# SOIL MECHANICS AND FOUNDATIONS

# SOIL MECHANICS AND FOUNDATIONS

**MUNI BUDHU**

*Professor, Department of Civil Engineering & Engineering Mechanics*
*University of Arizona*

**JOHN WILEY & SONS, INC.**
*New York / Chichester / Weinheim / Brisbane / Singapore / Toronto*

Editor    *Wayne Anderson*
Marketing Manager    *Katherine Hepburn*
Senior Production Manager    *Lucille Buonocore*
Production Editor    *Leslie Surovick*
Cover Designer    *Lynn Rogan*
Illustration Editor    *Sigmund Malinowski*
Illustration Studio    *Radiant Illustration & Design*
Cover Photo    *CORBIS/Roger Wood*

This book was set in 10/12 Times Ten by UG / GGS Information Services, Inc. and printed and bound by RR Donnelley/Willard. The cover was printed by Phoenix Color Corporation.

This book is printed on acid-free paper.

The paper in this book was manufactured by a mill whose forest management programs include sustained yield harvesting of its timberlands. Sustained yield harvesting principles ensure that the numbers of trees cut each year does not exceed the amount of new growth.

*Library of Congress Cataloging in Publication Data:*
Budhu, M.
    Soil mechanics and foundations / by Muni Budhu.
        p. cm.
    ISBN 0-471-25231-X (alk. paper)
    1. Soil mechanics. 2. Foundations. I. Title.

TA710.B765 1999
624.1'5136221                          99-050184

Printed in the United States of America

10 9 8 7 6 5 4 3

# PREFACE

This textbook is written for an undergraduate course in soil mechanics and foundations. It has three primary objectives. First, to present basic concepts and fundamental principles of soil mechanics and foundations in a simple pedagogy using the students' background in mechanics, physics, and mathematics. Second, to integrate modern learning principles, teaching techniques and learning aids to assist students in understanding the various topics in soil mechanics and foundations. Third, to provide a solid background knowledge to hopefully launch students in life-long learning of geotechnical engineering issues.

Topics are presented thoroughly and systematically to elucidate the basic concepts and fundamental principles without diluting technical rigor. The intention is to provide the student with a firm grounding in the principles behind the practice. Example problems have been solved to demonstrate or to provide further insights into the basic concepts and applications of fundamental principles. The solution of each example is preceded by a strategy, which is intended to teach students to think about possible solutions to a problem before they begin to solve it. Each solution provides a step by step procedure to guide the student in problem solving. Each chapter in the book is structured on modern teaching themes of gaining attention (reception), learning objectives (expectancy), recollecting of prior knowledge (working memory), presenting the information (selective perception), providing guidance (semantic encoding), key points (reinforcement), eliciting performance on key concepts through an interactive quiz on CD-ROM (retrieval and reinforcement), summarizing (retrieval and generalization), and assessing performance through problem solving. The text is written in colloquial English to try to engage students and make them feel that they are active participants in the learning process. The emphasis in this book is on the delivery and retention of the fundamental principles and concepts.

Today's students are surrounded by visual media and obtain much of their information from them. This book is accompanied by an interactive multimedia CD-ROM in which many of the concepts are animated. The author hopes that this would capture the visual learners and enhance learning. Various interactive tools such as interactive problem solving, virtual laboratories, quizzes, etc., are included to facilitate learning, retention, evaluation, and assessment.

With the proliferation and accessibility of computers, programmable calculators, and software, students will likely use these tools in their practice. Consequently, generalized equations which the students can program into their calculators, and computer program utilities are provided rather than charts.

I am grateful to the following reviewers who offered many valuable suggestions for improving this textbook:

- Professor Hilary I. Inyang, University of Massachusetts—Lowell
- Professor Derek Morris, Texas A&M University
- Professor Cyrus Aryani, California State University

- Professor Shobha K. Bhatia, Syracuse University
- Major Richard L. Shelton, United States Military Academy
- Professor Colby C. Swan, University of Iowa
- Professor Panos Kiousis, University of Arizona
- Professor Carlos Santamarina, Georgia Institute of Technology
- Dr. William Isenhower, Ensoft, Inc.

Mr. Wayne Anderson and his staff, and Leslie Surovick of John Wiley & Sons were particularly helpful in getting this book done. My heartfelt thanks goes to my wife and children who have contributed significantly to the completion of this book.

Additional resources are available online at *www.wiley.com/college/budhu*.

# *NOTES for Instructors*

I would like to present some guidance to assist you in using this book in undergraduate geotechnical engineering courses.

## DESCRIPTION OF CHAPTERS

The philosophy behind each chapter is to seek coherence and group topics that are directly related to each other. This is a rather difficult task in geotechnical engineering because topics are intertwined. Attempts have been made to group topics based on whether they relate directly to the physical characteristics of soils or mechanical behavior or are applications of concepts to analysis of geotechnical systems. The sequencing of the chapters is such that the preknowledge required in a chapter is covered in previous chapters.

Chapter 1 sets the introductory stage of informing the students of the importance of geotechnical engineering. Most of the topics related to the physical characteristics of soils are grouped in Chapter 2. Description of soils, soil constituents, index properties, soils classification, soil compaction, permeability, and soil investigations form the core.

Chapter 3 deals with stresses, strains, and elastic deformation of soils. Most of the material in this chapter builds on course materials that students would have encountered in their courses in statics and strength of materials. Often, elasticity is used in preliminary calculations in analyses and design of geotechnical systems. The use of elasticity to find stresses and settlement of soils is presented and discussed. Stress increases due to applied surface loads common to geotechnical problems are described. Students are introduced to stress and strain states and stress and strain invariants. The importance of effective stresses and seepage in soil mechanics is emphasized. Stress paths are introduced not only because of their importance in understanding the performance of a geotechnical structure but because they are connected to the presentation of stresses—stress paths being a graphical representation or vector representation of stresses. Drained and undrained conditions are introduced within the context of elasticity.

One-dimensional consolidation of soils is considered in Chapter 4. Here the basic concepts of consolidation are presented with methods to calculate consolidation settlement. The theory of one-dimensional consolidation is developed to show the students the theoretical framework from which soil consolidation settlement is interpreted and the parameter required to determine time rate of settlement. The oedometer test is described and procedures to determine the various parameters for settlement calculations are presented.

Chapter 5 deals with the shear strength of soils and the tests (laboratory and field) required for its determination. The Mohr–Coulomb failure criterion is discussed using the student's background in strength of materials (Mohr's circle) and in statics (dry friction). Soils are treated as a dilatant-frictional material rather than the conventional cohesive-frictional material. Typical stress–strain responses of sand and clay are presented and discussed. The implications of drained and undrained conditions on the shear strength of soils are discussed. Laboratory and field tests to determine the shear strength of soils are described.

Chapter 6 deviates from traditional undergraduate textbook topics that presents soil consolidation and strength as separate issues. In this chapter, deformation and strength are integrated within the framework of critical state soil mechanics using a simplified version of the modified cam-clay model. The emphasis is on understanding the mechanical behavior of soils rather than presenting the mathematical formulation of critical state soil mechanics and the modified cam-clay model. The amount of mathematics is kept to the minimum needed for understanding and clarification of important concepts. Projection geometry is used to illustrate the different responses of soils when the loading changes under drained and undrained loading. Although this chapter deals with a simplification and an idealization of real soils, the real benefit is a simple framework, which allows the student to think about possible soil responses if conditions change from those originally conceived, as is usual in engineering practice. It also allows them to better interpret soil test results.

Chapter 7 deals with bearing capacity and settlement of footings. Here bearing capacity and settlement are treated as a single topic. In the design of foundations, the geotechnical engineer must be satisfied that the bearing capacity is sufficient and the settlement at working load is tolerable. Indeed, for most shallow footings, it is settlement that governs the design, not bearing capacity. Limit equilibrium analysis is introduced to illustrate the method that has been used to find the popular bearing capacity equations and to make use of the student's background in statics (equilibrium) to introduce a simple but powerful analytical tool. Three sets of bearing capacity equations (Terzaghi as modified by Vesic, Meyerhof, and Skempton), the influence of groundwater level, and eccentric loads on bearing capacity are discussed. These equations are simplified by breaking them down into two categories—one relating to drained conditions, the other to undrained conditions. Elastic, one-dimensional consolidation, and Skempton and Bjerrum's method of determining settlement are presented. The elastic method of finding settlement is based on work done by Gazettas (1985), who described problems associated with the Janbu, Bjerrum, and Kjaernali (1956) method that is conventionally quoted in textbooks.

Pile foundations are described and discussed in Chapter 8. Methods for finding bearing capacity and settlement of single and group piles are presented.

Chapter 9 is about two-dimensional steady state flow through soils. Solutions to two-dimensional flow using flow nets and the finite difference technique are discussed. Emphases are placed on seepage, pore water pressure, and instability. This chapter normally comes early in most current textbooks. The reason for placing this chapter here is because two-dimensional flow influences the stability of earth structures (retaining walls and slopes), discussion of which follows in Chapters 10 and 11. A student would then be able to make the practical connection of two-dimensional flow and stability of geotechnical systems readily.

Lateral earth pressures and their use in the analysis of earth retaining systems and excavations are presented in Chapter 10. Gravity and flexible retaining walls, in addition to reinforced soil walls, are discussed. Guidance is provided as to what strength parameters to use in drained and undrained conditions.

Chapter 11 is about slope stability. Here stability conditions are described based on drained or undrained conditions.

An appendix (Appendix A) allows easy access to frequently used typical soil parameters and correlations.

## CHAPTER LAYOUT

The ***Introduction*** of each chapter attempts to capture the student's attention, to present the learning objectives, and to inform the student on what prior knowledge is needed to master the material. At the end of the introduction, a ***Sample Practical Situation*** is described. The intention is to give the student a feel for the kind of problem that he/she should be able to solve on completion of the chapter. At the end of the chapter, a problem similar to the sample practical situation is solved. This provides closure to the chapter.

***Definitions of Key Terms*** are presented to alert and introduce the students to new terms in the topics to be covered. A section on ***Questions to Guide Your Reading*** is intended to advise the students on key information that they should grasp and absorb. These questions form the core of the quiz on the CD-ROM.

Each topic is presented thoroughly with the intention of engaging the students and making them feel involved in the process of learning. At various stages, ***Key Points*** are summarized for reinforcement. ***Examples*** are solved at the end of each major topic to illustrate problem-solving techniques, to reinforce and to apply the basic concepts. A ***What's Next*** section serves as a link between articles and informs students about this connection. This prepares them for the next topic and serves as a break point for your lectures. A ***Summary*** at the end of each chapter reminds students, in a general way, of key information. The ***Exercises*** or problems are divided into three sections. The first section contains problems that are theoretically based, the second section contains problems suitable for problem solving, and the third section contains problems biased toward application. This gives you flexibility in setting problems based on the objectives of the course.

## CD ROM

With the advent of personal computers, learning has become more visual. Some studies have reported that visual images have improved learning by as much as 400%. This textbook is accompanied by a CD ROM that contains text, interactive animation, images, a glossary, notation, quizzes, notepads, and interactive problem solving. It should appeal, particularly, to visual learners.

A quiz is included in appropriate chapters on the CD ROM to elicit performance and provide feedback on key concepts. Interactive problem solving is used to help students solve problems similar to the problem-solving exercises. When an interactive problem is repeated, new values are automatically generated. Sounds are used to a limited extent. The CD ROM contains a virtual soils laboratory for the students to conduct geotechnical tests. These virtual tests are not intended to replace the necessary hands-on experience in a soil laboratory. Rather, they complement the hands-on experience, prepare the students for the real experience, test relevant prior knowledge of basic concepts for the interpretation of the test results, guide them through the evaluation and interpretation of the results, allow them to conduct tests that cannot otherwise be done during laboratory sessions, and allow them to use the results of their tests in practical applications.

## ABET REQUIREMENTS

The United States Accreditation Board for Engineering and Technology (ABET) has introduced new criteria for accreditation purposes. Each chapter in this book has the author's judgment on how it satisfies ABET's engineering science (ES) and engineering design (ED) criteria. You may adjust the recommended percentages allocated to ES and ED based on your own judgment.

## COURSE MATERIAL

I have used this book and CD ROM for teaching two undergraduate courses. One course is on soil mechanics in which I cover Chapters 1 through 6. The other course is an introduction to foundation engineering in which I cover Chapters 7 through 11. If you wish, you may consider Chapter 9, which deals with two-dimensional flow through porous media, in your first course in soil mechanics or geotechnical engineering. Visual learners may find the CD ROM more helpful than the textbook. However, the CD ROM does not cover the full range of topics in the textbook. In addition, the textbook contains a lot more detail.

## COURSE DELIVERY

I have used the CD ROM exclusively in my classroom to deliver my course. This significantly cuts down on lecture preparation time. Moreover, the interactivities and animations coded in the CD ROM make it easier for students to understand difficult topics. Before a new topic is introduced, I used the quiz on the CD ROM to evaluate how much of the content of the previous topic was retained. The interactive problem solving on the CD ROM is used to illustrate problem solving to the students.

# NOTES for Students and Instructors

## PURPOSES OF THIS BOOK

This book is intended to present the principles of soil mechanics and its application to foundation analyses. It will provide you with an understanding of the properties and behavior of soils, albeit not a perfect understanding. The design of safe and economical geotechnical structures or systems requires considerable experience and judgment, which cannot be obtained by reading this or any other textbook. It is hoped that the fundamental principles and guidance provided in this textbook will be a base for lifelong learning in the science and art of geotechnical engineering.

The goals of this textbook in a course on soil mechanics and foundation are as follows:

1. To understand the physical and mechanical properties of soils.
2. To determine parameters from soil testing to characterize soil properties, soil strength, and soil deformations.
3. To apply the principles of soil mechanics to analyze and design simple geotechnical systems.

## LEARNING OUTCOMES

When you complete studying this textbook you should be able to:

- Describe soils and determine their physical characteristics such as, grain size, water content, and void ratio
- Classify soils
- Determine compaction of soils
- Understand the importance of soil investigations and be able to plan a soil investigation
- Understand the concept of effective stress
- Determine total and effective stresses and pore water pressures
- Determine soil permeability
- Determine how surface stresses are distributed within a soil mass
- Specify, conduct, and interpret soil tests to characterize soils

- Determine soil strength and deformation parameters from soil tests, for example, Young's modulus, friction angle and undrained shear strength
- Discriminate between "drained" and "undrained" conditions
- Understand the effects of seepage on the stability of structures
- Estimate the bearing capacity and settlement of structures founded on soils
- Analyze and design simple foundations
- Determine the stability of earth structures, for example, retaining walls and slopes

Certain topics are ubiquitous. The table below provides a guide to the distribution of the major topics in this textbook. A solid circle denotes significant coverage of the topic while an open circle denotes some coverage of the topic or the use of it in the chapter. For example, total and effective stresses and pore water pressures are introduced and significantly covered in Chapter 3. However, they are used in every chapter thereafter. You should pay particular attention to the topics that have wide distributions in this textbook. Chapters 2 and 6 cover the fundamentals of soil mechanics, Chapters 7 and 11 cover analysis of foundations and earth structures.

**Distribution of Main Topics in This Textbook**

| Description | Chapter | | | | | | | | | | |
| --- | --- | --- | --- | --- | --- | --- | --- | --- | --- | --- | --- |
| | Soil mechanics | | | | | | Foundation and earth structures | | | | |
| | 1 | 2 | 3 | 4 | 5 | 6 | 7 | 8 | 9 | 10 | 11 |
| Physical characteristics and properties | I | ● | | ○ | ○ | ○ | ○ | ○ | ○ | ○ | ○ |
| Compaction | N | ● | | | | | ○ | | | | |
| Soil investigation | T | ● | | ○ | ○ | ○ | ○ | ○ | ○ | ○ | ○ |
| Total and effective stresses and pore water pressure | R | | ● | ○ | ○ | ○ | ○ | ○ | ○ | ○ | ○ |
| Permeability | O | ● | | ○ | | | | | ○ | | |
| Stresses in soils | D | | ● | ○ | ○ | ○ | ○ | ○ | ○ | ○ | ○ |
| Drained and undrained conditions | U | | ● | | ● | ● | ○ | ○ | | ○ | ○ |
| Settlement and deformation | C | | ○ | ● | | ○ | ● | ● | | ○ | |
| Shear strength | T | | | | ● | ● | ○ | ○ | ○ | ○ | ○ |
| Seepage | I | ● | | | | | | | ● | ○ | ○ |
| Bearing capacity and settlement of foundations | O | | | | | | ● | ● | | ○ | |
| Stability of earth structures | N | | | | | | | | ○ | ● | ● |

## ASSESSMENT

You will be assessed on how well you absorb and use the fundamentals of soil mechanics. Three areas of assessment are incorporated in the Exercise sections of this textbook. The first area called "Theory" is intended for you to demonstrate your knowledge of the theory and extend it to uncover new relationships. The questions under "Theory" will help you later in your career to address unconventional issues using fundamental principles. The second area called "Problem Solving" requires you to apply the fundamental principles and concepts to a wide variety of problems. These problems will test your understanding and use of the fundamental principles and concepts. The third area called "Practical" is intended to create practical scenarios for you to use not only the subject matter in the specific chapter but prior materials that you have encountered. These problems try to mimic some aspects of real situations and give you a feel for how the materials you have studied so far can be applied in practice. Communications are, at least, as important as the technical details. In many of these "Practical" problems you are placed in a situation to convince stakeholders of your technical competence. A quiz (multiple choice) on each chapter is included in the CD to test your general knowledge of the subject matter in that chapter. The questions on the quiz are related to the section "Questions to Guide Your Reading," included in each chapter.

## SUGGESTIONS FOR PROBLEM SOLVING

Engineering is, foremost, about problem solving. For most engineering problems, there is no unique method or procedure for finding solutions. Often, there is no unique solution to an engineering problem. A suggested problem-solving procedure is outlined below.

1. Read the problem carefully; note or write down what is given and what you are required to find.
2. Draw clear diagrams or sketches wherever possible.
3. Devise a strategy to find the solution. Determine what principles, concepts, and equations are needed to solve the problem.
4. Perform calculations making sure that you are using the correct units.
5. Check whether your results are reasonable.

The units of measurement used in this textbook follow the SI system. Engineering calculations are approximations and do not result in exact numbers. All calculations in this book are rounded, at the most, to two decimal places except in some exceptional cases, for example, void ratio.

## SUGGESTIONS FOR USING TEXTBOOK AND CD-ROM

This textbook is accompanied by and integrated with a CD-ROM. Not all sections of the textbook are covered in the CD-ROM. The textbook provides significantly more details on the subject matter than the CD-ROM. The CD-ROM provides animations, interactive problem solving, quizzes, virtual laboratories, special modules (for example, a computer program to find stresses within a soil), spreadsheets, videos, a notepad, a glossary, a list of notations, and a calculator.

CD icons in the textbook have inset numbers that are intended to alert you to special features present on the CD-ROM. The numbers have the following meaning:

1. Interactive animation
2. Virtual lab
3. Interactive problem solving
4. Spreadsheet
5. Video
6. Computer program utility

## PRESENTATION

Few engineering projects involve design and construction only. Often, you have to present your work, orally or in writing, to clients who do not have the technical knowledge to grasp technical lingo. In oral presentations, you are normally given a very short time to present your project to your client. On many projects, the awarding of a contract depends on those precious few minutes. Your task is to make the best, easily understood presentation in the shortest possible time. Because of the importance of communicating your ideas and solutions, problems are included in several chapters in this book in which you will have to make a short presentation of your solution to stakeholders.

The key to a good presentation is to tell your audience what you are going to say, say it (body of presentation), and then tell them what you told them (conclusion). There are several media available for making visual presentations. These include a computer-aided presentation using a laptop computer connected to an LCD projector, overhead transparencies, and slides. You should use one (or more) of these media that best suit your audience and the message you want to transmit. Your visuals should have a reasonable font size (24 points), should not be busy, and should have simple language and clear graphics. The visuals should highlight what you are going to present and you should not normally read from them.

A sample page of a presentation is shown below. Here the central idea is "Shear Strength of Soils" (font size: 30 points). In this visual, you would be defining shear strength and inform your audience that Coulomb's law is used to interpret the shear strength of soils.

# SHEAR STRENGTH OF SOILS
- Resistance to shear forces
- Coulomb's law

Now, let us start our study of soil mechanics and foundations.

# CONTENTS

**CHAPTER 3**    *STRESSES, STRAINS AND ELASTIC DEFORMATIONS OF SOILS*
        **79**

## CHAPTER 7   *BEARING CAPACITY OF SOILS AND SETTLEMENT OF SHALLOW FOUNDATIONS* **318**

# Notation

*Note:* A prime (′) after a notation for stress denotes effective stress.

| | | | |
|---|---|---|---|
| A | Area | $H_{dr}$ | Drainage path |
| B | Width | $i$ | Hydraulic gradient |
| $c_o$ | Cohesion | $I$ | Influence factor |
| $C_c$ | Compression index | $I_s$ | Settlement influence factor |
| $C_r$ | Recompression index | $I_L$ | Liquidity index |
| $C_h$ | Horizontal coefficient of consolidation | $I_p$ | Plasticity index |
| | | $k$ | Hydraulic conductivity or coefficient of permeability |
| $C_v$ | Vertical coefficient of consolidation | $K_a$ | Active lateral earth pressure coefficient |
| $C_\alpha$ | Secondary compression index | $K_p$ | Passive lateral earth pressure coefficient |
| CC | Coefficient of curvature | | |
| CSM | Critical state model | $K_0$ | Lateral earth pressure coefficient at rest |
| D | Diameter | | |
| $D_f$ | Embedment depth | $L$ | Length |
| $D_r$ | Relative density | $m_v$ | Modulus of volume compressibility |
| $D_{10}$ | Effective particle size | | |
| $D_{50}$ | Average particle diameter | $N$ | Standard penetration number |
| $e$ | Void ratio | $N_c, N_q, N_\gamma$ | Bearing capacity factors |
| $E$ | Modulus of elasticity | $n$ | Porosity |
| $E_i$ | Initial tangent modulus | NCL | Normal consolidation line |
| $E_p$ | Modulus of elasticity of pile | OCR | Overconsolidation ratio |
| $E_s$ | Secant modulus | $p$ | Mean stress |
| $E_{so}$ | Modulus of elasticity of soil | $q$ | Deviatoric stress or shear stress |
| $E_t$ | Tangent modulus | | |
| $f_b$ | Ending bearing stress | $q_a$ | Allowable bearing capacity |
| $f_s$ | Skin friction | $q_s$ | Surface stress |
| FS | Factor of safety | $q_{ult}$ | Ultimate bearing capacity |
| $F_\phi$ | Mobilization factor for $\phi$ | $q_v$ | Flow rate |
| $F_u$ | Mobilization factor for $s_u$ | $Q$ | Flow, quantity of flow |
| $G$ | Shear modulus | $Q_b$ | End bearing or point resistance |
| $G_s$ | Specific gravity | | |
| $h_p$ | Pressure head | $Q_f$ | Skin or shaft friction |
| $h_z$ | Elevation head | $Q_p$ | Point bearing resistance |
| $H$ | Head | $Q_{ult}$ | Ultimate load capacity |
| $H_o$ | Height | $(Q_{ult})_g$ | Ultimate group load capacity |

| | | | |
|---|---|---|---|
| $R_T$ | Temperature correction factor | $\delta$ | Deflection or settlement |
| $R_x$ | Resultant lateral force | $\varepsilon$ | Normal strain |
| $R_o$ | Overconsolidation ratio with respect to stress invariants | $\varepsilon_p$ | Volumetric strain |
| | | $\varepsilon_q$ | Deviatoric strain |
| $s_u$ | Undrained shear strength | $\phi'$ | Generic friction angle |
| $S$ | Degree of saturation | $\phi'_{cs}$ | Critical state friction angle |
| $S_t$ | Sensitivity | $\phi'_p$ | Peak friction angle |
| $SPT$ | Standard penetration test | $\phi'_r$ | Residual friction angle |
| $T$ | Sliding force or resistance | $\gamma$ | Bulk unit weight |
| $T_v$ | Time factor | $\gamma'$ | Effective unit weight |
| $u$ | Pore water pressure | $\gamma_{sat}$ | Saturated unit weight |
| $U$ | Average degree of consolidation | $\gamma_d$ | Dry unit weight |
| | | $\gamma_{d(max)}$ | Maximum dry unit weight |
| UC | Uniformity coefficient | $\gamma_w$ | Unit weight of water |
| URL | Unloading/reloading line | $\gamma_{zx}$ | Shear strain |
| $v$ | Velocity | $\kappa$ | Recompression index |
| $v_s$ | Seepage velocity | $\lambda$ | Compression index |
| $v_{sh}$ | Shear wave velocity | $\mu$ | Viscosity |
| $V$ | Volume | $\mu_s$ | Shape coefficient |
| $V'$ | Specific volume | $\mu_{emb}$ | Embedment coefficient |
| $V_a$ | Volume of air | $\mu_{wall}$ | Wall friction coefficient |
| $V_s$ | Volume of solid | $\nu$ | Poisson's ratio |
| $V_w$ | Volume of water | $\rho_e$ | Elastic settlement |
| $w$ | Water content | $\rho_{pc}$ | Primary consolidation |
| $w_{LL}$ | Liquid limit | $\rho_{sc}$ | Secondary consolidation settlement |
| $w_{opt}$ | Optimum water content | | |
| $w_{PL}$ | Plastic limit | $\sigma$ | Normal stress |
| $w_{SL}$ | Shrinkage limit | $\tau$ | Shear stress |
| $W$ | Weight | $\tau_{cs}$ | Critical state shear strength |
| $W_a$ | Weight of air | $\tau_f$ | Shear strength at failure |
| $W_s$ | Weight of solid | $\tau_p$ | Peak shear strength |
| $W_w$ | Weight of water | $\tau_r$ | Residual shear strength |
| $z$ | Depth | $\xi$ | Velocity potential |
| $\alpha$ | Dilation angle | $\psi$ | Rotation of principal plane to the horizontal |
| $\alpha_s$ | Slope angle | | |
| $\alpha_u$ | Adhesion factor | $\psi_p$ | Plastification angle for piles |
| $\beta$ | Skin friction coefficient for drained condition | $\psi_s$ | Stream potential |

## CHAPTER 1

# INTRODUCTION TO SOIL MECHANICS AND FOUNDATIONS

## 1.0 INTRODUCTION

Soil is the oldest and most complex engineering material. Our ancestors used soils as construction material to build burial sites, flood protection, and shelters. Western civilization credits the Romans for recognizing the importance of soils in the stability of structures. Roman engineers, especially Vitruvius, who served during the reign of Emperor Augustus in the first century B.C., paid great attention to soil types (sand, gravel, etc.) and to the design and construction of solid foundations. There was no theoretical basis for design; experience from trial and error was relied upon.

Coulomb (1773) is credited as the first person to use mechanics to solve soil problems. He was a member of the French Royal Engineers, who were interested in protecting old fortresses that fell easily from cannon fire. To protect the fortresses from artillery attack, sloping masses of soil were placed in front of them (Fig. 1.1). The enemy had to tunnel below the soil mass and the fortress to attack. Of course, the enemy then became an easy target. The mass of soil applies a lateral force to the fortress that could cause it to topple over or could cause it to slide away from the soil mass. Coulomb attempted to determine the lateral force so that he could evaluate the stability of the fortress. He postulated that a wedge of soil *ABC* (Fig. 1.1) would fail along a slip plane BC and this wedge would push the wall out or overtopple it as it moves down the slip plane.

Movement of the wedge along the slip plane would occur only if the soil resistance along the wedge were overcome. Coulomb assumed that the soil resistance is provided by friction between the particles and the problem became one of a wedge sliding on a rough (frictional) plane, which you may have analyzed in your physics or mechanics course. Coulomb has tacitly defined a failure criterion for soils. Today, Coulomb's failure criterion and method of analysis still prevail.

From the early 20th century, the rapid growth of cities, industry, and commerce required a myriad of building systems: for example, skyscrapers, large public buildings, dams for electric power generation and reservoirs for water supply and irrigation, tunnels, roads and railroads, port and harbor facilities, bridges, airports and runways, mining activities, hospitals, sanitation systems, drainage systems, and towers for communication systems. These building systems require stable and economic foundations and new questions about soils were asked. For example, what is the state of stress in a soil mass, how can one design safe and economic foundations, how much would a building settle, and what is

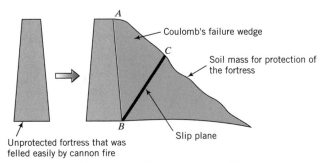

**FIGURE 1.1**   Unprotected and protected fortress.

the stability of structures founded on or within soil? We continue to ask these questions and to try to find answers as new issues have confronted us. Some of these new issues include removing toxic compounds from soil and water, designing foundations and earth structures to mitigate damage from earthquakes and other natural hazards, and designing systems to protect the environment.

To answer these questions, we need the help of some rational method and, consequently, soil mechanics was born. Karl Terzaghi (1883–1963) is the undisputed father of soil mechanics. The publication of his book *Erdbaumechanik* in 1925 laid the foundation for soil mechanics and brought recognition to the importance of soils in engineering activities. Soil mechanics, also called geotechnique or geotechnics or geomechanics, is the application of engineering mechanics to the solution of problems dealing with soils as a foundation and a construction material. Engineering mechanics is used to understand and interpret the properties, behavior, and performance of soils.

Soil mechanics is a subset of geotechnical engineering, which involves the application of soil mechanics, geology, and hydraulics to the analysis and design of geotechnical systems such as dams, embankments, tunnels, canals and waterways, foundations for bridges, roads, buildings, and solid waste disposal systems. Every application of soil mechanics involves uncertainty because of the variability of soils—their stratification, composition, and engineering properties. Thus, engineering mechanics can provide only partial solutions to soil problems. Experience and approximate calculations are essential for the successful application of soil mechanics to practical problems. Many of the calculations in this textbook are approximations.

Stability and economy are two tenets of engineering design. In geotechnical engineering, the uncertainties of the performance of soils, the uncertainties of the applied loads, and the vagaries of natural forces nudge us to compromise between sophisticated analyses and simple analyses or approximate methods. Stability should never be compromised for economy. An unstable structure compromised to save a few dollars can result in death and destruction.

# 1.1  MARVELS OF CIVIL ENGINEERING— THE HIDDEN TRUTH

The work that geotechnical engineers do is often invisible once construction is completed. For example, four marvelous structures—the Sears Tower (Fig. 1.2), the Empire State Building (Fig. 1.3), the Taj Mahal (Fig. 1.4), and the Hoover

**FIGURE 1.2**   Sears Tower. (© Bill Bachmann/Photo Researchers.)

**FIGURE 1.3**   Empire State Building. (© Rafael Macia/Photo Researchers.)

**FIGURE 1.4**   Taj Mahal. (© Will & Deni McIntyre/Photo Researchers.)

**FIGURE 1.5**  Hoover Dam. (Courtesy Bureau of Reclamation, U.S. Department of the Interior. Photo by E.E. Hertzog.)

Dam (Fig. 1.5)—grace us with their engineering and architectural beauty. However, if the foundations, which are invisible, on which these structures stand were not satisfactorily designed then these structures would not exist. A satisfactory foundation design requires the proper application of soil mechanics principles, accumulated experience and good judgment.

The stability and life of any structure—a building, an airport, a road, dams, levees, natural slopes, power plants—depend on the stability, strength, and deformation of soils. If the soil fails, structures founded on or within it will fail or be impaired, regardless of how well these structures are designed. Thus, successful civil engineering projects are heavily dependent on geotechnical engineering.

## 1.2  GEOTECHNICAL LESSONS FROM FAILURES

All structures that are founded on earth rely on our ability to design safe and economic foundations. Because of the natural vagaries of soils, failures do occur. Some failures have been catastrophic and caused severe damage to lives and properties; others have been insidious. Failures occur because of inadequate site and soil investigations; unforeseen soil and water conditions; natural hazards; poor engineering analysis, design, construction, and quality control; postconstruction activities; and usage outside the design conditions. When failures are investigated thoroughly, we obtain lessons and information that will guide us to prevent similar types of failure in the future. Some types of failure caused by natural hazards (earthquakes, hurricanes, etc.) are difficult to prevent and our efforts must be directed toward solutions that mitigate damages to lives and properties.

One of the earliest failures that was investigated and contributed to our knowledge of soil behavior is the failure of the Transcona Grain Elevator in 1913 (Fig. 1.6). Within 24 hours after loading the grain elevator at a rate of about 1 m of grain height per day, the bin house began to tilt and settle. Fortunately, the structural damage was minimal and the bin house was later restored. No borings were done to identify the soils and to obtain information on their strength.

**FIGURE 1.6** Failure of the Transcona Grain Elevator. (Photo courtesy of Parrish and Heimbecker Limited.)

Rather, an open pit about 4 m deep was made for the foundations and a plate was loaded to determine the bearing strength of the soil.

The information gathered from the Transcona Grain Elevator failure and the subsequent detailed soil investigation was used (Peck and Bryant, 1953; Skempton, 1951) to verify the theoretical soil bearing strength. Peck and Bryant (1953) found that the applied pressure from loads imposed by the bin house and the grains was nearly equal to the calculated maximum pressure that the soil can withstand, thereby verifying the theory for calculating the bearing strength of soils. We also learn from this failure the importance of soil investigations, soils tests, and the effects of rate of loading.

The Transcona Grain Elevator was designed at a time when soil mechanics was not born. One eyewitness (White, 1953) wrote: "Soil Mechanics as a special science had hardly begun at that time. If as much had been known then as is now about the shear strength and behaviour of soils, adequate borings would have been taken and tests made and these troubles would have been avoided. We owe more to the development of this science than is generally recognized."

We have come a long way in understanding soil behavior since its fatherhood by Terzaghi in 1925. We continue to learn more daily though research on and experience from failures and your contribution to understanding soil behavior is needed. Join me on a journey of learning the fundamentals of soil mechanics and its applications to practical problems so that we can avoid failures or, at least, reduce the probability of their occurrence.

## CHAPTER 2

# PHYSICAL CHARACTERISTICS OF SOILS AND SOIL INVESTIGATIONS

| ABET | ES | ED |
|------|----|----|
|      | 90 | 10 |

## 2.0 INTRODUCTION

The purpose of this chapter is to introduce you to soils. You will learn some basic descriptions of soils and some fundamental physical soil properties that you should retain for future use in this text and in geotechnical engineering practice. Soils, derived from the weathering of rocks, are very complex materials and vary widely. There is no certainty that a soil will have the same properties within a few centimeters of its current location.

One of the primary tasks of a geotechnical engineer is to collect, classify, and investigate the physical properties of soils. In this chapter, we will deal with descriptions of soils, tests to determine the physical properties of soils, soil classification, one-dimensional flow of water through soils, and methods of soil investigations. Usually soils investigations are conducted only on a fraction of a proposed site because it would be prohibitively expensive to conduct an extensive investigation of a whole site. We then have to make estimates and judgments based on information from a limited set of observations and field and laboratory test data.

In engineering, we disassemble complex systems into parts and then study each part and its relationship to the whole. We will do the same for soils. Soils will be dismantled into three constituents and we will examine how the proportions of each constituent characterize soils. When you complete this chapter, you should be able to:

- Describe and classify soils
- Determine particle size distribution in a soil mass
- Determine the proportions of the main constituents in a soil
- Determine index properties of soils
- Determine the rate of flow of water through soils
- Determine maximum dry unit weight and optimum water content
- Plan a soil investigation

*Sample Practical Situation*   A highway is proposed to link the city of Noscut to the village of Windsor Forest. The highway route will pass through a terrain that is relatively flat and is expected to be flooded by a 100 year storm event.

**FIGURE 2.1**   A highway under construction.

The highway will be supported on an embankment constructed from soils trucked to the site from two possible pits. You, the geotechnical engineer, are to advise a stakeholders' committee, consisting of engineers, farmers, community representatives, and lawyers, on the estimated cost of the embankment. In arriving at a cost, you must consider the suitability of the soil in each of the two pits for the embankment, the amount of soil, and the number of truckloads required. A highway under construction is shown in Fig. 2.1.

## 2.1   DEFINITIONS OF KEY TERMS

*Soils* are materials that are derived from the weathering of rocks.

*Water content ($w$)* is the ratio of the weight of water to the weight of solids.

*Void ratio ($e$)* is the ratio of the volume of void space to the volume of solids.

*Porosity ($n$)* is the ratio of the volume of void to the total volume of soil.

*Degree of saturation ($S$)* is the ratio of the volume of water to the volume of void.

*Bulk unit weight ($\gamma$)* is the weight density, that is, the weight of a soil per unit volume.

*Saturated unit weight ($\gamma_{sat}$)* is the weight of a saturated soil per unit volume.

*Dry unit weight ($\gamma_d$)* is the weight of a dry soil per unit volume.

*Effective unit weight ($\gamma'$)* is the weight of soil solids in a submerged soil per unit volume.

*Relative density ($D_r$)* is an index that quantifies the degree of packing between the loosest and densest state of coarse-grained soils.

*Effective particle size ($D_{10}$)* is the average particle diameter of the soil at 10 percentile; that is, 10% of the particles are smaller than this size (diameter).

*Average particle diameter ($D_{50}$)* is the average particle diameter of the soil.

*Liquid limit ($w_{LL}$)* is the water content at which a soil changes from a plastic state to a liquid state.

*Plastic limit ($w_{PL}$)* is the water content at which a soil changes from a semisolid to a plastic state.

*Shrinkage limit ($w_{SL}$)* is the water content at which a soil changes from a solid to a semisolid state without further change in volume.

*Groundwater* is water under gravity in excess of that required to fill the soil pores.

*Head ($H$)* is the mechanical energy per unit weight.

*Coefficient of permeability ($k$)* is a proportionality constant to determine the flow velocity of water through soils.

*Maximum dry unit weight ($\gamma_{d(max)}$)* is the maximum unit weight that a soil can attain using a specified means of compaction.

*Optimum water content ($w_{opt}$)* is the water content required to allow a soil to attain its maximum dry unit weight.

## 2.2   QUESTIONS TO GUIDE YOUR READING

1. How is soil described and classified?
2. What are the main minerals in soils?
3. What are the physical parameters that characterize soils?
4. What are the differences between coarse-grained and fine-grained soils?
5. What are some of the basic soil tests required to characterize soils?
6. What are the Atterberg limits?
7. What is the purpose of classifying soils?
8. What is a grading curve?
9. What causes flow of water through soils?
10. What law describes the flow of water through soils?
11. What is permeability and how is it determined?
12. What is a soil investigation?
13. How do you plan a soil investigation?
14. What are the effects of water on the unit weight of soils?
15. What factors affect the compaction of soils?

## 2.3   COMPOSITION OF SOILS

### 2.3.1 Soil Formation

Soils are formed from the physical and chemical weathering of rocks. Physical weathering involves reduction of size without any change in the original composition of the parent rock. The main agents responsible for this process are exfoliation, unloading, erosion, freezing, and thawing. Chemical weathering causes both reductions in size and chemical alteration of the original parent rock. The main agents responsible for chemical weathering are hydration, carbonation, and oxidation. Often, chemical and physical weathering take place in concert.

Soils that remain at the site of weathering are called residual soils. These soils retain many of the elements that comprise the parent rock. Alluvial soils, also called fluvial soils, are soils that were transported by rivers and streams. The composition of these soils depends on the environment under which they were transported and is often different from the parent rock. The profile of alluvial soils usually consists of layers of different soils. Much of our construction activities has been and is occurring in and on alluvial soils. Glacial soils are soils that were transported and deposited by glaciers. Marine soils are soils deposited in a marine environment.

## 2.3.2 Soil Types

Common descriptive terms such as gravels, sands, silts, and clays are used to identify specific textures in soils. We will refer to these soil textures as soil types; that is, sand is one soil type, clay is another. Texture refers to the appearance or feel of a soil. Sands and gravels are grouped together as coarse-grained soils. Clays and silts are fine-grained soils. Coarse-grained soils feel gritty and hard. Fine-grained soils feel smooth. The coarseness of soils is determined from knowing the distribution of particle sizes, which is the primary means of classifying coarse-grained soils. To characterize fine-grained soils, we need further information on the types of minerals present and their contents. The response of fine-grained soils to loads, known as the mechanical behavior, depends on the type of predominant minerals present.

Currently, many soil descriptions and soil types are in usage. A few of these are listed below.

- *Calcareous soil* contains calcium carbonate and effervesces when treated with hydrochloric acid.
- *Caliche* consists of gravel, sand, and clay cemented together by calcium carbonate.
- *Expansive soils* are clays that undergo large volume changes from cycles of wetting and drying.
- *Glacial soils* are mixed soils consisting of rock debris, sand, silt, clays, and boulders.
- *Glacial till* is a soil that consists mainly of coarse particles.
- *Glacial clays* are soils that were deposited in ancient lakes and subsequently frozen. The thawing of these lakes reveals a soil profile of neatly stratified silt and clay, sometimes called varved clay. The silt layer is light in color and was deposited during summer periods while the thinner, dark clay layer was deposited during winter periods.
- *Gypsum* is calcium sulphate formed under heat and pressure from sediments in ocean brine.
- *Lateritic* soils are residual soils that are cemented with iron oxides and are found in tropical regions.

- *Loam* is a mixture of sand, silt, and clay that may contain organic material.
- *Loess* is a wind blown, uniform fine-grained soil.
- *Mud* is clay and silt mixed with water into a viscous fluid.

### 2.3.3 Clay Minerals

Minerals are crystalline materials and make up the solids constituent of a soil. The mineral particles of fine-grained soils are platy. Minerals are classified according to chemical composition and structure. Most minerals of interest to geotechnical engineers are composed of oxygen and silicon—two of the most abundant elements on earth. Silicates are a group of minerals with a structural unit called the silica tetraheron. A central silica cation (positively charged ion) is surrounded by four oxygen anions (negatively charged ions), one at each corner of the tetrahedron (Fig. 2.2a). The charge on a single tetrahedron is $-4$ and to achieve a neutral charge, cations must be added or single tetrahedrons must be linked to each other sharing oxygen ions. Silicate minerals are formed by addition of cations and interactions of tetrahedrons. Silica tetrahedrons combine to form sheets, called silicate sheets, which are thin layers of silica tetrahedrons in which three oxygen ions are shared between adjacent tetrahedrons (Fig. 2.2b). Silicate sheets may contain other structural units such as alumina sheets. Alumina sheets are formed by combination of alumina minerals, which consists of an aluminum ion surrounded by six oxygen or hydroxyl atoms in an octahedron (Fig. 2.2c,d).

The main groups of crystalline materials that make up clays are the minerals: kaolinite, illite, and montmorillonite. Kaolinite has a structure that consists

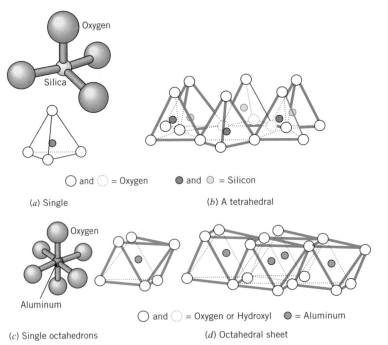

$\bigcirc$ and $\bigcirc$ = Oxygen        $\bullet$ and $\bigcirc$ = Silicon

(*a*) Single                (*b*) A tetrahedral

$\bigcirc$ and $\bigcirc$ = Oxygen or Hydroxyl    $\bullet$ = Aluminum

(*c*) Single octahedrons              (*d*) Octahedral sheet

**FIGURE 2.2**   (a) Silica tetrahedrons, (b) silica sheets, (c) single aluminum octahedrons, and (d) aluminum sheets.

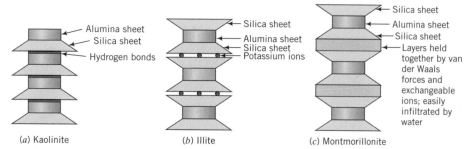

**FIGURE 2.3**    Structure of kaolinite, illite, and montmorillonite.

of one silica sheet and one alumina sheet bonded together into a layer about 0.72 nm thick and stacked repeatedly (Fig. 2.3a). The layers are held together by hydrogen bonds. Tightly stacked layers results from numerous hydrogen bonds. Kaolinite is common in clays in humid tropical regions. Illite consists of repeated layers of one alumina sheet sandwiched by two silicate sheets (Fig. 2.3b). The layers, each of thickness 0.96 nm, are held together by potassium ions.

Montmorillonite has a structure similar to illite, but the layers are held together by weak van der Waals forces and exchangeable ions (Fig. 2.3c). Water can easily enter the bond and separate the layers in montmorillonite, causing swelling. Montmorillonite is often called a swelling or expansive clay.

### 2.3.4 Surface Forces and Adsorbed Water

If we subdivide a body, the ratio of its surface area to its volume increases. For example, a cube of sides 1 cm has a surface area of 6 cm$^2$. If we subdivide this cube into smaller cubes of sides 1 mm, the original volume is unchanged but the surface area increases to 60 cm$^2$. The surface area per unit mass (specific surface) of sands is typically 0.01 m$^2$ per gram, while for clays, it is as high as 1000 m$^2$ per gram (montmorillonite). The specific surface of kaolinite ranges from 10 to 20 m$^2$ per gram while that of illite ranges from 65 to 100 m$^2$ per gram. The surface area of 45 grams of illite is equivalent to the area of a football field. Because of their large surfaces, surface forces significantly influence the behavior of fine-grained soils compared to coarse-grained soils.

The surface charges on fine-grained soils are negative (anions). These negative surface charges attract cations and the positively charged side of water molecules from surrounding water. Consequently, a thin film or layer of water, called adsorbed water, is bonded to the mineral surfaces. The thin film or layer of water is known as the diffuse double layer (Fig. 2.4). The largest concentration of cations occurs at the mineral surface and decreases exponentially with distance away from the surface (Fig. 2.4).

Drying of most soils, with the exception of gypsum, using an oven for which the standard temperature is 105 ± 5°C, cannot remove the adsorbed water. The adsorbed water influences the way a soil behaves. For example, plasticity, which we will deal with in Section 2.6, in soils is attributed to the adsorbed water. Toxic chemicals that seep into the ground contaminate soil and groundwater. The surface chemistry of fine-grained soils is important in understanding the migration, sequestration, re-release, and ultimate removal of toxic compounds from soils.

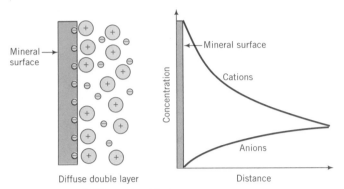

**FIGURE 2.4** Diffuse double layer.

Our main concern in this book is on the physical and mechanical properties of soils. Accordingly, we will not deal with the surface chemistry of fine-grained soils. You may refer to Mitchell (1993) for further information on the surface chemistry of fine-grained soils that are of importance to geotechnical and geoenvironmental engineers.

### 2.3.5 Soil Fabric

Soil particles are assumed rigid. During deposition, the mineral particles are arranged into structural frameworks that we call soil fabric (Fig. 2.5). Each particle is in random contact with neighboring particles. The environment under which deposition occurs influences the structural framework that is formed. In particular, the electrochemical environment has the greatest influence on the kind of soil fabric that is formed during deposition.

Two common types of soil fabric—flocculated and dispersed—are formed during soil deposition as shown schematically in Fig. 2.5. A flocculated structure, formed under a saltwater environment, results when many particles tend to orient

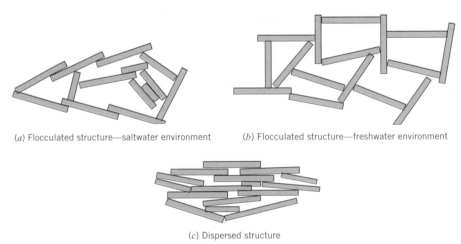

(a) Flocculated structure—saltwater environment      (b) Flocculated structure—freshwater environment

(c) Dispersed structure

**FIGURE 2.5** Soil fabric.

parallel to each other. A flocculated structure, formed under a freshwater environment, results when many particles tend to orient perpendicular to each other. A dispersed structure is the result when a majority of the particles orient parallel to each other.

Any loading (tectonic or otherwise) during or after deposition permanently alters the soil fabric or structural arrangement in a way that is unique to that particular loading condition. Consequently, the history of loading and changes in the environment is imprinted in the soil fabric. The soil fabric is the brain; it retains the memory of the birth of the soil and subsequent changes that occur.

The spaces between the mineral particles are called voids, which may be filled with liquids (essentially water) and gases (essentially air). Voids occupy a large proportion of the soil volume. Interconnected voids form the passageway through which water flows in and out of soils. If we change the volume of voids, we will cause the soil to either compress (settle) or expand (dilate). Loads applied by a building, for example, will cause the mineral particles to be forced closer together, reducing the volume of voids and changing the orientation of the structural framework. Consequently, the building settles. The amount of settlement depends on how much we compress the volume of voids. The rate at which the settlement occurs depends on the interconnectivity of the voids. Free water, not the adsorbed water, and/or air trapped in the voids must be forced out for settlement to occur. The decrease in volume, which results in settlement of buildings and other structures, is usually very slow in fine-grained soils and almost ceaseless because of their (fine-grained soils) large surface area compared with coarse-grained soils. The larger surface area of fine-grained soils compared with coarse-grained soils provides greater resistance to the flow of water through the voids.

### 2.3.6 Comparison of Coarse-Grained and Fine-Grained Soils for Engineering Use

Coarse-grained soils have good load-bearing capacities and good drainage qualities, and their strength and volume change characteristics are not significantly affected by change in moisture conditions. They are practically incompressible when dense, but significant volume changes can occur when they are loose. Vibrations accentuate volume changes in loose coarse-grained soils by rearranging the soil fabric into a dense configuration.

Fine-grained soils have poor load-bearing capacities compared with coarse-grained soils. Fine-grained soils are practically impermeable, change volume and strength with variations in moisture conditions, and are frost susceptible. The engineering properties of coarse-grained soils are controlled mainly by the grain size of the particles and their structural arrangement. The engineering properties of fine-grained soils are controlled by mineralogical factors rather than grain size. Thin layers of fine-grained soils, even within thick deposits of coarse-grained soils, have been responsible for many geotechnical failures and therefore you need to pay special attention to fine-grained soils.

In this book, we will deal with soil as a construction and a foundation material. We will not consider soils containing organic material or the parent material of soils—rock. We will label our soils as engineering soils to distinguish our consideration of soils from geologists, agronomists, and soil scientists, who have additional interests in soils not related to construction activities.

*The* **essential points** *are:*

1. *Soils are derived from the weathering of rocks and are commonly described by textural terms such as gravels, sands, silts, and clays.*

2. *Particle size is used to distinguish various soil textures.*

3. *Clays are composed of three main types of mineral—kaolinite, illite, and montmorillonite.*

4. *The clay minerals consist of silica and alumina sheets that are combined to form layers. The bonds between layers play a very important role in the mechanical behavior of clays. The bond between the layers in montmorillonite is very weak compared with kaolinite and illite. Water can easily enter between the layers in montmorillonite, causing swelling.*

5. *A thin layer of water is bonded to the mineral surfaces of soils and significantly influences the physical and mechanical characteristics of fine-grained soils.*

6. *Fine-grained soils have much larger surface areas than coarse-grained soils and are responsible for the major physical and mechanical differences between coarse-grained and fine-grained soils.*

7. *The engineering properties of fine-grained soils depend mainly on mineralogical factors.*

*What's next . . .* You should now have a general idea of engineering soil. Next, we will describe the constituents of soils and develop fundamental relationships between them. These relationships should be memorized. Many of these relationships can be presented in different forms. The most popular form of each is presented. When you work on these relationships, think about a bread dough in which you have to reconstruct the amount of the constituent ingredients, for example, the amount of flour or water. If you add too much water to a bread dough it becomes softer and more malleable. The same phenomenon occurs in fine-grained soils.

## 2.4  PHASE RELATIONSHIPS

Soil is composed of solids, liquids, and gases (Fig. 2.6a). The solid phase may be mineral, organic matter, or both. As mentioned before, we will not deal with organic matter in this textbook. The spaces between the solids (soil particles) are called voids. Water is often the predominant liquid and air is the predominant gas. We will use the terms water and air instead of liquids and gases. The soil water is commonly called pore water and it plays a very important role in the behavior of soils under load. If all the voids are filled by water, the soil is saturated. Otherwise, the soil is unsaturated. If all the voids are filled with air, the soil is said to be dry.

We can idealize the three phases of soil as shown in Fig. 2.6b. The physical properties of soils are influenced by the relative proportions of each of these

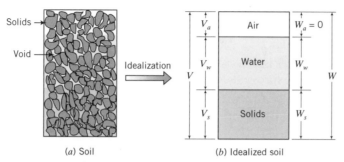

(a) Soil          (b) Idealized soil

FIGURE 2.6  **Soil phases.**

phases. The total volume of the soil is the sum of the volume of solids ($V_s$), volume of water ($V_w$), and volume of air ($V_a$): that is,

$$V = V_s + V_w + V_a = V_s + V_v \tag{2.1}$$

where

$$V_v = V_w + V_a$$

is the volume of voids. The weight of the soil is the sum of the weight of solids ($W_s$) and the weight of water ($W_w$). The weight of air is negligible. Thus,

$$W = W_s + W_w \tag{2.2}$$

The following definitions have been established to describe the proportion of each constituent in soil. Each equation can be presented with different variables. The most popular and convenient forms are given. You should try to memorize these definitions and equations.

**1.** *Water content* ($w$) is the ratio, often expressed as a percentage, of the weight of water to the weight of solids:

$$w = \frac{W_w}{W_s} \times 100\% \tag{2.3}$$

The water content of a soil is found by weighing a sample of the soil and then placing it in an oven at $110 \pm 5°C$ until the weight of the sample remains constant; that is, all the absorbed water is driven out. For most soils, a constant weight is achieved in about 24 hours. The soil is removed from the oven, cooled, and then weighed. Example 2.2 illustrates the measurements and calculations required to determine the water content.

**2.** *Void ratio* ($e$) is the ratio of the volume of void space to the volume of solids. Void ratio is usually expressed as a decimal quantity.

$$e = \frac{V_v}{V_s} \tag{2.4}$$

**3.** *Specific volume* ($V'$) is the volume of soil per unit volume of solids:

$$V' = \frac{V}{V_s} = 1 + e \tag{2.5}$$

This equation is useful in relating volumes as shown in Example 2.3 and in the calculation of volumetric strains (Chapter 3).

**4. Porosity** ($n$) is the ratio of the volume of voids to the total volume. Porosity is usually expressed as a percentage.

$$n = \frac{V_v}{V} \qquad (2.6)$$

Porosity and void ratio are related by the expression

$$n = \frac{e}{1 + e} \qquad (2.7)$$

Let us prove Eq. (2.7). We will start with the basic definition [Eq. (2.6)] and then algebraically manipulate it to get Eq. (2.7). The total volume will be decomposed into the volume of solids and the volume of voids and then both the numerator and denominator will be divided by the volume of solids; that is,

$$n = \frac{V_v}{V} = \frac{V_v}{V_s + V_v} = \frac{V_v/V_s}{V_s/V_s + V_v/V_s} = \frac{e}{1 + e}$$

**5. Specific gravity** ($G_s$) is the ratio of the weight of the soil solids to the weight of water of equal volume:

$$G_s = \frac{W_s}{V_s \gamma_w} \qquad (2.8)$$

where $\gamma_w = 9.81$ kN/m$^3$ is the unit weight of water. We will use $\gamma_w = 9.8$ kN/m$^3$ in this book. The specific gravity of soils ranges from approximately 2.6 to 2.8. For most problems, $G_s$ can be assumed, with little error, to be equal to 2.7.

Two types of container are used to determine the specific gravity. One is a pycnometer, which is used for coarse-grained soils. The other is a 50 mL density bottle, which is used for fine-grained soils. The container is weighed and a small quantity of dry soil is placed in it. The mass of the container and the dry soil is determined. De-aired water is added to the soil in the container. The container is then agitated to remove air bubbles. When all air bubbles have been removed, the container is filled with de-aired water. The mass of container, soil, and water is determined. The contents of the container are discarded and the container is thoroughly cleaned. De-aired water is added to fill the container and the mass of the container and water is determined.

Let $m_1$ be the mass of the container; $m_2$ be the mass of the container and dry soil; $m_3$ be the mass of the container, soil, and water; and $m_4$ be the mass of the container and water. The mass of dry soil is $m_s = m_2 - m_1$, the mass of water displaced by the soil particles is $m_5 = m_4 - m_3 + m_s$, and $G_s = m_s/m_5$.

**6. Degree of saturation** ($S$) is the ratio, often expressed as a percentage, of the volume of water to the volume of voids:

$$S = \frac{V_w}{V_v} = \frac{wG_s}{e} \qquad \text{or} \qquad Se = wG_s \qquad (2.9)$$

If $S = 1$ or 100%, the soil is saturated. If $S = 0$, the soil is bone dry. It is practically impossible to obtain a soil with $S = 0$.

**TABLE 2.1 Typical Values of Unit Weight for Soils**

| Soil type | $\gamma_{sat}$ (kN/m³) | $\gamma_d$ (kN/m³) |
|-----------|------------------------|---------------------|
| Gravel | 20–22 | 15–17 |
| Sand | 18–20 | 13–16 |
| Silt | 18–20 | 14–18 |
| Clay | 16–22 | 14–21 |

**7.** *Unit weight* is the weight of a soil per unit volume. We will use the term *bulk unit weight*, $\gamma$, to denote unit weight:

$$\gamma = \frac{W}{V} = \left(\frac{G_s + Se}{1 + e}\right)\gamma_w \qquad (2.10)$$

**Special Cases**

**(a)** Saturated unit weight ($S = 1$):

$$\gamma_{sat} = \left(\frac{G_s + e}{1 + e}\right)\gamma_w \qquad (2.11)$$

**(b)** Dry unit weight:

$$\gamma_d = \frac{W_s}{V} = \left(\frac{G_s}{1 + e}\right)\gamma_w = \frac{\gamma}{1 + w} \qquad (2.12)$$

**(c)** Effective or buoyant unit weight is the weight of a saturated soil, surrounded by water, per unit volume of soil:

$$\gamma' = \gamma_{sat} - \gamma_w = \left(\frac{G_s - 1}{1 + e}\right)\gamma_w \qquad (2.13)$$

Typical values of unit weight of soils are given in Table 2.1.

**8.** *Relative density* ($D_r$) is an index that quantifies the degree of packing between the loosest and densest possible state of coarse-grained soils as determined by experiments:

$$D_r = \frac{e_{max} - e}{e_{max} - e_{min}} \qquad (2.14)$$

where $e_{max}$ is the maximum void ratio (loosest condition), $e_{min}$ is the minimum void ratio (densest condition), and $e$ is the current void ratio.

The maximum void ratio is found by pouring dry sand, for example, into a mold of volume ($V$) 2830 cm³ using a funnel. The sand that fills the mold is weighed. If the weight of the sand is $W$, then by combining Eqs. (2.10) and (2.12) we get $e_{max} = G_s\gamma_w(V/W) - 1$. The minimum void ratio is determined by vibrating the sand with a weight imposing a vertical stress of 13.8 kPa on top of the sand. Vibration occurs for 8 minutes at a frequency of 3600 Hz and amplitude of 0.064

**TABLE 2.2   Description Based on Relative Density**

| $D_r$ (%) | Description |
|-----------|-------------|
| 0–15      | Very loose  |
| 15–35     | Loose       |
| 35–65     | Medium dense |
| 65–85     | Dense       |
| 85–100    | Very dense  |

mm. From the weight of the sand ($W_1$) and the volume ($V_1$) occupied by it after vibration, we can calculate the minimum void ratio using $e_{min} = G_s\gamma_w(V_1/W_1) - 1$.

The relative density correlates very well with the strength of coarse-grained soils—denser soils being stronger than looser soils. A description of sand based on relative density is given in Table 2.2.

*What's next . . .*Four examples will be used to illustrate how to solve a variety of problems involving the constituents of soils. In the first example, we will derive some of the equations describing relationships between the soil constituents.

**EXAMPLE 2.1**

Prove the following relationships:

(a) $S = \dfrac{wG_s}{e}$

(b) $\gamma_d = \dfrac{\gamma}{1 + w}$

(c) $\gamma = \left(\dfrac{G_s + Se}{1 + e}\right)\gamma_w = \dfrac{G_s\gamma_w\,(1 + w)}{1 + e}$

*Strategy*   The proofs of these equations are algebraic manipulations. Start with the basic definition and then manipulate the basic equation algebraically to get the desired form.

**Solution 2.1**

(a) For this relationship, we proceed as follows:

**Step 1:**  Write down the basic equation,

$$S = \frac{V_w}{V_v}$$

**Step 2:**  Manipulate the basic equation to get the desired equation.
You want to get $e$ in the denominator and you have $V_v$. You know that $V_v = eV_s$ and $V_w$ is the weight of water divided by the unit

weight of water. From the definition of water content, the weight of water is $wW_s$. Here is the algebra:

$$V_v = eV_s; \quad V_w = \frac{W_w}{\gamma_w} = \frac{wW_s}{\gamma_w}$$

$$S = \frac{wW_s}{e\gamma_w V_s} = \frac{G_s w}{e}$$

(b) For this relationship, we proceed as follows:

**Step 1:** Write down the basic equation,

$$\gamma_d = \frac{W_s}{V}$$

**Step 2:** Manipulate the basic equation to get the new form of the equation.

$$\gamma_d = \frac{W_s}{V} = \frac{W - W_w}{V} = \frac{W}{V} - \frac{wW_s}{V} = \gamma - w\gamma_d$$

$$\therefore \gamma_d + w\gamma_d = \gamma$$

$$\gamma_d = \frac{\gamma}{1 + w}$$

(c) For this relationship, we proceed as follows:

**Step 1:** Start with the basic equation,

$$\gamma = \frac{W}{V}$$

**Step 2:** Manipulate the basic equation to get the new form of the equation.

$$\gamma = \frac{W}{V} = \frac{W_s + W_w}{V_s + V_v} = \frac{W_s + wW_s}{V_s + V_v}$$

Substituting $w = Se/G_s$ and $V_v = eV_s$, we obtain

$$\gamma = \frac{W_s(1 + Se/G_s)}{V_s(1 + e)}$$

$$= \frac{G_s\gamma_w(1 + Se/G_s)}{1 + e} = \frac{G_s\gamma_w(1 + w)}{1 + e}$$

or

$$\gamma = \left(\frac{G_s + Se}{1 + e}\right)\gamma_w$$

∎

## EXAMPLE 2.2

A sample of saturated clay was placed in a container and weighed. The weight was 6 N. The clay in its container was placed in an oven for 24 hours at 105°C. The weight reduced to a constant value of 5 N. The weight of the container is 1 N. If $G_s = 2.7$, determine the (a) water content, (b) void ratio, (c) bulk unit weight, (d) dry unit weight, and (e) effective unit weight.

***Strategy***   Write down what is given and then use the appropriate equations to find the unknowns. You are given the weight of the natural soil, sometimes called the wet weight, and the dry weight of the soil. The difference between these will give the weight of water and you can find the water content by using Eq. (2.3). You are also given a saturated soil, which means that $S = 1$.

## Solution 2.2

**Step 1:**   Write down what is given.

Weight of sample + container = 6 N.
Weight of dry sample + container = 5 N.

**Step 2:**   Determine the weight of water and the weight of dry soil:

Weight of water:   $W_w = 6 - 5 = 1$ N
Weight of dry soil:   $W_s = 5 - 1 = 4$ N

**Step 3:**   Determine the water content.

$$w = \frac{W_w}{W_s} \times 100 = \frac{1}{4} \times 100 = 25\%$$

**Step 4:**   Determine the void ratio.

$$e = \frac{wG_s}{S} = \frac{0.25 \times 2.7}{1} = 0.675$$

**Step 5:**   Determine the bulk unit weight.

$$\gamma = \frac{W}{V} = \frac{G_s\gamma_w(1 + w)}{1 + e} \quad \text{(see Example 2.1)}$$

$$\gamma = \frac{2.7 \times 9.8(1 + 0.25)}{1 + 0.675} = 19.7 \text{ kN/m}^3$$

In this case the soil is saturated, so that the bulk unit weight is equal to the saturated unit weight.

**Step 6:**   Determine the dry unit weight.

$$\gamma_d = \frac{W_s}{V} = \left(\frac{G_s}{1 + e}\right)\gamma_w = \frac{2.7}{1 + 0.675} \times 9.8 = 15.8 \text{ kN/m}^3$$

or

$$\gamma_d = \frac{\gamma}{(1 + w)} = \frac{19.7}{1 + 0.25} = 15.8 \text{ kN/m}^3$$

**Step 7:**   Determine the effective unit weight.

$$\gamma' = \left(\frac{G_s - 1}{1 + e}\right)\gamma_w = \left(\frac{2.7 - 1}{1 + 0.675}\right) \times 9.8 = 9.9 \text{ kN/m}^3$$

or

$$\gamma' = \gamma_{\text{sat}} - \gamma_w = 19.7 - 9.8 = 9.9 \text{ kN/m}^3 \qquad \blacksquare$$

## EXAMPLE 2.3

An embankment for a highway is to be constructed from a soil compacted to a dry unit weight of 18 kN/m³. The clay has to be trucked to the site from a borrow pit. The bulk unit weight of the soil in the borrow pit is 17 kN/m³ and its natural water content is 5%. Calculate the volume of clay from the borrow pit required for 1 cubic meter of embankment. Assume $G_s = 2.7$.

***Strategy*** This problem can be solved in many ways. We will use two of these ways. One way is direct; the other a bit longer. In the first way, we are going to use the ratio of the dry unit weights of the compacted soil to the borrow pit soil to determine the volume. In the second way, we will use the specific volume. In this case, we need to find the void ratio for the borrow pit clay and the desired void ratio for the embankment. We can then relate the specific volumes of the embankment and the borrow pit clay.

## Solution 2.3

**Step 1:** Find the dry unit weight of the borrow pit soil.

$$\gamma_d = \frac{\gamma}{1 + w} = \frac{17}{1 + 0.05} = 16.2 \text{ kN/m}^3$$

**Step 2:** Find the volume of borrow pit soil required.

Volume of borrow pit soil required per m³

$$= \frac{(\gamma_d)_{\text{compacted soil}}}{(\gamma_d)_{\text{borrow pit soil}}} = \frac{18}{16.2} = 1.11 \text{ m}^3$$

*Alternatively:*

**Step 1:** Define parameters for the borrow pit and embankment. Let

$V_1'$, $e_1$ = specific volume and void ratio, respectively, of borrow pit clay
$V_2'$, $e_2$ = volume and void ratio, respectively, of compacted clay

**Step 2:** Determine $e_1$ and $e_2$.

$$\gamma_d = \frac{\gamma}{1 + w} = \frac{17}{1 + 0.05} = 16.2 \text{ kN/m}^3$$

But

$$\gamma_d = \frac{G_s}{1 + e_1} \gamma_w$$

and therefore

$$e_1 = G_s \frac{\gamma_w}{\gamma_d} - 1 = 2.7\left(\frac{9.8}{16.2}\right) - 1 = 0.633$$

Similarly,

$$e_2 = G_s \frac{\gamma_w}{\gamma_d} - 1 = 2.7\left(\frac{9.8}{18}\right) - 1 = 0.47$$

**Step 3:** Determine the volume of borrow pit material.

$$\frac{V_1'}{V_2'} = \frac{1 + e_1}{1 + e_2}$$

Therefore

$$V_1' = V_2' \frac{1 + e_1}{1 + e_2} = 1\left(\frac{1 + 0.633}{1 + 0.47}\right) = 1.11 \text{ m}^3 \quad \blacksquare$$

### EXAMPLE 2.4

If the borrow soil in Example 2.3 were to be compacted to attain a dry unit weight of 18 kN/m³ at a water content of 7%, determine the amount of water required per cubic meter of embankment, assuming no loss of water during transportation.

**Strategy**   Since water content is related to the weight of solids and not the total weight, we need to use the data given to find the weight of solids.

### Solution 2.4

**Step 1:** Determine the weight of solids per unit volume of borrow pit soil.

$$W_s = \frac{\gamma}{1 + w} = \frac{17}{1 + 0.05} = 16.2 \text{ kN/m}^3$$

**Step 2:** Determine the amount of water required.

Additional water = $7 - 5 = 2\%$

Weight of water = $W_w = wW_s = 0.02 \times 16.2 = 0.32$ kN

$$V_w = \frac{W_w}{\gamma_w} = \frac{0.32}{9.8} = 0.033 \text{ m}^3 = 33 \text{ liters} \quad \blacksquare$$

*What's next . . .*In most soils, there is a distribution of particles with various sizes. The distribution of particle size influences the response of soils to loads and to flow of water. We will describe methods used in the laboratory to find particle sizes in soils.

## 2.5   DETERMINATION OF PARTICLE SIZE OF SOILS

### 2.5.1 Particle Size of Coarse-Grained Soils

The distribution of particle sizes or average grain diameter of coarse-grained soils—gravels and sands—is obtained by screening a known weight of the soil through a stack of sieves of progressively finer mesh size. A typical stack of sieves is shown in Fig. 2.7.

Each sieve is identified by a number that corresponds to the number of square holes per linear inch of mesh. The particle diameter in the screening process, often called sieve analysis, is the maximum particle dimension to pass through the square hole of a particular mesh. A known weight of dry soil is placed on the largest sieve (the top sieve) and the nest of sieves is then placed on a vibrator, called a sieve shaker, and shaken. The nest of sieves is dismantled, one sieve at a time. The soil retained on each sieve is weighed and the percentage of soil retained on each sieve is calculated. The results are plotted on a graph of

**FIGURE 2.7** Stack of sieves.

percent of particles finer than a given sieve (not the percent retained) as the ordinate versus the logarithm of the particle sizes as shown in Fig. 2.8. The resulting plot is called a particle size distribution curve(s) or, simply, the gradation curve(s). Engineers have found it convenient to use a logarithm scale for particle size because the ratio of particle sizes from the largest to the smallest in a soil can be greater than $10^4$.

Let $W_i$ be the weight of soil retained on the $i$th sieve from the top of the nest of sieves and $W$ be the total soil weight. The percent weight retained is

$$\% \text{ Retained on } i\text{th sieve} = \frac{W_i}{W} \times 100 \qquad (2.15)$$

The percent finer is

$$\% \text{ Finer than } i\text{th sieve} = 100 - \sum_{i=1}^{i} (\% \text{ Retained on } i\text{th sieve}) \qquad (2.16)$$

You can use mass instead of weight. The unit of mass is grams or kilograms.

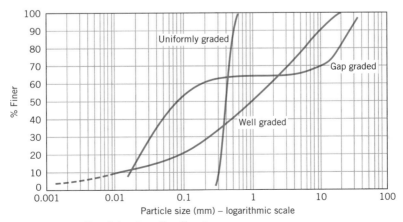

**FIGURE 2.8** Particle size distribution curves.

## 2.5.2 Particle Size of Fine-Grained Soils

The screening process cannot be used for fine-grained soils—silts and clays—because of their extremely small size. The common laboratory method to determine the size distribution of fine-grained soils is a hydrometer test (Fig. 2.9). The hydrometer test involves mixing a small amount of soil into a suspension and observing how the suspension settles in time. Larger particles will settle quickly followed by smaller particles. When the hydrometer is lowered into the suspension, it will sink into the suspension until the buoyancy force is sufficient to balance the weight of the hydrometer. The length of the hydrometer projecting above the suspension is a function of the density, so it is possible to calibrate the hydrometer to read the density of the suspension at different times. The calibration of the hydrometer is affected by temperature and the specific gravity of the suspended solids. You must then apply a correction factor to your hydrometer reading based on the test temperatures.

Typically, a hydrometer test is conducted by taking a small quantity of a dry fine-grained soil (approximately 10 grams) and thoroughly mixing it with distilled water to form a paste. The paste is placed in a 1 liter glass cylinder and distilled water is added to bring the level to the 1 liter mark. The glass cylinder is then repeatedly shaken and inverted before being placed in a constant-temperature bath. A hydrometer is placed in the glass cylinder and a clock is simultaneously started. At different times, the hydrometer is read. The diameter $D$ of the particle at time $t_D$ is calculated from Stokes's law as

$$D = \sqrt{\frac{18\mu z}{(G_s - 1)\gamma_w t_D}} \tag{2.17}$$

where $\mu$ is the viscosity of water (10.09 millipoises at 20°C), $z$ is the depth, $\gamma_w$ is the unit weight of water, and $G_s$ is the specific gravity.

There are several assumptions made in Stokes's law that are not fulfilled by the hydrometer test. In Stokes's law, the particles are assumed to be free-falling spheres with no collision, but the mineral particles of clays are plate-like and collision of particles during sedimentation is unavoidable. However, the results of this test suffice for most geotechnical engineering needs. For more accurate size distribution measurements in fine-grained soils, other more

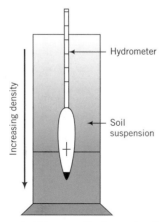

**FIGURE 2.9** Hydrometer in soil-water suspension.

sophisticated methods are available (e.g., light scattering methods). The dashed line in Fig. 2.8 shows a typical particle size distribution for fine-grained soils.

### 2.5.3 Characterization of Soils Based on Particle Size

The grading curve is used for textural classification of soils. Various classification systems have evolved over the years to describe soils based on their particle size. Each system was developed for a specific engineering purpose. In this book, we will use the Unified Soil Classification System (USCS). The USCS separates soils into two categories. One category is coarse-grained soils that are delineated if more than 50% of the soil is greater than 0.075 mm (No. 200 sieve). The other category is fine-grained soils that are delineated if more than 50% of the soil is finer than 0.075 mm. Coarse-grained soils are subdivided into gravels and sands while fine-grained soils are divided into silts and clays. Each soil type—gravel, sand, silt, and clay—is identified by grain size as shown in Table 2.3. The USCS does not differentiate silts from clays. Clays have particle sizes less than 0.002 mm. We will discuss the USCS in more detail in Section 2.8.

Real soils consist of a mixture of particle sizes. The selection of a soil for a particular use may depend on the assortment of particles it contains. Two coefficients have been defined to provide guidance on distinguishing soils based on the distribution of the particles. One of these is a numerical measure of uniformity, called the *uniformity coefficient*, UC, defined as

$$\text{UC} = \frac{D_{60}}{D_{10}} \tag{2.18}$$

where $D_{60}$ is the diameter of the soil particles for which 60% of the particles are finer, and $D_{10}$ is the diameter of the soil particles for which 10% of the particles are finer. Both of these diameters are obtained from the grading curve.

The other coefficient is the *coefficient of curvature*, CC (other terms used are the coefficient of gradation and the coefficient of concavity), defined as

$$\text{CC} = \frac{(D_{30})^2}{D_{10}D_{60}} \tag{2.19}$$

where $D_{30}$ is the diameter of the soil particles for which 30% of the particles are finer.

### TABLE 2.3 Soil Types, Descriptions, and Average Grain Sizes According to USCS

| Soil type | Description | Average grain size |
|---|---|---|
| Gravel | Rounded and/or angular bulky hard rock | Coarse: 75 mm to 19 mm<br>Fine: 19 mm to 4.75 mm |
| Sand | Rounded and/or angular bulky hard rock | Coarse: 4.75 mm to 1.7 mm<br>Medium: 1.7 mm to 0.380 mm<br>Fine: 0.380 mm to 0.075 mm |
| Silt | Particles smaller than 0.075 mm, exhibit little or no strength when dried | 0.075 mm to 0.002 mm |
| Clay | Particles smaller than 0.002 mm, exhibit significant strength when dried; water reduces strength | <0.002 mm |

A soil that has a uniformity coefficient of <4 contains particles of uniform size (approximately one size). The minimum value of UC is 1 and corresponds to an assemblage of particles of the same size. The gradation curve for a uniform soil is almost vertical (Fig. 2.8). Humps in the gradation curve indicate two or more uniform soils. Higher values of uniformity coefficient (>4) indicate a wider assortment of particle sizes. A soil that has a uniformity coefficient of >4 is described as a well-graded soil and is indicated by a flat curve (Fig. 2.8). The coefficient of curvature is between 1 and 3 for well-graded soils. The absence of certain grain sizes, termed gap-graded, is diagnosed by a coefficient of curvature outside the range 1 to 3 and a sudden change of slope in the particle size distribution curve as shown in Fig. 2.8.

Uniform soils are sorted by water (e.g., beach sands) or by wind. Gap-graded soils are also sorted by water but certain sizes were not transported. Well-graded soils are produced by bulk transport processes (e.g., glacial till). The uniformity coefficient and the coefficient of concavity are strictly applicable to coarse-grained soils.

The diameter $D_{10}$ is called the effective size of the soil and was established by Allen Hazen (1893) in connection with his work on soil filters. The effective size is the diameter of an artificial sphere that will approximately produce the same effect of an irregular shaped particle. The effective size is particularly important in regulating the flow of water through soils and can dictate the mechanical behavior of soils since the coarser fractions may not be in effective contact with each other; that is, they float in a matrix of finer particles. The higher the $D_{10}$ value, the coarser the soil and the better the drainage characteristics. The diameter of the finer particle sizes, in particular $D_{15}$, has been used to develop criteria for soil filters. Terzaghi and Peck (1948), for example, proposed the following set of criteria for an effective soil filter:

$$\frac{D_{15(F)}}{D_{85(BS)}} < 4 \quad \text{(to prevent the filter soil from being washed out)}$$

and

$$\frac{D_{15(F)}}{D_{15(BS)}} > 4 \quad \text{(to ensure a high rate of flow of water)}$$

where F denotes filter and BS is the base soil. The average diameter of a soil is given as $D_{50}$.

Particle size analyses have many uses in engineering. They are used to select aggregates for concrete, soils for the construction of dams and highways, soils as filters, and material for grouting and chemical injection. In Section 2.8, you will learn about how the particle size distribution is used with other physical properties of soils in a classification system designed to help you select soils for particular applications.

> *The* **essential points** *are:*
> 1. *A sieve analysis is used to determine the grain size distribution of coarse-grained soils.*
> 2. *For fine-grained soils, a hydrometer analysis is used to find the particle size distribution.*

3. *Particle size distribution is represented on a semilogarithmic plot of % finer (ordinate, arithmetic scale) versus particle size (abscissa, logarithm scale).*

4. *The particle size distribution plot is used to delineate the different soil textures (percentages of gravel, sand, silt, and clay) in a soil.*

5. *The effective size, $D_{10}$, is the diameter of the particles of which 10% of the soil is finer. $D_{10}$ is an important value in regulating flow through soils and can significantly influence the mechanical behavior of soils.*

6. *$D_{50}$ is the average grain size diameter of the soil.*

7. *Two coefficients—the uniformity coefficient and the coefficient of curvature—are used to characterize the particle size distribution. Uniform soils have uniformity coefficients <4 and steep gradation curves. Well-graded soils have uniformity coefficients >4, coefficients of curvature between 1 and 3, and flat gradation curves. Gap-graded soils have coefficients of curvature <1 or >3, and one or more humps on the gradation curves.*

## EXAMPLE 2.5

A sample of a dry coarse-grained material of mass 500 grams was shaken through a nest of sieves and the following results were obtained:

| Sieve no. | Opening (mm) | Mass retained (grams) |
|---|---|---|
| 4 | 4.75 | 0 |
| 10 | 2.00 | 14.8 |
| 20 | 0.85 | 98 |
| 40 | 0.425 | 90.1 |
| 100 | 0.15 | 181.9 |
| 200 | 0.075 | 108.8 |
| Pan | | 6.1 |

(a) Plot the particle size distribution curve.

(b) Determine (1) the effective size, (2) the average particle size, (3) the uniformity coefficient, and (4) the coefficient of curvature.

(c) Determine the textural composition of the soil (i.e., the amount of gravel, sand, etc.).

(d) Describe the gradation curve.

**Strategy** The best way to solve this type of problem is to make a table to carry out the calculations and then plot a gradation curve. Total mass of dry sample (M) used is 500 grams but on summing the masses of the retained soil in column 2 we obtain 499.7 grams. The reduction in mass is due to losses mainly from a small quantity of soil that gets stuck in the meshes of the sieves. You should use the "after sieving" total mass of 499.7 grams in the calculations.

## Solution 2.5

**Step 1:**   Tabulate data to obtain % finer.

| Sieve no. | Mass retained (grams) $M_r$ | % Retained $(M_r/M) \times 100$ | Σ(% Retained) | % Finer |
|---|---|---|---|---|
| 4 | 0 | 0 | 0 | $100 - 0 = 100$ |
| 10 | 14.8 | 3.0 | add   3.0 | $100 - 3.0 = 97.0$ |
| 20 | 98.0 | 19.6 | 22.6 | $100 - 22.6 = 77.4$ |
| 40 | 90.1 | 18.0 | 40.6 | $100 - 40.6 = 59.4$ |
| 100 | 181.9 | 36.4 | 77.0 | $100 - 77 = 23.0$ |
| 200 | 108.8 | 21.8 | 98.8 | $100 - 98.8 = 1.2$ |
| Pan | 6.1 | 1.2 | check | |
| Total mass $M = 499.7$ | | 100 | | |

**Step 2:**   Plot the gradation curve.
See Fig. E2.5 for a plot of the gradation curve.

**Step 3:**   Extract the effective size.

$$\text{Effective size} = D_{10} = 0.1 \text{ mm}$$

**Step 4:**   Extract percentages of gravel, sand, silt, and clay.

Gravel = 0%

Sand = 98.8%

Silt and clay = 1.2%

**Step 5:**   Calculate UC and CC.

$$UC = \frac{D_{60}}{D_{10}} = \frac{0.45}{0.1} = 4.5$$

$$CC = \frac{(D_{30})^2}{D_{10}D_{60}} = \frac{0.18^2}{0.1 \times 0.45} = 0.72$$

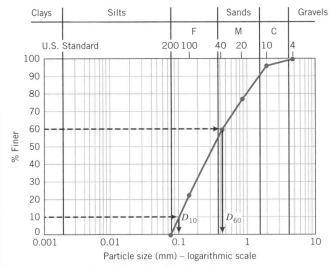

**FIGURE E2.5**   Particle size distribution curve.

***What's next*** . . .Particle size is adequate to classify coarse-grained soils. To classify fine-grained soils, we need additional information on the surface forces and mineral contents. We will discuss laboratory tests that are used to account for the surface forces and mineral content of a soil. In a conventional geotechnical course, you will be required to conduct the tests described next.

## 2.6 PHYSICAL STATES AND INDEX PROPERTIES OF FINE-GRAINED SOILS

The physical and mechanical behavior of fine-grained soils is linked to four distinct states: solid, semisolid, plastic, and liquid in order of increasing water content. Let us consider a soil initially in a liquid state that is allowed to dry uniformly. If we plot a diagram of volume versus water content as shown in Fig. 2.10, we can locate the original liquid state as point $A$. As the soil dries, its water content reduces and consequently its volume (see Fig. 2.6b).

At point $B$, the soil becomes so stiff that it can no longer flow as a liquid. The boundary water content at point $B$ is called the liquid limit; it is denoted by $w_{LL}$. As the soil continues to dry, there is a range of water content at which the soil can be molded into any desired shape without rupture. The soil at this state is said to exhibit plastic behavior—the ability to deform continuously without rupture. But if drying is continued beyond the range of water content for plastic behavior, the soil becomes a semisolid. The soil cannot be molded now without visible cracks appearing. The water content at which the soil changes from a plastic to a semisolid is known as the plastic limit, denoted by $w_{PL}$. The range of water contents over which the soil deforms plastically is known as the plasticity index, $I_p$:

$$\boxed{I_p = w_{LL} - w_{PL}}$$

(2.20)

As the soil continues to dry, it comes to a final state called the solid state. At this state, no further volume change occurs since nearly all the water in the soil has been removed. The water content at which the soil changes from a semisolid to a solid is called the shrinkage limit, denoted by $w_{SL}$. The shrinkage limit is useful for the determination of the swelling and shrinking capacity of soils. The liquid and plastic limits are called the Atterberg limits named after their originator, Swedish soil scientist, A. Atterberg (1911).

We have changed the states of fine-grained soils by changing the water

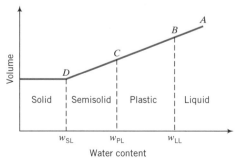

**FIGURE 2.10** Changes in soil states as a function of soil volume and water content.

**Table 2.4 Description of Soil Strength Based on Liquidity Index**

| Values of $I_L$ | Description of soil strength |
|---|---|
| $I_L < 0$ | Semisolid state—high strength, brittle (sudden) fracture is expected |
| $0 < I_L < 1$ | Plastic state—intermediate strength, soil deforms like a plastic material |
| $I_L > 1$ | Liquid state—low strength, soil deforms like a viscous fluid |

content. Since engineers are interested in the strength and deformation of materials, we can associate specific strength characteristics to each of the soil states. At one extreme, the liquid state, the soil has the lowest strength and the largest deformation. At the other extreme, the solid state, the soil has the largest strength and the lowest deformation. A measure of soil strength using the Atterberg limits is known as the liquidity index ($I_L$) and is expressed as

$$I_L = \frac{w - w_{PL}}{I_p} \tag{2.21}$$

The liquidity index is the ratio of the difference in water content between the natural or in situ water content of a soil and its plastic limit to its plasticity index. Table 2.4 shows a description of soil strength based on values of $I_L$

Typical values for the Atterberg limits for soils are shown in Table 2.5. The Atterberg limits depend on the type of predominant mineral in the soil. If montmorillonite is the predominant mineral, the liquid limit can exceed 100%. Why? Recall that the bond between the layers in montmorillonite is weak and large amounts of water can easily infiltrate the spaces between the layers. In the case of kaolinite, the layers are held relatively tightly and water cannot easily infiltrate between the layers in comparison with montmorillonite. Therefore, you can expect the Atterberg limits for kaolinite to be, in general, much lower than either montmorillonite or illite.

Skempton (1953) showed that for soils with a particular mineralogy, the plasticity index is linearly related to the amount of the clay fraction. He coined a term called activity ($A$) to describe the importance of the clay fractions on the plasticity index. The equation for $A$ is

$$A = \frac{I_p}{\text{Clay fraction (\%)}}$$

You should recall that the clay fraction is the amount of particles less than 2 μm.

**TABLE 2.5 Typical Atterberg Limits for Soils**

| Soil type | $w_{LL}$ (%) | $w_{PL}$ (%) | $I_p$ (%) |
|---|---|---|---|
| Sand | | Nonplastic | |
| Silt | 30–40 | 20–25 | 10–15 |
| Clay | 40–150 | 25–50 | 15–100 |

## 2.7 DETERMINATION OF THE LIQUID, PLASTIC, AND SHRINKAGE LIMITS

### 2.7.1 Casagrande Cup Method

The liquid limit is determined from an apparatus (Fig. 2.11) that consists of a semispherical brass cup that is repeatedly dropped onto a hard rubber base from a height of 10 mm by a cam-operated mechanism. The apparatus was developed by A. Casagrande (1932) and the procedure for the test is called the Casagrande cup method.

A dry powder of the soil is mixed with distilled water into a paste and placed in the cup to a thickness of about 12.5 mm. The soil surface is smoothed and a groove is cut into the soil using a standard grooving tool. The crank operating the cam is turned at a rate of 2 revolutions per second and the number of blows required to close the groove over a length of 12.5 mm is counted and recorded. A specimen of soil within the closed portion is extracted for determination of the water content. The liquid limit is defined as the water content at which the groove cut into the soil will close over a distance of 12.5 mm following 25 blows. This is difficult to achieve in a single test. Four or more tests at different water contents are usually required for terminal blows (number of blows to close the groove over a distance of 12.5 mm) ranging from 10 to 40. The results are presented in a plot of water content (ordinate, arithmetic scale) versus terminal blows (abscissa, logarithm scale) as shown in Fig. 2.12.

The best-fit straight line to the data points, usually called the flow line, is drawn. We will call this line the liquid state line to distinguish it from flow lines used in describing the flow of water through soils. The liquid limit is read from the graph as the water content on the liquid state line corresponding to 25 blows.

The cup method of determining the liquid limit has many shortcomings. Two of these are:

1. The tendency of soils of low plasticity to slide and to liquefy with shock in the cup rather than to flow plastically.

2. Sensitivity to operator and to small differences in apparatus.

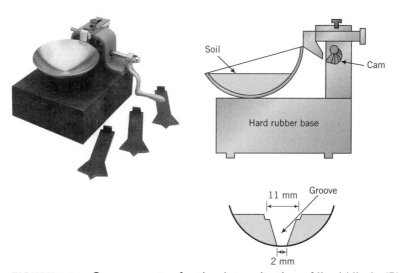

**FIGURE 2.11** Cup apparatus for the determination of liquid limit. (Photo courtesy of Geotest.)

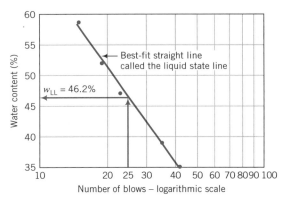

**FIGURE 2.12**   Typical liquid limit results from the Casagrande cup method.

## 2.7.2 Plastic Limit Test

The plastic limit is determined by rolling a small clay sample into threads and finding the water content at which threads approximately 3 mm in diameter will just start to crumble. Two or more determinations are made and the average water content is reported as the plastic limit.

## 2.7.3 Fall Cone Method to Determine Liquid and Plastic Limits

A fall cone test, popular in Europe and Asia, appears to offer a more accurate (less prone to operator's errors) method of determining both the liquid and plastic limits. In the fall cone test (Fig. 2.13), a cone with an apex angle of 30° and total mass of 80 grams is suspended above, but just in contact with, the soil sample. The cone is permitted to fall freely for a period of 5 seconds. The water content corresponding to a cone penetration of 20 mm defines the liquid limit.

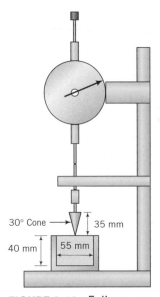

**FIGURE 2.13**   Fall cone apparatus.

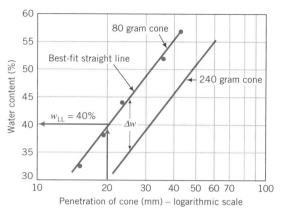

**FIGURE 2.14**  Typical fall cone test results.

The sample preparation is similar to the cup method except that the sample container in the fall cone test has a different shape and size (Fig. 2.13).

Four or more tests at different water contents are also required because of the difficulty of achieving the liquid limit from a single test. The results are plotted as water content (ordinate, arithmetic scale) versus penetration (abscissa, logarithm scale) and the best-fit straight line (liquid state line) linking the data points is drawn (Fig. 2.14). The liquid limit is read from the plot as the water content on the liquid state line corresponding to a penetration of 20 mm. The plastic limit is found by repeating the test with a cone of similar geometry, but with a mass of 240 grams. The penetration depth in the soil for the bigger cone mass at given water content will be larger than the smaller cone mass of 80 grams. Thus, the liquid state line for the 240 gram cone will be below the liquid state line for the 80 gram cone and parallel to it. The plastic limit is given as

$$w_{PL} = w_{LL} - \frac{2\Delta w}{\log_{10}(M_2/M_1)} = w_{LL} - 4.2\Delta w \tag{2.22}$$

where $\Delta w$ is the separation in terms of water content between the liquid state lines (Fig. 2.14) of the two cones, $M_1$ is the mass of 80 gram cone, and $M_2$ is the mass of the 240 gram cone.

## 2.7.4 Shrinkage Limit

The shrinkage limit is determined as follows. A mass of wet soil, $m_1$, is placed in a porcelain dish 44.5 mm in diameter and 12.5 mm high and then oven-dried. The volume of oven-dried soil is determined by using mercury to occupy the vacant spaces caused by shrinkage. The mass of the mercury is determined and the volume decrease caused by shrinkage can be calculated from the known density of mercury. The shrinkage limit is calculated from

$$w_{SL} = \left( \frac{m_1 - m_2}{m_2} - \frac{V_1 - V_2}{m_2} \frac{\gamma_w}{g} \right) \times 100 \tag{2.23}$$

where $m_1$ is the mass of the wet soil, $m_2$ is the mass of the oven-dried soil, $V_1$ is the volume of wet soil, $V_2$ is the volume of the oven-dried soil, and $g$ is the acceleration due to gravity (9.8 m/s$^2$).

*The essential points are:*

1. *Fine-grained soils can exist in one of four states: solid, semisolid, plastic, and liquid.*
2. *Water is the agent that is responsible for changing the states of soils.*
3. *A soil gets weaker if its water content increases.*
4. *Three limits are defined based on the water content that causes a change of state. These are the liquid limit—the water content that caused the soil to change from a liquid to a plastic state; the plastic limit—the water content that caused the soil to change from a plastic to a semisolid; and the shrinkage limit—the water content that caused the soil to change from a semisolid to a solid state. All these limiting water contents are found from laboratory tests.*
5. *The plasticity index defines the range of water content for which the soil behaves like a plastic material.*
6. *The liquidity index gives a measure of strength.*

## EXAMPLE 2.6

A liquid limit test conducted on a soil sample in the cup device gave the following results:

| Number of blows | 10 | 19 | 23 | 27 | 40 |
|---|---|---|---|---|---|
| Water content (%) | 60.0 | 45.2 | 39.8 | 36.5 | 25.2 |

Two determinations for the plastic limit gave water contents of 20.3% and 20.8%. Determine (a) the liquid limit and plastic limit, (b) the plasticity index, (c) the liquidity index if the natural water content is 27.4%, and (d) the void ratio at the liquid limit, if $G_s = 2.7$. If the soil were to be loaded to failure, would you expect a brittle failure?

**Strategy**   To get the liquid limit, you must make a semi-logarithm plot of water content versus number of blows. Use the data to make your plot, then extract the liquid limit (water content on the liquid state line corresponding to 25 blows). Two determinations of the plastic limit were made and the differences in the results are small. So, use the average value of water content as the plastic limit.

## Solution 2.6

**Step 1:**   Plot the data.
See Fig. E2.6.

**Step 2:**   Extract the liquid limit.
The water content on the liquid state line corresponding to a terminal blow of 25 gives the liquid limit.

$$w_{LL} = 38\%$$

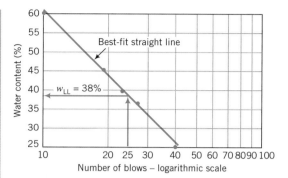

**FIGURE E2.6**   Plot of the liquid state line for the liquid limit by the Casagrande cup method.

**Step 3:**   Calculate plastic limit.
The plastic limit is

$$w_{PL} = \frac{20.3 + 20.8}{2} = 20.6\%$$

**Step 4:**   Calculate $I_p$.

$$I_p = w_{LL} - w_{PL} = 38 - 20.6 = 17.4\%$$

**Step 5:**   Calculate $I_L$.

$$I_L = \frac{(w - w_{PL})}{I_p} = \frac{27.4 - 20.6}{17.4} = 0.39$$

**Step 6:**   Calculate the void ratio.
Assume the soil is saturated at the liquid limit. For a saturated soil, $e = wG_s$. Thus,

$$e_{LL} = w_{LL}G_s = 0.38 \times 2.7 = 1.03$$

Brittle failure is not expected as the soil is in a plastic state $(0 < I_L < 1)$.

■

## EXAMPLE 2.7

The results of a fall cone test are shown in the table below.

| Parameter | 80 gram cone | | | | | 240 gram cone | | | |
|---|---|---|---|---|---|---|---|---|---|
| Penetration (mm) | 5.5 | 7.8 | 14.8 | 22 | 32 | 8.5 | 15 | 21 | 35 |
| Water content (%) | 39.0 | 44.8 | 52.5 | 60.3 | 67 | 36.0 | 45.1 | 49.8 | 58.1 |

Determine (a) the liquid limit, (b) the plastic limit, (c) the plasticity index, and (d) the liquidity index if the natural water content is 36%.

***Strategy***   Adopt the same strategy as in Example 2.6. Make a semilogarithm plot of water content versus penetration. Use the data to make your plot, then extract the liquid limit (water content on the liquid state line corresponding to 20 mm). Find the water content difference between the two liquid state lines at any fixed penetration. Use this value to determine the plastic limit.

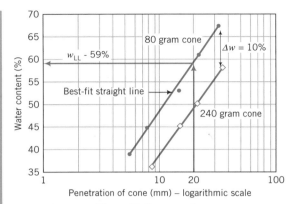

**FIGURE E2.7** Plot of fall cone data.

## Solution 2.7

**Step 1:** Plot the data.
See Fig. E2.7.

**Step 2:** Extract the liquid limit.

$$w_{LL} = 59\%$$

**Step 3:** Determine the plastic limit.

$$\Delta w = 10\%$$
$$w_{PL} = w_{LL} - 4.2\Delta w = 59 - 4.2 \times 10 = 17\%$$

**Step 4:** Determine $I_p$.

$$I_p = w_{LL} - w_{PL} = 59 - 17 = 42\%$$

**Step 5:** Determine $I_L$.

$$I_L = \frac{w - w_{PL}}{I_p} = \frac{36 - 17}{42} = 0.45$$

∎

*What's next . . .* We now know how to obtain some basic soil data—particle size and indices—from quick, simple tests. The question that arises is: What do we do with these data? Engineers would like to use the data to get a first impression on the use and possible performance of the soil for a particular purpose such as a construction material for an embankment. This is currently achieved by classification systems. We will study one such system next.

## 2.8   SOIL CLASSIFICATION SCHEMES

A classification scheme provides a method of identifying soils in a particular group that would likely exhibit similar characteristics. Soil classification is used to specify a certain soil type that is best suitable for a given application. There are several classification schemes available. Each was devised for a specific use. For example, the American Association of State Highway and Transportation

Officials (AASHTO) developed one scheme that classifies soils according to their usefulness in roads and highways while the Unified Soil Classification System (USCS) was originally developed for use in airfield construction but was later modified for general use.

We will study only the USCS because it is neither too elaborate nor too simplistic. The USCS uses symbols for the particle size groups. These symbols and their representations are: G—gravel, S—sand, M—silt, C—clay. These are combined with other symbols expressing gradation characteristics—W for well graded and P for poorly graded—and plasticity characteristics—H for high and L for low, and a symbol, O, indicating the presence of organic material. A typical classification of CL means a clay soil with low plasticity, while SP means a poorly graded sand. The flowcharts shown in Figs. 2.15a,b provide systematic means of classifying a soil according to the USCS.

Experimental results from soils tested from different parts of the world were plotted on a graph of plasticity index (ordinate) versus liquid limit (abscissa). It was found that clays, silts, and organic soils lie in distinct regions of the graph. A line defined by the equation

$$I_p = 0.73(w_{LL} - 20) \%$$

(2.24)

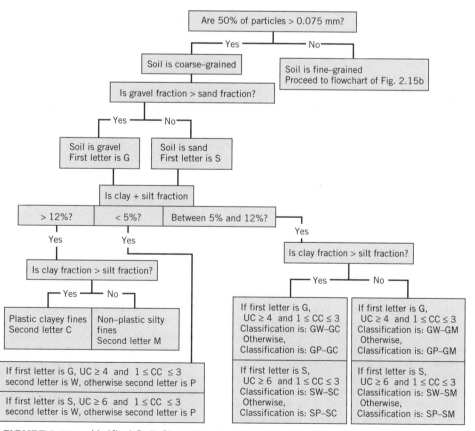

**FIGURE 2.15a** Unified Soil Classification flowchart for coarse-grained soils.

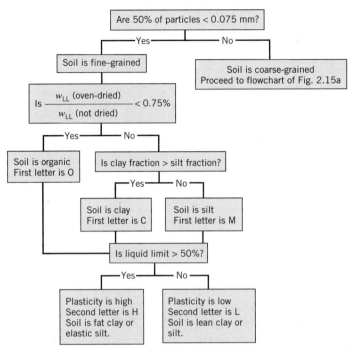

**FIGURE 2.15b**   Unified Soil Classification flowchart for fine-grained soils.

called the "A-line," delineates the boundaries between clays (above the line) and silts and organic soils (below the line) as shown in Fig. 2.16. A second line, the U-line expressed as $I_p = 0.9(w_{LL} - 8)$, defines the upper limit of the correlation between plasticity index and liquid limit. If the results of your soil tests fall above the U-line, you should be suspicious of your results and repeat your tests.

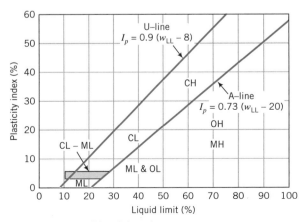

**FIGURE 2.16**   Plasticity chart.

## 2.9   ENGINEERING USE CHART

You may ask: "How do I use a soil classification to select a soil for a particular type of construction, for example, a dam?" Geotechnical engineers have prepared charts based on experience to assist you in selecting a soil for a particular construction purpose. One such chart is shown in Table 2.6. The numerical values 1 to 9 are ratings with No. 1 the best. The chart should only be used to provide guidance and to make a preliminary assessment of the suitability of a soil for a particular use. You should not rely on such descriptions as "excellent" shear strength or "negligible" compressibility to make final design and construction decisions. We will deal later (Chapters 4 and 5) with more reliable methods to determine strength and compressibility properties.

### EXAMPLE 2.8

Particle size analyses were carried out on two soils—Soil A and Soil B—and the particle size distribution curves are shown in Fig. E2.8. The Atterberg limits for the two soils are:

| Soil | $w_{LL}$ | $w_{PL}$ |
|------|------|------|
| A | 26 | 18 |
| B | Nonplastic | |

**(a)** Classify these soils according to the Unified Soil Classification Scheme.

**(b)** Is either of the soils organic?

**(c)** In a preliminary assessment, which of the two soils is a better material for the core of a rolled earth dam?

*Strategy*   If you examine the flowcharts of Figs. 2.15a,b, you will notice that you need to identify the various soil types based on texture: for example, the percentage of gravel or sand. Use the particle size distribution curve to extract

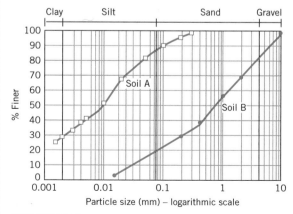

**FIGURE E2.8**

**TABLE 2.6   Engineering Use Chart (After Wagner, 1957)**

| Typical names of soil groups | Group symbols | Important properties | | | |
| --- | --- | --- | --- | --- | --- |
| | | Permeability when compacted | Shearing strength when compacted and saturated | Compressibility when compacted and saturated | Workability as a construction material |
| Well-graded gravels, gravel–sand mixtures, little or no fines | GW | Pervious | Excellent | Negligible | Excellent |
| Poorly graded gravels, gravel–sand mixtures, little or no fines | GP | Very pervious | Good | Negligible | Good |
| Silty gravels, poorly graded gravel–sand–silt mixtures | GM | Semipervious to impervious | Good | Negligible | Good |
| Clayey gravels, poorly graded gravel–sand–clay mixtures | GC | Impervious | Good to fair | Very low | Good |
| Well-graded sands, gravelly sands, little or no fines | SW | Pervious | Excellent | Negligible | Excellent |
| Poorly graded sands, gravelly sands, little or no fines | SP | Pervious | Good | Very low | Fair |
| Silty sands, poorly graded sand–silt mixtures | SM | Semipervious to impervious | Good | Low | Fair |
| Clayey sands, poorly graded sand–clay mixtures | SC | Impervious | Good to fair | Low | Good |
| Inorganic silts and very fine sands, rock flour, silty or clayey fine sands with slight plasticity | ML | Semipervious to impervious | Fair | Medium | Fair |
| Inorganic clays of low to medium plasticity, gravelly clays, sandy clays, silky clays, lean clays | CL | Impervious | Fair | Medium | Good to fair |
| Organic silts and organic silt–clays of low plasticity | OL | Semipervious to impervious | Poor | Medium | Fair |
| Inorganic silts, micaceous or diatomaceous fine sandy or silty soils, elastic silts | MH | Semipervious to impervious | Fair to poor | High | Poor |
| Inorganic clays of high plasticity, fat clays | CH | Impervious | Poor | High | Poor |
| Organic clays of medium to high plasticity | OH | Impervious | Poor | High | Poor |
| Peat and other highly organic soils | Pt | — | — | — | — |

the different percentages of each soil type and then follow the flowchart. To determine whether your soil is organic or inorganic, plot your Atterberg limits on the plasticity chart and check whether the limits fall within an inorganic or organic soil region.

## Solution 2.8

**Step 1:**   Determine the percentages of each soil type from the particle size distribution curve.

| Relative desirability for various uses | | | | | | | | | |
|---|---|---|---|---|---|---|---|---|---|
| Rolled earth dams | | | Canal sections | | Foundations | | Roadways | | |
| | | | | | | | Fills | | |
| Homogeneous embankment | Core | Shell | Erosion resistance | Compacted earth lining | Seepage important | Seepage not important | Frost heave not possible | Frost heave possible | Surfacing |
| — | — | 1 | 1 | — | — | 1 | 1 | 1 | 3 |
| — | — | 2 | 2 | — | — | 3 | 3 | 3 | — |
| 2 | 4 | — | 4 | 4 | 1 | 4 | 4 | 9 | 5 |
| 1 | 1 | — | 3 | 1 | 2 | 6 | 5 | 5 | 1 |
| — | — | 3 if gravelly | 6 | — | — | 2 | 2 | 2 | 4 |
| — | — | 4 if gravelly | 7 if gravelly | — | — | 5 | 6 | 4 | — |
| 4 | 5 | — | 8 if gravelly | 5 erosion critical | 3 | 7 | 8 | 10 | 6 |
| 3 | 2 | — | 5 | 2 | 4 | 8 | 7 | 6 | 2 |
| 6 | 6 | — | — | 6 erosion critical | 6 | 9 | 10 | 11 | — |
| 5 | 3 | — | 9 | 3 | 5 | 10 | 9 | 7 | 7 |
| 8 | 8 | — | — | 7 erosion critical | 7 | 11 | 11 | 12 | — |
| 9 | 9 | — | — | — | 8 | 12 | 12 | 13 | — |
| 7 | 7 | — | 10 | 8 volume change critical | 9 | 13 | 13 | 8 | — |
| 10 | 10 | — | — | — | 10 | 14 | 14 | 14 | — |
| — | — | — | — | — | — | — | — | — | — |

| Constituent | Soil A | Soil B |
|---|---|---|
| Percent of particle greater than 0.075 mm | 12 | 80 |
| Gravel fraction (%) | 0 | 16 |
| Sand fraction (%) | 12 | 64 |
| Silt fraction (%) | 59 | 20 |
| Clay fraction (%) | 29 | 0 |

**Step 2:** Use the flowchart.
Following the flowchart, Soil A is ML and Soil B is SM.

**Step 3:** Plot the Atterberg limits on the plasticity chart.

$$\text{Soil A:} \quad I_p = 26 - 18 = 8\%$$

The point (26, 8) falls above the A-line; the soil is inorganic.

$$\text{Soil B:} \quad \text{Nonplastic and inorganic}$$

**Step 4:** Use Table 2.6 to make a preliminary assessment.
Soil B with a rating of 5 is better than Soil B with a rating of 6 for the dam core. ∎

*What's next . . .*We have discussed particle sizes and index properties and used these to classify soils. You know that water changes the soil states in fine-grained soils; the greater the water content in a soil the weaker it is ($I_L$ increases). Soils are porous materials much like sponges. Water can flow between the interconnected voids. Particle sizes and the structural arrangement of the particles influence the rate of flow. In Table 2.6, you should have noticed that one of the important soil properties is permeability. Next, we will discuss soil permeability by considering one-dimensional flow of water through soils. Water can cause instability and many geotechnical structures have failed because of instability from flow of water. We will discuss instability from two-dimensional flow of water in Chapter 9.

## 2.10  ONE-DIMENSIONAL FLOW OF WATER THROUGH SOILS

### 2.10.1 Groundwater

We will be discussing gravitational flow of water under a steady state condition. You may ask: "What is a steady state condition?" Gravitational flow can only occur if there is a gradient. Flow takes place downhill. The steady state condition occurs if neither the flow nor the pore water pressures change with time. Pore water pressure is the water pressure within the voids.

If you dig a hole into a soil mass that has all the voids filled with water (fully saturated), you will observe water in the hole up to a certain level. This water level is called groundwater level or groundwater table and exists under a hydrostatic condition. A hydrostatic condition occurs when the flow is zero. The top of the groundwater level is under atmospheric pressure. We will denote groundwater level by the symbol ▼.

### 2.10.2 Head

Darcy's law governs the flow of water through soils. But before we delve into Darcy's law, we will discuss an important principle in fluid mechanics—Bernoulli's principle—which is essential in understanding flow through soils.

If you cap one end of a tube, fill the tube with water, and then rest it on your table (Fig. 2.17), the height of water with reference to your table is called the pressure head ($h_p$). Head refers to the mechanical energy per unit weight. If you raise the tube above the table, the mechanical energy or total head increases.

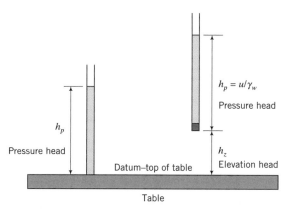

**FIGURE 2.17**   Illustration of elevation and pressure heads.

You now have two components of total head—the pressure head ($h_p$) and the elevation head ($h_z$). If water were to flow through the tube with a velocity $v$, under steady state condition, then we have an additional head due to the velocity given as $v^2/2g$. The total head (sometimes called piezometric head), $H$, according to Bernoulli's principle is

$$H = h_z + h_p + \frac{v^2}{2g} \qquad (2.25)$$

The elevation or potential head is referenced to an arbitrary datum and the total head will change depending on the choice of the datum position. Therefore, it is essential that you identify your datum position in solutions to flow problems. Pressures are defined relative to atmospheric pressure. The velocity of flow through soils is generally small (<1 cm/s) and we usually neglect the velocity head. The total head in soils is then

$$H = h_z + h_p = h_z + \frac{u}{\gamma_w} \qquad (2.26)$$

where $u = h_p \gamma_w$ is the pore water pressure.

Consider a cylinder containing a soil mass with water flowing through it at a constant rate as depicted in Fig. 2.18. If we connect two tubes, A and B, called piezometers, at a distance $l$ apart, the water will rise to different heights in each of the tubes. The height of water in tube B near the exit is lower than A. Why? As the water flows through the soil, energy is dissipated through friction with the soil particles, resulting in a loss of head. The head loss between A and B, assum-

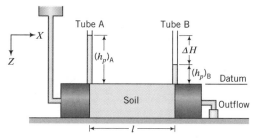

**FIGURE 2.18**   Head loss due to flow of water through soil.

ing decrease in head is positive and our datum is arbitrarily selected at the top of the cylinder, is $\Delta H = |(h_p)_B - (h_p)_A|$.

### 2.10.3 Darcy's Law

Darcy (1856) proposed that average flow velocity through soils is proportional to the gradient of the total head. The flow in any direction, $j$, is

$$v_j = k_j \frac{dH}{dx_j} \tag{2.27}$$

where $v$ is the average flow velocity, $k$ is a coefficient of proportionality called the coefficient of permeability or hydraulic conductivity, and $dH$ is the change in total head over a distance $dx$. The unit of measurement for $k$ is length/time, that is, cm/s. With reference to Fig. 2.18, Darcy's law becomes

$$v_x = k_x \frac{\Delta H}{l} = k_x i \tag{2.28}$$

where $i = \Delta H/l$ is the hydraulic gradient. Darcy's law is valid for all soils if the flow is laminar (Reynolds number $< 1$).

The average velocity, $v$, calculated from Eq. (2.28) is for the cross-sectional area normal to the direction of flow. Flow through soils, however, occurs only through the interconnected voids. The velocity through the void spaces is called seepage velocity ($v_s$) and is obtained by dividing the average velocity by the porosity of the soil:

$$v_s = \frac{k_j}{n} i \tag{2.29}$$

The volume rate of flow, $q_v$, or, simply, flow rate is the product of the average velocity and the cross-sectional area:

$$q_v = v_j A = A k_j i \tag{2.30}$$

The unit of measurement for $q_v$ is m³/s or cm³/s. The conservation of flow (law of continuity) stipulates that the volume rate of inflow $(q_v)_{\text{in}}$ into a soil element must equal the volume rate of outflow, $(q_v)_{\text{out}}$, or, simply, inflow must equal outflow: $(q_v)_{\text{in}} = (q_v)_{\text{out}}$.

The coefficient of permeability depends on the soil type, the particle size distribution, the structural arrangement of the grains or void ratio, and the wholeness (homogeneity, layering, fissuring, etc.) of the soil mass. Typical value ranges of $k_z$ for various soil types are shown in Table 2.7.

Homogeneous clays are practically impervious. Two popular uses of "im-

**TABLE 2.7    Coefficient of Permeability for Common Soil Types**

| Soil type | $k_z$ (cm/s) |
|---|---|
| Clean gravel | $>1.0$ |
| Clean sands, clean sand and gravel mixtures | $1.0$ to $10^{-3}$ |
| Find sands, silts, mixtures comprising sands, silts, and clays | $10^{-3}$ to $10^{-7}$ |
| Homogeneous clays | $<10^{-7}$ |

pervious" clays are in dam construction to curtail the flow of water through the dam and as barriers in landfills to prevent migration of effluent to the surrounding area. Clean sands and gravels are pervious and can be used as drainage materials or soil filters.

### 2.10.4 Empirical Relationships for *k*

For a homogeneous soil, the coefficient of permeability depends predominantly on its void ratio. You should recall that the void ratio is dependent on the soil fabric or structural arrangement of the soil grains. A number of empirical relationships have been proposed linking $k$ to void ratio and grain size for coarse-grained soils. Hazen (1930) proposed one of the early relationships as

$$k_z = CD_{10}^2 \text{ cm/s} \tag{2.31}$$

where $C$ is a constant varying between 0.4 and 1.2 if the unit of measurement of $D_{10}$ is mm. Typically, $C = 1.0$. Other relationships were proposed for coarse- and fine-grained soils by Samarasinghe et al. (1982), Kenny et al. (1984), and others. One has to be extremely cautious in using empirical relationships for $k$ because it is very sensitive to changes in void ratio and the wholeness of your soil mass.

> *The* **essential points** *are:*
> 1. *The flow of water through soils is governed by Darcy's law, which states that the average flow velocity is proportional to the hydraulic gradient.*
> 2. *The proportionality coefficient in Darcy's law is called the coefficient of permeability or hydraulic conductivity,* **k.**
> 3. *The value of* **k** *is influenced by the void ratio, particle size distribution, and the wholeness of the soil mass.*
> 4. *Homogeneous clays are practically impervious while sands and gravels are pervious.*

### EXAMPLE 2.9

A soil sample 10 cm in diameter is placed in a tube 1 m long. A constant supply of water is allowed to flow into one end of the soil at A and the outflow at B is collected by a beaker (Fig. E2.9). The average amount of water collected is 1 cm$^3$ for every 10 seconds. The tube is inclined as shown in Fig. E2.9. Determine the

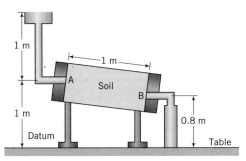

**FIGURE E2.9**

(a) hydraulic gradient, (b) flow rate, (c) average velocity, (d) seepage velocity, if $e = 0.6$, and (e) coefficient of permeability.

***Strategy***   In flow problems, you must define a datum position. So your first task is to define the datum position and then find the difference in total head between A and B. Use the head difference to calculate the hydraulic gradient and use Eqs. (2.28) to (2.30) to solve the problem.

## Solution 2.9

**Step 1:**   Define the datum position. Select the top of the table as the datum.

**Step 2:**   Find the total heads at A (inflow) and B (outflow).

$$H_A = (h_p)_A + (h_z)_A = 1 + 1 = 2 \text{ m}$$
$$H_B = (h_p)_B + (h_z)_B = 0 + 0.8 = 0.8 \text{ m}$$

**Step 3:**   Find the hydraulic gradient.

$$\Delta H = |H_B - H_A| = |0.8 - 2| = 1.2 \text{ m}$$
$$l = 1 \text{ m}$$
$$i = \frac{\Delta H}{l} = \frac{1.2}{1} = 1.2$$

If you were to select the outflow, point B, as the datum, then $H_A = 1 \text{ m} + 0.2 \text{ m} = 1.2 \text{ m}$ and $H_B = 0$. The head loss is $\Delta H = 1.2$ m, which is the same value we obtained using the table's top as the datum. It is often simpler, for calculation purposes, to select the exit flow position as the datum.

**Step 4:**   Determine the flow rate.

Volume of water collected, $Q = 1 \text{ cm}^3$, $t = 10$ seconds

$$q_v = \frac{Q}{t} = \frac{1}{10} = 0.1 \text{ cm}^3/\text{s}$$

**Step 5:**   Determine the average velocity.

$$q_v = Av$$
$$A = \frac{\pi \times (\text{diam})^2}{4} = \frac{\pi \times 10^2}{4} = 78.5 \text{ cm}^2$$
$$v = \frac{q_v}{A} = \frac{0.1}{78.5} = 0.0013 \text{ cm/s}$$

**Step 6:**   Determine seepage velocity.

$$v_s = \frac{v}{n}$$
$$n = \frac{e}{1 + e} = \frac{0.6}{1 + 0.6} = 0.38$$
$$v_s = \frac{0.0013}{0.38} = 0.0034 \text{ cm/s}$$

**Step 7:**   Determine the coefficient of permeability. From Darcy's law $v = ki$.

$$\therefore k = \frac{v}{i} = \frac{0.0013}{1.2} = 10.8 \times 10^{-4} \text{ cm/s}$$   ∎

## EXAMPLE 2.10

A drainage pipe (Fig. E2.10a) became completely blocked during a storm by a plug of sand, 1.5 m long, followed by another plug of a mixture of clays, silts, and sands, 0.5 m long. When the storm was over, the water level above ground was 1 m. The coefficient of permeability of the sand is 2 times that of the mixture of clays, silts, and sands.

(a) Plot the variation of pressure, elevation, and total head over the length of the pipe.

(b) Calculate the pore water pressure at (1) the center of the sand plug and (2) the center of the mixture of clays, silts, and sands.

(c) Find the average hydraulic gradients in the sand and in the mixture of clays, silts and sands.

***Strategy*** You need to select a datum. From the information given, you can calculate the total head at A and B. The difference in head is the head loss over both plugs but you do not know how much head is lost in the sand and in the mixture of clays, silts, and sands. The continuity equation provides the key to finding the head loss over each plug.

## Solution 2.10

**Step 1:** Select a datum.
Select the exit at B along the centerline of the drainage pipe as the datum.

**Step 2:** Determine heads at A and B.
$$(h_z)_A = 0 \text{ m}, \quad (h_p)_A = 0.3 + 2 + 1 = 3.3 \text{ m}, \quad H_A = 0 + 3.3 = 3.3 \text{ m}$$
$$(h_z)_B = 0 \text{ m}, \quad (h_p)_B = 0 \text{ m}, \quad H_B = 0 \text{ m}$$

**Step 3:** Determine the head loss in each plug.
Head loss between A and B $= |H_B - H_A| = 3.3 \text{ m}$ (decrease in head taken as positive). Let $\Delta H_1$, $L_1$, $k_1$, and $q_1$ be the head loss, length, coefficient of permeability, and

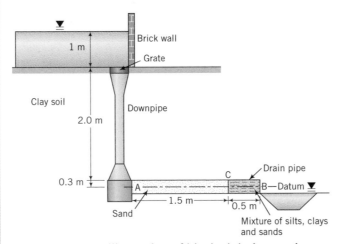

**FIGURE E2.10a** Illustration of blocked drainage pipe.

flow in the sand; let $\Delta H_2$, $L_2$, $k_2$, and $q_2$ be the head loss, length, coefficient of permeability, and flow in the mixture of clays, silts, and sands. Now,

$$q_1 = Ak_1 \frac{\Delta H_1}{L_1} = A \times 2k_2 \frac{\Delta H_1}{L_1}$$

$$q_2 = Ak_2 \frac{\Delta H_2}{L_2} = A \times k_2 \frac{\Delta H_2}{L_2}$$

From the continuity equation, $q_1 = q_2$.

$$\therefore A \times 2k_2 \frac{\Delta H_1}{L_1} = Ak_2 \frac{\Delta H_2}{L_2}$$

Solving, we get

$$\frac{\Delta H_1}{\Delta H_2} = \frac{L_1}{2L_2} = \frac{1.5}{2 \times 0.5} = 1.5$$

$$\Delta H_1 = 1.5\ \Delta H_2 \qquad\qquad (1)$$

However, we know that

$$\Delta H_1 + \Delta H_2 = \Delta H = 3.3 \text{ m} \qquad\qquad (2)$$

Solving for $\Delta H_1$ and $\Delta H_2$ from Eqs. (1) and (2), we obtain

$$\Delta H_1 = 1.98 \text{ m} \quad \text{and} \quad \Delta H_2 = 3.3 - 1.98 = 1.32 \text{ m}$$

**Step 4:** Calculate heads at the junction of the two plugs.

$$\text{Total head at } C = H_C = H_A - \Delta H_1 = 3.3 - 1.98 = 1.32 \text{ m}$$
$$(h_z)_C = 0$$
$$(h_p)_C = H_C - (h_z)_C = 1.32 \text{ m}$$

**Step 5:** Plot distribution of heads.
See Fig. E2.10b.

**Step 6:** Calculate pore water pressures.
Let D be the center of the sand.

$$(h_p)_D = \frac{(h_p)_A + (h_p)_C}{2} = \frac{3.3 + 1.32}{2} = 2.31 \text{ m}$$

$$u_D = 2.31 \times \gamma_w = 2.31 \times 9.8 = 22.6 \text{ kPa}$$

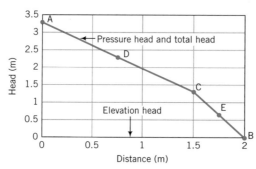

**FIGURE E2.10b**  Variation of elevation, pressure, and total heads along pipe.

Let E be the center of the mixture of clays, silts, and sands.

$$(h_p)_E = \frac{(h_p)_C + (h_p)_B}{2} = \frac{1.32 + 0}{2} = 0.66 \text{ m}$$

$$u_E = 0.66 \times 9.8 = 6.5 \text{ kPa}$$

**Step 7:**  Find the average hydraulic gradients.

$$i_1 = \frac{\Delta H_1}{l_1} = \frac{1.98}{1.5} = 1.32$$

$$i_2 = \frac{\Delta H_2}{l_2} = \frac{1.32}{0.5} = 2.64$$ ■

*What's next . . .*We have considered flow only through homogeneous soils. In reality, soils are stratified or layered with different soil types. In calculating flow through layered soils, an average or equivalent permeability representing the whole soil mass is determined from the permeability of each layer. Next, we will consider flow of water through layered soil masses: one flow occurs parallel to the layers, the other flow occurs normal to the layers.

### 2.10.5 Flow Parallel to Soil Layers

When the flow is parallel to the soil layers (Fig. 2.19), the hydraulic gradient is the same at all points. The flow through the soil mass as a whole is equal to the sum of the flow through each of the layers. There is a parallel here with the flow of electricity through resistors in parallel. If we consider a unit width (in the $y$ direction) of flow and use Eq. (2.30), we obtain

$$q_v = Av = (1 \times H_o)k_{x(eq)}i = (1 \times z_1)k_{x1}i + (1 \times z_2)k_{x2}i + \cdots + (1 \times z_n)k_{xn}i \quad (2.32)$$

where $H_o$ is the total thickness of the soil mass, $k_{x(eq)}$ is the equivalent permeability in the horizontal ($x$) direction, $z_1$ to $z_n$ are the thickness of the first to the $n$th layers, and $k_{x1}$ to $k_{xn}$ are the horizontal permeabilities of the first to the $n$th layer. Solving Eq. (2.32) for $k_{x(eq)}$, we get

$$k_{x(eq)} = \frac{1}{H_o}(z_1 k_{x1} + z_2 k_{x2} + \cdots + z_n k_{xn}) \quad (2.33)$$

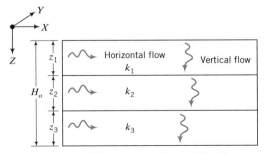

**FIGURE 2.19**  Flow through stratified layers.

### 2.10.6 Flow Normal to Soil Layers

For flow normal to the soil layers, the head loss in the soil mass is the sum of the head losses in each layer:

$$\Delta H = \Delta h_1 + \Delta h_2 + \cdots + \Delta h_n \tag{2.34}$$

where $\Delta H$ is the total head loss, and $\Delta h_1$ to $\Delta h_n$ are the head losses in each of the $n$ layers. The velocity in each layer is the same. The analogy to electricity is flow of current through resistors in series. From Darcy's law, we obtain

$$k_{z(eq)} \frac{\Delta H}{H_o} = k_{z1} \frac{\Delta h_1}{z_1} = k_{z2} \frac{\Delta h_2}{z_2} = \cdots = k_{zn} \frac{\Delta h_n}{z_n} \tag{2.35}$$

where $k_{z(eq)}$ is the equivalent permeability in the vertical ($z$) direction and $k_{z1}$ to $k_{zn}$ are the vertical permeabilities of the first to the $n$th layer. Solving Eqs. (2.34) and (2.35) leads to

$$k_{z(eq)} = \frac{H_o}{\dfrac{z_1}{k_{z1}} + \dfrac{z_2}{k_{z2}} + \cdots + \dfrac{z_n}{k_{zn}}} \tag{2.36}$$

Values of $k_{z(eq)}$ are generally less than $k_{x(eq)}$—sometimes as much as 10 times less.

### EXAMPLE 2.11

A canal is cut into a soil with a stratigraphy shown in Fig. E2.11. Assuming flow takes place laterally and vertically through the sides of the canal and vertically below the canal, determine the equivalent permeability in the horizontal and vertical directions. Calculate the ratio of the equivalent horizontal permeability to the equivalent vertical permeability for flow through the sides of the canal.

**Strategy**   Use Eq. (2.33) to find the equivalent horizontal permeability over the depth of the canal (3.0 m) and then use Eq. (2.36) to find the equivalent vertical permeability below the canal. To make the calculations easier, convert all exponential quantities to a single exponent.

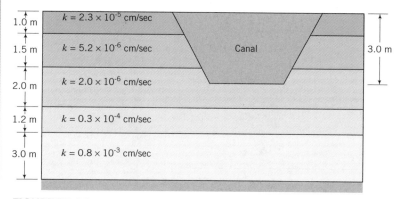

**FIGURE E2.11**

**Solution 2.11**

**Step 1:** Find $k_{x(eq)}$ and $k_{z(eq)}$ for flow through the sides of the canal.

$$H_o = 3 \text{ m}$$

From Eq. (2.33),

$$k_{x(eq)} = \frac{1}{H_o}(z_1 k_{x1} + z_2 k_{x2} + \cdots + z_n k_{xn})$$

$$= \frac{1}{3}(1 \times 0.23 \times 10^{-6} + 1.5 \times 5.2 \times 10^{-6} + 0.5 \times 2 \times 10^{-6})$$

$$= 3 \times 10^{-6} \text{ cm/s}$$

From Eq. (2.36),

$$k_{z(eq)} = \frac{H_o}{\dfrac{z_1}{k_{z1}} + \dfrac{z_2}{k_{z2}} + \cdots + \dfrac{z_n}{k_{zn}}}$$

$$= \frac{3}{\dfrac{1}{10^{-6}}\left(\dfrac{1}{0.23} + \dfrac{1.5}{5.2} + \dfrac{0.5}{2}\right)} = 0.61 \times 10^{-6} \text{ cm/s}$$

**Step 2:** Find $k_{x(eq)}/k_{z(eq)}$ ratio.

$$\frac{k_{x(eq)}}{k_{z(eq)}} = \frac{3 \times 10^{-6}}{0.61 \times 10^{-6}} = 4.9$$

**Step 3:** Find $k_{z(eq)}$ below the bottom of the canal.

$$H_o = 1.5 + 1.2 + 3.0 = 5.7 \text{ m}$$

$$k_{z(eq)} = \frac{H_o}{\dfrac{z_1}{k_{z1}} + \dfrac{z_2}{k_{z2}} + \cdots + \dfrac{z_n}{k_{zn}}} = \frac{5.7}{\dfrac{1.5}{2 \times 10^{-6}} + \dfrac{1.2}{30 \times 10^{-6}} + \dfrac{3}{800 \times 10^{-6}}}$$

$$= 7.2 \times 10^{-6} \text{ cm/s}$$

∎

*What's next . . .* In order to calculate flow, we need to know the coefficient of permeability $k$. We will discuss how this coefficient is determined in the laboratory and in the field.

## 2.11    DETERMINATION OF THE COEFFICIENT OF PERMEABILITY

### 2.11.1 Constant-Head Test

The constant-head test is used to determine the coefficient of permeability of coarse-grained soils. A typical constant-head apparatus is shown in Fig. 2.20. Water is allowed to flow through a cylindrical sample of soil under a constant head ($h$). The outflow ($Q$) is collected in a graduated cylinder at a convenient duration ($t$).

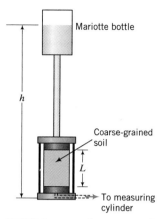

**FIGURE 2.20**   A constant-head apparatus.

With reference to Fig. 2.20,

$$\Delta H = h \quad \text{and} \quad i = \frac{\Delta H}{L} = \frac{h}{L}$$

The flow rate through the soil is $q_v = Q/t$, where $Q$ is the total quantity of water collected in the measuring cylinder over time $t$.

From Eq. (2.30),

$$k_z = \frac{q_v}{Ai} = \frac{QL}{tAh} \tag{2.37}$$

where $k_z$ is the coefficient of permeability in the vertical direction.

The viscosity of the fluid, which is a function of temperature, influences the value of $k$. The experimental value ($k_{T°C}$) is corrected to a baseline temperature of 20°C using

$$k_{20°C} = k_{T°C} \frac{\mu_{T°C}}{\mu_{20°C}} = k_{T°C} R_T \tag{2.38}$$

where $\mu$ is the viscosity of water, $T$ is the temperature in °C at which the measurement was made, and $R_T = \mu_{T°C}/\mu_{20°C}$ is the temperature correction factor that can be calculated from

$$R_T = 2.42 - 0.475 \ln(T) \tag{2.39}$$

### 2.11.2 Falling-Head Test

The falling-head test is used for fine-grained soils because the flow of water through these soils is too slow to get reasonable measurements from the constant-head test. A compacted soil sample or a sample extracted from the field is placed in a metal or acrylic cylinder (Fig. 2.21). Porous stones are positioned at the top and bottom faces of the sample to prevent its disintegration and to allow water to percolate through it. Water flows through the sample from a standpipe attached to the top of the cylinder. The head of water ($h$) changes with time as flow occurs through the soil. At different times, the head of water is recorded.

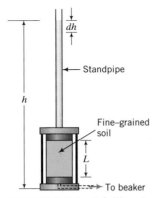

**FIGURE 2.21**   A falling-head apparatus.

Let $dh$ be the drop in head over a time period $dt$. The velocity or rate of head loss in the tube is

$$v = -\frac{dh}{dt}$$

and the inflow of water to the soil is

$$(q_v)_{\text{in}} = av = -a\frac{dh}{dt}$$

where $a$ is the cross-sectional area of the tube. We now appeal to Darcy's law to get the outflow:

$$(q_v)_{\text{out}} = Aki = Ak\frac{h}{L}$$

where $A$ is the cross-sectional area, $L$ is the length of the soil sample, and $h$ is the head of water at any time $t$. The continuity condition requires that $(q_v)_{\text{in}} = (q_v)_{\text{out}}$. Therefore,

$$-a\frac{dh}{dt} = Ak\frac{h}{L}$$

By separating the variables ($h$ and $t$) and integrating between the appropriate limits, the last equation becomes

$$\frac{Ak}{aL}\int_{t_1}^{t_2} dt = -\int_{h_1}^{h_2}\frac{dh}{h}$$

and the solution for $k$ in the vertical direction is

$$k = k_z = \frac{aL}{A(t_2 - t_1)}\ln\left(\frac{h_1}{h_2}\right) \tag{2.40}$$

*The* essential points *are:*

1. **The constant-head test is used to determine the coefficient of permeability of coarse-grained soils.**
2. **The falling-head test is used to determine the coefficient of permeability of fine-grained soils.**

## EXAMPLE 2.12

A sample of sand, 5 cm in diameter and 15 cm long, was prepared at a porosity of 60% in a constant-head apparatus. The total head was kept constant at 30 cm and the amount of water collected in 5 seconds was 40 cm³. The test temperature was 20°C. Calculate the coefficient of permeability and the seepage velocity.

**Strategy**  From the data given, you can readily apply Darcy's law to find $k$.

## Solution 2.12

**Step 1:**  Calculate the sample cross-sectional area, hydraulic gradient, and flow.

$$D = 5 \text{ cm}$$

$$A = \frac{\pi \times D^2}{4} = \frac{\pi \times 5^2}{4} = 19.6 \text{ cm}^2$$

$$\Delta H = 30 \text{ cm}$$

$$i = \frac{\Delta H}{L} = \frac{30}{15} = 2$$

$$Q = 40 \text{ cm}^3$$

$$q_v = \frac{Q}{t} = \frac{40}{5} = 8 \text{ cm}^3/\text{s}$$

**Step 2:**  Calculate $k$.

$$k = \frac{q_v}{Ai} = \frac{8}{19.6 \times 2} = 0.2 \text{ cm/s}$$

**Step 3:**  Calculate the seepage velocity.

$$v_s = \frac{ki}{n} = \frac{0.2 \times 2}{0.6} = 0.67 \text{ cm/s}$$  ∎

## EXAMPLE 2.13

The data from a falling-head test on a silty clay are:

Cross-sectional area of soil = 80 cm²

Length of soil = 10 cm

Initial head = 90 cm

Final head = 84 cm

Duration of test = 15 minutes

Diameter of tube = 6 mm

Temperature = 22°C

Determine $k$.

**Strategy**  Since this is a falling-head test, you should use Eq. (2.40). Make sure you are using consistent units.

**Solution 2.13**

**Step 1:** Calculate the parameters required in Eq. (2.40).

$$a = \frac{\pi \times (6/10)^2}{4} = 0.28 \text{ cm}^2$$

$$A = 80 \text{ cm}^2 \quad \text{(given)}$$

$$t_2 - t_1 = 15 \times 60 = 900 \text{ seconds}$$

**Step 2:** Calculate $k$.

$$k = \frac{aL}{A(t_2 - t_1)} \ln\left(\frac{h_1}{h_2}\right) = \frac{0.28 \times 10}{80 \times 900} \ln\left(\frac{90}{84}\right) = 2.7 \times 10^{-6} \text{ cm/s}$$

From Eq. (2.39), $R_T = 2.42 - 0.475 \ln(22) = 0.95$

$$k_{20^\circ C} = kR_T = 2.7 \times 10^{-6} \times 0.95 = 2.6 \times 10^{-6} \text{ cm/s} \quad \blacksquare$$

*What's next . . .*In the constant-head test and the falling-head test, we determined the coefficient of permeability of only a small volume of soil at a specific location in a soil mass. In some cases, we have to use remolded or disturbed soil samples. In addition, if field samples are used, they are invariably disturbed by sampling processes (see Section 2.11). The coefficient of permeability is sensitive to alteration in the fabric of the soil and, consequently, there are doubts about the accuracy of representing the in situ soil conditions using laboratory permeability tests. There are several field methods to determine the coefficient of permeability. Next, we will discuss one popular method.

### 2.11.3 Pumping Test to Determine the Coefficient of Permeability

One common method of determining the coefficient of permeability in the field is by pumping water at a constant flow rate from a well and measuring the decrease in groundwater level at observation wells (Fig. 2.22). The equation, called the simple well formula, is derived using the following assumptions.

1. The pumping well penetrates through the water-bearing stratum and is perforated only at the section that is below the groundwater level.
2. The soil mass is homogeneous, isotropic, and of infinite size.
3. Darcy's law is valid.
4. Flow is radial toward the well.
5. The hydraulic gradient at any point in the water-bearing stratum is constant and is equal to the slope of groundwater surface (Dupuit's assumptions).

Let $dz$ be the drop in total head over a distance $dr$. Then according to Dupuit's assumption the hydraulic gradient is

$$i = \frac{dz}{dr}$$

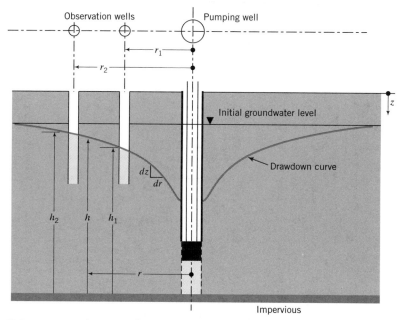

**FIGURE 2.22**   Layout of a pump test to determine $k$.

The area of flow at a radial distance $r$ from the center of the pumping well is

$$A = 2\pi r z$$

where $z$ is the thickness of an elemental volume of the pervious soil layer.
From Darcy's law, the flow is

$$q_v = 2\pi r \, z \, k \, \frac{dz}{dr}$$

We need to rearrange the above equation and integrate it between the limits $r_1$ and $r_2$, and $h_1$ and $h_2$:

$$q_v \int_{r_1}^{r_2} \frac{dr}{r} = 2k\pi \int_{h_1}^{h_2} z \, dz$$

Completing the integration leads to

$$k = \frac{q_v \ln(r_2/r_1)}{\pi(h_2^2 - h_1^2)} \tag{2.41}$$

With measurements of $r_1$, $r_2$, $h_1$, $h_2$, and $q_v$ (flow rate of the pump), $k$ can be calculated from Eq. (2.41). This test is only practical for coarse-grained soils.

Pumping tests lower the groundwater, which then causes stress changes in the soil. Since the groundwater is not lowered uniformly as shown by the drawdown curve in Fig. 2.22, the stress changes in the soil will not be even. Consequently, pumping tests near existing structures can cause them to settle unevenly. You should consider the possibility of differential settlement on existing structures when you plan a pumping test. Also, it is sometimes necessary to temporarily lower the groundwater level for construction. The process of lowering the groundwater is called dewatering.

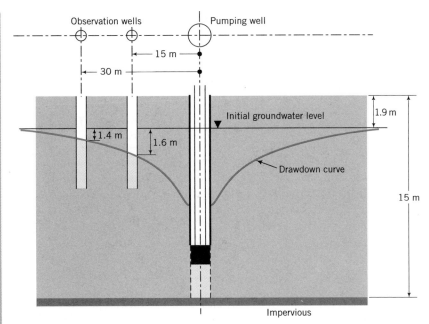

**FIGURE E2.14**

**EXAMPLE 2.14** A pumping test was carried out in a soil bed of thickness 15 m and the following measurements were recorded. Rate of pumping was 10.6 $\times 10^{-3} \text{m}^3/\text{s}$; drawdowns in observation wells located at 15 m and 30 m from the center of the pumping well were 1.6 m and 1.4 m, respectively, from the initial groundwater level. The initial groundwater level was located at 1.9 m below ground level.

*Strategy* You are given all the measurements to directly apply Eq. (2.41) to find $k$. You should draw a sketch of the pump test to identify the values to be used in Eq. (2.41).

## Solution 2.14

**Step 1:** Draw a sketch of the pump test with the appropriate dimensions—see Fig. E2.14.

**Step 2:** Substitute given values in Eq. (2.41) to find $k$.

$$r_2 = 30 \text{ m}, \quad r_1 = 15 \text{ m}, \quad h_2 = 15 - (1.9 + 1.4) = 11.7 \text{ m},$$
$$h_1 = 15 - (1.9 + 1.6) = 11.5 \text{ m}$$
$$k = \frac{q_v \ln(r_2/r_1)}{\pi(h_2^2 - h_1^2)} = \frac{10.6 \times 10^{-3} \ln(30/15)}{\pi(11.7^2 - 11.5^2)10^4} = 5.0 \times 10^{-2} \text{ cm/s} \quad \blacksquare$$

*What's next . . .*Water, although regarded as the "foe" in geotechnical engineering, can be used to improve soil strength, reduce soil deformations under loads, and reduce the permeability. Next, we will study how water can assist in the improvement of soils.

## 2.12 DRY UNIT WEIGHT–WATER CONTENT RELATIONSHIP

### 2.12.1 Basic Concept

Let's examine Eq. (2.12) for dry unit weight, that is,

$$\gamma_d = \left(\frac{G_s}{1 + e}\right)\gamma_w = \left(\frac{G_s}{1 + wG_s/S}\right)\gamma_w \qquad (2.42)$$

The extreme right-hand side term was obtained by replacing $e$ by $e = wG_s/S$. How can we increase the dry unit weight? Examination of Eq. (2.42) reveals that we have to reduce the void ratio; that is, $w/S$ must be reduced. The theoretical maximum dry unit weight is obtained when $S = 1$; that is,

$$e_{min} = wG_s \qquad (2.43)$$

### 2.12.2 Proctor Compaction Test

A laboratory test, called the Proctor test, was developed to deliver a standard amount of mechanical energy (compactive effort) to determine the maximum dry unit weight of a soil. In the standard Proctor test, a dry soil specimen is mixed with water and compacted in a cylindrical mold of volume of $9.44 \times 10^{-4}$ m$^3$ (standard Proctor mold) by repeated blows from the mass of a hammer, 2.5 kg, falling freely from a height of 305 mm (Fig. 2.23). The soil is compacted in three layers, each of which is subjected to 25 blows.

A modified Proctor test was developed for compaction of airfields to support heavy aircraft loads. In the modified Proctor test, a hammer with a mass 4.54 kg falls freely from a height of 457 mm. The soil is compacted in five layers with 25 blows per layer in the standard Proctor mold.

Four or more tests are conducted on the soil using different water contents. The last test is identified when additional water causes the bulk unit weight of the soil to decrease. The results are plotted as dry unit weight (ordinate) versus water content (abscissa). Typical dry unit weight–water content plots are shown in Fig. 2.24.

Clays usually yield bell-shaped curves. Sands do not often yield a clear bell;

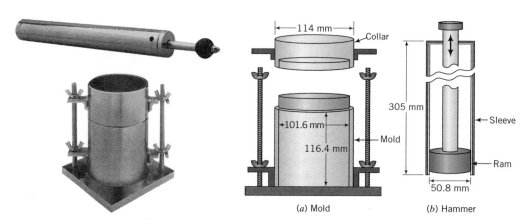

(a) Mold          (b) Hammer

**FIGURE 2.23** Compaction apparatus. (Photo courtesy of Geotest.)

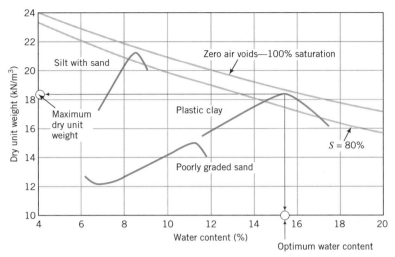

**FIGURE 2.24** Dry unit weight–water content curves.

rather, they sometimes show an initial decrease in dry unit weight, attributed to capillary tension that restrains the free movement of soil particles, and then yield the bell-shaped curve. Some soils, those with liquid limit less than 30%, may produce one or more humps before the maximum dry unit weight is achieved.

The water content at which the maximum dry unit weight, $(\gamma_d)_{max}$, is achieved is called the optimum water content ($w_{opt}$). At water contents below optimum (dry of optimum), air is expelled and water facilitates the rearrangement of soil grains into a denser configuration—the number of soil grains per unit volume of soil increases. At water contents just above optimum (wet of optimum), the compactive effort cannot expel more air and additional water displaces soil grains, thus decreasing the number of soil grains per unit volume of soil. Consequently, the dry unit weight decreases.

### 2.12.3 Zero Air Voids Curve

The soil is invariably unsaturated at the maximum dry unit weight, that is, $S < 1$. We can determine the degree of saturation at the maximum dry unit weight using Eq. (2.42). We know $\gamma_d = (\gamma_d)_{max}$ and $w = w_{opt}$ from our Proctor test results. If $G_s$ is known, we can solve Eq. (2.42) for $S$. If $G_s$ is not known, you can substitute a value of 2.7 with little resulting error in most cases. Equation (2.42) can be used to plot a series of theoretical curves of dry unit weight versus water content for different degrees of saturation (lines of constant degree of saturation) as shown in Fig. 2.24 for $S = 100\%$ and $S = 80\%$. You plot these curves as follows:

1. Assume a fixed value of $S$, say, $S = 1$ (100% saturation).

2. Substitute arbitrarily chosen values of $w$, approximately within the range of water content on your graph.

3. With the fixed value of $S$ and either an estimated value of $G_s$ ($= 2.7$) or a known value, find $\gamma_d$ for each value of $w$ using Eq. (2.42) and plot the results of $\gamma_d$ versus $w$.

4. Repeat for a different value of $S$.

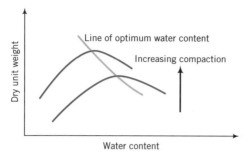

**FIGURE 2.25**   Effect of increasing compaction efforts on the dry unit weight–water content relationship.

The curve corresponding to $S = 1$ is known as the saturation line or zero air voids line. This line represents the minimum void ratio ($e_{min}$) attainable at a given water content [Eq. (2.43)].

The achievement of zero air voids by compaction is rare. The modified Proctor test, using higher levels of compaction energy, achieves a higher maximum dry unit weight at a lower optimum water content than the standard test (Fig. 2.25). The degree of saturation is also lower at higher levels of compaction than the standard compaction test.

### 2.12.4 Importance of Compaction

Knowledge of the optimum water content and the maximum dry unit weight of soils is very important for construction specifications of soil improvement by compaction. Specifications for earth structures (embankments, footings, etc.) usually call for a minimum of 95% of Proctor maximum dry unit weight. This level of compaction can be attained at two water contents: one before the attainment of the maximum dry unit weight or dry of optimum, the other after attainment of the maximum dry unit weight or wet of optimum. Normal practice is to compact the soil at the lower water content value except for swelling (expansive) soils.

Compaction increases the strength, lowers the compressibility, and reduces the permeability of a soil by rearranging its fabric. The soil fabric is forced into a denser configuration by the mechanical effort used in compaction. Compaction is the most popular technique of improving soils.

### 2.12.5 Field Compaction

A variety of mechanical equipment is used to compact soils in the field. You may have seen various types of rollers being used in road construction. Each type of roller has special mechanical systems to effectively compact a particular soil type. For example, a sheepsfoot roller (Fig. 2.26a) is generally used to compact fine-grained soils while a drum type roller (Fig. 2.26b) is generally used to compact coarse-grained soils.

### 2.12.6 Compaction Quality Control

A geotechnical engineer needs to check that field compaction meets specifications. Various types of equipment are available to check the amount of compac-

(a)

(b)

**FIGURE 2.26**  Two types of machinery for field compaction. (Photos courtesy of Vibromax America, Inc.)

tion achieved in the field. Three popular apparatuses are (1) the sand cone, (2) the balloon, and (3) nuclear density meters.

***2.12.6.1 Sand Cone***   A sand cone apparatus is shown in Fig. 2.27. It consists of a glass or plastic jar with a funnel attached to the neck of the jar.

The procedure for a sand cone test is as follows:

1. Fill the jar with a standard sand—a sand with known density—and determine the weight of the sand cone apparatus with the jar filled with sand

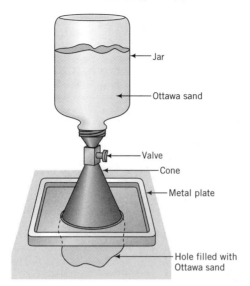

**FIGURE 2.27** A sand cone apparatus.

($W_1$). The American Society for Testing and Materials (ASTM) recommends Ottawa sand as the standard.

2. Determine the weight of sand to fill the cone ($W_2$).

3. Excavate a small hole in the soil and determine the weight of the excavated soil ($W_3$).

4. Determine the water content of the excavated soil ($w$).

5. Fill the hole with the standard sand by inverting the sand cone apparatus over the hole and opening the valve.

6. Determine the weight of the sand cone apparatus with the remaining sand in the jar ($W_4$).

7. Calculate the unit weight of the soil as follows:

Weight of sand to fill hole: $W_s = W_1 - (W_2 + W_4)$

Volume of hole: $V = \dfrac{W_s}{(\gamma_d)_{\text{Ottawa sand}}}$

Weight of dry soil: $W_d = \dfrac{W_3}{1 + w}$

Dry unit weight: $\gamma_d = \dfrac{W_d}{V}$

***2.12.6.2 Balloon Test*** The balloon test apparatus (Fig. 2.28) consists of a graduated cylinder with a centrally placed balloon. The cylinder is filled with water. The procedure for the balloon test is as follows:

1. Fill the cylinder with water and record its volume, $V_1$.

2. Excavate a small hole in the soil and determine the weight of the excavated soil ($W$).

3. Determine the water content of the excavated soil ($w$).

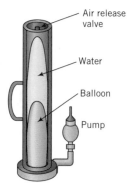

**FIGURE 2.28**  Balloon test device.

**4.** Use the pump to invert the balloon to fill the hole.

**5.** Record the volume of water remaining in the cylinder, $V_2$.

**6.** Calculate the unit weight of the soil as follows:

$$\gamma = \frac{W}{V_1 - V_2}; \quad \gamma_d = \frac{\gamma}{1 + w}$$

***2.12.6.3 Nuclear Density Meter***    The nuclear density apparatus (Fig. 2.29) is a versatile device to rapidly obtain the unit weight and water content of the soil nondestructively. Soil particles cause radiation to scatter to a detector tube and the amount of scatter is counted. The scatter count rate is inversely proportional to the unit weight of the soil. If water is present in the soil, the hydrogen in water scatters the neutrons and the amount of scatter is proportional to the water content. The radiation source is either radium or radioactive isotopes of cesium and americium. The nuclear density apparatus is first calibrated using the manufacturer's reference blocks. This calibration serves as a reference to determine the unit weight and water content of a soil at a particular site.

**FIGURE 2.29**  Nuclear density meter. (Photo courtesy of Seaman Nuclear Corp.)

*The* essential points *are:*

1. *Compaction is the densification of a soil by the expulsion of air and the rearrangement of soil particles.*
2. *The Proctor test is used to determine the maximum dry unit weight and the optimum water content and serves as the reference for field specifications of compaction.*
3. *Higher compactive effort increases the maximum dry unit weight and reduces the optimum water content.*
4. *Compaction increases strength, lowers compressibility, and reduces the permeability of soils.*
5. *A variety of field equipment is used to check the dry unit weights achieved in the field. Popular field equipment includes the sand cone apparatus, the balloon apparatus, and the nuclear density meter.*

## EXAMPLE 2.15

The results of a standard compaction test are shown in the table below. Determine the maximum dry unit weight and optimum water content.

| Water content (%) | 6.2 | 8.1 | 9.8 | 11.5 | 12.3 | 13.2 |
|---|---|---|---|---|---|---|
| Bulk unit weight (kN/m³) | 16.9 | 18.7 | 19.5 | 20.5 | 20.4 | 20.1 |

(a) What is the dry unit weight and water content at 95% standard compaction?

(b) Determine the degree of saturation at the maximum dry density.

(c) Plot the zero air voids line.

**Strategy**   Compute $\gamma_d$ and then plot the results of $\gamma_d$ versus $w$ (%). Then extract the required information.

## Solution 2.15

**Step 1:**   Use a table or a spreadsheet program to tabulate $\gamma_d$.
Use Eq. (2.12) to find the dry unit weight and Eq. (2.42) to calculate the dry unit weight for $S = 1$.

| Water content (%) | Bulk unit weight (kN/m³) | Dry unit weight (kN/m³) $\gamma_d = \dfrac{\gamma}{1 + w}$ | Zero air voids | |
|---|---|---|---|---|
| | | | Water content (%) | Dry unit weight (kN/m³) $\gamma_d = \left(\dfrac{G_s}{1 + wG_s/S}\right)\gamma_w;$ $S = 1$ |
| 6.2 | 16.9 | 15.9 | 6 | 22.8 |
| 8.1 | 18.7 | 17.3 | 8 | 21.8 |
| 9.8 | 19.5 | 17.8 | 10 | 20.8 |
| 11.5 | 20.5 | 18.4 | 12 | 20.0 |
| 12.3 | 20.4 | 18.2 | 14 | 19.2 |
| 13.2 | 20.1 | 17.8 | | |

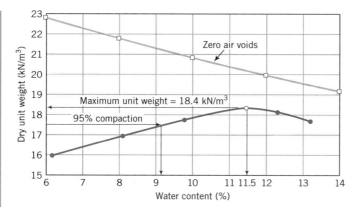

**FIGURE E2.15** Compaction test results.

**Step 2:** Plot graphs as shown in Fig. E2.15.

**Step 3:** Extract the desired values.

$$(\gamma_d)_{\max} = 18.4 \text{ kN/m}^3, \quad w_{\text{opt}} = 11.5\%$$

At 95% compaction, $\gamma_d = 18.4 \times 0.95 = 17.5 \text{ kN/m}^3$ and $w = 9.2\%$ (from graph).

**Step 4:** Calculate the degree of saturation at maximum dry unit weight. Algebraically manipulate Eq. (2.42) to find $S$ as

$$S = \frac{w G_s (\gamma_d)_{\max}/\gamma_w}{G_s - (\gamma_d)_{\max}/\gamma_w} = \frac{0.115 \times 2.7 \times (18.4/9.8)}{2.7 - 18.4/9.8} = 0.71 = 71\% \quad \blacksquare$$

*What's next . . .*We have discussed soils with the tacit assumption that we have observed and recovered samples of soils from the field on which we conducted tests as have been described. Soils are observed and recovered during a soil investigation of a proposed site. A soil investigation is an essential part of the design and construction of a proposed structural system (buildings, dams, roads and highways, etc.). You will now be given a brief introduction to a site or soil investigation.

## 2.13   SOIL INVESTIGATION

### 2.13.1 Purposes of a Soil Investigation

A soil investigation program is necessary to provide information for design and construction and for environmental assessment. The purposes of a soil investigation are:

1. To evaluate the general suitability of the site for the proposed project.
2. To enable an adequate and economical design to be made.
3. To disclose and make provision for difficulties that may arise during construction due to ground and other local conditions.

## 2.13.2 Phases of a Soil Investigation

The scope of a soil investigation depends on the type, size, and importance of the structure, the client, the engineer's familiarity with the soils at the site, and local building codes. Structures that are sensitive to settlement such as machine foundations and high-use buildings usually require a thorough soil investigation compared to a foundation for a house. A client may wish to take a greater risk than normal to save money and set limits on the type and extent of the site investigation. If the geotechnical engineer is familiar with a site, he/she may undertake a very simple soil investigation to confirm his/her experience. Some local building codes have provisions that set out the extent of a site investigation. It is mandatory that a visit be made to the proposed site.

In the early stages of a project, the available information is often inadequate to allow a detailed plan to be made. A site investigation must be developed in phases.

**Phase I.** Collection of available information such as a site plan, type, size, and importance of the structure, loading conditions, previous geotechnical reports, topographic maps, air photographs, geologic maps, and newspaper clippings.

**Phase II.** Preliminary reconnaissance or a site visit to provide a general picture of the topography and geology of the site. It is necessary that you take with you on the site visit all the information gathered in Phase I to compare with the current conditions of the site.

**Phase III.** Detailed soil exploration. The objectives of a detailed soil exploration are:

1. To determine the geological structure, which should include the thickness, sequence, and extent of the soil strata.
2. To determine the groundwater conditions.
3. To obtain disturbed and undisturbed samples for laboratory tests.
4. To conduct in situ tests.

**Phase IV.** Write a report. The report must contain a clear description of the soils at the site, methods of exploration, soil profile, test methods and results, and the location of the groundwater. You should include information and/or explanations of any unusual soil, water-bearing stratum, and soil and groundwater condition that may be troublesome during construction.

## 2.13.3 Soil Exploration Program

A soil exploration program usually involves test pits and/or soil borings (boreholes). During the site visit (Phase II), you should work out most of the soil exploration program. A detailed soil exploration consists of:

1. Preliminary location of each borehole and/or test pits.
2. Numbering of the boreholes or test pits.

3. Planned depth of each borehole or test pit.

4. Methods and procedures for advancing the boreholes.

5. Sampling instructions for at least the first borehole. The sampling instructions must include the number of samples and possible locations. Changes in the sampling instructions often occur after the first borehole.

6. Requirements for groundwater observations.

## 2.13.4 Soil Exploration Methods

Access to the soil may be obtained by the following methods:

- Trial pits or test pits
- Hand or powered augers
- Wash boring
- Rotary rigs

The advantages and disadvantages of each of these methods are shown in Table 2.8.

## 2.13.5 Soil Identification in the Field

In the field, the predominant soil types based on texture are identified by inspection. Gravels and sands are gritty and the individual particles are visible. Silts easily crumble and water migrates to the surface on application of pressure. Clays fail this water migration test since water flows very slowly through clays. Clays feel smooth, greasy, and sticky to the touch when wet but are very hard and strong when dry.

## 2.13.6 Depth of Boreholes

In compressible soils such as clays, the borings should penetrate either 1.5 to 2 times the least dimension of the foundation or until the stress increment due to the foundation loads is less than 10%, whichever is greater. Borings should penetrate at least 1 m into rock. In very stiff clays, borings should penetrate 5 m to 7 m to prove that the thickness of the strata is adequate.

## 2.13.7 Soil Sampling

The objective of soil sampling is to obtain soils of satisfactory size with minimum disturbance for observations and laboratory tests. Soil samples are usually obtained by attaching an open-ended thin-walled tube—called a Shelby tube or, simply, a sampling tube—to drill rods and forcing it down into the soil.

The tube is carefully withdrawn, hopefully, with the soil inside it. Soil disturbances occur from several sources during sampling, such as friction between

**TABLE 2.8   Advantages and Disadvantages of Soil Exploration Methods**

| Method | Advantages | Disadvantages |
|---|---|---|
| **Test pits**<br>A pit is dug either by hand or by a backhoe. | • Cost effective<br>• Provide detailed information of stratigraphy<br>• Large quantities of disturbed soils are available for testing<br>• Large blocks of undisturbed samples can be carved out from the pits<br>• Field tests can be conducted at the bottom of the pit | • Depth limited to about 6 m<br>• Deep pits uneconomical<br>• Excavation below groundwater and into rock difficult and costly<br>• Too many pits may scar site and require backfill soils |
| **Hand augers**<br>The auger is rotated by turning and pushing down on the handlebar. | • Cost effective<br>• Not dependent on terrain<br>• Portable<br>• Low headroom required<br>• Used in uncased holes<br>• Groundwater location can easily be identified and measured | • Depth limited to about 6 m<br>• Labor intensive<br>• Undisturbed samples can be taken only for soft clay deposit<br>• Cannot be used in rock, stiff clays, dry sand, or caliche soils. |
| **Power augers**<br>Truck mounted and equipped with continuous flight augers that bore a hole 100 to 250 mm in diameter. Augers can have a solid or hollow stem. | • Quick<br>• Used in uncased holes<br>• Undisturbed samples can be obtained quite easily<br>• Drilling mud not used<br>• Groundwater location can easily be identified | • Depth limited to about 15 m. At greater depth drilling becomes difficult and expensive<br>• Site must be accessible to motorized vehicle |
| **Wash boring**<br>Water is pumped to bottom of borehole and soil washings are returned to surface. A drill bit is rotated and dropped to produce a chopping action. | • Can be used in difficult terrain<br>• Low equipment costs<br>• Used in uncased holes | • Depth limited to about 30 m<br>• Slow drilling through stiff clays and gravels<br>• Difficulty in obtaining accurate location of groundwater level<br>• Undisturbed soil samples cannot be obtained |
| **Rotary drills**<br>A drill bit is pushed by the weight of the drilling equipment and rotated by a motor. | • Quick<br>• Can drill through any type of soil or rock<br>• Can drill to depths of 7500 m<br>• Undisturbed samples can easily be recovered | • Expensive equipment<br>• Terrain must be accessible to motorized vehicle<br>• Difficulty in obtaining location of groundwater level<br>• Additional time required for setup and cleanup |

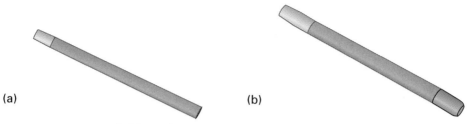

(a)                                    (b)

FIGURE 2.30   A thin-walled tube and a split tube sampler.

the soil and the sampling tube, the wall thickness of the sampling tube, the sharpness of the cutting edge, and care and handling during transportation of the sample tube. To minimize friction, the sampling tube should be pushed instead of driven into the ground.

Sampling tubes that are in common use have been designed to minimize sampling disturbances. One measure of the effects of sampler wall thickness is the recovery ratio defined as $L/z$, where $L$ is the length of the sample and $z$ is the distance that the sampler was pushed. Higher wall thickness leads to a greater recovery ratio and greater sampling disturbance.

One common type of soil sampler is the "Shelby tube," which is a thin-walled seamless steel tube of diameter 50 or 75 mm and length of 600–900 mm (Fig. 2.30a). Another popular sampler is the "standard" sampler, popularly known as the split spoon sampler, which has an inside diameter of 35 mm and an outside diameter of 50 mm (Fig. 2.30b). The sampler has a split barrel that is held together using a screw-on driving shoe at the bottom end and a cap at the upper end. The thicker wall of the standard sampler permits higher driving stresses than the Shelby tube but does so at the expense of higher levels of soil disturbances. Split spoon samples are disturbed. They are used for visual examination and for classification tests.

### 2.13.8 Boring Log

During soil exploration all pertinent details are recorded and presented in a boring log. A typical boring log is shown in Fig. 2.31. Additional information consisting mainly of laboratory and field test results is added to complete the boring log.

> *The* **essential points** *are:*
>
> *1. A site investigation is necessary to determine the nature of the soils at a proposed site for design and construction.*
>
> *2. A soil investigation needs careful planning and is usually done in phases.*
>
> *3. A number of tools are available for soil exploration. You need to use judgment as to the type appropriate for a given project.*

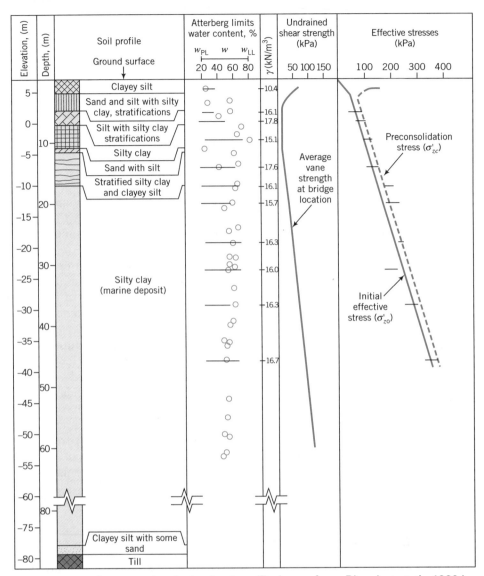

**FIGURE 2.31** An example of a boring log. (Redrawn from Blanchet et al., 1980.)

## 2.14 SUMMARY

We have dealt with a large body of basic information on the physical characteristics of soils and methods of soil exploration. A brief summary of what we covered follows. Soils are derived from the weathering of rocks and consist of gravels, sands, silts, and clays in decreasing order of particle size. Soils are conveniently idealized as three-phase materials: solids, water, and air. The physical properties of a soil depend on the relative proportion of these constituents in a given mass. Soils are classified into groups through their particle sizes and Atterberg limits. Soils within the same group are likely to have similar mechanical behavior and construction use. Flow of water through soils is governed by Darcy's law. A soil mass can be made denser by removing the air constituents through mechanical effort (compaction). The main physical parameters for soils are the particle sizes, void ratio, liquid limit, plastic limit, shrinkage limit, plasticity and liquidity indices, and the coefficient of permeability. Water can significantly change the characteristics of soils.

*Practical Examples*

### EXAMPLE 2.16

An embankment for a highway 30 m wide and 1.5 m in compacted thickness is to be constructed from a sandy soil trucked from a borrow pit. The water content of the sandy soil in the borrow pit is 15% and its void ratio is 0.69. The specification requires the embankment be compacted to a dry unit weight of 18 kN/m$^3$. Determine, for 1 km length of embankment, the following:

**(a)** The weight of sandy soil from the borrow pit required to construct the embankment.

**(b)** The number of 10.0 m$^3$ truckloads of sandy soil required for the construction.

**(c)** The weight of water per truckload of sandy soil.

**(d)** The degree of saturation of the sandy soil in situ.

*Strategy*   The strategy is similar to that adopted in Example 2.3.

### Solution 2.16

**Step 1:**   Calculate $\gamma_d$ for the borrow pit material.

$$\gamma_d = \frac{G\gamma_w}{1 + e} = \frac{2.7 \times 9.8}{1 + 0.69} = 15.7 \text{ kN/m}^3$$

**Step 2:**   Determine the volume of borrow pit soil required.

Volume of finished embankment:   $V = 30 \times 1.5 \times 1 = 45 \text{ m}^3$

Volume of borrow pit soil required:   $\dfrac{(\gamma_d)_{\text{required}}}{(\gamma_d)_{\text{borrow pit}}} \times V = \dfrac{18}{15.7} \times 45 \times 10^3$

$$= 51.6 \times 10^3 \text{ m}^3$$

**Step 3:**   Determine the number of trucks required.

$$\text{Number of trucks} = \frac{51.6 \times 10^3}{10} = 5160$$

**Step 4:**   Determine the weight of water required.

Weight of dry soil in one truckload:   $W_d = 10 \times 15.7 = 157 \text{ kN}$

Weight of water:   $wW_d = 0.15 \times 157 = 23.6 \text{ kN}$

**Step 5:**   Determine the degree of saturation.

$$S = \frac{wG}{e} = \frac{0.15 \times 2.7}{0.69} = 0.59 = 59\%$$

■

### EXAMPLE 2.17

An earth dam requires 1 million cubic meters of soil compacted to a void ratio of 0.8. In the vicinity of the proposed dam, three borrow pits were identified as having suitable materials. The cost of purchasing the soil and the cost of excavation are the same for each borrow pit. The only cost difference is transportation cost. The table below provides the void ratio and the transportation cost for each borrow pit. Which borrow pit would be the most economical?

| Borrow pit | Void ratio | Transportation cost ($/m³) |
|---|---|---|
| #1 | 1.8 | $0.60 |
| #2 | 0.9 | $1.00 |
| #3 | 1.5 | $0.75 |

**Strategy**   The specific volume is very useful in this problem to find the desired volume of borrow pit material.

## Solution 2.17

**Step 1:**   Define parameters and set up relevant equations. Let $V_o$, $e_o$ be the specific volume and void ratio of the compacted soil in the dam; and $V_i$, $e_i$ be the specific volume and void ratio of the soil from the borrow pits, where $i = 1, 2, 3$. Now

$$\frac{V_i}{V_o} = \frac{1 + e_i}{1 + e_o}$$

and

$$V_i = V_o\left(\frac{1 + e_i}{1 + e_o}\right) = \frac{1 \times 10^6}{1 + 0.8}(1 + e_i)$$

**Step 2:**   Determine the volume of soil from each borrow pit. Substituting the void ratio from the table into the last equation, we obtain

$$V_1 = 1,555,555 \text{ m}^3$$
$$V_2 = 1,055,555 \text{ m}^3$$
$$V_3 = 1,388,888 \text{ m}^3$$

**Step 3:**   Determine transportation costs.

Transportation cost = Volume $\times$ $/m³

Pit 1 = $933,333

Pit 2 = $1,055,555

Pit 3 = $1,041,667

Borrow pit 1 is therefore the most economical.   ∎

# EXERCISES

## Theory

**2.1**   Prove the following relations:

**(a)** $\gamma_d = \dfrac{G_s\gamma_w}{1 + e}$

**(b)** $S = \dfrac{wG_s(1 - n)}{n}$

**2.2** Show that

$$D_r = \frac{\gamma_d - (\gamma_d)_{min}}{(\gamma_d)_{max} - (\gamma_d)_{min}} \left\{ \frac{(\gamma_d)_{max}}{\gamma_d} \right\}$$

**2.3** Tests on a soil gave the following results: $G_s = 2.7$ and $e = 1.96$. Make a plot of degree of saturation versus water content for this soil.

**2.4** A spherical soil particle of radius $r$ is settling in water, which is at rest, at a constant velocity, $v$. If the drag force for a sphere is $6\pi r \eta v$, show that

$$r = \sqrt{\frac{9\eta}{2(\gamma_s - \gamma_w)}} \sqrt{v}$$

where $\gamma_s$ is the unit weight of the spherical soil particle, $\gamma_w$ is the unit weight of water and $\eta$ is the dynamic or absolute viscosity of water. [*Hint:* Draw a free body diagram (see Fig. P2.4) and use statics to solve the problem. You are actually deriving Stokes's equation.]

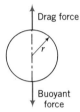

Drag force

$r$

Buoyant force

**FIGURE P2.4**

**2.5** A pump test is carried out to determine the coefficient of permeability of a confined water-bearing stratum shown in Fig. P2.5. Show that the equation for $k$ is

$$k = \frac{q \ln(r_1/r_2)}{2\pi H(h_1 - h_2)}$$

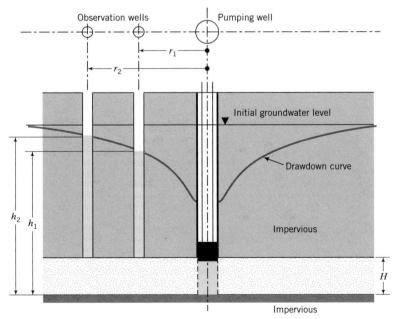

**FIGURE P2.5**

## Problem Solving

Assume $G_s = 2.7$, where necessary, for solving the following problems.

2.6  The mass of a sample of saturated soil is 520 grams. The dry mass, after oven drying, is 405 grams. Determine the (a) water content, (b) void ratio, (c) saturated unit weight, and (d) effective unit weight.

2.7  A soil sample has a bulk unit weight of 19.6 kN/m³ at a water content of 10%. Determine the void ratio, percentage air in the voids (air voids), and the degree of saturation of this sample.

2.8  A saturated silty clay encountered in a deep excavation is found to have a water content of 23%. Determine its porosity and bulk unit weight.

2.9  A wet sand sample has a volume of $4.64 \times 10^{-4}$ m³ and weighs 8 N. After oven drying, the weight reduces to 7.5 N. Calculate the water content, void ratio, and degree of saturation.

2.10  A sandy soil has a natural water content of 27.5% and bulk unit weight of 19.5 kN/m³. The void ratios corresponding to the densest and loosest state of this soil are 0.51 and 0.87. Find the relative density and degree of saturation for this soil.

2.11  The void ratio of a soil is 0.85. Determine the bulk and effective unit weights for the following degrees of saturation: (a) 75%, (b) 95%, and (c) 100%. What is the percentage error in the bulk unit weight if the soil were 95% saturated but assumed to be 100% saturated?

2.12  A particle size analysis on a soil sample yields the following data. Plot the particle size distribution curve and classify the soil using the Unified Soil Classification System. Clay to silt ratio is 3:1.

| Sieve no. | 4 | 10 | 20 | 60 | 200 | Pan |
|---|---|---|---|---|---|---|
| Sieve size (mm) | 4.75 | 2.0 | 0.84 | 0.25 | 0.074 | — |
| Weight retained (N) | 3.1 | 5.8 | 3.8 | 2.6 | 6.8 | 2.1 |

2.13  The following results were obtained from sieve analyses and Atterberg limits of three soils—A, B, and C. Hydrometer tests were conducted on the material that passed the 200 sieve for Soil A and Soil C. The ratio of clay to silt is 4:1 for Soil A and 1.5:1 for Soil C. Classify each of these soils according to the USCS.

| Sieve no. | Opening (mm) | Mass retained (grams) | | |
|---|---|---|---|---|
| | | Soil A | Soil B | Soil C |
| 4 | 4.75 | 0 | 0 | 0 |
| 10 | 2.00 | 20.2 | 48.2 | 15 |
| 20 | 0.85 | 25.7 | 19.6 | 98 |
| 40 | 0.425 | 40.4 | 60.3 | 90 |
| 100 | 0.15 | 18.1 | 37.2 | 182 |
| 200 | 0.075 | 27.2 | 22.1 | 109 |
| Pan | | 68.2 | | 6 |

2.14 The following results were obtained from a liquid limit test on a clay using the Casagrande cup device.

| Number of blows | 6 | 12 | 20 | 28 | 32 |
|---|---|---|---|---|---|
| Water content (%) | 52.5 | 47.1 | 43.2 | 38.6 | 37.0 |

(a) Determine the liquid limit of this clay.

(b) If the natural water content is 38% and the plastic limit is 23%, calculate the liquidity index.

(c) Do you expect a brittle type of failure for this soil? Why?

2.15 A fall cone test was carried out on a soil to determine its liquid and plastic limit using cones of masses 80 grams and 240 grams. The following results were obtained:

| | 80 gram cone | | | | 240 gram cone | | | |
|---|---|---|---|---|---|---|---|---|
| Penetration (mm) | 8 | 15 | 19 | 28 | 9 | 18 | 22 | 30 |
| Water content (%) | 43.1 | 52.0 | 56.1 | 62.9 | 37.0 | 47.5 | 51.0 | 55.1 |

Determine the liquid and plastic limits.

2.16 Determine the pressure head, elevation head, and total head at A, B, and C for the arrangement shown in Fig. P2.16. Take the water level at exit as datum. *Hint:* You need to convert the pressure 10 kPa to head.

2.17 In a constant-head permeability test, a sample of sand 12 cm long and 8 cm in diameter discharged $1.5 \times 10^{-3}$ m³ of water in 10 minutes. The head difference in two manometers A and B located at 10 mm and 110 mm, respectively, from the bottom of the sample is 20 mm. Determine the coefficient of permeability of the sand.

2.18 A constant-head test was conducted on a sample of soil 15 cm long and 60 cm² in cross-sectional area. The quantity of water collected was 40.5 cm³ in 15 seconds under a head difference of 24 cm. Calculate the coefficient of permeability. If the porosity of the sand were 55%, calculate the average velocity and the seepage velocity. Estimate the coefficient of permeability of a similar soil with a porosity of 35% from the results of this test.

2.19 A falling-head permeability test was carried out on a clay soil of diameter 100 mm and length 150 mm. In 4 minutes the head in the standpipe of diameter 5 mm dropped from 680 to 530 mm. Calculate the coefficient of permeability of this clay.

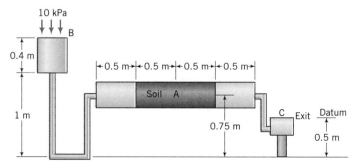

**FIGURE P2.16**

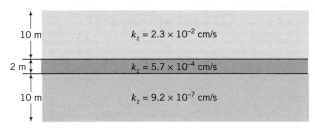

**FIGURE P2.20**

**2.20**  Calculate the equivalent coefficient of permeability in the vertical direction for the soil profile shown in Fig. P2.20.

**2.21**  A pumping test was carried out to determine the average permeability of a sand deposit 20 m thick overlying impermeable clay. The discharge from the pumping well was $12 \times 10^{-3}$ m³/s. Drawdowns in the observation wells located 15 m and 30 m from the centerline of the pumping well were 2.1 m and 1.6 m, respectively. Groundwater table was reached at 3.2 m below the ground surface. Determine the permeability of the sand. Estimate the effective grain size using Hazen's equation.

**2.22**  The following data were obtained from a Proctor test:

| Water content (%) | 12 | 13 | 14 | 16 | 19 |
|---|---|---|---|---|---|
| Weight of wet sample (N) | 17 | 18 | 18.7 | 19.3 | 18.9 |

The volume of the mold was $9.44 \times 10^{-4}$ m³.

**(a)** Determine the optimum water content and maximum dry unit weight.

**(b)** Plot the zero air voids line.

**(c)** Determine the degree of saturation at the maximum dry unit weight.

**2.23**  A 200,000 m³ clay embankment is to be constructed at an optimum water content of 16%. It is expected that 95% saturation will be attained. Calculate the volume of soil that will be needed if the void ratio of the fill material is 1.3.

## Practical

**2.24**  A Proctor compaction test in the laboratory on a borrow pit soil gives a maximum dry unit weight of 19 kN/m³ and optimum water content of 11.5%. The bulk unit weight and water content of the soil in the borrow pit are 17.2 kN/m³ and 8.2%, respectively. A highway fill is to be constructed using this soil. The specifications require the fill to be compacted to 95% Proctor compaction.

**(a)** How many cubic meters of borrow pit material is needed for 1 cubic meter of highway fill?

**(b)** The water content at 95% compaction is 11% on the dry side and 12% on the wet side of the compaction curve. What is the minimum additional weight of water per unit volume to achieve 95% Proctor compaction?

**(c)** If the soil were expansive, what is the additional weight of water per unit volume to achieve 95% Proctor compaction?

**(d)** How many truckloads of soil will be required for a 100,000 m³ highway embankment? Each truck has a capacity of 10 m³.

**(e)** Determine the cost for 100,000 m³ of compacted soil based on the following:

purchase and load borrow pit material at site, haul 2 km round trip, and spread with 200 HP dozer = $25/m³; extra mileage charge for each km = $3.10/m³; round trip distance = 10 km; compaction = $1.02/m³.

**2.25**  The highway embankment from Noscut to Windsor Forest, described in the sample practical situation, is 10 km long. The average cross section of the embankment is shown in Fig. P2.25a. The gradation curves for the soils at the two borrow pits are shown in Fig. P2.25b. Pit 1 is located 5 km from the start of the embankment while pit 2 is 3 km away. Estimated costs for various earthmoving operations are shown in the table below. You are given 10 minutes by the stakeholder's committee to present your recommendations. Prepare your presentation. The available visual aid equipment consists of an overhead projector and an LCD projector.

|  | Cost | |
| --- | --- | --- |
| **Operation** | **Pit 1** | **Pit 2** |
| Purchase and load borrow pit material at site, haul 2 km round trip, and spread with 200 HP dozer | $15/m³ | $18/m³ |
| Extra mileage charge for each km | $0.50/m³ | $0.55/m³ |
| Compaction | $1.02/m³ | $1.26 |
| Miscellaneous | $1.10/m³ | $0.85/m³ |

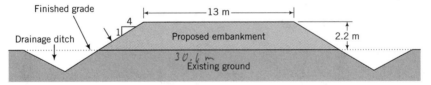

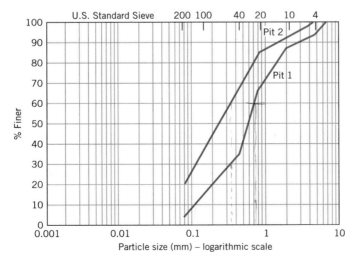

**FIGURE P2.25a**

**FIGURE P2.25b**

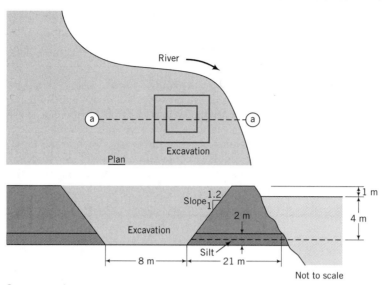

Cross section a–a.
**FIGURE P2.26**

2.26   An excavation is proposed for a square area near the bend of a river as shown in Fig. P2.26. It is expected that the flow of water into the excavation will come through the silt layer. Pumping tests reveal an average horizontal permeability of $5.2 \times 10^{-5}$ cm/s in the silt layer. The excavation has to be kept dry. Determine the flow ($q$) into the excavation.

## CHAPTER 3

# STRESSES, STRAINS, AND ELASTIC DEFORMATIONS OF SOILS

| ABET | ES | ED |
|------|----|----|
|      | 90 | 10 |

## 3.0 INTRODUCTION

You would have studied in mechanics the stresses imposed on homogeneous, elastic, rigid bodies by external forces. Soils are not homogeneous, elastic, rigid bodies, so the determination of stresses and strains in soils is a particularly difficult task. You may ask: "If soils are not elastic materials, then why do I have to study elastic methods of analysis?" Here are some reasons why a knowledge of elastic analysis is advantageous.

An elastic analysis of an isotropic material involves only two constants—Young's modulus and Poisson's ratio—and thus if we assume that soils are isotropic elastic materials then we have a powerful, but simple, analytical tool to predict a soil's response under loading. We will have to determine only the two elastic constants from our laboratory or field tests.

A geotechnical engineer must ensure that a geotechnical structure must not collapse under any anticipated loading condition and that settlement under working load (a fraction of the collapse load) must be within tolerable limits. We would prefer the settlement under working loads to be elastic so that no permanent settlement would occur. To calculate the elastic settlement, we have to use an elastic analysis. For example, in designing foundations on coarse-grained soils, we normally assume that the settlement is elastic and we then use elastic analysis to calculate the settlement.

An important task of a geotechnical engineer is to determine the stresses and strains that are imposed on a soil mass by external loads. It is customary to assume that the strains in the soils are small and this assumption allows us to apply our knowledge of mechanics of elastic bodies to soils. Small strains mean infinitesimal strains. For a realistic description of soils, elastic analysis is not satisfactory. We need soil models that can duplicate the complexity of soil behavior. However, even for complex soil models, an elastic analysis is a first step.

In this chapter, we will review some fundamental principles of mechanics and strength of materials and apply these principles to soils as elastic porous materials. This chapter contains a catalog of a large number of equations for soil stresses and strains. You may become weary of these equations but they are necessary for the analyses of the mechanical behavior of soils. You do not have to memorize these equations except the fundamental ones.

When you complete this chapter, you should be able to:

- Calculate stresses and strains in soils (assuming elastic behavior) from external loads
- Determine stress states
- Determine effective stresses
- Determine stress paths
- Calculate elastic settlement

You will use the following principles learned from statics and strength of materials:

- Stresses and strains
- Mohr's circle
- Elasticity—Hooke's law

***Sample Practical Situation***   Two storage tanks are to be founded on a deep layer of stiff saturated clay. Your client and the mechanical engineer, who is designing the pipe works, need an estimate of the settlement of the tanks when they are completely filled. Because of land restrictions, your client desires that

**FIGURE 3.1**   The "kissing" silos (Bozozuk, 1976, permission from National Research Council of Canada).

the tanks be as close as possible to each other. If two separate foundations are placed too close to each other, the stresses in the soil induced by each foundation overlap and cause intolerable tilting of the structures and their foundations. An example of tilting of structures caused by stress overlap is shown in Fig. 3.1.

These silos tilted toward each other at the top because stresses in the soil overlap at and near the internal edges of their foundations. The foundations are too close to each other.

## 3.1   DEFINITIONS OF KEY TERMS

*Stress* or intensity of loading is the load per unit area. The fundamental definition of stress is the ratio of the force $\Delta P$ acting on a plane $\Delta S$ to the area of the plane $\Delta S$ when $\Delta S$ tends to zero; $\Delta$ denotes a small quantity.

*Effective stress* $(\sigma')$ is the stress carried by the soil particles.

*Total stress* $(\sigma)$ is the stress carried by the soil particles and the liquids and gases in the voids.

*Strain* or intensity of deformation is the ratio of the change in a dimension to the original dimension or the ratio of change in length to the original length.

*Stress (strain) state* at a point is a set of stress (strain) vectors corresponding to all planes passing through that point. Mohr's circle is used to graphically represent stress (strain) state for two-dimensional bodies.

*Mean stress, p,* is the average stress on a body or the average of the orthogonal stresses in three dimensions.

*Deviatoric stress, q,* is the shear or distortional stress or stress difference on a body.

*Pore water pressure, u,* is the pressure of the water held in the soil pores.

*Stress path* is a graphical representation of the locus of stresses on a body.

*Isotropic* means the same material properties in all directions and also the same loading in all directions.

*Anisotropic* means the material properties are different in different directions and also the loadings are different in different directions.

*Elastic materials* are materials that return to their original configuration on unloading and obey Hooke's law.

## 3.2   QUESTIONS TO GUIDE YOUR READING

1. What are normal and shear stresses?
2. What is stress state and how is it determined?
3. Is soil an elastic material?
4. What are the limitations in analyzing soils based on the assumption that they (soils) are elastic materials?

5. What are shear strains, vertical strains, volumetric strains, and deviatoric strains?

6. How do I use elastic analysis to estimate the elastic settlement of soils and what are the limitations?

7. What are mean and deviatoric stresses?

8. What are the differences between plane strain and axisymmetric conditions?

9. How do I determine the stresses and strains/displacements imposed on a soil mass by external loads?

10. What is effective stress?

11. Is deformation a function of effective or total stress?

12. What is a stress path and its significance in practical problems?

## 3.3 STRESSES AND STRAINS

### 3.3.1 Normal Stresses and Strains

Consider a cube of dimensions $x = y = z$ that is subjected to forces $P_x$, $P_y$, $P_z$, normal to three adjacent sides as shown in Fig. 3.2. The normal stresses are

$$\sigma_z = \frac{P_z}{xy}, \quad \sigma_x = \frac{P_x}{yz}, \quad \sigma_y = \frac{P_y}{xz} \tag{3.1}$$

Let us assume that under these forces the cube compressed by $\Delta x$, $\Delta y$, and $\Delta z$ in the $X$, $Y$, and $Z$ directions. The strains in these directions, assuming they are small (infinitesimal), are

$$\varepsilon_z = \frac{\Delta z}{z}, \quad \varepsilon_x = \frac{\Delta x}{x}, \quad \varepsilon_y = \frac{\Delta y}{y} \tag{3.2}$$

### 3.3.2 Volumetric Strain

The volumetric strain is

$$\varepsilon_p = \varepsilon_x + \varepsilon_y + \varepsilon_z \tag{3.3}$$

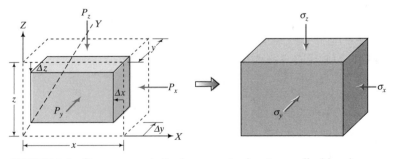

**FIGURE 3.2** Stresses and displacements due to applied loads.

### 3.3.3 Shear Stresses and Shear Strains

Let us consider, for simplicity, the $XZ$ plane and apply a force $F$ that causes the square to distort into a parallelogram as shown in Fig. 3.3. The force $F$ is a shearing force and the shear stress is

$$\tau = \frac{F}{xy} \tag{3.4}$$

Simple shear strain is a measure of the angular distortion of a body by shearing forces. If the horizontal displacement is $\Delta x$, the shear strain or simple shear strain, $\gamma_{zx}$, is

$$\gamma_{zx} = \tan^{-1}\frac{\Delta x}{z}$$

For small strains, $\tan \gamma_{zx} = \gamma_{zx}$ and therefore

$$\gamma_{zx} = \frac{\Delta x}{z} \tag{3.5}$$

If the shear stress on a plane is zero, the normal stress on that plane is called a principal stress. We will discuss principal stresses later. In geotechnical engineering, compressive stresses in soils are assumed to be positive. Soils cannot sustain any appreciable tensile stresses and we normally assume that the tensile strength of soils is negligible. Strains can be compressive or tensile.

> *The essential points are:*
> 1. *A normal stress is the load per unit area on a plane normal to the direction of the load.*
> 2. *A shear stress is the load per unit area on a plane parallel to the direction of the shear force.*
> 3. *Normal stresses compress or elongate a material; shear stresses distort a material.*
> 4. *A normal strain is the change in length divided by the original length in the direction of the original length.*
> 5. *Principal stresses are normal stresses on planes of zero shear stress.*
> 6. *Soils can only sustain compressive stresses.*

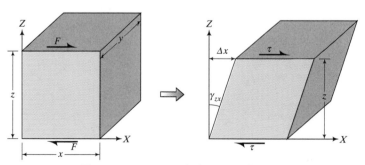

**FIGURE 3.3** Shear stresses and shear strains.

*What's next* . . .What happens when we apply stresses to a deformable material? From the last section, you may answer that the material deforms, and you are absolutely correct. Different materials respond differently to applied loads. Next, we will examine some typical responses of deformable materials to applied loads to serve as a base to characterize the loading responses of soils.

## 3.4 IDEALIZED STRESS–STRAIN RESPONSE AND YIELDING

### 3.4.1 Material Responses to Normal Loading and Unloading

If we apply an incremental vertical load, $\Delta P$, to a deformable cylinder (Fig. 3.4) of cross-sectional area $A$, the cylinder will compress by, say, $\Delta z$ and the radius will increase by $\Delta r$. The loading condition we apply here is called uniaxial loading. The change in vertical stress is

$$\Delta\sigma_z = \frac{\Delta P}{A} \tag{3.6}$$

The vertical and radial strains are, respectively,

$$\Delta\varepsilon_z = \frac{\Delta z}{H_o} \tag{3.7}$$

$$\Delta\varepsilon_r = \frac{\Delta r}{r_o} \tag{3.8}$$

where $H_o$ is the original length and $r_o$ is the original radius. The ratio of the radial (or lateral) strain to the vertical strain is called Poisson's ratio, $v$, defined as

$$v = \frac{-\Delta\varepsilon_r}{\Delta\varepsilon_z} \tag{3.9}$$

Typical values of Poisson's ratio for soil are listed in Table 3.1.

We can plot a graph of $\sigma_z = \Sigma \Delta\sigma_z$ versus $\varepsilon_z = \Sigma \Delta\varepsilon_z$. If for equal increments of $\Delta P$, we get the same value of $\Delta z$, then we will get a straight line in the graph of $\sigma_z$ versus $\varepsilon_z$ as shown by $OA$ in Fig. 3.5. If at some stress point, say, at $A$ (Fig. 3.5), we unload the cylinder and it returns to its original configuration, the ma-

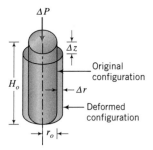

**FIGURE 3.4** Forces and displacements on a cylinder.

**TABLE 3.1 Typical Values of Poisson's Ratio**

| Soil type | Description | $\nu^a$ |
|---|---|---|
| Clay | Soft | 0.35–0.40 |
| | Medium | 0.30–0.35 |
| | Stiff | 0.20–0.30 |
| Sand | Loose | 0.15–0.25 |
| | Medium | 0.25–0.30 |
| | Dense | 0.25–0.35 |

[a]These values are effective values, $\nu'$ (see discussions later in this chapter, Sections 3.12 and 3.13).

terial comprising the cylinder is called a *linearly elastic* material. Suppose for equal increments of $\Delta P$ we get different values of $\Delta z$, but on unloading the cylinder it returns to its original configuration. Then a plot of the stress–strain relationship will be a curve as illustrated by $OB$ in Fig. 3.5. In this case, the material comprising the cylinder is called a *nonlinearly elastic* material. If we apply a load $P_1$ that causes a displacement $\Delta z_1$ on an elastic material and a second load $P_2$ that causes a displacement $\Delta z_2$, then the total displacement is $\Delta z = \Delta z_1 + \Delta z_2$. Elastic materials obey the principle of superposition. The order in which the load is applied is not important; we could apply $P_2$ first and then $P_1$ but the final displacement would be the same.

Some materials, soil is one of them, do not return to their original configurations after unloading. They exhibit a stress–strain relationship similar to that depicted in Fig. 3.6, where $OA$ is the loading response, $AB$ the unloading response, and $BC$ the reloading response. The strains that occur during loading, $OA$, consist of two parts—an elastic or recoverable part, $BD$, and a plastic or unrecoverable part, $OB$. Such material behavior is called *elastoplastic*. Part of the loading response is elastic, the other plastic.

As engineers, we are particularly interested in the plastic strains since these are the result of permanent deformations of the material. But to calculate the permanent deformation, we must know the elastic deformation. Here, elastic

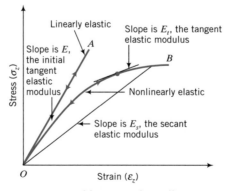

**FIGURE 3.5** Linear and nonlinear stress–strain curves of an elastic material.

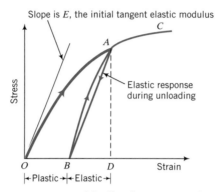

**FIGURE 3.6**   Idealized stress–strain curves of an elastoplastic material.

analyses become useful. The stress at which permanent deformation initiates is called the yield stress.

The *elastic modulus* or *initial tangent elastic modulus* (E) is the slope of the stress–strain line for linear isotropic material (Fig. 3.5). For a nonlinear elastic material either the tangent modulus ($E_t$) or the secant modulus ($E_s$) or both are determined from the stress–strain relationship (Fig. 3.5). The *tangent elastic modulus* is the slope of the tangent to the stress–strain point under consideration. The *secant elastic modulus* is the slope of the line joining the origin $(0, 0)$ to some desired stress–strain point. For example, some engineers prefer to determine the secant modulus by using a point on the stress–strain curve corresponding to the maximum stress while others prefer to use a point on the stress–strain curve corresponding to a certain level of strain, for example, 1%. The tangent elastic modulus and the secant elastic modulus are not constants. These moduli tend to decrease as shear strains increase. It is customary to determine the *initial tangent elastic modulus* for an elastoplastic material by unloading it and calculating the initial slope of the unloading line as the initial tangent elastic modulus (Fig. 3.6).

### 3.4.2 Material Response to Shear Forces

Shear forces distort materials. A typical response of an elastoplastic material to simple shear is shown in Fig. 3.7. The initial shear modulus ($G_i$) is the slope of the initial straight portion of the $\tau_{zx}$ versus $\gamma_{zx}$ curve. The secant shear modulus

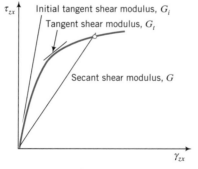

**FIGURE 3.7**   Shear stress–shear strain response of an elastoplastic material.

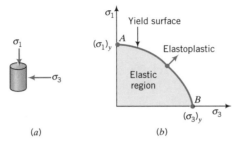

**FIGURE 3.8** Elastic, yield, and elastoplastic stress states.

(G) is the slope of a line from the desired shear stress–shear strain point to the origin of the $\tau_{zx}$ versus $\gamma_{zx}$ plot (Fig. 3.7).

### 3.4.3 Yield Surface

Let us consider a more complex situation than the uniaxial loading of a cylinder (Fig. 3.8a). In this case, we are going to apply increments of vertical and radial stresses. Since we are not applying any shear stresses, the axial stresses and radial stresses are principal stresses: $\sigma_z = \sigma_1 = \Sigma \, \Delta\sigma_z$ and $\sigma_r = \sigma_3 = \Sigma \, \Delta\sigma_r$, respectively. Let us, for example, set $\sigma_3$ to zero and increase $\sigma_1$. The material will yield at some value of $\sigma_1$, which we will call $(\sigma_1)_y$, and plots as point $A$ in Fig. 3.8b. If, alternatively, we set $\sigma_1 = 0$ and increase $\sigma_3$, the material will yield at $(\sigma_3)_y$ and is represented by point $B$ in Fig. 3.8b. We can then subject the cylinder to various combinations of $\sigma_1$ and $\sigma_3$ and plot the resulting yield points. Linking the yield points results in a curve, $AB$, which is called the *yield curve* or *yield surface* as shown in Fig. 3.8b. A material subjected to a combination of stresses that lies below this curve will respond elastically (recoverable deformation). If loading is continued beyond the yield stress, the material will respond elastoplastically (irrecoverable or permanent deformations occur). If the material is isotropic, the yield surface will be symmetrical about the $\sigma_1$, $\sigma_3$ axes.

*The essential points are:*

1. *An elastic material recovers its original configuration on unloading; an elastoplastic material undergoes both elastic (recoverable) and plastic (permanent) deformation during loading.*

2. *Soils are elastoplastic materials.*

3. *At small strains, soils behave like an elastic material and thereafter like an elastoplastic material.*

4. *The locus of the stresses at which a soil yields is called a yield surface. Stresses below the yield stress cause the soil to respond elastically; stresses beyond the yield stress cause the soil to respond elastoplastically.*

**What's next . . .** In the next two sections, we will write the general expression for Hooke's law, which is the fundamental law for linear elastic materials, and then consider two loading cases appropriate to soils.

## 3.5   HOOKE'S LAW

### 3.5.1 General State of Stress

Stresses and strains for a linear, isotropic, elastic soil are related through Hooke's law. For a general state of stress (Fig. 3.9), Hooke's law is

$$
\begin{Bmatrix} \varepsilon_x \\ \varepsilon_y \\ \varepsilon_z \\ \gamma_{xy} \\ \gamma_{yz} \\ \gamma_{zx} \end{Bmatrix} = \frac{1}{E}
\begin{bmatrix}
1 & -\nu & -\nu & 0 & 0 & 0 \\
-\nu & 1 & -\nu & 0 & 0 & 0 \\
-\nu & -\nu & 1 & 0 & 0 & 0 \\
0 & 0 & 0 & 2(1+\nu) & 0 & 0 \\
0 & 0 & 0 & 0 & 2(1+\nu) & 0 \\
0 & 0 & 0 & 0 & 0 & 2(1+\nu)
\end{bmatrix}
\begin{Bmatrix} \sigma_x \\ \sigma_y \\ \sigma_z \\ \tau_{xy} \\ \tau_{yz} \\ \tau_{zx} \end{Bmatrix}
\tag{3.10}
$$

where $E$ is the elastic or Young's modulus and $\nu$ is Poisson's ratio. Equation (3.10) is called the elastic equations or elastic stress–strain constitutive equations. From Eq. (3.10), we have, for example,

$$
\gamma_{zx} = \frac{2(1+\nu)}{E}\,\tau_{zx} = \frac{\tau_{zx}}{G}
\tag{3.11}
$$

where

$$
G = \frac{E}{2(1+\nu)}
\tag{3.12}
$$

is the shear modulus. We will call $E$, $G$, and $\nu$ the elastic parameters. Only two of these parameters—either $E$ or $G$ and $\nu$—are required to solve problems dealing with isotropic, elastic materials. We can calculate $G$ from Eq. (3.12), if $E$ and $\nu$ are known. Poisson's ratio for soils is not easy to determine and a direct way to obtain $G$ is to subject the material to shearing forces as described in Section 3.4.2. For nonlinear elastic materials, the tangent modulus or the secant modulus is used in Eq. (3.10) and the calculations are done incrementally for small increments of stress.

The elastic and shear moduli for soils depend on the stress history, direction of loading, and the magnitude of the applied strains. In Chapter 5, we will study a few tests that are used to determine $E$ and $G$ and, in Chapter 6, explore the details of the use of $E$ and $G$ in soil analyses. Typical values of $E$ and $G$ are shown in Table 3.2.

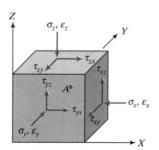

**FIGURE 3.9**   General state of stress.

**TABLE 3.2    Typical Values of *E* and *G***

| Soil type | Description | $E^a$ (MPa) | $G*$ (MPa) |
|-----------|-------------|-------------|------------|
| Clay | Soft | 1–15 | 0.5–5 |
| | Medium | 15–30 | 5–15 |
| | Stiff | 30–100 | 15–40 |
| Sand | Loose | 10–20 | 5–10 |
| | Medium | 20–40 | 10–15 |
| | Dense | 40–80 | 15–35 |

[a]These are average secant elastic moduli for the drained condition (see discussions later in this chapter, Sections 3.12 and 3.13).

### 3.5.2 Principal Stresses

If the stresses applied to a soil are principal stresses, then Hooke's law reduces to

$$\begin{Bmatrix} \varepsilon_1 \\ \varepsilon_2 \\ \varepsilon_3 \end{Bmatrix} = \frac{1}{E} \begin{bmatrix} 1 & -\nu & -\nu \\ -\nu & 1 & -\nu \\ -\nu & -\nu & 1 \end{bmatrix} \begin{Bmatrix} \sigma_1 \\ \sigma_2 \\ \sigma_3 \end{Bmatrix} \tag{3.13}$$

The matrix on the right-hand side of Eq. (3.13) is called the compliance matrix. The inverse of Eq. (3.13) is

$$\begin{Bmatrix} \sigma_1 \\ \sigma_2 \\ \sigma_3 \end{Bmatrix} = \frac{E}{(1 + \nu)(1 - 2\nu)} \begin{bmatrix} 1 - \nu & \nu & \nu \\ \nu & 1 - \nu & \nu \\ \nu & \nu & 1 - \nu \end{bmatrix} \begin{Bmatrix} \varepsilon_1 \\ \varepsilon_2 \\ \varepsilon_3 \end{Bmatrix} \tag{3.14}$$

The matrix on the right-hand side is called the stiffness matrix. If you know the stresses and the material parameters $E$ and $\nu$, you can use Eq. (3.13) to calculate the strains; or if you know the strains, $E$, and $\nu$, you can use Eq. (3.14) to calculate the stresses.

### 3.5.3 Displacements from Strains and Forces from Stresses

The displacements and forces are obtained by integration. For example, the vertical displacement, $\Delta z$, is

$$\Delta z = \int \varepsilon_z \, dz \tag{3.15}$$

and the axial force is

$$P_z = \int \Delta \sigma_z \, dA \tag{3.16}$$

where $dz$ is the height or thickness of the element and $dA$ is the elemental area.

> *The* **essential points** *are:*
> 1. *Hooke's law applies to a linearly elastic material.*
> 2. *As a first approximation, you can use Hooke's law to calculate stresses, strains, and elastic settlement of soils.*
> 3. *For nonlinear materials, Hooke's law is used with an approximate elastic modulus (tangent modulus or secant modulus) and the calculations are done for incremental increases in stresses or strains.*

*What's next . . .*The stresses and strains in three dimensions become complicated when applied to real problems. For practical purposes, many geotechnical problems can be solved using two-dimensional stress and strain parameters. In the next section, we will discuss two conditions that simplify the stress and strain states of soils.

## 3.6   PLANE STRAIN AND AXIAL SYMMETRIC CONDITIONS

### 3.6.1 Plane Strain

There are two conditions of stresses and strains that are common in geotechnical engineering. One is the *plane strain* condition in which the strain in one direction is zero. As an example of a plane strain condition, let us consider an element of soil, $A$, behind a retaining wall (Fig. 3.10). Because the displacement that is likely to occur in the $Y$ direction ($\Delta y$) is small compared with the length in this direction, the strain tends to zero; that is, $\varepsilon_y = \Delta y/y \cong 0$. We can then assume that soil element $A$ is under a plane strain condition. Since we are considering principal stresses, we will map the $X$, $Y$, and $Z$ directions as 3, 2, and 1 directions. In the case of the retaining wall the $Y$ direction (2 direction) is the zero strain direction and therefore $\varepsilon_2 = 0$ in Eq. (3.13). Hooke's law for a plane strain condition is

$$\varepsilon_1 = \frac{1 + v}{E} \left[ (1 - v)\sigma_1 - v\sigma_3 \right] \tag{3.17}$$

$$\varepsilon_3 = \frac{1 + v}{E} \left[ (1 - v)\sigma_3 - v\sigma_1 \right] \tag{3.18}$$

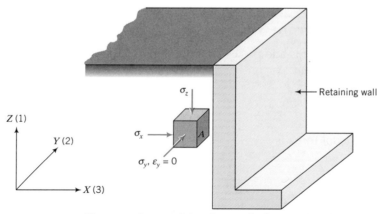

**FIGURE 3.10**  Plane strain condition in a soil element behind a retaining wall.

and

$$\sigma_2 = \nu(\sigma_1 + \sigma_3)$$ (3.19)

In matrix form, Eqs. (3.17) and (3.18) become

$$\left\{\begin{matrix} \varepsilon_1 \\ \varepsilon_3 \end{matrix}\right\} = \frac{1 + \nu}{E} \left[\begin{matrix} 1 - \nu & -\nu \\ -\nu & 1 - \nu \end{matrix}\right] \left\{\begin{matrix} \sigma_1 \\ \sigma_3 \end{matrix}\right\}$$ (3.20)

The inverse of Eq. (3.20) gives

$$\left\{\begin{matrix} \sigma_1 \\ \sigma_3 \end{matrix}\right\} = \frac{E}{(1 + \nu)(1 - 2\nu)} \left[\begin{matrix} 1 - \nu & \nu \\ \nu & 1 - \nu \end{matrix}\right] \left\{\begin{matrix} \varepsilon_1 \\ \varepsilon_3 \end{matrix}\right\}$$ (3.21)

### 3.6.2 Axisymmetric Condition

The other condition that occurs in practical problems is *axial symmetry* or the *axisymmetric* condition where two stresses are equal. Let us consider a water tank or an oil tank founded on a soil mass as illustrated in Fig. 3.11.

The radial stresses ($\sigma_r$) and circumferential stresses ($\sigma_\theta$) on a cylindrical element of soil directly under the center of the tank are equal because of axial symmetry. The oil tank will apply a uniform vertical (axial) stress at the soil surface and the soil element will be subjected to an increase in axial stress, $\Delta\sigma_z = \Delta\sigma_1$, and an increase in radial stress, $\Delta\sigma_r = \Delta\sigma_\theta = \Delta\sigma_3$. Will a soil element under the edge of the tank be under an axisymmetric condition? The answer is no, since the stresses at the edge of the tank are all different; there is no symmetry.

Hooke's law for the axisymmetric condition is

$$\varepsilon_1 = \frac{1}{E} [\sigma_1 - 2\nu\sigma_3]$$ (3.22)

$$\varepsilon_3 = \frac{1}{E} [(1 - \nu)\sigma_3 - \nu\sigma_1]$$ (3.23)

or, in matrix form,

$$\left\{\begin{matrix} \varepsilon_1 \\ \varepsilon_3 \end{matrix}\right\} = \frac{1}{E} \left[\begin{matrix} 1 & -2\nu \\ -\nu & 1 - \nu \end{matrix}\right] \left\{\begin{matrix} \sigma_1 \\ \sigma_3 \end{matrix}\right\}$$ (3.24)

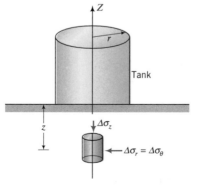

**FIGURE 3.11** Axisymmetric condition on a soil element under the center of a tank.

The inverse of Eq. (3.24) gives

$$\begin{Bmatrix} \sigma_1 \\ \sigma_3 \end{Bmatrix} = \frac{E}{(1 + v)(1 - 2v)} \begin{bmatrix} 1 - v & 2v \\ v & 1 \end{bmatrix} \begin{Bmatrix} \varepsilon_1 \\ \varepsilon_3 \end{Bmatrix} \tag{3.25}$$

*The essential points are:*

1. *A plane strain condition is one in which the strain in one or more directions is zero or small enough to be neglected.*

2. *An axisymmetric condition is one in which two stresses are equal.*

## EXAMPLE 3.1

A retaining wall moves outward causing a lateral strain of 0.1% and a vertical strain of 0.05% on a soil element located 3 m below ground level. Assuming the soil is a linear, isotropic, elastic material with $E = 5000$ kPa and $v = 0.3$, calculate the increase in stresses imposed. If the retaining wall is 6 m high and the stresses you calculate are the average stresses, determine the lateral force increase per unit length of wall.

**Strategy**   You will have to make a decision whether to use the plane strain or axisymmetric condition and then use the appropriate equation. You are asked to find the increase in stresses, so it is best to write the elastic equations in terms of increment. The retaining wall moves outward, so the lateral strain is tensile ($-$) while the vertical strain is compressive ($+$). The increase in lateral force is found by integration of the average lateral stress increase.

## Solution 3.1

**Step 1:**   Determine the appropriate stress condition and write the appropriate equation.

The soil element is likely to be under the plane strain condition ($\varepsilon_2 = 0$); use Eq. (3.21).

$$\begin{Bmatrix} \Delta\sigma_1 \\ \Delta\sigma_3 \end{Bmatrix} = \frac{5000}{(1 + 0.3)(1 - 2 \times 0.3)} \begin{bmatrix} 1 - 0.3 & 0.3 \\ 0.3 & 1 - 0.3 \end{bmatrix} \begin{Bmatrix} 0.0005 \\ -0.001 \end{Bmatrix}$$

**Step 2:**   Solve the equation.

$$\Delta\sigma_1 = 9615.4\{(0.7 \times 0.0005) + [0.3 \times (-0.001)]\} = 0.5 \text{ kPa}$$
$$\Delta\sigma_3 = 9615.4\{(0.3 \times 0.0005) + [0.7 \times (-0.001)]\} = -5.3 \text{ kPa};$$
the negative sign means reduction

**Step 3:**   Calculate the lateral force per unit length.

$$\Delta\sigma_3 = \Delta\sigma_x$$
$$\Delta P_x = \int_0^6 \Delta\sigma_x \, dA = -\int_0^6 5.3 \, (dx \times 1) = -[5.3x]_0^6 = -31.8 \text{ kN/m} \quad \blacksquare$$

## EXAMPLE 3.2

An oil tank is founded on a layer of medium sand 5 m thick underlain by a deep deposit of dense sand. The geotechnical engineer assumed, based on experience,

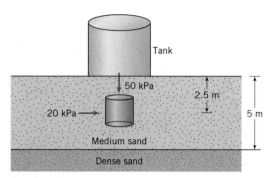

**FIGURE E3.2**

that the settlement of the tank would occur from settlement in the medium sand. The vertical and lateral stresses at the middle of the medium sand directly under the center of the tank are 50 kPa and 20 kPa, respectively. The values of $E$ and $\nu$ are 20 MPa and 0.3, respectively. Assuming a linear, isotropic, elastic material behavior, calculate the strains imposed on the medium sand and the vertical settlement.

***Strategy*** You have to decide on the stress conditions on the soil element directly under the center of the tank. Once you make your decision, use the appropriate equations to find the strains and then integrate the vertical strains to calculate the settlement. Draw a diagram illustrating the problem.

### Solution 3.2

**Step 1:** Draw a diagram of the problem—see Fig. E3.2.

**Step 2:** Decide on a stress condition.
The element is directly under the center of the tank, so the axisymmetric condition prevails.

**Step 3:** Choose the appropriate equations and solve.
Use Eq. (3.24).

$$\begin{Bmatrix} \Delta\varepsilon_1 \\ \Delta\varepsilon_3 \end{Bmatrix} = \frac{1}{20 \times 10^3} \begin{bmatrix} 1 & -0.6 \\ -0.3 & 0.7 \end{bmatrix} \begin{Bmatrix} 50 \\ 20 \end{Bmatrix}$$

Using algebra, we get

$$\Delta\varepsilon_1 = \frac{1}{20 \times 10^3} [1 \times 50 - 0.6 \times 20] = 1.9 \times 10^{-3}$$

$$\Delta\varepsilon_3 = \frac{1}{20 \times 10^3} [-0.3 \times 50 + 0.7 \times 20] = -5 \times 10^{-5}$$

**Step 4:** Calculate vertical displacement.

$$\Delta\varepsilon_1 = \Delta\varepsilon_z$$

$$\Delta z = \int_0^5 \Delta\varepsilon_z \, dz = [1.9 \times 10^{-3} \, z]_0^5 = 9.5 \times 10^{-3} \text{ m} = 9.5 \text{ mm} \quad \blacksquare$$

***What's next*** . . .We have used the elastic equations to calculate stresses, strains, and displacements in soils assuming that soils are linear, isotropic, elastic materials.

Soils, in general, are not linear, isotropic, elastic materials. We will briefly discuss anisotropic, elastic materials in the next section.

## 3.7    ANISOTROPIC ELASTIC STATES

Anisotropic materials have different elastic parameters in different directions. Anisotropy in soils results from essentially two causes.

1. The manner in which the soil is deposited. This is called structural anisotropy and it is the result of the kind of soil fabric that is formed during deposition. You should recall (Chapter 2) that the soil fabric produced is related to the history of the environment in which the soil is formed. A special form of structural anisotropy occurs when the horizontal plane is a plane of isotropy. We call this form of structural anisotropy, transverse anisotropy.

2. The difference in stresses in the different directions. This is known as stress-induced anisotropy.

Transverse anisotropy, also called cross anisotropy, is the most prevalent type of anisotropy in soils. If we were to load the soil in the vertical direction ($Z$ direction) and repeat the same loading in the horizontal direction, say, the $X$ direction, the soil will respond differently; its stress–strain characteristics and strength would be different in these directions. However, if we were to load the soil in the $Y$ direction, the soil's response would be similar to the response obtained in the $X$ direction. The implication is that a soil mass will, in general, respond differently depending on the direction of the load. For transverse anisotropy, the elastic parameters are the same in the lateral directions ($X$ and $Y$ directions) but are different from the vertical direction.

To fully describe anisotropic soil behavior, we need 21 elastic constants (Love, 1927) but for transverse anisotropy, we need only 5 elastic constants; these are $E_z$, $E_x$, $v_{xx}$, $v_{zx}$, and $v_{zz}$. The first letter in the double subscripts denotes the direction of loading and the second letter denotes the direction of measurement. For example, $v_{zx}$ means Poisson's ratio determined from the ratio of the strain in the lateral direction ($X$ direction) to the strain in the vertical direction ($Z$ direction) with the load applied in the vertical direction ($Z$ direction).

In the laboratory, the direction of loading of soil samples taken from the field is invariably vertical. Consequently, we cannot determine the five desired elastic parameters from conventional laboratory tests. Graham and Houlsby (1983) suggested a method to overcome the lack of knowledge of the five desired elastic parameters in solving problems on transverse anisotropy. However, their method is beyond the scope of this book.

For axisymmetric conditions, the transverse anisotropic, elastic equations are

$$
\begin{Bmatrix} \Delta\varepsilon_z \\ \Delta\varepsilon_r \end{Bmatrix} = \begin{bmatrix} \dfrac{1}{E_z} & \dfrac{-2v_{rz}}{E_r} \\ \dfrac{-v_{zr}}{E_z} & \dfrac{(1-v_{rr})}{E_r} \end{bmatrix} \begin{Bmatrix} \Delta\sigma_z \\ \Delta\sigma_r \end{Bmatrix}
\tag{3.26}
$$

where the subscript $z$ denotes vertical and $r$ denotes radial. By superposition, $v_{rz}/v_{zr} = E_r/E_z$.

> *The* **essential points** *are:*
>
> 1. *Two forms of anisotropy are present in soils. One is structural anisotropy, which is related to the history of loading and environmental conditions during deposition, and the other is stress-induced anisotropy, which results from differences in stresses in different directions.*
> 2. *The prevalent form of structural anisotropy in soils is transverse anisotropy; the soil properties and the soil response in the lateral directions are the same but are different from those in the vertical direction.*
> 3. *You need to find the elastic parameters in different directions of a soil mass to determine elastic stresses, strains, and displacements.*

## EXAMPLE 3.3

Redo Example 3.2 but now the soil under the oil tank is an anisotropic elastic material with $E_z = 20$ MPa, $E_r = 25$ MPa, $v_{rz} = 0.15$, and $v_{rr} = 0.3$.

**Strategy**  The solution of this problem is a straightforward application of Eq. (3.26).

## Solution 3.3

**Step 1:**  Determine $v_{zr}$ (by superposition).

$$\frac{v_{rz}}{v_{zr}} = \frac{E_r}{E_z}$$

$$v_{zr} = \frac{20}{25} \times 0.15 = 0.12$$

**Step 2:**  Find the strains.
Use Eq. (3.26).

$$\begin{Bmatrix} \Delta\varepsilon_z \\ \Delta\varepsilon_r \end{Bmatrix} = 10^{-3} \begin{bmatrix} \dfrac{1}{20} & \dfrac{-2 \times 0.15}{25} \\ \dfrac{-0.12}{20} & \dfrac{(1-0.3)}{25} \end{bmatrix} \begin{Bmatrix} 50 \\ 20 \end{Bmatrix}$$

The solution is $\varepsilon_z = 2.26 \times 10^{-3} = 0.23\%$ and $\varepsilon_r = 0.26 \times 10^{-3} = 0.03\%$.

**Step 3:**  Determine vertical displacement.

$$\Delta z = \int_0^5 \varepsilon_z \, dz = [2.26 \times 10^{-3} z]_0^5 = 11.3 \times 10^{-3} \text{ m} = 11.3 \text{ mm}$$

The vertical displacement in the anisotropic case is about 19% more than in the isotropic case (Example 3.2). Also, the radial strain is tensile for the isotropic case but compressive in the anisotropic case for this problem.  ∎

***What's next . . .***We now know how to calculate stresses and strains in soils if we assume soils are elastic, homogeneous materials. One of the important tasks for engineering works is to determine strength or failure of materials. We can draw an analogy of the strength of materials with the strength of a chain. The chain is only as strong as its weakest link. For soils, failure may be initiated at a point within a soil mass and then propagate through it; this is known as progressive failure. The stress state at a point in a soil mass due to applied boundary forces may be equal to the strength of the soil, thereby initiating failure. Therefore, as engineers, we need to know the stress state at a point due to applied loads. We will discuss stress state next using your knowledge in strength of materials.

##  3.8   STRESS AND STRAIN STATES

### 3.8.1 Mohr's Circle for Stress States

Suppose a cuboidal sample of soil is subjected to the stresses shown in Fig. 3.9. We would like to know what the stresses are at a point, say, $A$, within the sample due to the applied stresses. One approach to find the stresses at $A$, called the stress state at $A$, is to use Mohr's circle. The stress state at a point is the set of stress vectors corresponding to all planes passing through that point. For simplicity, we will consider a two-dimensional element with stresses as shown in Fig. 3.12a. Let us draw Mohr's circle. First, we have to choose a sign convention. We have already decided that compressive stresses are positive for soils. We will assume clockwise shear is positive and $\sigma_z > \sigma_x$. The two coordinates of the circle are $(\sigma_z, \tau_{zx})$ and $(\sigma_x, \tau_{xz})$. Recall from your strength of materials course that, for equilibrium, $\tau_{xz} = -\tau_{zx}$; these are called complementary shear stresses and are orthogonal to each other. Plot these two coordinates on a graph of shear stress (ordinate) and normal stress (abscissa) as shown by $A$ and $B$ in Fig. 3.12b. Draw a circle with $AB$ as the diameter. The circle crosses the normal stress axis at 1

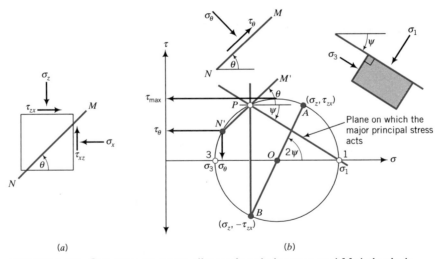

$(a)$          $(b)$

**FIGURE 3.12**   Stresses on a two-dimensional element and Mohr's circle.

and 3. The stresses at these points are the major principal stress, $\sigma_1$, and the minor principal stress, $\sigma_3$.

The principal stresses are related to the stress components $\sigma_z$, $\sigma_x$, $\tau_{zx}$ by

$$\sigma_1 = \frac{\sigma_z + \sigma_x}{2} + \sqrt{\left(\frac{\sigma_z - \sigma_x}{2}\right)^2 + \tau_{zx}^2} \tag{3.27}$$

$$\sigma_3 = \frac{\sigma_z + \sigma_x}{2} - \sqrt{\left(\frac{\sigma_z - \sigma_x}{2}\right)^2 + \tau_{zx}^2} \tag{3.28}$$

The angle between the major principal stress plane and the horizontal plane ($\psi$) is

$$\tan \psi = \frac{\tau_{zx}}{\sigma_1 - \sigma_x} \tag{3.29}$$

The stresses on a plane oriented at an *angle $\theta$ to the major principal stress plane* are

$$\sigma_\theta = \frac{\sigma_1 + \sigma_3}{2} + \frac{\sigma_1 - \sigma_3}{2} \cos 2\theta \tag{3.30}$$

$$\tau_\theta = \frac{\sigma_1 - \sigma_3}{2} \sin 2\theta \tag{3.31}$$

The stresses on a plane oriented at an *angle $\theta$ to the horizontal plane* are

$$\sigma_\theta = \frac{\sigma_z + \sigma_x}{2} + \frac{\sigma_z - \sigma_x}{2} \cos 2\theta + \tau_{zx} \sin 2\theta \tag{3.32}$$

$$\tau_\theta = \tau_{zx} \cos 2\theta - \frac{\sigma_z - \sigma_x}{2} \sin 2\theta \tag{3.33}$$

In the above equations, $\theta$ is positive for clockwise orientation.

The maximum shear stress is at the top of the circle with magnitude

$$\tau_{max} = \frac{\sigma_1 - \sigma_3}{2} \tag{3.34}$$

For the stresses shown in Fig. 3.9, we would get three circles but we have simplified the problem by plotting one circle for stresses on all planes perpendicular to one principal direction.

The stress $\sigma_z$ acts on the horizontal plane and the stress $\sigma_x$ acts on the vertical plane for our case. If we draw these planes in Mohr's circle, they intersect at a point, $P$. Point $P$ is called the pole of the stress circle. It is a special point because any line passing through the pole will intersect Mohr's circle at a point that represents the stresses on a plane parallel to the line. Let us see how this works. Suppose we want to find the stresses on a plane inclined at an angle $\theta$ to the horizontal plane as depicted by $MN$ in Fig. 3.12a. Once we locate the pole, $P$, we can draw a line parallel to $MN$ through $P$ as shown by $M'N'$ in Fig. 3.12b. The line $M'N'$ intersects the circle at $N'$ and the coordinates of $N'$, ($\sigma_\theta$, $\tau_\theta$), represent the normal and shear stresses on $MN$.

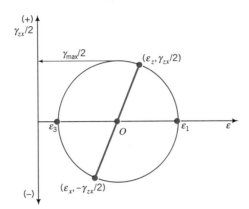

**FIGURE 3.13** Mohr's circle of strain.

## 3.8.2 Mohr's Circle for Strain States

So far, we have studied stress states. The strain state is found in a similar manner to the stress state. With reference to Fig. 3.13, the principal strains are

Major principal strain:
$$\varepsilon_1 = \frac{\varepsilon_z + \varepsilon_x}{2} + \sqrt{\left(\frac{\varepsilon_z - \varepsilon_x}{2}\right)^2 + \left(\frac{\gamma_{zx}}{2}\right)^2} \qquad (3.35)$$

Minor principal strain:
$$\varepsilon_3 = \frac{\varepsilon_z + \varepsilon_x}{2} - \sqrt{\left(\frac{\varepsilon_z - \varepsilon_x}{2}\right)^2 + \left(\frac{\gamma_{zx}}{2}\right)^2} \qquad (3.36)$$

where $\gamma_{zx}$ is called the engineering shear strain or simple shear strain.

The maximum simple shear strain is

$$\gamma_{max} = \varepsilon_1 - \varepsilon_3 \qquad (3.37)$$

In soils, strains can be compressive or tensile. There is no absolute reference strain. For stresses, we can select atmospheric pressure as the reference but not so for strains. Usually, we deal with changes or increments of strains resulting from stress changes.

*The essential points are:*

1. *Mohr's circle is used to find the stress state or strain state from a two-dimensional set of stresses or strains on a soil.*
2. *The pole on a Mohr's circle identifies a point through which any plane passing through it will intersect the Mohr's circle at a point that represents the stresses on that plane.*

## EXAMPLE 3.4

A sample of soil is subjected to the forces shown in Fig. E3.4a. Determine (a) $\sigma_1$, $\sigma_3$, and $\Psi$; (b) the maximum shear stress, and (c) the stresses on a plane oriented at 30° clockwise to the major principal stress plane.

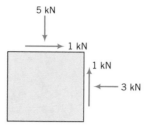

**FIGURE E3.4a**

***Strategy*** There are two approaches to solve this problem. You can either use Mohr's circle or the appropriate equations. Both approaches will be used here.

## Solution 3.4

**Step 1:** Find the area.

$$\text{Area:} \quad A = 100 \times 100 = 10^4 \text{ mm}^2 = 10^{-2} \text{ m}^2$$

**Step 2:** Calculate the stresses.

$$\sigma_z = \frac{\text{Force}}{\text{Area}} = \frac{5}{10^{-2}} = 500 \text{ kPa}$$

$$\sigma_x = \frac{3}{10^{-2}} = 300 \text{ kPa}$$

$$\tau_{zx} = \frac{1}{10^{-2}} = 100 \text{ kPa}; \quad \tau_{xz} = -\tau_{zx} = -100 \text{ kPa}$$

**Step 3:** Draw Mohr's circle and extract $\sigma_1$, $\sigma_3$, and $\tau_{max}$. Mohr's circle is shown in Fig. E3.4b.

$$\sigma_1 = 540 \text{ kPa}, \quad \sigma_3 = 260 \text{ kPa}, \quad \tau_{max} = 140 \text{ kPa}$$

**Step 4:** Draw the pole on Mohr's circle. The pole of Mohr's circle is shown by point $P$ in Fig. E3.4b.

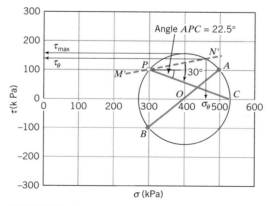

**FIGURE E3.4b**

**Step 5:** Determine $\Psi$.
Draw a line from $P$ to $\sigma_1$ and measure the angle between the horizontal plane and this line.

$$\Psi = 22.5°$$

Alternatively, the angle $AOC = 2\Psi = 45°$

$$\therefore \Psi = 22.5°$$

**Step 6:** Determine the stresses on a plane inclined at 30° to the major principal stress plane.
Draw a line $M^1N^1$ through $P$ with an inclination of 30° to the major principal stress plane, angle $CPN'$. The coordinate at point $N'$ is (470, 120).

*Alternatively:*

**Step 1:** Use Eqs. (3.27) to (3.29) and (3.34) to find $\sigma_1$, $\sigma_3$, $\Psi$, and $\tau_{max}$.

$$\sigma_1 = \frac{500 + 300}{2} + \sqrt{\left(\frac{500 - 300}{2}\right)^2 + 100^2} = 541.4 \text{ kPa}$$

$$\sigma_3 = \frac{500 + 300}{2} - \sqrt{\left(\frac{500 - 300}{2}\right)^2 + 100^2} = 258.6 \text{ kPa}$$

$$\tan \Psi = \frac{\tau_{yx}}{\sigma_1 - \sigma_x} = \frac{100}{541.4 - 300} = 0.414$$

$$\therefore \Psi = 22.5°$$

$$\tau_{max} = \frac{\sigma_1 - \sigma_3}{2} = \frac{541.4 - 258.6}{2} = 141.4 \text{ kPa}$$

**Step 2:** Use Eqs. (3.30) and (3.31) to find $\sigma_\theta$ and $\tau_\theta$.

$$\sigma_\theta = \frac{541.4 + 258.6}{2} + \frac{541.4 - 258.6}{2} \cos(2 \times 30) = 470.7 \text{ kPa}$$

$$\tau_\theta = \frac{541.4 - 258.6}{2} \sin(2 \times 30) = 122.5 \text{ kPa} \qquad \blacksquare$$

*What's next* . . .The stresses we have calculated are for soils as solid elastic materials. We have not accounted for the pressure within the soil pore spaces. In the next section, we will discuss the principle of effective stresses that accounts for the pressures within the soil pores. This principle is the most important principle in soil mechanics.

## 3.9 TOTAL AND EFFECTIVE STRESSES

### 3.9.1 The Principle of Effective Stress

The deformations of soils are similar to the deformations of structural framework such as a truss. The truss deforms from changes in loads carried by each member. If the truss is loaded in air or submerged in water, the deformations under a given

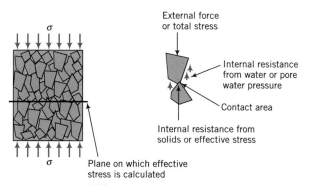

**FIGURE 3.14** Effective stress.

load will remain unchanged. Deformations of the truss are independent of hydrostatic pressure. The same is true for soils.

Let us consider an element of a *saturated soil* subjected to a normal stress, $\sigma$, applied on the horizontal boundary as shown in Fig. 3.14. The stress $\sigma$ is called the *total stress* and for equilibrium (Newton's third law) the stresses in the soil must be equal and opposite to $\sigma$. The resistance or reaction to $\sigma$ is provided by a combination of the stresses from the solids, called *effective stress* ($\sigma'$), and from water in the pores, called *pore water pressure* ($u$). We will denote effective stresses by a prime ($'$) following the symbol for normal stress, usually $\sigma$. The equilibrium equation is

$$\sigma = \sigma' + u \qquad (3.38)$$

so that

$$\sigma' = \sigma - u \qquad (3.39)$$

Equation (3.39) is called the *principle of effective stress* and was first recognized by Terzaghi (1883–1963) in the mid-1920s during his research into soil consolidation (Chapter 4). ***The principle of effective stress is the most important principle in soil mechanics. Deformations of soils are a function of effective stresses not total stresses. The principle of effective stresses applies only to normal stresses and not to shear stresses.*** The pore water cannot sustain shear stresses and therefore the soil solids must resist the shear forces. Thus $\tau = \tau'$, where $\tau$ is the total shear stress and $\tau'$ is the effective shear stress. The effective stress is not the contact stress between the soil solids. Rather, it is the average stress on a plane through the soil mass.

Soils cannot sustain tension. Consequently, the effective stress cannot be less than zero. Pore water pressures can be positive or negative. The latter is sometimes called suction or suction pressure.

For unsaturated soils, the effective stress (Bishop et al., 1960) is

$$\sigma' = \sigma - u_a + \chi(u_a - u_w) \qquad (3.40)$$

where $u_a$ is the pore air pressure, $u_w$ is the pore water pressure, and $\chi$ is a factor depending on the degree of saturation. For dry soil, $\chi = 0$; for saturated soil, $\chi = 1$. Values of $\chi$ for a silt are shown in Fig. 3.15.

To determine the effective stress in a soil mass, the pore water pressure must be known. The pore water pressure at a particular point in a soil mass is

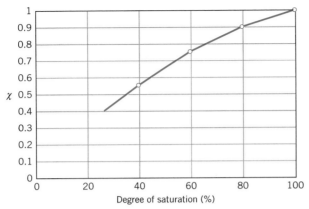

**FIGURE 3.15**    Values of $\chi$ for a silt at different degrees of saturation.

the depth of water above that point multiplied by the unit weight of water. Pore water pressures are measured by pore water pressure transducers (Fig. 3.16) or by piezometers (Fig. 3.17). In a pore water pressure transducer, water passes through a porous material and pushes against a metal diaphragm to which a strain gauge is attached. The strain gauge is usually wired into a Wheatstone bridge. The pore water pressure transducer is calibrated by applying known pressures and measuring the electrical voltage output from the Wheatstone bridge. Piezometers are porous tubes that allow the passage of water. In a simple piezometer, you can measure the height of water in the tube from a fixed elevation and then calculate the pore water pressure by multiplying the height of water by the unit weight of water (Chapter 2). A borehole cased to a certain depth acts like a piezometer. Modern piezometers are equipped with pore water pressure transducers for electronic reading and data acquisition.

### 3.9.3 Effective Stresses Due to Geostatic Stress Fields

The effective stress in a soil mass not subjected to external loads is found from the unit weight of the soil and the depth of groundwater. Consider a soil element

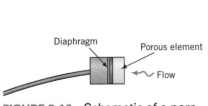

**FIGURE 3.16**    Schematic of a pore water pressure transducer.

**FIGURE 3.17**    Piezometers.

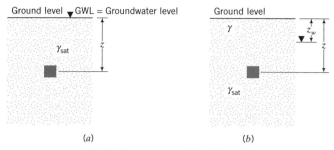

**FIGURE 3.18** Soil element at a depth $z$ with groundwater level at (a) ground level and (b) below ground level.

at a depth $z$ below the ground surface and the groundwater level (GWL) is at ground surface (Fig. 3.18a). The total vertical stress is

$$\sigma = \gamma_{\text{sat}} z \tag{3.41}$$

The pore water pressure is

$$u = \gamma_w z \tag{3.42}$$

and the effective stress is

$$\sigma' = \sigma - u = \gamma_{\text{sat}} z - \gamma_w z = (\gamma_{\text{sat}} - \gamma_w) z = \gamma' z \tag{3.43}$$

If the GWL is at a depth $z_w$ below ground level (Fig. 3.18b), then

$$\sigma = \gamma z_w + \gamma_{\text{sat}}(z - z_w) \quad \text{and} \quad u = \gamma_w(z - z_w)$$

The effective stress is

$$\sigma' = \sigma - u = \gamma z_w + \gamma_{\text{sat}}(z - z_w) - \gamma_w(z - z_w)$$
$$= \gamma z_w + (\gamma_{\text{sat}} - \gamma_w)(z - z_w) = \gamma z_w + \gamma'(z - z_w)$$

### 3.9.3 Effects of Capillarity

In silts and fine sands, the soil above the groundwater can be saturated by capillary action. You would have encountered capillary action in your physics course when you studied menisci. We can get an understanding of capillarity in soils by idealizing the continuous void spaces as capillary tubes. Consider a single idealized tube as shown in Fig. 3.19. The height at which water will rise in the tube can be found from statics. Summing forces vertically (upward forces are negative), we get

$$\Sigma F_z = \text{weight of water} - \text{the tension forces from capillary action}$$

that is,

$$\frac{\pi d^2}{4} z_c \gamma_w - \pi \, dT \cos \alpha = 0 \tag{3.44}$$

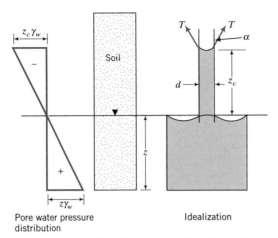

**FIGURE 3.19**    Capillary simulation in soils.

Solving for $z_c$, we get

$$z_c = \frac{4T \cos \alpha}{d\gamma_w} \qquad (3.45)$$

where $T$ is the surface tension (force per unit length), $\alpha$ is the contact angle, $z_c$ is the height of capillary rise, and $d$ is the diameter of the tube representing the diameter of the void space. The surface tension of water is 0.073 N/m and the contact angle of water with a clean glass surface is 0. Since $T$, $\alpha$, and $\gamma_w$ are constants,

$$z_c \propto \frac{1}{d} \qquad (3.46)$$

The interpretation of Eq. (3.46) is that the smaller the soil pores, the higher the capillary zone. The capillary zone in fine sands will be larger than for medium or coarse sands.

The pore water pressure due to capillarity is negative (suction) as shown in Fig. 3.19 and is a function of the size of the soil pores and the water content. At the groundwater level, the pore water pressure is zero and decreases (becomes negative) as you move up the capillary zone. The effective stress increases because the pore water pressure is negative. For example, for the capillary zone, $z_c$, the pore water pressure at the top is $-z_c\gamma_w$ and the effective stress is $\sigma' = \sigma - (-z_c\gamma_w) = \sigma + z_c\gamma_w$.

The approach we have taken to interpret capillary action in soils is simple but it is sufficient for most geotechnical applications. For a comprehensive treatment of capillary action you can refer to Adamson (1982).

### 3.9.4 Effects of Seepage

In Section 2.9, we discussed one-dimensional flow of water through soils. As water flows through soil it exerts a frictional drag on the soil particles resulting in head losses. The frictional drag is called seepage force in soil mechanics. It is often convenient to define seepage as the seepage force per unit volume (it has

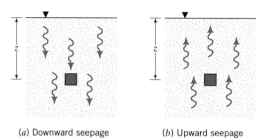

(a) Downward seepage          (b) Upward seepage

**FIGURE 3.20**   Seepage in soils.

units similar to unit weight), which we will denote by $j_s$. If the head loss over a flow distance, $L$, is $\Delta h$, the seepage force is

$$j_s = \frac{\Delta h\, \gamma_w}{L} = i\gamma_w \tag{3.47}$$

If seepage occurs downward (Fig. 3.20a), then the seepage stresses are in the same direction as the gravitational effective stresses. From static equilibrium the resultant vertical effective stress is

$$\sigma'_z = \gamma'z + iz\gamma_w = \gamma'z + j_s z \tag{3.48}$$

If seepage occurs upwards (Fig. 3.20b), then the seepage stresses are in the opposite direction to the gravitational effective stresses. From static equilibrium the resultant vertical effective stress is

$$\sigma'_z = \gamma'z - iz\gamma_w = \gamma'z - j_s z \tag{3.49}$$

Seepage forces play a very important role in destabilizing geotechnical structures. For example, a cantilever retaining wall, shown in Fig. 3.21, depends on the depth of embedment for its stability. The retained soil (left side of wall) applies an outward lateral pressure to the wall, which is resisted by an inward lateral resistance from the soil on the right side of the wall. If a steady quantity of water is available on the left side of the wall, for example, from a busted water pipe, then water will flow from the left side to the right side of the wall. The path followed by a particle of water is depicted by $AB$ in Fig. 3.21 and as water flows from $A$ to $B$, head loss occurs. The seepage stresses on the left side of the wall

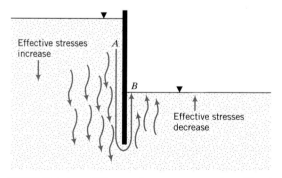

**FIGURE 3.21**   Effects of seepage on the effective stresses near a retaining wall.

are in the direction of the gravitational stresses. The effective stress increases and, consequently, an additional outward lateral force is applied on the left side of the wall. On the right side of the wall, the seepage stresses are upward and the effective stress decreases. The lateral resistance provided by the embedment is reduced. Seepage stresses in this problem play a double role (increase the lateral disturbing force and reduce the lateral resistance) in reducing the stability of a geotechnical structure. In Chapters 9 through 11, you will study the effects of seepage on the stability of several types of geotechnical structures.

> *The essential points are:*
>
> 1. *The effective stress represents the average stress carried by the soil solids and is the difference between the total stress and the pore water pressure.*
> 2. *The effective stress principle applies only to normal stresses and not to shear stresses.*
> 3. *Deformations of soils are due to effective not total stress.*
> 4. *Soils, especially silts and fine sands, can be affected by capillary action.*
> 5. *Capillary action results in negative pore water pressures and increases the effective stresses.*
> 6. *Downward seepage increases the resultant effective stress; upward seepage decreases the resultant effective stress.*

## EXAMPLE 3.5

Calculate the effective stress for a soil element at depth 5 m in a uniform deposit of soil as shown in Fig. E3.5.

**Strategy**  You need to get unit weights from the given data and you should note that the soil above the groundwater level is not saturated.

## Solution 3.5

**Step 1:**   Calculate unit weights.

**Above groundwater level**

$$\gamma = \left(\frac{G_s + Se}{1 + e}\right)\gamma_w = \frac{G_s(1 + w)}{1 + e}\,\gamma_w$$

$$Se = wG_s, \quad \therefore e = \frac{0.3 \times 2.7}{0.6} = 1.35$$

$$\gamma = \frac{2.7(1 + 0.3)}{1 + 1.35} \times 9.8 = 14.6 \text{ kN/m}^3$$

**Below groundwater level**

Soil is saturated, $S = 1$

$$e = wG_s = 0.4 \times 2.7 = 1.08$$

$$\gamma_{sat} = \left(\frac{G_s + e}{1 + e}\right)\gamma_w = \left(\frac{2.7 + 1.08}{1 + 1.08}\right)9.8 = 17.8 \text{ kN/m}^3$$

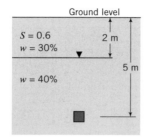

**FIGURE E3.5**

**Step 2:** Calculate the effective stress.

Total stress: $\sigma = 2\gamma + 3\gamma_{sat} = 2 \times 14.6 + 3 \times 17.8 = 82.6$ kPa

Pore water pressure: $u = 3\gamma_w = 3 \times 9.8 = 29.4$ kPa

Effective stress: $\sigma' = \sigma - u = 82.6 - 29.4 = 53.2$ kPa

Alternatively,

$$\sigma' = 2\gamma + 3(\gamma_{sat} - \gamma_w) = 2\gamma + 3\gamma' = 2 \times 14.6 + 3(17.8 - 9.8) = 53.2 \text{ kPa}$$

■

## EXAMPLE 3.6

A borehole at a site reveals the soil profile shown in Fig. E3.6a. Plot the distribution of total and effective stresses with depth.

***Strategy*** From the data given, you will have to find the unit weight of each soil layer to calculate the stresses. You are given that the 1.0 m of fine sand above the groundwater level is saturated by capillary action. Therefore, the pore water pressure in this 1.0 m zone is negative.

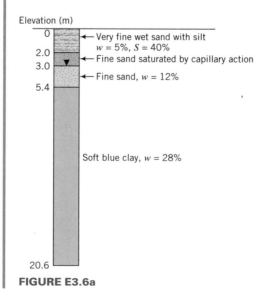

**FIGURE E3.6a**

## Solution 3.6

**Step 1:** Calculate the unit weights.

**0–2 m**

$$S = 40\% = 0.4; \quad w = 0.05$$

$$e = \frac{wG_s}{S} = \frac{0.05 \times 2.7}{0.4} = 0.34$$

$$\gamma = \frac{G_s(1 + w)}{1 + e}\,\gamma_w = \frac{2.7(1 + 0.05)}{1 + 0.34}\,9.8 = 20.7 \text{ kN/m}^3$$

**2–5.4 m**

$$S = 1; \quad w = 0.12$$

$$e = wG_s = 0.12 \times 2.7 = 0.32$$

$$\gamma_{sat} = \left(\frac{G_s + e}{1 + e}\right)\gamma_w = \left(\frac{2.7 + 0.32}{1 + 0.32}\right)9.8 = 22.4 \text{ kN/m}^3$$

**5.4–20.6 m**

$$S = 1; \quad w = 0.28$$

$$e = wG_s = 0.28 \times 2.7 = 0.76$$

$$\gamma_{sat} = \left(\frac{2.7 + 0.76}{1 + 0.76}\right)9.8 = 19.3 \text{ kN/m}^3$$

**Step 2:** Calculate the stresses using a table or use a spreadsheet program.

| Depth (m) | Thickness (m) | σ (kPa) | u (kPa) | σ' = σ − u (kPa) |
|---|---|---|---|---|
| 0 | 0 | 0 | 0 | 0 |
| 2 | 2 | 20.7 × 2 = 41.4 | −1 × 9.8 = −9.8 | 51.6 |
| 3 | 1 | 41.4 + 22.4 × 1 = 63.8 | 0 | 63.8 |
| 5.4 | 2.4 | 63.8 + 22.4 × 2.4 = 117.6 | 2.4 × 9.8 = 23.5 | 94.1 |
| 20.6 | 15.2 | 117.6 + 19.3 × 15.2 = 411 | 23.5 + 15.2 × 9.8 = 172.5 or 17.6 × 9.8 = 172.5 | 238.5 |

**Step 3:** Plot the stresses versus depth—see Fig. E3.6b.

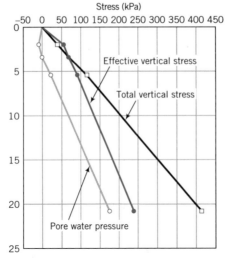

**FIGURE E3.6b**

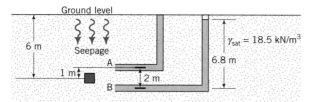

**FIGURE E3.7**

## EXAMPLE 3.7

Water is seeping downward through a soil layer as shown in Fig. E3.7. Two piezometers (A and B) located 2 m apart (vertically) showed a head loss of 0.2 m. Calculate the resultant vertical effective stress for a soil element at a depth of 6 m as shown in Fig. E3.7.

**Strategy**  You have to calculate the seepage stress. But to obtain this you must know the hydraulic gradient, which you can find from the data given.

## Solution 3.7

**Step 1:**  Find the hydraulic gradient.

$$\Delta H = 0.2 \text{ m}; \ L = 2 \text{ m}; \quad i = \frac{\Delta H}{L} = \frac{0.2}{2} = 0.1$$

**Step 2:**  Determine the effective stress.
Assume the hydraulic gradient is the average for the soil mass; then

$$\sigma'_z = (\gamma_{sat} - \gamma_w)z + i\gamma_w z = (18.5 - 9.8)6 + 0.1 \times 9.8 \times 6 = 58.1 \text{ kPa} \quad \blacksquare$$

**What's next . . .**We have only considered vertical stresses. But an element of soil in the ground is also subjected to lateral stresses. Next, we will introduce an equation that relates the vertical and lateral effective stresses.

## 3.10  LATERAL EARTH PRESSURE AT REST

The ratio of the horizontal principal effective stress to the vertical principal effective stress is called the lateral earth pressure coefficient at rest ($K_o$), that is,

$$K_o = \frac{\sigma'_3}{\sigma'_1} \tag{3.50}$$

The at-rest condition implies that no deformation occurs. We will revisit the at-rest coefficient in Chapters 5, 6, and 10. You must remember that $K_o$ applies only to effective not total stresses. To find the lateral total stress, you must add the pore water pressure. Remember that the pore water pressure is hydrostatic and, at any given depth, the pore water pressures in all directions are equal.

### EXAMPLE 3.8

Calculate the lateral effective stress and the lateral total stress for the soil element at 5 m in Example 3.5 if $K_o = 0.5$.

***Strategy***    The stresses on the horizontal and vertical planes on the soil element are principal stresses (no shear stress occurs on these planes). You need to apply $K_o$ to the effective stress and then add the pore water pressure to get the lateral total stress.

### Solution 3.8

**Step 1:**    Calculate the lateral effective stress.

$$K_o = \frac{\sigma_3'}{\sigma_1'} = \frac{\sigma_x'}{\sigma_z'}; \quad \sigma_x' = K_o\sigma_z' = 0.5 \times 53.2 = 26.6 \text{ kPa}$$

**Step 2:**    Calculate the lateral total stress.

$$\sigma_x = \sigma_x' + u = 26.6 + 29.4 = 56 \text{ kPa} \quad \blacksquare$$

***What's next***...The stresses we considered so far are called geostatic stresses and when we considered elastic deformation of soils, the additional stresses imposed on the soil were given. But in practice, we have to find these additional stresses from applied loads located either on the ground surface or within the soil mass. We will use elastic analysis to find these additional stresses. Next, we will consider increases in stresses from a number of common surface loads. You will encounter a myriad of equations. You are not expected to remember these equations but you are expected to know how to use them.

## 3.11 STRESSES IN SOIL FROM SURFACE LOADS

Various types of surface loads are applied to soils. For example, an oil tank will impose a uniform, circular, vertical stress on the surface of the soil while an unsymmetrical building may impose a nonuniform vertical stress. We would like to know how the surface stresses are distributed within the soil mass and the resulting deformations. The distribution of surface stresses within a soil is determined by assuming that the soil is a semi-infinite, homogeneous, linear, isotropic, elastic material. A semi-infinite mass is bounded on one side and extends infinitely in all other directions; this is also called an "elastic half-space." For soils, the horizontal surface is the bounding side. Equations and charts for several types of surface loads based on the above assumptions are presented. A computer program to compute stresses from surface loads is available in the CD ROM.

Most soils exist in layers with finite thicknesses. The solution based on a semi-infinite soil mass will not be accurate for these layered soils. In Appendix B, you will find selected graphs and tables for vertical stress increases in one-layer and two-layer soils. A comprehensive set of equations for a variety of loading situations is available in Poulos and Davis (1974).

### 3.11.1 Point Load

Boussinesq (1885) presented a solution for the distribution of stresses for a point load applied on the soil surface. An example of a point load is the vertical load transferred to the soil from an electric power line pole.

The increases in stresses on a soil element located at point $A$ (Fig. 3.22a) due to a point load, $Q$, are

$$\Delta\sigma_z = \frac{3Q}{2\pi z^2}\left(\frac{1}{1 + (r/z)^2}\right)^{5/2} \tag{3.51}$$

$$\Delta\sigma_r = \frac{Q}{2\pi}\left(\frac{3r^2 z}{(r^2 + z^2)^{5/2}} - \frac{1 - 2v}{r^2 + z^2 + z(r^2 + z^2)^{1/2}}\right) \tag{3.52}$$

$$\Delta\sigma_\theta = -\frac{Q}{2\pi}(1 - 2v)\left(\frac{z}{(r^2 + z^2)^{3/2}} - \frac{1}{r^2 + z^2 + z(r^2 + z^2)^{1/2}}\right) \tag{3.53}$$

$$\Delta\tau_{rz} = \frac{3Q}{2\pi}\left[\frac{rz^2}{(r^2 + z^2)^{5/2}}\right] \tag{3.54}$$

where $v$ is Poisson's ratio. Most often, the increase in vertical stress is needed in practice. Equation (3.51) can be written as

$$\sigma_z = \frac{Q}{z^2} I \tag{3.55}$$

where $I$ is an influence factor, and

$$I = \frac{3}{2\pi}\left(\frac{1}{1 + (r/z)^2}\right)^{5/2} \tag{3.56}$$

The distributions of the increase in vertical stress from Eq. 3.55 reveal that the increase in vertical stress decreases with depth (Fig. 3.22b) and radial distance (Fig. 3.22c).

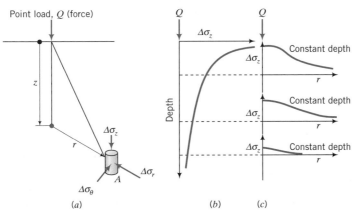

**FIGURE 3.22** Point load and vertical stress distribution with depth and radial distance.

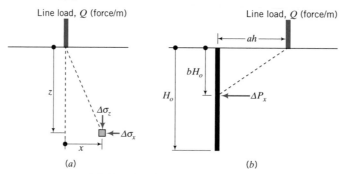

**FIGURE 3.23** (a) Line load and (b) line load near a retaining wall.

### 3.11.2 Line Load

With reference to Fig. 3.23a, the increases in stresses due to a line load, $Q$ (force/length), are

$$\Delta\sigma_z = \frac{2Qz^3}{\pi(x^2 + z^2)^2} \qquad (3.57)$$

$$\Delta\sigma_x = \frac{2Qx^2z}{\pi(x^2 + z^2)^2} \qquad (3.58)$$

$$\Delta\tau_{zx} = \frac{2Qxz^2}{\pi(x^2 + z^2)^2} \qquad (3.59)$$

A practical example of a line load is the load from a long brick wall.

### 3.11.3 Line Load Near a Buried Earth Retaining Structure

The increase in lateral stress on a buried earth retaining structure (Fig. 3.23b) due to a line load of intensity $Q$ (force/length) is

$$\Delta\sigma_x = \frac{4Qa^2b}{\pi H_o(a^2 + b^2)^2} \qquad (3.60)$$

The increase in lateral force is

$$\Delta P_x = \frac{2Q}{\pi(a^2 + 1)} \qquad (3.61)$$

### 3.11.4 Strip Load

A strip load is the load transmitted by a structure of finite width and infinite length on a soil surface. Two types of strip loads are common in geotechnical engineering. One is a load that imposes a uniform stress on the soil, for example, the middle section of a long embankment (Fig. 3.24a). The other is a load that induces a triangular stress distribution over an area of width $B$ (Fig. 3.24b). An example of a strip load with a triangular stress distribution is the stress under the side of an embankment.

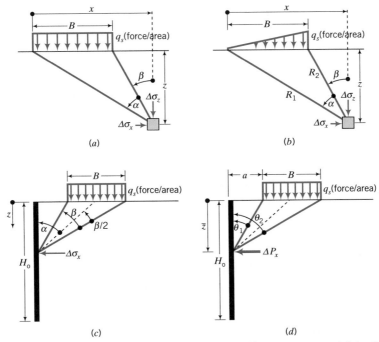

**FIGURE 3.24** Strip load imposing (a) a uniform stress and (b) a linear varying stress. (c) Strip load near a retaining wall and (d) lateral force on a retaining wall from a strip load.

The increases in stresses due to a surface stress $q_s$ (force/area) are as follows:

(a) **Area transmitting a uniform stress (Fig. 3.24a)**

$$\Delta\sigma_z = \frac{q_s}{\pi} [\alpha + \sin\alpha\cos(\alpha + 2\beta)] \tag{3.62}$$

$$\Delta\sigma_x = \frac{q_s}{\pi} [\alpha - \sin\alpha\cos(\alpha + 2\beta)] \tag{3.63}$$

$$\Delta\tau_{zx} = \frac{q_s}{\pi} [\sin\alpha\sin(\alpha + 2\beta)] \tag{3.64}$$

where $q_s$ is the applied surface stress.

(b) **Area transmitting a triangular stress (Fig. 3.24b)**

$$\Delta\sigma_z = \frac{q_s}{\pi} \left( \frac{x}{B}\alpha - \tfrac{1}{2}\sin 2\beta \right) \tag{3.65}$$

$$\Delta\sigma_x = \frac{q_s}{\pi} \left( \frac{x}{B}\alpha - \frac{z}{B}\ln\frac{R_1^2}{R_2^2} + \tfrac{1}{2}\sin 2\beta \right) \tag{3.66}$$

$$\Delta\tau_{zx} = \frac{q_s}{2\pi} \left( 1 + \cos 2\beta - 2\frac{z}{B}\alpha \right) \tag{3.67}$$

### (c) Area transmitting a uniform stress near a retaining wall (Fig. 3.24c,d)

$$\Delta\sigma_x = \frac{2q_s}{\pi}(\beta - \sin\beta\cos 2\alpha) \qquad (3.68)$$

The lateral force and its location were derived by Jarquio (1981) and are

$$\Delta P_x = \frac{q_s}{90}[H_o(\theta_2 - \theta_1)] \qquad (3.69)$$

$$\bar{z} = \frac{H_o^2(\theta_2 - \theta_1) - (R_1 - R_2) + 57.3BH_o}{2H_o(\theta_2 - \theta_1)} \qquad (3.70)$$

where

$$\theta_1 = \tan^{-1}\left(\frac{a}{H_o}\right), \quad \theta_2 = \tan^{-1}\left(\frac{a + B}{H_o}\right),$$
$$R_1 = (a + B)^2(90 - \theta_2), \quad \text{and} \quad R_2 = a^2(90 - \theta_1)$$

### 3.11.5 Uniformly Loaded Circular Area

An example of a circular area that transmits stresses to a soil mass is a circular foundation of an oil or water tank. The increases of vertical and radial stresses under the center of a circular area of radius $r_o$ are

$$\Delta\sigma_z = q_s\left[1 - \left(\frac{1}{1 + (r_o/z)^2}\right)^{3/2}\right] = q_sI_c \qquad (3.71)$$

where

$$I_c = \left[1 - \left(\frac{1}{1 + (r_o/z)^2}\right)^{3/2}\right]$$

is an influence factor and

$$\Delta\sigma_r = \Delta\sigma_\theta = \frac{q_s}{2}\left[(1 + 2v) - \frac{2(1 + v)}{[1 + (r_o/z)^2]^{1/2}} + \frac{1}{[1 + (r_o/z)^2]^{3/2}}\right] \qquad (3.72)$$

The vertical elastic settlement at the surface due to a circular flexible loaded area is

Below center of loaded area: $\quad \Delta z = \dfrac{q_sD(1 - v^2)}{E} \qquad (3.73)$

Below edge: $\quad \Delta z = \dfrac{2}{\pi}\dfrac{q_sD(1 - v^2)}{E} \qquad (3.74)$

where $D = 2r_o$ is the diameter of the loaded area.

## 3.11.6 Uniformly Loaded Rectangular Area

Many structural foundations are rectangular or approximately rectangular in shape. The increases in stresses below the *corner* of a rectangular area of width $B$ and length $L$ are

$$\Delta\sigma_z = \frac{q_s}{2\pi}\left[\tan^{-1}\frac{LB}{zR_3} + \frac{LBz}{R_3}\left(\frac{1}{R_1^2} + \frac{1}{R_2^2}\right)\right] \tag{3.75}$$

$$\Delta\sigma_x = \frac{q_s}{2\pi}\left[\tan^{-1}\frac{LB}{zR_3} - \frac{LBz}{R_1^2 R_3}\right] \tag{3.76}$$

$$\Delta\sigma_y = \frac{q_s}{2\pi}\left[\tan^{-1}\frac{LB}{zR_3} - \frac{LBz}{R_2^2 R_3}\right] \tag{3.77}$$

$$\Delta\tau_{zx} = \frac{q_s}{2\pi}\left[\frac{B}{R_2} - \frac{z^2 B}{R_1^2 R_3}\right] \tag{3.78}$$

where $R_1 = (L^2 + z^2)^{1/2}$, $R_2 = (B^2 + z^2)^{1/2}$, and $R_3 = (L^2 + B^2 + z^2)^{1/2}$.

These equations can be written as

$$\Delta\sigma_z = q_s I_z \tag{3.79}$$

$$\Delta\sigma_x = q_s I_x \tag{3.80}$$

$$\Delta\sigma_y = q_s I_y \tag{3.81}$$

$$\tau_{zx} = q_s I_\tau \tag{3.82}$$

where $I$ denotes the influence factor. The influence factor for the vertical stress is

$$I_z = \frac{1}{4\pi}\left[\frac{2mn\sqrt{m^2 + n^2 + 1}}{m^2 + n^2 + m^2 n^2 + 1}\left(\frac{m^2 + n^2 + 2}{m^2 + n^2 + 1}\right) + \tan^{-1}\left(\frac{2mn\sqrt{m^2 + n^2 + 1}}{m^2 + n^2 - m^2 n^2 + 1}\right)\right] \tag{3.83}$$

where $m = B/z$ and $n = L/z$. You can program your calculator or use a spreadsheet program to find $I_z$. You must be careful in the last term ($\tan^{-1}$) in programming. If $m^2 + n^2 + 1 < m^2 n^2$, then you have to add $\pi$ to the bracketed quantity in the last term. A chart for $I_z$ produced by a spreadsheet program is shown in Fig. 3.25. You would have to calculate $m = B/z$ and $n = L/z$ and read $I_z$ from the chart; $m$ and $n$ are interchangeable. In general, the vertical stress increase is less than 10% of the surface stress when $z > 3B$.

The vertical elastic settlement at the ground surface under a rectangular flexible surface load is

$$\Delta z = \frac{q_s B(1 - v^2)}{E} I_s \tag{3.84}$$

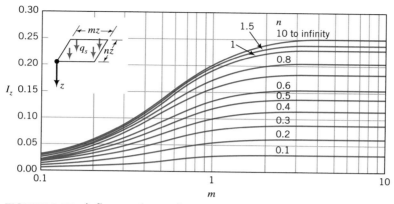

**FIGURE 3.25** Influence factor for calculating the vertical stress increase under the corner of a rectangle.

where $I_s$ is a settlement influence factor that is a function of the $L/B$ ratio ($L$ is length and $B$ is width). Setting $\xi_s = L/B$, the equations for $I_s$ are

At center of a rectangle (Giroud, 1968):

$$I_s = \frac{2}{\pi}\left[\ln(\xi_s + \sqrt{1 + \xi_s^2}) + \xi_s \ln\frac{1 + \sqrt{1 + \xi_s^2}}{\xi_s}\right]$$

At corner of a rectangle (Giroud, 1968):

$$I_s = \frac{1}{\pi}\left[\ln(\xi_s + \sqrt{1 + \xi_s^2}) + \xi_s \ln\frac{1 + \sqrt{1 + \xi_s^2}}{\xi_s}\right]$$

The above equations can be simplified to the following for $\xi_s \geq 1$:

At center of a rectangle: $\boxed{I_s \cong 0.62 \ln(\xi_s) + 1.12}$

At corner of a rectangle: $\boxed{I_s \cong 0.31 \ln(\xi_s) + 0.56}$

### 3.11.7 Approximate Method for Rectangular Loads

In preliminary analyses of vertical stress increases under the center of rectangular loads, geotechnical engineers often use an approximate method (sometimes

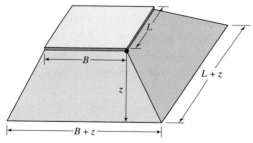

**FIGURE 3.26** Dispersion of load for approximate increase in vertical stress under a rectangle.

called the 2:1 method). The surface load on an area, $B \times L$, is dispersed at a depth $z$ over an area $(B + z) \times (L + z)$ as illustrated in Fig. 3.26. The vertical stress increase under the center of the load is

$$\Delta\sigma_z = \frac{q_sBL}{(B + z)(L + z)} \tag{3.85}$$

The approximate method is reasonably accurate (compared with Boussinesq's elastic solution) when $z > B$.

### 3.11.8 Vertical Stress Below Arbitrarily Shaped Areas

Newmark (1942) developed a chart to determine the increase in vertical stress due to a uniformly loaded area of any shape. The chart consists of concentric circles divided by radial lines (Fig. 3.27). The area of each segment represents an equal proportion of the applied surface stress at a depth $z$ below the surface. If there are 10 concentric circles (only 9 are shown because the 10th extends to infinity) and 20 radial lines, the stress on each circle is $q_s/10$ and on each segment is $q_s/(10 \times 20)$. The radius to depth ratio of the first (inner) circle is found by setting $\Delta\sigma_z = 0.1q_s$ in Eq. (3.71), that is,

$$0.1q_s = q_s\left[1 - \left\{\frac{1}{1 + (r_o/z)^2}\right\}^{3/2}\right]$$

from which $r/z = 0.27$. For the other circles, substitute the appropriate value for $\Delta\sigma_z$; for example, for the second circle $\Delta\sigma_z = 0.2q_s$, and find $r/z$. The chart is normalized to the depth; that is, all dimensions are scaled by a factor initially determined for the depth. Every chart should show a scale and an influence factor $I_N$, which for our case is $1/(10 \times 20) = 0.005$.

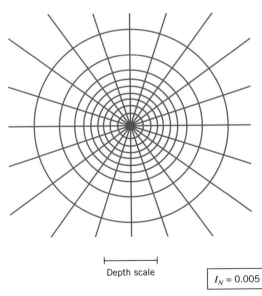

Depth scale

$I_N = 0.005$

**FIGURE 3.27** Newmark's chart for increase in vertical stress.

The procedure for using Newmark's chart is as follows:

1. Set the scale, shown on the chart, equal to the depth at which the increase in vertical stress is required. We will call this the depth scale.
2. Identify the point on the loaded area below which the stress is required. Let us say this point is $A$.
3. Plot the loaded area using the depth scale with point $A$ at the center of the chart.
4. Count the number of segments ($N_s$) covered by the scaled loaded area. If certain segments are not fully covered, you can estimate what fraction is covered.
5. Calculate the increase in vertical stress as $\Delta\sigma_z = q_s I_N N_s$.

*The essential points are:*

1. *The increases in stresses below a surface load are found by assuming the soil is an elastic, semi-infinite mass.*
2. *Various equations are available for the increases in stresses from surface loading.*
3. *The stress increase at any depth depends on the shape and distribution of the surface load.*
4. *A stress applied at the surface of a soil mass by a loaded area decreases with depth and lateral distance away from the center of the loaded area.*
5. *The vertical stress increases are generally less than 10% of the surface stress when the depth to width ratio is greater than 3.*

### EXAMPLE 3.9

A pole carries a vertical load of 200 kN. Determine the vertical stress increase at a depth 5 m (a) directly below the pole and (b) at a radial distance of 2 m.

*Strategy*   The first step is to determine the type of surface load. The load carried by the pole can be approximated to a point load. You can then use the equation for the vertical stress increase for a point load.

### Solution 3.9

**Step 1:**   Determine the load type.
   Assume the load from the pole can be approximated by a point load.

**Step 2:**   Use the equation for a point load. Use Eq. (3.55):

$$z = 5 \text{ m}, \quad Q = 200 \text{ kN}; \quad \text{Under load, } r = 0, \quad \therefore \frac{r}{z} = 0$$

$$\text{From Eq. (3.56):} \quad \frac{r}{z} = 0, \quad I = 0.48$$

$$\Delta\sigma_z = \frac{Q}{z^2} I = \frac{200}{5^2} \times 0.48 = 3.8 \text{ kPa}$$

**Step 3:**   Determine the vertical stress at the radial distance.

$$r = 2 \text{ m}, \quad \frac{r}{z} = \frac{2}{5} = 0.4, \quad I = 0.33$$

$$\Delta\sigma_z = \frac{200}{5^2} \times 0.33 = 2.6 \text{ kPa} \qquad \blacksquare$$

## EXAMPLE 3.10

A rectangular concrete slab, 3 m × 4.5 m, rests on the surface of a soil mass. The load on the slab is 2025 kN. Determine the vertical stress increase at a depth of 3 m (a) under the center of the slab, point $A$ (Fig. E3.10a), (b) under point $B$ (Fig. E3.10a), and (c) at a distance of 1.5 m from a corner, point $C$ (Fig. E3.10a).

***Strategy***  The slab is rectangular and the equations for a uniformly loaded rectangular area are for the corner of the area. You should divide the area so that the point of interest is a corner of a rectangle(s). You may have to extend the loaded area if the point of interest is outside it (loaded area). The extension is fictitious so you have to subtract the fictitious increase in stress for the extended area.

## Solution 3.10

**Step 1:**   Identify the loading type.
It is a uniformly loaded rectangle.

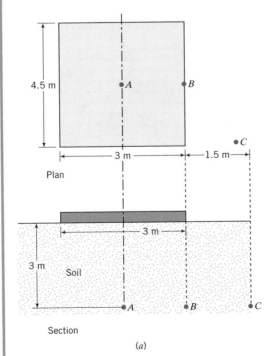

**FIGURE E3.10a**

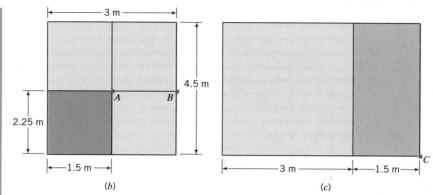

**FIGURE E3.10b,c**

**Step 2:**   Divide the rectangle so that the center is a corner.
In this problem, all four rectangles after the subdivision are equal (Fig. E3.10b), so you only need to find the stress increase for one rectangle of size $B = 1.5$ m, $L = 2.25$ m and multiply the results by 4.

$$m = \frac{B}{z} = \frac{1.5}{3} = 0.5; \quad n = \frac{L}{z} = \frac{2.25 \text{ m}}{3} = 0.75$$

From the chart in Fig. 3.25, $I_z = 0.105$.

**Step 3:**   Find the stress increase at the center of the slab (point $A$, Fig. E3.10b).

$$q_s = \frac{2025}{3 \times 4.5} = 150 \text{ kPa}$$

$$\Delta\sigma_z = 4q_s I_z = 4 \times 150 \times 0.105 = 63 \text{ kPa}$$

Note: The approximate method [Eq. (3.85)] gives

$$\Delta\sigma_z = \frac{q_s BL}{(B + z)(L + z)} = \frac{2025}{(3 + 3)(4.5 + 3)} = 45 \text{ kPa}$$

which is about 30% less than the elastic solution.

**Step 4:**   Find the stress increase for point $B$.
Point $B$ is at the corner of two rectangles each of width 3 m and length 2.25 m. You need to find the stress increase for one rectangle and multiply the result by 2.

$$m = \frac{3}{3} = 1; \quad n = \frac{2.25}{3} = 0.75$$

From the chart in Fig. 3.25, $I_z = 0.158$.

$$\Delta\sigma_z = 2 \times 150 \times 0.158 = 47.4 \text{ kPa}$$

You should note that the vertical stress increase at $B$ is lower than at $A$, as expected.

**Step 5:** Find the stress increase for point $C$.

Stress point $C$ is outside the rectangular slab. You have to extend the rectangle to $C$ (Fig. E3.10c) and find the stress increase for the large rectangle of width $B = 4.5$, length $L = 4.5$ m and then subtract the stress increase for the smaller rectangle of width $B = 1.5$ m and length $L = 4.5$ m.

**Large rectangle**

$$m = \frac{4.5}{3} = 1.5, \quad n = \frac{4.5}{3} = 1.5; \quad \text{from chart in Fig. 3.25, } I_z = 0.22$$

**Small rectangle**

$$m = \frac{1.5}{3} = 0.5, \quad n = \frac{4.5}{3} = 1.5; \quad \text{from chart in Fig. 3.25, } I_z = 0.13$$

$$\Delta\sigma_z = 150 \times (0.22 - 0.13) = 13.5 \text{ kPa} \qquad \blacksquare$$

## EXAMPLE 3.11

The plan of a foundation of uniform thickness for a building is shown in Fig. E3.11a. Determine the vertical stress increase at a depth of 4 m below the centroid. The foundation applies a vertical stress of 200 kPa on the soil surface.

***Strategy*** You need to locate the centroid of the foundation, which you can find using the given dimensions. The shape of the foundation does not fit neatly into one of the standard shapes (e.g., rectangles or circles) discussed. The convenient method to use for this (odd) shape foundation is Newmark's chart.

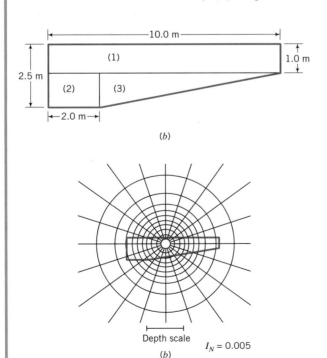

(b)

Depth scale $I_N = 0.005$

(b)

**FIGURE E3.11a,b**

### Solution 3.11

**Step 1:**   Find the centroid.

$$\bar{x} = \frac{\begin{array}{c}1.0 \times 10.0 \times 5.0 + 1.5 \times 2.0 \times 1.0 \\ + \frac{1}{2} \times 8.0 \times 1.5 \times (2.0 + \frac{1}{3} \times 8.0)\end{array}}{(1.0 \times 10.0) + (1.5 \times 2.0) + \frac{1}{2} \times 8.0 \times 1.5} = \frac{81}{19} = 4.26 \text{ m}$$

$$\bar{y} = \frac{\begin{array}{c}1.0 \times 10.0 \times 0.5 + 1.5 \times 2.0 \times (1.0 + 1.5/2) \\ + \frac{1}{2} \times 8.0 \times 1.5 \times (1.0 + 1.5/3)\end{array}}{(1.0 \times 10.0) + (1.5 \times 2.0) + \frac{1}{2} \times 8.0 \times 1.5} = \frac{16.25}{19} = 0.86 \text{ m}$$

**Step 2:**   Scale and plot the foundation on a Newmark's chart.
The scale on the chart is set equal to the depth. The centroid is located at the center of the chart and the foundation is scaled using the depth scale (Fig. E3.11b).

**Step 3:**   Count the number of segments covered by the foundation.

$$N_s = 61$$

**Step 4:**   Calculate the vertical stress increase.

$$\Delta \sigma_z = 200 \times 0.005 \times 61 = 61 \text{ kPa}$$ ∎

*What's next . . .* The stresses and strains discussed in previous sections are all dependent on the axis system chosen. We have arbitrarily chosen the Cartesian coordinate and the cylindrical coordinate systems. We could, however, define a set of stresses and strains that are independent of the axis system. Such a system, which we will discuss next, will allow us to use generalized stress and strain parameters to analyze and interpret soil behavior. In particular, we would be able to represent a three-dimensional system of stresses and strains by a two-dimensional system.

## 3.12   STRESS AND STRAIN INVARIANTS

Stress and strain invariants are measures that are independent of the axis system. We will define stress invariants that provide measures of (1) mean stress and (2) deviatoric or distortional or shear stress, and strain invariants that provide measures of (1) volumetric strains and (2) deviatoric or distortional or shear strains.

### 3.12.1 Mean Stress

$$p = \frac{\sigma_1 + \sigma_2 + \sigma_3}{3} = \frac{\sigma_x + \sigma_y + \sigma_z}{3} \tag{3.86}$$

On a graph with orthogonal principal stress axes, $\sigma_1$, $\sigma_2$, $\sigma_3$, the mean stress is the space diagonal, that is, a line oriented at equal angles to all three axes (Fig. 3.28). Mean stress causes volume changes.

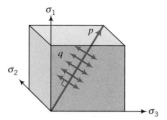

**FIGURE 3.28**  Mean and deviatoric stresses.

### 3.12.2 Deviatoric or Shear Stress

$$q = \frac{1}{\sqrt{2}}[(\sigma_1 - \sigma_2)^2 + (\sigma_2 - \sigma_3)^2 + (\sigma_3 - \sigma_1)^2]^{1/2} \tag{3.87}$$

A line normal to the mean stress as illustrated in Fig. 3.28 represents the deviatoric stress. Deviatoric stress causes distortions or shearing of a soil mass.

### 3.12.3 Volumetric Strain

$$\varepsilon_p = \varepsilon_1 + \varepsilon_2 + \varepsilon_3 = \varepsilon_x + \varepsilon_y + \varepsilon_z \tag{3.88}$$

### 3.12.4 Deviatoric or Distortional or Shear Strain

$$\varepsilon_q = \frac{\sqrt{2}}{3}[(\varepsilon_1 - \varepsilon_2)^2 + (\varepsilon_2 - \varepsilon_3)^2 + (\varepsilon_3 - \varepsilon_1)^2]^{1/2} \tag{3.89}$$

In this book, we will be dealing with two conditions—axisymmetric and plane strain. For these two conditions, the stress and strain invariants are as follows.

### 3.12.5 Axisymmetric Condition, $\sigma_2' = \sigma_3'$ or $\sigma_2 = \sigma_3$; $\varepsilon_2 = \varepsilon_3$

$$p' = \frac{\sigma_1' + 2\sigma_3'}{3} \quad \text{and} \quad p = \frac{\sigma_1 + 2\sigma_3}{3} \tag{3.90}$$

$$p' = p - u$$

$$q = \sigma_1 - \sigma_3; \quad q' = \sigma_1' - \sigma_3' = (\sigma_1 - \Delta u) - (\sigma_3 - \Delta u) = \sigma_1 - \sigma_3 \tag{3.91}$$

Therefore, $q = q'$; shear is unaffected by pore water pressures.

$$\varepsilon_p = \varepsilon_1 + 2\varepsilon_3 \tag{3.92}$$

$$\varepsilon_q = \tfrac{2}{3}(\varepsilon_1 - \varepsilon_3) \tag{3.93}$$

### 3.12.6 Plane Strain, $\varepsilon_2 = 0$

$$p' = \frac{\sigma_1' + \sigma_2' + \sigma_3'}{3} \quad \text{and} \quad p = \frac{\sigma_1 + \sigma_2 + \sigma_3}{3} \tag{3.94}$$

$$p' = p - u \tag{3.95}$$

$$q' = q = \frac{1}{\sqrt{2}} [(\sigma_1' - \sigma_2')^2 + (\sigma_2' - \sigma_3')^2 + (\sigma_3' - \sigma_1')^2]^{1/2} \tag{3.96}$$

or

$$q = q' = \frac{1}{\sqrt{2}} [(\sigma_1 - \sigma_3)^2 + (\sigma_2 - \sigma_3)^2 + (\sigma_3 - \sigma_1)^2]^{1/2} \tag{3.97}$$

$$\varepsilon_p = \varepsilon_1 + \varepsilon_3 \tag{3.97}$$

$$\varepsilon_q = \tfrac{2}{3}(\varepsilon_1^2 + \varepsilon_3^2 - \varepsilon_1\varepsilon_3)^{1/2} \tag{3.98}$$

### 3.12.7 Hooke's Law Using Stress and Strain Invariants

The stress and strain invariants for an elastic material are related as follows:

$$\varepsilon_p^e = \frac{1}{K'} p' \tag{3.99}$$

where

$$K' = \frac{p'}{\varepsilon_p^e} = \frac{E'}{3(1 - 2v')} \tag{3.100}$$

is the effective bulk modulus and the superscript $e$ denotes elastic.

$$\varepsilon_q^e = \frac{1}{3G} q \tag{3.101}$$

where

$$G = G' = \frac{E'}{2(1 + v')} \tag{3.102}$$

is called the shear modulus. Hooke's law in terms of the stress and strain invariants is

$$\begin{Bmatrix} p' \\ q \end{Bmatrix} = \begin{bmatrix} K' & 0 \\ 0 & 3G \end{bmatrix} \begin{Bmatrix} \varepsilon_p^e \\ \varepsilon_q^e \end{Bmatrix} \tag{3.103}$$

Equation (3.103) reveals that for a linear, isotropic, elastic material, shear stresses do not cause volume changes and mean effective stresses do not cause shear deformation.

We can generate a generalized Poisson's ratio by eliminating $E'$ from Eqs. (3.100) and (3.102) as follows:

$$\text{Equation (3.100):} \quad E' = 3K'(1 - 2v')$$
$$\text{Equation (3.102):} \quad E' = 2G(1 + v')$$
$$\therefore \frac{3K'(1 - 2v')}{2G(1 + v')} = 1$$

and

$$v' = \frac{3K' - 2G}{2G + 6K'} \tag{3.104}$$

*The* essential points *are:*

1. *Stress and strain invariants are independent of the chosen axis system.*
2. *Stress and strain invariants are convenient measures to determine the effects of a general state of stresses and strains on soils.*
3. *Mean stress represents the average stress on a soil while deviatoric stress represents the average shear or distortional stress.*

## EXAMPLE 3.12

A cylindrical sample of soil 50 mm in diameter and 100 mm long is subjected to an axial effective principal stress of 400 kPa and a radial effective principal stress of 100 kPa. The axial and radial displacements are 0.5 mm and −0.04 mm, respectively. Assuming the soil is an isotropic, elastic material, calculate (a) the mean and deviatoric stresses, (b) the volumetric and shear (distortional) strains, and (c) the shear, bulk, and elastic moduli.

*Strategy*  This is a straightforward problem. You only need to apply the equations given in the previous section. The negative sign for the radial displacement indicates an expansion.

## Solution 3.12

**Step 1:**  Calculate the mean and deviatoric stresses.

$$\sigma_1' = \sigma_z' = 400 \text{ kPa}, \quad \sigma_3' = \sigma_r' = 100 \text{ kPa}$$

**(a)** $p' = \dfrac{\sigma_z' + 2\sigma_r'}{3} = \dfrac{400 + 2 \times 100}{3} = 200 \text{ kPa}$

$q = q' = \sigma_z' - \sigma_r' = 400 - 100 = 300 \text{ kPa}$

**Step 2:**  Calculate the volumetric and shear strains.

$$\Delta z = 0.5 \text{ mm}, \quad \Delta r = -0.04 \text{ mm}, \quad r = 50/2 = 25 \text{ mm}, \quad L = 100 \text{ mm}$$

**(b)** $\varepsilon_z = \varepsilon_1 = \dfrac{\Delta z}{L} = \dfrac{0.5}{100} = 0.005$

$\varepsilon_r = \varepsilon_3 = \dfrac{\Delta r}{r} = \dfrac{-0.04}{25} = -0.0016$

$\varepsilon_p^e = \varepsilon_z + 2\varepsilon_r = 0.005 - 2 \times 0.0016 = 0.0018 = 0.18\%$

$\varepsilon_q^e = \frac{2}{3}(\varepsilon_z - \varepsilon_r) = \frac{2}{3}(0.005 + 0.0016) = 0.0044 = 0.44\%$

**Step 3:** Calculate the moduli.

**(c)** $K' = \dfrac{p'}{\varepsilon_p^e} = \dfrac{200}{0.0018} = 111{,}111 \text{ kPa}$

$G = \dfrac{q}{3\varepsilon_q^e} = \dfrac{300}{3 \times 0.0044} = 22{,}727 \text{ kPa}$

but

$G = \dfrac{E'}{2(1 + \nu')}$ and $\nu' = \dfrac{3K' - 2G}{2G + 6K'} = \dfrac{3 \times 111{,}111 - 2 \times 22{,}727}{2 \times 22{,}727 + 6 \times 111{,}111} = 0.4$

$\therefore E' = 2G(1 + \nu') = 2 \times 22{,}727(1 + 0.4) = 63{,}636 \text{ kPa}$ ∎

## EXAMPLE 3.13

A sample of soil was subjected to simple shear, which consists of deforming a cuboidal sample into a parallelepiped (Fig. E3.13). The strains in the $X$ and $Y$ directions are zero. The sample size is 100 mm × 100 mm × 20 mm high. The maximum lateral displacement applied at the top is 0.2 mm and results in a vertical displacement of 0.04 mm and a shear force of 2 kN. Calculate the principal, volumetric, and deviatoric strains.

**Strategy** You are given the displacement. You can use this information to calculate strains. The sample is deformed under the plane strain condition. Therefore, the appropriate equations to use are Eqs. (3.97) and (3.98).

## Solution 3.13

**Step 1:** Calculate $\varepsilon_1$ and $\varepsilon_3$.

$\Delta z = 0.04, \quad h = 20 \text{ mm}, \quad \Delta x = 0.2 \text{ mm}$

$\varepsilon_z = \dfrac{\Delta z}{h} = \dfrac{0.04}{20} = 0.002, \quad \varepsilon_x = \varepsilon_y = 0, \quad \gamma_{zx} = \dfrac{\Delta x}{h} = \dfrac{0.2}{20} = 0.01$

Equation (3.35): $\varepsilon_1 = \dfrac{0.002 + 0}{2} + \sqrt{\left(\dfrac{0.002 - 0}{2}\right)^2 + \left(\dfrac{0.01}{2}\right)^2} = 0.006$

Equation (3.36): $\varepsilon_3 = \dfrac{0.002 + 0}{2} - \sqrt{\left(\dfrac{0.002 - 0}{2}\right)^2 + \left(\dfrac{0.01}{2}\right)^2} = -0.004$

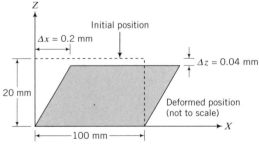

**FIGURE E3.13**

**Step 2:** Calculate the volumetric and deviatoric strains.

$$\varepsilon_p = \varepsilon_1 + \varepsilon_3 = 0.006 - 0.004 = 0.002$$

Also,

$$\varepsilon_p = \varepsilon_x + \varepsilon_y + \varepsilon_z = 0 + 0 + 0.002 = 0.002$$

$$\varepsilon_q = \tfrac{2}{3}(\varepsilon_1^2 + \varepsilon_3^2 - \varepsilon_1\varepsilon_3)^{1/2} = \tfrac{2}{3}[0.006^2 + (-0.004)^2 - (0.006)(-0.004)]^{1/2} = 0.0058$$

∎

*What's next . . .* We have examined how applied surface stresses are distributed in soils as if soils were linear, isotropic, elastic materials. Different structures will impose different stresses and cause the soil to respond differently. For example, an element of soil under the center of an oil tank will experience a continuous increase/decrease in vertical stress while the tank is being filled/emptied. However, the soil near a retaining earth structure will suffer a reduction in lateral stress if the wall moves out. If the soil were the same for both structures, the different loading conditions would cause the soil to respond differently. Therefore, we need to trace the history of stress increases/decreases in soils to evaluate possible soil responses and to conduct tests that replicate the loading history of the in situ soil. In the next section, a method of keeping track of the loading history of a soil is described.

## 3.13 STRESS PATHS

### 3.13.1 Basic Concept

Consider two marbles representing two particles of a coarse-grained soil. Let us fix one marble in a hemispherical hole and stack the other on top of it (Fig. 3.29a). We are constructing a one-dimensional system in which relative displacement of the two marbles will occur at the contact. Let us incrementally apply a vertical, concentric force, $F_z$, on the top marble. We will call this loading "*A*". The forces at the contact are equal to the applied loads and the marbles are forced together vertically. No relative displacement between the marbles occurs. For the system to become unstable or fail, the applied forces must crush the marbles. We can make a plot of our loading by arbitrarily choosing an axis system. Let us choose a Cartesian system with the $X$ axis representing the horizontal force and the $Z$ axis representing the vertical forces. We can represent loading "*A*" by a line $OA$ as shown in Fig. 3.29c. The line $OA$ is called a load path or a force path.

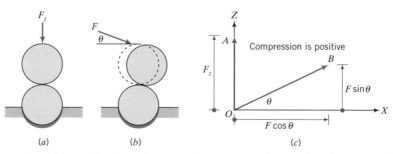

**FIGURE 3.29** Effects of force paths on a one-dimensional system of marbles.

Let us now apply the same force at an angle $\theta$ to the $X$ axis in the $ZX$ plane (Fig. 3.29b) and call this loading "$B$". There are now two components of force. One component is $F_x = F \cos \theta$ and the other is $F_z = F \sin \theta$. If the frictional resistance at the contacts of the two marbles is less than the horizontal force, the top marble will slide relative to the bottom. You should recall from your mechanics or physics course that the frictional resistance is $\mu F_z$ (Coulomb's law), where $\mu$ is the coefficient of friction at the contact between the two marbles. Our one-dimensional system now has two modes of instability or failure—one due to relative sliding and the other due to crushing of the marbles. The force path for loading "$B$" is represented by $OB$ in Fig. 3.29c. The essential point or principle is that the response, stability, and failure of the system depend on the force path.

Soils, of course, are not marbles but the underlying principle is the same. The soil fabric can be thought of as a space frame with the soil particles representing the members of the frame and the particle contacts representing the joints. The response, stability, and failure of the soil fabric or the space frame depend on the stress path.

Stress paths are presented in a plot showing the relationship between stress parameters and provide a convenient way to allow a geotechnical engineer to study the changes in stresses in a soil caused by loading conditions. We can, for example, plot a two-dimensional graph of $\sigma_1$ versus $\sigma_3$ or $\sigma_2$, which will give us a relationship between these stress parameters. However, the stress invariants, being independent of the axis system, are more convenient to use.

### 3.13.2 Plotting Stress Paths

We will explore the stress paths for a range of loading conditions. We will use a cylindrical soil sample for illustrative purposes and subject it to several loading conditions. Let us apply equal increments of axial and radial stresses ($\Delta \sigma_z = \Delta \sigma_r = \Delta \sigma$) to an initially stress-free sample as illustrated in the inset figure labeled "1" in Fig. 3.30. Since we are not applying any shearing stresses on the horizontal and vertical boundaries, the axial and radial stresses are principal stresses: that is, $\Delta \sigma_z = \Delta \sigma_1$ and $\Delta \sigma_r = \Delta \sigma_3$.

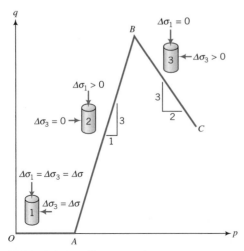

**FIGURE 3.30**   Stress paths.

The loading condition we are applying is called isotropic compression; that is, the stresses in all directions are equal ($\Delta\sigma_1 = \Delta\sigma_2 = \Delta\sigma_3$). We will call this loading condition, loading "1." It is often convenient to work with increments of stresses in determining stress paths. Consequently, we are going to use the incremental form of the stress invariants. The stress invariants for isotropic compression are

$$\Delta p_1 = \frac{\Delta\sigma_1 + 2\Delta\sigma_3}{3} = \frac{\Delta\sigma_1 + 2\Delta\sigma_1}{3} = \Delta\sigma_1$$

$$\Delta q_1 = \Delta\sigma_1 - \Delta\sigma_3 = \Delta\sigma_1 - \Delta\sigma_1 = 0$$

The subscript 1 on $p$ and $q$ denotes loading "1."

Let us now prepare a graph with axes $p$ (abscissa) and $q$ (ordinate), as depicted in Fig. 3.30. We will call this graph the $q$-$p$ plot. The initial stresses on the soil sample are zero; that is, $p_o = 0$ and $q_o = 0$. The stresses at the end of loading "1" are

$$p_1 = p_o + \Delta p_1 = 0 + \Delta\sigma_1 = \Delta\sigma_1$$

$$q_1 = q_o + \Delta q_1 = 0 + 0 = 0$$

and are shown as coordinate $A$ in Fig. 3.30. The line $OA$ is called the stress path for isotropic compression. The slope of $OA$ is

$$\frac{\Delta q_1}{\Delta p_1} = 0$$

Let us now apply loading "2" by keeping $\sigma_3$ constant, that is, $\Delta\sigma_3 = 0$, but continue to increase $\sigma_1$, that is, $\Delta\sigma_1 > 0$ (inset figure labeled "2" in Fig. 3.30). Increases in the stress invariants for loading "2" are

$$\Delta p_2 = \frac{\Delta\sigma_1 + 2 \times 0}{3} = \frac{\Delta\sigma_1}{3}$$

$$\Delta q_2 = \Delta\sigma_1 - 0 = \Delta\sigma_1$$

and the stress invariants at the end of loading "2" are

$$p_2 = p_1 + \Delta p_2 = \Delta\sigma_1 + \frac{\Delta\sigma_1}{3} = \tfrac{4}{3}\Delta\sigma_1$$

$$q_2 = q_1 + \Delta q_2 = 0 + \Delta\sigma_1 = \Delta\sigma_1$$

Point $B$ in Fig. 3.30 represents ($q_2$, $p_2$) and the line $AB$ is the stress path for loading "2." The slope of $AB$ is

$$\frac{\Delta q_2}{\Delta p_2} = \frac{\Delta\sigma_1}{(\Delta\sigma_1/3)} = 3$$

Let us make another change to the loading conditions. We will now keep $\sigma_1$ constant ($\Delta\sigma_1 = 0$) and then increase $\sigma_3$ ($\Delta\sigma_3 > 0$) as illustrated by the inset figure labeled "3" in Fig. 3.30. The increases in stress invariants are

$$\Delta p_3 = \frac{0 + 2\Delta\sigma_3}{3} = \frac{2\Delta\sigma_3}{3}$$

$$\Delta q_3 = 0 - \Delta\sigma_3 = -\Delta\sigma_3$$

The stress invariants at the end of loading "3" are

$$p_3 = p_2 + \Delta p_3 = \tfrac{4}{3}\Delta\sigma_1 + \tfrac{2}{3}\Delta\sigma_3$$
$$q_3 = q_2 + \Delta q_3 = \Delta\sigma_1 - \Delta\sigma_3$$

The stress path for loading "3" is shown as $BC$ in Fig. 3.30. The slope of $BC$ is

$$\frac{\Delta q_3}{\Delta p_3} = \frac{-\Delta\sigma_3}{\tfrac{2}{3}\Delta\sigma_3} = -\frac{3}{2}$$

You should note that $q$ decreases but $p$ increases for stress path $BC$.

So far, we have not discussed whether the soil was allowed to drain or not. You will recall that the soil solids and the pore water (Section 3.9) must carry the applied increase in stresses in a saturated soil. If the soil pore water is allowed to drain from the soil sample, the increase in stress carried by the pore water, called excess pore water pressure ($\Delta u$), will continuously decrease to zero and the soil solids will have to support all of the increase in applied stresses. We will assume that during loading "1," the excess pore water was allowed to drain— this is called the drained condition in geotechnical engineering. The type of loading imposed by loading "1" is called isotropic consolidation. In Chapter 5, we will discuss isotropic consolidation further. Since the excess pore water pressure ($\Delta u_1$) dissipates as the pore water drains from the soil, the mean effective stress at the end of each increment of loading "1" is equal to the mean total stress; that is,

$$\Delta p_1' = \Delta p_1 - \Delta u_1 = \Delta p_1 - 0 = \Delta p$$

The effective stress path (ESP) and the total stress path (TSP) are the same and represented by $OA$ in Fig. 3.31. You should note that we have used dual labels, $p'$, $p$, for the horizontal axis in Fig. 3.31. This dual labeling allows us to use one plot to represent both the effective and total stress paths.

We will assume that for loadings "2" and "3" the excess pore water pressures were prevented from draining out of the soil. In geotechnical engineering, the term undrained is used to denote a loading situation in which the excess pore

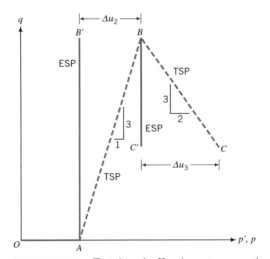

**FIGURE 3.31**   Total and effective stress paths.

water cannot drain from the soil. The implication is that the volume of our soil sample remains constant. In Chapter 5, we will discuss drained and undrained loading conditions in more detail. For loading "2," the total stress path is $AB$. In this book, we will represent total stress paths by dashed lines.

If our soil were an isotropic, elastic material, then according to Eq. (3.99), written in incremental form

$$\Delta\varepsilon_p^e = \frac{\Delta p'}{K'} = 0 \tag{3.105}$$

The solution of Eq. (3.105) leads to either $\Delta p' = 0$ or $K' = \infty$. There is no reason why $K'$ should be $\infty$. The act of preventing the drainage of the excess pore water cannot change the (effective) bulk modulus of the soil solids. Remember the truss analogy we used for effective stresses. The same analogy is applicable here. The only tenable solution is $\Delta p' = 0$. We can also write Eq. (3.105) in terms of total stresses; that is,

$$\Delta\varepsilon_p^e = \frac{\Delta p}{K} = 0 \tag{3.106}$$

where $K = E_u/3(1 - 2v_u)$ and the subscript $u$ denotes undrained condition. In this case, $\Delta p$ cannot be zero since this is the change in mean total stress from the applied loading. Therefore, the only tenable solution is $K = K_u = \infty$, which leads to $v_u = 0.5$. The implications of Eqs. (3.105) and (3.106) for a linear, isotropic, elastic soil under undrained conditions are:

1. The change in mean effective stress is zero and, consequently, the effective stress path is vertical.
2. The undrained bulk modulus is $\infty$ and $v_u = 0.5$.

The deviatoric stress is unaffected by pore water pressure changes. We can write Eq. (3.102) in terms of total stress parameters as

$$G = G_u = \frac{E_u}{2(1 + v_u)}$$

Since $G = G_u = G'$ then

$$\frac{E_u}{2(1 + v_u)} = \frac{E'}{2(1 + v')}$$

and, by substituting $v_u = 0.5$, we obtain

$$E_u = \frac{1.5E'}{(1 + v')} \tag{3.107}$$

For many soils, $v' \cong \frac{1}{3}$ and, as a result, $E_u \cong 1.1E'$; that is, the undrained elastic modulus is about 10% greater than the effective elastic modulus.

The effective stress path for loading "2," assuming our soil sample behaves like an isotropic, elastic material, is represented by $AB'$ (Fig. 3.31); the coordinates of $B'$ are

$$p_2' = p_1' + \Delta p_2' = p_1' + 0 = \Delta\sigma_1$$
$$q_2 = q_1 + \Delta q_2 = 0 + \Delta\sigma_1 = \Delta\sigma_1$$

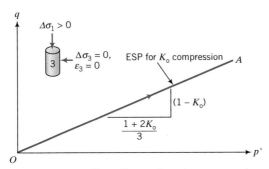

**FIGURE 3.32**   One-dimensional compression stress path.

The difference in mean stress between the TSP and the ESP at a fixed value of $q$ is the change in excess pore water pressure. That is, the magnitude of a horizontal line between the TSP and ESP is the change in excess pore water pressure. The maximum change in excess pore water pressure at the end of loading "2" is

$$\Delta u_2 = p_2 - p_2' = \tfrac{4}{3}\Delta\sigma_1 - \Delta\sigma_1 = \tfrac{1}{3}\Delta\sigma_1$$

For loading "3," the ESP for an elastic soil is $BC'$ and the maximum change in excess pore water pressure is denoted by $CC'$ (Fig. 3.31).

Soils only behave as elastic materials over a small range of strains and therefore the condition $\Delta p' = 0$ under undrained loading has only limited application. Once the soil yields, the ESP tends to bend. In Chapter 6, we will discuss how soil yielding affects the ESP.

You can use the above procedure to determine the stress paths for any loading condition. For example, let us confine our soil sample laterally, that is, we are keeping the diameter constant, $\Delta\varepsilon_r = 0$, and incrementally increase $\sigma_1$ under drained conditions (Fig. 3.32). The loading condition we are imposing on our sample is called one-dimensional compression.

The increase in lateral effective stress for an increment of vertical stress $\Delta\sigma_1$ under the drained condition is given by Eq. (3.50) as $\Delta\sigma_3 = \Delta\sigma_3' = K_o\,\Delta\sigma_1'$. The stress invariants are

$$\Delta p' = \frac{\Delta\sigma_1' + 2\Delta\sigma_3'}{3} = \frac{\Delta\sigma_1' + 2K_o\,\Delta\sigma'}{3} = \Delta\sigma_1'\left(\frac{1 + 2K_o}{3}\right)$$

$$\Delta q = \Delta q' = \Delta\sigma_1' - \Delta\sigma_3' = \Delta\sigma_1' - K_o\,\Delta\sigma_1' = \Delta\sigma_1'(1 - K_o)$$

The slope of the TSP is equal to the slope of the ESP; that is

$$\frac{\Delta q}{\Delta p} = \frac{\Delta q}{\Delta p'} = \frac{3(1 - K_o)}{1 + 2K_o}$$

The one-dimensional compression stress path is shown in Fig. 3.32.

---

*The essential points are:*

*1. A stress path is a graphical representation of stresses in stress space. For convenience, stress paths are plotted as deviatoric stress (q) on the ordinate versus mean effective stress (p′) and/or mean total stress (p) on the abscissa.*

> 2. *The effective stress path for a linear, elastic soil under the undrained condition is vertical; that is,* $\Delta p' = 0$.
> 3. *The mean stress difference between the total stress path and the effective stress path is the excess pore water pressure.*
> 4. *The response, stability, and failure of soils depend on stress paths.*

### 3.13.3 Procedure for Plotting Stress Paths

A summary of the procedure for plotting stress paths is as follows:

1. Determine the loading conditions drained or undrained or both.
2. Calculate the initial loading values of $p'_o$, $p_o$, and $q_o$.
3. Set up a graph of $p'$ (and $p$, if you are going to also plot the total stress path) as the abscissa and $q$ as the ordinate. Plot the initial values of $(p'_o, q_o)$ and $(p_o, q_o)$.
4. Determine the increase in stresses, $\Delta\sigma_1$, $\Delta\sigma_2$, and $\Delta\sigma_3$. These stresses can be negative.
5. Calculate the increase in stress invariants, $\Delta p'$, $\Delta p$, and $\Delta q$. These stress invariants can be negative.
6. Calculate the current stress invariants as $p' = p'_o + \Delta p'$, $p = p_o + \Delta p$, and $q = q_o + \Delta q$. The current value of $p'$ cannot be negative but $q$ can be negative.
7. Plot the current stress invariants $(p', q)$ and $(p, q)$.
8. Connect the points identifying effective stresses and do the same for total stresses.
9. Repeat items 5 to 8 for the next loading condition.
10. The pore water pressure at a desired level of deviatoric stress is the mean stress difference between the total stress path and the effective stress path.

Remember that for a drained loading condition, ESP = TSP, and for an undrained condition, the ESP for a linear, elastic soil is vertical.

### EXAMPLE 3.14

Two cylindrical specimens, A and B, of a soil were loaded as follows. Both specimens were isotropically loaded by a stress of 200 kPa under drained conditions. Subsequently, the radial stress applied on specimen A was held constant and the axial stress was incrementally increased to 440 kPa under undrained conditions. The axial stress on specimen B was held constant and the radial stress incrementally reduced to 50 kPa under drained conditions. Plot the total and effective stress paths for each specimen assuming the soil is a linear, isotropic, elastic material. Calculate the maximum excess pore water pressure in specimen A.

*Strategy*  The loading conditions on both specimens are axisymmetric. The easiest approach is to write the mean stress and deviatoric stress equations in terms of increments and make the necessary substitutions.

## Solution 3.14

**Step 1:** Determine loading condition.
Loading is axisymmetric and both drained and undrained conditions are specified.

**Step 2:** Calculate initial stress invariants for isotropic loading path.
For axisymmetric, isotropic loading under drained conditions, $\Delta u = 0$,

$$\Delta p' = \frac{\Delta \sigma_a' + 2\Delta \sigma_r'}{3} = \frac{\Delta \sigma_1' + 2\Delta \sigma_1'}{3} = \Delta \sigma_1' = 200 \text{ kPa}$$

$p_o = p_o' = 200$ kPa, since the soil specimens were loaded from a stress-free state under drained conditions.

$$q_o = q_o' = 0$$

**Step 3:** Set up graph and plot initial stress points.
Create a graph with axes $p'$ and $p$ as the abscissa and $q$ as the ordinate and plot the isotropic stress path with coordinates $(0, 0)$ and $(200, 0)$ as shown by $OA$ in Fig. E3.14.

**Step 4:** Determine the increases in stresses.

**Specimen A**
We have (1) an undrained condition, $\Delta u$ is not zero and (2) no change in the radial stress but the axial stress is increased to 440 kPa. Therefore,

$$\Delta \sigma_3 = 0, \quad \Delta \sigma_1 = 440 - 200 = 240 \text{ kPa}$$

**Specimen B**
Drained loading ($\Delta u = 0$); therefore, TSP = ESP.
Axial stress held constant, $\Delta \sigma_1 = \Delta \sigma_1' = 0$; radial stress decreases to 50 kPa; that is,

$$\Delta \sigma_3 = \Delta \sigma_3' = 50 - 200 = -150 \text{ kPa}$$

**Step 5:** Calculate the increases in stress invariants.

**Specimen A**

$$\Delta p = \frac{\Delta \sigma_1 + 2\Delta \sigma_3}{3} = \frac{240 + 2 \times 0}{3} = 80 \text{ kPa}$$

$$\Delta q = \Delta \sigma_1 - \Delta \sigma_3 = 240 - 0 = 240 \text{ kPa}$$

$$\text{Slope of total stress path} = \frac{\Delta q}{\Delta p} = \frac{240}{80} = 3$$

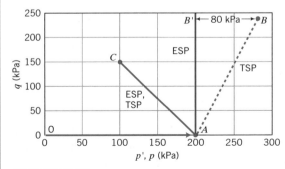

**FIGURE E3.14**

**Specimen B**

$$\Delta p = \Delta p' = \frac{\Delta \sigma_1' + 2\Delta \sigma_3'}{3} = \frac{0 + 2 \times (-150)}{3} = -100 \text{ kPa}$$

$$\Delta q = \Delta \sigma_1 - \Delta \sigma_3 = 0 - (-150) = 150 \text{ kPa}$$

$$\text{Slope of ESP (or TSP)} = \frac{\Delta q}{\Delta p'} = \frac{150}{-100} = -1.5$$

**Step 6:** Calculate the current stress invariants.

**Specimen A**

$p = p_o + \Delta p = 200 + 80 = 280 \text{ kPa}, \quad q = q' = q_o + \Delta q = 0 + 240 = 240 \text{ kPa}$

$p' = p_o + \Delta p' = 200 + 0 = 200 \text{ kPa}$ (elastic soil)

**Specimen B**

$$p = p' = p_o + \Delta p = 200 - 100 = 100 \text{ kPa},$$

$$q = q_o + \Delta q = 0 + 150 = 150 \text{ kPa}$$

**Step 7:** Plot the current stress invariants.

**Specimen A**

Plot point B as (280, 240); plot point B' as (200, 240).

**Specimen B**

Plot point C as (100, 150).

**Step 8:** Connect the stress points.

**Specimen A**

$AB$ in Fig. E3.14 shows the total stress path and $AB'$ shows the effective stress path.

**Specimen B**

$AC$ in Fig. E3.14 shows the ESP and TSP.

**Step 9:** Determine the excess pore water pressure.

**Specimen A**

$BB'$ shows the maximum excess pore water pressure. The mean stress difference is $280 - 200 = 80$ kPa. ∎

## 3.14 SUMMARY

Elastic theory provides a simple, first approximation to calculate the deformation of soils at small strains. You are cautioned that the elastic theory cannot adequately describe the behavior of most soils and more involved theories are required. The most important principle in soil mechanics is the principle of effective stress. Soil deformation is due to effective not total stresses. Applied surface stresses are distributed such that their magnitudes decrease with depth and distance away from their points of application.

Stress paths provide a useful means through which the history of loading of a soil can be followed. The mean effective stress changes for a linear, isotropic, elastic soil are zero under undrained loading and the effective stress path is a vector parallel to the deviatoric stress axis with the $q$ ordinate equal to the corresponding state on the total stress path. The difference in mean stress between the total stress path and the effective stress path gives the excess pore water pressure at a desired value of deviatoric stress.

*Practical Example*

### EXAMPLE 3.15

A building foundation of width 10 m and length 40 m transmits a load of 80 MN to a deep deposit of stiff saturated clay (Fig. E3.15a). The elastic modulus of the clay varies with depth (Fig. E3.15b) and $v' = 0.32$. Estimate the elastic settlement of the clay under the center of the foundation assuming (1) drained condition and (2) undrained condition.

*Strategy*   The major decision in this problem is what depth to use to determine an appropriate elastic modulus. One option is to use an average elastic modulus over a depth of 3B. Beyond a depth of 3B, the vertical stress increase is less than 10%.

### Solution 3.15

**Step 1:**   Find the applied vertical surface stress.

$$q_s = \frac{80 \times 10^3}{10 \times 40} = 200 \text{ kPa}$$

**Step 2:**   Determine the elastic modulus.
Assume an effective depth of 3B = 3 × 10 = 30 m.
The average value of $E'$ is 34.5 MPa.

From Eq. (3.107) with $v_u = 0.5$, $E_u = 1.1\,E' = 1.1 \times 34.5 = 38$ MPa

**Step 3:**   Calculate the vertical settlement.

Use Eq. (3.84):   $\Delta z = \dfrac{q_s B (1 - v^2)}{E} I_s$

$\dfrac{L}{B} = \dfrac{40}{10} = 4,$   $I_s = 0.62 \ln\left(\dfrac{L}{B}\right) + 1.12 = 0.62 \ln(4) + 1.12 = 1.98$

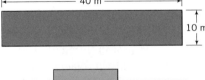

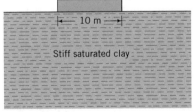

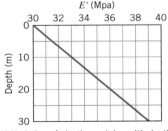

(a) Foundation plan and section          (b) Variation of elastic modulus with depth

**FIGURE E3.15**

**(1)** Drained condition.

$$\Delta z = \frac{200 \times 5 \times (1 - 0.32^2)}{34.5 \times 10^6} \, 1.98 = 51.5 \times 10^{-6} \, \text{m} = 51.5 \times 10^{-3} \, \text{mm}$$

**(2)** Undrained condition.

$$\Delta z = \frac{200 \times 5 \times (1 - 0.5^2)}{38 \times 10^6} \, 1.98 = 39.1 \times 10^{-6} \, \text{m} = 39.1 \times 10^{-3} \, \text{mm} \quad \blacksquare$$

# EXERCISES

## Theory

**3.1** An elastic soil is confined laterally and is axially compressed under drained conditions. In soil mechanics, the loading imposed on the soil is called $K_o$ compression or consolidation. Show that under the $K_o$ condition,

$$\frac{\sigma'_x}{\sigma'_z} = \frac{v'}{1 - v'}$$

**3.2** The increase in pore water pressure in a saturated soil is given by $\Delta u = \Delta \sigma_3 + A(\Delta \sigma_1 - \Delta \sigma_3)$. Show that if the soil is a linear, isotropic, elastic material, $A = \frac{1}{3}$ for the axisymmetric condition.

**3.3** If the axial stress on a cylindrical sample of soil is decreased and the radial stress is increased by twice the decrease in axial stress, show that the stress path has a slope $q/p = -3$.

**3.4** The initial mean effective stress on a soil is $p'_o$ and the deviatoric stress $q = 0$. If the soil is a linear, isotropic, elastic material, plot the total and effective stress paths for the following axisymmetric undrained loading condition: (a) $\Delta \sigma_3 = \frac{1}{2} \Delta \sigma_1$ and (b) $\Delta \sigma_3 = -\frac{1}{2} \Delta \sigma_1$.

## Problem Solving

**3.5** A long embankment is located on a soil profile consisting of 2 m of medium clay followed by 8 m of medium to dense sand on top of bedrock. A vertical settlement of 5 mm at the center of the embankment was measured during construction. Assuming all the settlement is elastic and occurs in the medium clay, determine the average stresses imposed on the medium clay under the center of the embankment using the elastic equations. The elastic parameters are $E_u = 15$ MPa and $v_u = 0.45$. [*Hint:* Assume the lateral strain is zero.]

**3.6** An element of soil (sand) behind a retaining wall is subjected to an increase in vertical stress of 60 kPa and an increase in lateral stress of 25 kPa. Determine the increase in vertical and lateral strains, assuming the soil is a linearly elastic material with $E' = 20$ MPa and $v' = 0.3$.

**3.7** A cylindrical specimen of soil is compressed by an axial principal stress of 500 kPa and a radial principal stress of 200 kPa. Plot Mohr's circle of stress and determine (a) the maximum shear stress and (b) the normal and shear stresses on a plane inclined at 30° clockwise to the horizontal.

**3.8** A soil specimen (100 mm × 100 mm × 100 mm) is subjected to the forces shown in Fig. P3.8. Determine (a) the magnitude of the principal stresses, (b) the orientation of the

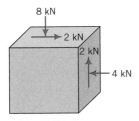

**FIGURE P3.8**

principal stress plane to the horizontal, (c) the maximum shear stress, and (d) the normal and shear stresses on a plane inclined at 20° counterclockwise to the horizontal.

3.9    The initial principal stresses at a certain depth in a clay soil are 200 kPa on the horizontal plane and 100 kPa on the vertical plane. Construction of a surface foundation induces additional stresses consisting of a vertical stress of 45 kPa, a lateral stress of 20 kPa, and a clockwise (with respect to the horizontal plane) shear stress of 40 kPa. Plot Mohr's circle (1) for the initial state of the soil and (2) after construction of the foundation. Determine (a) the change in magnitude of the principal stresses, (b) the change in maximum shear stress, and (c) the change in orientation of the principal stress plane resulting from the construction of the foundation.

3.10   Plot the distribution of total stress, effective stress, and pore water pressure with depth for the soil profile shown in Fig. P3.10. Neglect capillary action.

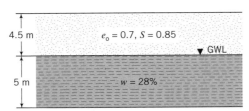

**FIGURE P3.10**

3.11   At what depth would the vertical effective stress in a deep deposit of clay be 100 kPa, if $e = 1.1$; groundwater level is at 1 m below ground surface and $S = 95\%$ above the groundwater level?

3.12   A culvert is to be constructed in a bed of sand ($e = 0.5$) for drainage purposes. The roof of the culvert will be located 3 m below ground surface. Currently, the groundwater level is at ground surface. But, after installation of the culvert, the groundwater level is expected to drop to 2 m below ground surface. Calculate the change in vertical effective stress on the roof of the culvert after installation. You can assume the sand above the groundwater level is saturated.

3.13   The effective size of a fine sand is 0.15 mm. Estimate the distance in the sand, above the groundwater level, that would be saturated by capillary action. Determine the vertical effective stress in the sand at depths of 1 m above and 3 m below the groundwater level for the following cases: (a) groundwater level 3 m below ground surface and (b) groundwater level 1.5 m below ground surface. The void ratio of the sand is 0.6. Assume that the degree of saturation is 90% for soil layers that are unsaturated.

3.14   A soil profile consists of a clay layer underlain by a sand layer as shown in Fig. P3.14. If a tube is inserted into the bottom sand layer (Fig. P3.14) and the water level rises to 1 m

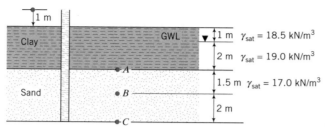

**FIGURE P3.14**

above the ground surface, determine the vertical effective stresses and pore water pressures at $A$, $B$, and $C$. If $K_o$ is 0.5, determine the effective and total lateral stresses at $A$, $B$, and $C$.

**3.15** A rectangular foundation 4 m $\times$ 6 m (Fig. P3.15) transmits a stress of 150 kPa on the surface of a soil deposit. Plot the distribution of increases of vertical stresses with depth under points $A$, $B$, and $C$ up to a depth of 20 m. At what depth is the increase in vertical stress below $A$ less than 10% of the surface stress?

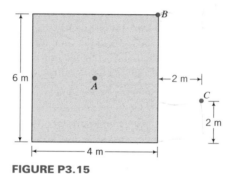

**FIGURE P3.15**

**3.16** Determine the increase in vertical stress at a depth of 5 m below the centroid of the foundation shown in Fig. P3.16.

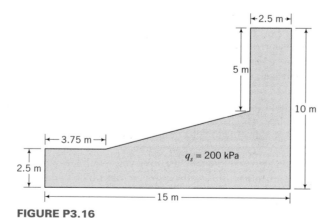

**FIGURE P3.16**

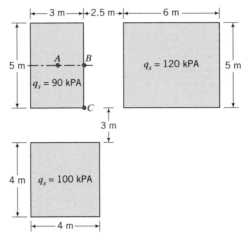

**FIGURE P3.17**

**3.17** Three foundations are located next to each other (Fig. P3.17). Determine the stress increases at $A$, $B$, and $C$ at a depth of 2 m below the ground surface.

**3.18** A cylindrical sample of soil is isotropically compressed under a drained condition with a vertical stress of 100 kPa and a radial stress of 100 kPa. Subsequently, the axial stress was held constant and the radial stress was increased to 300 kPa under an undrained condition. Plot the total and effective stress paths (assume the soil is a linear, isotropic, elastic material). Determine the slopes of the stress paths and the maximum excess pore water pressure.

**3.19** The initial stresses on a soil element at a certain depth in a soil are $\sigma_1 = 80$ kPa, $\sigma_2 = 40$ kPa, and $\sigma_3 = 40$ kPa. The changes in stresses resulting from a sudden outward movement of a retaining wall are as follows: $\Delta\sigma_1 = 0$ kPa, $\Delta\sigma_2 = -10$ kPa, and $\Delta\sigma_3 = -40$ kPa. Plot the total stress path. Plot the effective stress path assuming that the soil is a linearly elastic material, and determine the maximum excess pore water pressure.

## Practical

**3.20** You are the geotechnical engineer for a proposed office building in a densely clustered city. The office building will be constructed adjacent to an existing office complex. The soil at the site is a deposit of very dense sand with $E' = 45$ MPa and $\nu' = 0.3$. The sand rests on a deep deposit of dense gravel. The existing high rise complex is founded on a concrete slab, 100 m $\times$ 120 m located at 2 m below ground surface, that transmits a load of 2400 MN to the soil. Your office foundation is 50 m $\times$ 80 m and transmits a load of 1000 MN. You also intend to locate your foundation at 2 m below ground level. The front of your building is aligned with the existing office complex and the side distance is 0.5 m. The lesser dimension of each building is the frontal dimension. The owners of the existing building are concerned about possible settlement of their building due to your building. You are invited to a meeting with your client, the owners of the existing building, and their technical staff. You are expected to determine what effects your office building would have on the existing building. You only have one hour to make the preliminary calculations and you are expected to present the increase in stresses and the amount of settlement of the existing office complex due to the construction of your office building. Prepare your analysis and presentation.

## CHAPTER 4

# ONE-DIMENSIONAL CONSOLIDATION SETTLEMENT OF FINE-GRAINED SOILS

| ABET | ES | ED |
|------|----|----|
|      | 85 | 15 |

## 4.0  INTRODUCTION

In this chapter, we will consider one-dimensional consolidation settlement of fine-grained soils. We will restrict settlement consideration to the vertical direction. Under load, all soils will settle, causing settlement of structures founded on or within them. If the settlement is not kept to a tolerable limit, the desired use of the structure may be impaired and the design life of the structure may be reduced. Structures may settle uniformly or nonuniformly. The latter is called differential settlement and is often the crucial design consideration.

The total settlement usually consists of three parts—immediate or elastic compression, primary consolidation, and secondary compression. We have considered elastic settlement in Chapter 3 and we will consider some modifications to the elastic analysis for practical applications in Chapter 7. Here we will develop the basic concepts of consolidation and show how these concepts can be applied to calculate the consolidation settlement from applied loads. After that, we will formulate the theory of one-dimensional consolidation and use it to predict the time rate of settlement.

After you have studied this chapter, you should:

- Have a basic understanding of the consolidation of soils under vertical loads
- Be able to calculate one-dimensional consolidation settlement and time rate of settlement

You will make use of the following concepts learned from the previous chapters and your courses in mechanics and mathematics.

- Stresses in soils—effective stresses and vertical stress increases from surface loads (Chapter 3)
- Strains in soils—vertical and volumetric strains (Chapter 3)
- Elasticity (Chapter 3)
- Flow of water through soils (Chapter 2, Darcy's law)
- Solutions of partial differential equations

**FIGURE 4.1** The leaning tower of Pisa. (© Photo Disc.)

*Sample Practical Situation*   The foundation for an office building is to be constructed on a sand deposit. Below the sand deposit is a layer of soft clay followed by a thick layer of sand and gravel. The office building is asymmetrical and the loads are not uniformly distributed. The allowable total settlement is 50 mm and the differential settlement is limited to 6 mm. You are expected to determine the total and differential settlement of the building, the rate of settlement, and the time to achieve 90% consolidation settlement.

The leaning tower of Pisa is the classic example of differential settlement (Fig. 4.1). Construction of the tower started in 1173 and by the end of 1178 when two-thirds of the tower was completed it tilted. Since then the tower has been settling differentially. The foundation of the tower is located about 3 m into a bed of silty sand that is underlain by 30 m of clay on a deposit of sand. A sand layer approximately 5 m thick intersects the clay. The structure of the tower is intact but its function is impaired by differential settlement.

## 4.1   DEFINITIONS OF KEY TERMS

*Consolidation* is the time-dependent settlement of soils resulting from the expulsion of water from the soil pores.

*Primary consolidation* is the change in volume of a fine-grained soil caused by the expulsion of water from the voids and the transfer of load from the excess pore water pressure to the soil particles.

*Secondary compression* is the change in volume of a fine-grained soil caused by the adjustment of the soil fabric (internal structure) after primary consolidation has been completed.

*Excess pore water pressure, $\Delta u$,* is the pore water pressure in excess of the current equilibrium pore water pressure. For example, if the pore water pressure in a soil is $u_o$ and a load is applied to the soil so that the existing pore water pressure increases to $u_1$, then the excess pore water pressure is $\Delta u = u_1 - u_o$.

*Drainage path, $H_{dr}$,* is the longest vertical path that a water particle will take to reach the drainage surface.

*Preconsolidation stress* or past maximum effective stress, $\sigma'_{zc}$, is the maximum vertical effective stress that a soil was subjected to in the past.

*Normally consolidated soil* is one that has never experienced vertical effective stresses greater than its current vertical effective stress ($\sigma'_{zo} = \sigma'_{zc}$).

*Overconsolidated soil* is one that has experienced vertical effective stresses greater than its existing vertical effective stress ($\sigma'_{zo} < \sigma'_{zc}$).

*Overconsolidation ratio, OCR,* is the ratio by which the current vertical effective stress in the soil was exceeded in the past (OCR $= \sigma'_{zc}/\sigma'_{zo}$).

*Compression index, $C_c$,* is the slope of the normal consolidation line in a plot of the logarithm of vertical effective stress versus void ratio.

*Unloading/reloading index* or *recompression index, $C_r$,* is the average slope of the unloading/reloading curves in a plot of the logarithm of vertical effective stress versus void ratio.

*Modulus of volume compressibility, $m_v$,* is the slope of the curve between two stress points in a plot of vertical effective stress versus vertical strain.

## 4.2   QUESTIONS TO GUIDE YOUR READING

1. What is the process of soil consolidation?
2. What is the difference between consolidation and compaction?
3. What is the governing equation in one-dimensional consolidation theory?
4. How is the excess pore water pressure distributed within the soil when a load is applied and after various elapsed times?
5. What factors determine the consolidation settlement of soils?
6. What are the average degree of consolidation, time factor, modulus of volume compressibility, and compression and recompression indices?
7. What is the difference between primary consolidation and secondary compression?
8. What is the drainage path for single drainage and double drainage?
9. Why do we need to carry out consolidation tests, how are they conducted, and what parameters are deduced from the test results?

   **10.** How is time rate of settlement and consolidation settlement calculated?

   **11.** Are there significant differences between the calculated settlements and field settlements?

## 4.3   BASIC CONCEPTS

In our development of the various ideas on consolidation settlement, we will assume:

   • A homogenous, saturated soil
   • The soil particles and the water to be incompressible
   • Vertical flow of water
   • The validity of Darcy's law
   • Small strains

   We will conduct a simple experiment to establish the basic concepts of the one-dimensional consolidation settlement of fine-grained soils. Let us take a thin, soft, saturated sample of clay and place it between porous stones in a rigid, cylindrical container whose inside wall is frictionless (Fig. 4.2a). The porous stones are used to facilitate drainage of the pore water from the top and bottom faces of the soil. The top half of the soil will drain through the top porous stone and the bottom half of the soil will drain through the bottom porous stone. A platen on the top porous stone transmits applied loads to the soil. Expelled water is transported by plastic tubes to a burette. A valve is used to control the flow of the expelled water into the burette. Three pore water pressure transducers are mounted in the side wall of the cylinder to measure the excess pore water pressure near the porous stone at the top $(A)$, at a distance of one-quarter the height $(B)$, and at midheight of the soil $(C)$. Excess pore water pressure is the additional pore water pressure induced in a soil mass by loads. A displacement gauge with its stem on the platen keeps track of the vertical settlement of the soil.

   We will assume that the pore water and the soil particles are incompressible, and the initial pore water pressure is zero. The volume of excess pore water that drains from the soil is then a measure of the volume change of the soil resulting

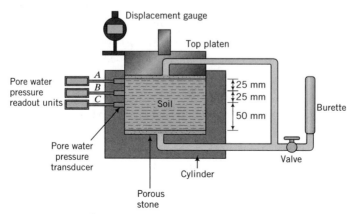

**FIGURE 4.2a**   Experimental setup for illustrating basic concepts on consolidation.

from the applied loads. Since the side wall of the container is rigid, no radial displacement can occur. The lateral and circumferential strains are then equal to zero ($\varepsilon_r = \varepsilon_\theta = 0$), and the volumetric strain ($\varepsilon_p = \varepsilon_z + \varepsilon_\theta + \varepsilon_r$) is equal to the vertical strain, $\varepsilon_z = \Delta z/H_o$, where $\Delta z$ is the change in height or thickness and $H_o$ is the initial height or thickness of the soil.

### 4.3.1 Instantaneous Load

Let us now apply a load $P$ to the soil through the load platen and keep the valve closed. Since no excess pore water can drain from the soil, the change in volume of the soil is zero ($\Delta V = 0$) and no load or stress is transferred to the soil particles ($\Delta\sigma_z' = 0$). The pore water carries the total load. The initial excess pore water pressure in the soil ($\Delta u_o$) is then equal to the change in applied vertical stress, $\Delta\sigma_z = P/A$, where $A$ is the cross-sectional area of the soil, or more appropriately the change in mean total stress, $\Delta p = (\Delta\sigma_z + 2\Delta\sigma_r)/3$, where $\Delta\sigma_r$ is the change in radial stress. For our thin soil layer, we will assume that the initial excess pore water pressure will be distributed uniformly with depth so that at every point in the soil layer, the initial excess pore water pressure is equal to the applied vertical stress. For example, if $\Delta\sigma_z = 100$ kPa, then $\Delta u_o = 100$ kPa as shown in Fig. 4.2b.

### 4.3.2 Consolidation Under a Constant Load—Primary Consolidation

Let us now open the valve and allow the initial excess pore water to drain. The total volume of soil at time $t_1$ decreases by the amount of excess pore water that drains from it as indicated by the change in volume of water in the burette (Fig. 4.2c). At the top and bottom of the soil sample, the excess pore water pressure is zero because these are the drainage boundaries. The decrease of initial excess pore water pressure at the middle of the soil (position $C$) is the slowest because a water particle must travel from the middle of the soil to either the top or bottom boundary to exit the system.

You may have noticed that the settlement of the soil ($\Delta z$) with time $t$ (Fig. 4.2c) is not linear. Most of the settlement occurs shortly after the valve was opened. The rate of settlement, $\Delta z/t$, is also much faster soon after the valve was opened compared with later times. Before the valve was opened, an initial hydraulic head, $\Delta u_o/\gamma_w$, was created by the applied vertical stress. When the valve was opened, the initial excess pore water was forced out of the soil by this initial

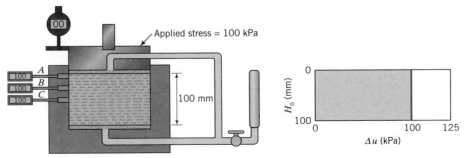

**FIGURE 4.2b** Instantaneous or initial excess pore water pressure when a vertical load is applied.

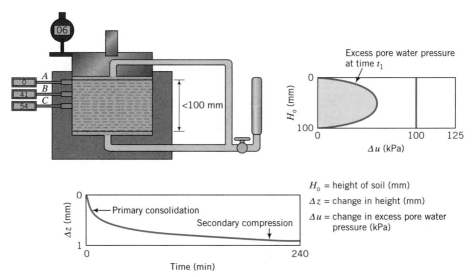

**FIGURE 4.2c**  Excess pore water pressure distribution and settlement during consolidation.

hydraulic head. With time, the initial hydraulic head decreases and, consequently, smaller amounts of excess pore water are forced out. An analogy can be drawn with a pipe containing pressurized water that is ruptured. A large volume of water gushes out as soon as the pipe is ruptured, but soon after the flow becomes substantially reduced. We will call the initial settlement response, soon after the valve was opened, the early time response or primary consolidation. Primary consolidation is the change in volume of the soil caused by the expulsion of water from the voids and the transfer of load from the excess pore water pressure to the soil particles.

### 4.3.3 Secondary Compression

Primary consolidation ends when $\Delta u_o = 0$. The later time settlement response is called secondary compression or creep. Secondary compression is the change in volume of a fine-grained soil caused by the adjustment of the soil fabric (internal structure) after primary consolidation has been completed. The term consolidation is reserved for the process in which settlement of a soil occurs from changes in effective stresses resulting from decreases in excess pore water pressure. The rate of settlement from secondary compression is very slow compared with primary consolidation.

We have separated primary consolidation and secondary compression. In reality, the distinction is not clear because secondary compression occurs as part of the primary consolidation phase especially in soft clays. The mechanics of consolidation is still not fully understood and to make estimates of settlement, it is convenient to separate primary consolidation and secondary compression.

### 4.3.4 Drainage Path

The distance of the longest vertical path taken by a particle to exit the soil is called the length of the drainage path. Because we allowed the soil to drain on the top and bottom faces (double drainage), the length of the drainage path, $H_{dr}$, is

$$H_{dr} = \frac{H_{av}}{2} = \frac{H_o + H_f}{4} \qquad (4.1)$$

where $H_{av}$ is the average height and $H_o$ and $H_f$ are the initial and final heights, respectively, under the current loading. If drainage were permitted from only one face of the soil, then $H_{dr} = H_{av}$. Shorter drainage paths will cause the soil to complete its settlement in a shorter time than a longer drainage path. You will see later that, for single drainage, our soil sample will take four times longer to reach a particular settlement than for double drainage.

### 4.3.5 Rate of Consolidation

The rate of consolidation for a homogeneous soil depends on the soil's permeability, the thickness, and the length of the drainage path. A soil with a coefficient of permeability lower than that of our current soil will take longer to drain the initial excess pore water and settlement will proceed at a slower rate.

### 4.3.6 Effective Stress Changes

Since the applied vertical stress (total stress) remains constant, then according to the principle of effective stress ($\Delta\sigma'_z = \Delta\sigma_z - \Delta u$), any reduction of the initial excess pore water pressure must be balanced by a corresponding increase in vertical effective stress. Increases in vertical effective stresses lead to soil settlement caused by changes to the soil fabric. As time increases, the initial excess pore water continues to dissipate and the soil continues to settle (Fig. 4.2c).

After some time, usually within 24 hours for many soils tested in the laboratory, the initial excess pore water pressure in the middle of the soil reduces to approximately zero and the rate of decrease of the volume of the soil becomes very small. Since the initial excess pore water pressure becomes zero, then from the principle of effective stress all of the applied vertical stress is transferred to the soil; that is, the vertical effective stress is equal to the vertical total stress ($\Delta\sigma'_z = \Delta\sigma_z$).

> *The **essential points** are:*
>
> 1. *When a load is applied to a saturated soil, all of the applied stress is supported initially by the pore water (initial excess pore water pressure); that is, at $t = 0$, $\Delta u_o = \Delta\sigma_z$ or $\Delta u_o = \Delta p$. The change in effective stress is zero ($\Delta\sigma'_z = 0$).*
>
> 2. *If drainage of pore water is permitted, the initial excess pore water pressure decreases and soil settlement ($\Delta z$) increases with time; that is, $\Delta u(t) < \Delta u_o$ and $\Delta z > 0$. The change in effective stress is $\Delta\sigma'_z = \Delta\sigma_z - \Delta u(t)$.*
>
> 3. *When $t \to \infty$, the change in volume and the change in excess pore water pressure of the soil approach zero; that is, $\Delta V \to 0$ and $\Delta u_o \to 0$. The change in vertical effective stress is $\Delta\sigma'_z = \Delta\sigma_z$.*

> 4. *Soil settlement is not linearly related to time except very early in the consolidation process.*
> 5. *The change in volume of the soil is equal to the volume of initial excess pore water expelled.*
> 6. *The rate of settlement depends on the permeability of the soil.*

### 4.3.7 Void Ratio and Settlement Changes Under a Constant Load

The initial volume (specific volume) of a soil is $V = 1 + e_o$ (Chapter 2), where $e_o$ is the initial void ratio. Any change in volume of the soil ($\Delta V$) is equal to the change in void ratio ($\Delta e$). We can calculate the volumetric strain from the change in void ratio as

$$\varepsilon_p = \frac{\Delta V}{V} = \frac{\Delta e}{1 + e_o} \tag{4.2}$$

Since for one-dimensional consolidation, $\varepsilon_z = \varepsilon_p$, we can write a relationship between settlement and the change in void ratio as

$$\varepsilon_z = \frac{\Delta z}{H_o} = \frac{\Delta e}{1 + e_o} \tag{4.3}$$

where $H_o$ is the initial height of the soil. We can rewrite Eq. (4.3) as

$$\Delta z = H_o \frac{\Delta e}{1 + e_o} \tag{4.4}$$

We are going to use $\rho_{pc}$ to denote primary consolidation settlement rather than $\Delta z$, so

$$\boxed{\rho_{pc} = H_o \frac{\Delta e}{1 + e_o}} \tag{4.5}$$

The void ratio at the end of the consolidation under load $P$ is

$$\boxed{e = e_o - \Delta e = e_o - \frac{\Delta z}{H_o}(1 + e_o)} \tag{4.6}$$

### 4.3.8 Effects of Vertical Stresses on Primary Consolidation

We can apply additional loads to the soil and for each load increment we can calculate the final void ratio from Eq. (4.6) and plot the results as shown by segment $AB$ in Fig. 4.3. Three types of graph are shown in Fig. 4.3 to illustrate three different ways of plotting the data from our test. Figure 4.3a is an arithmetic plot of the void ratio versus vertical effective stress. Figure 4.3b is a similar plot except the vertical effective stress is plotted on a logarithmic scale. Figure 4.3c is an arithmetic plot of the vertical strain ($\varepsilon_z$) versus vertical effective stress. The segment $AB$ in Figs. 4.3a,c is not linear because the settlement that occurs for each increment of loading brings the soil to a denser state from its initial state

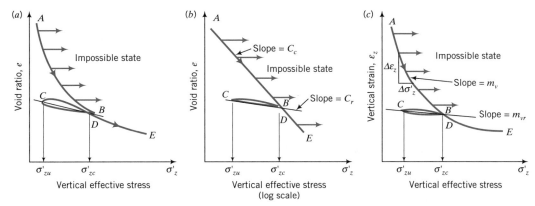

**FIGURE 4.3**  Three plots of settlement data from soil consolidation.

and the soil's permeability decreases. Therefore, doubling the load from a previous increment, for example, would not cause a twofold increase in settlement. The segment $AB$ (Fig. 4.3a–c) is called the virgin consolidation line or normal consolidation line (NCL). In a plot of log $\sigma'_z$ versus $e$, the NCL is approximately a straight line.

At some value of vertical effective stress, say, $\sigma'_{zc}$, let us unload the soil incrementally. Each increment of unloading is carried out only after the soil reaches equilibrium under the previous loading step. When an increment of load is removed, the soil will start to swell by absorbing water from the burette. The void ratio increases but the increase is much less than the decrease in void ratio for the same magnitude of loading that was previously applied.

Let us reload the soil after unloading it to, say, $\sigma'_{zu}$. The reloading path $CD$ is convex compared with the concave unloading path, $BC$. The average slopes of the unloading path and the reloading path are not equal but the difference is generally small for many soils. We will represent the unloading–reloading path by an average slope $BC$ and refer to it as the recompression line or the unloading–reloading line (URL).

A comparison of the soil's response with typical material responses to loads as shown in Figs. 3.5 and 3.6 reveals that soils can be considered to be an elastoplastic material (see Fig. 3.6). The path $BC$ represents the elastic response while the path $AB$ represents the elastoplastic response of the soil. Loads that cause the soil to follow path $BC$ will produce elastic settlement (recoverable settlement of small magnitude). Loads that cause the soil to follow path $AB$ will produce settlements that have both elastic and plastic (permanent) components.

Once the past maximum vertical effective stress, $\sigma'_{zc}$, is exceeded the slope of the path followed by the soil, $DE$, is the same as that of the initial loading path $AB$. Unloading and reloading the soil at any subsequent vertical effective stress would result in a soil's response similar to paths $BCDE$.

### 4.3.9 Primary Consolidation Parameters

The primary consolidation settlement of the soil (settlement that occurs along path $AB$ in Fig. 4.3) can be expressed through the slopes of the curves in Fig. 4.3. We are going to define two slopes for primary consolidation. One is called the coefficient of compression or compression index, $C_c$, and is obtained from the plot of $e$ versus $\log \sigma_z'$ (Fig. 4.3b) as

$$C_c = -\frac{e_2 - e_1}{\log \dfrac{(\sigma_z')_2}{(\sigma_z')_1}} = \frac{|\Delta e|}{\log \dfrac{(\sigma_z')_2}{(\sigma_z')_1}} \quad \text{(no units)} \tag{4.7}$$

where the subscripts 1 and 2 denote two arbitrarily selected points on the NCL.

The other is called the modulus of volume compressibility, $m_v$, and is obtained from the plot of $\varepsilon_z$ versus $\sigma_z'$ (Fig. 4.3c) as

$$m_v = -\frac{(\varepsilon_z)_2 - (\varepsilon_z)_1}{(\sigma_z')_2 - (\sigma_z')_1} = \frac{|\Delta \varepsilon_z|}{(\sigma_z')_2 - (\sigma_z')_1} \left(\frac{\text{m}^2}{\text{kN}}\right) \tag{4.8}$$

where the subscripts 1 and 2 denote two arbitrarily selected points on the NCL.

Similarly, we can define the slope $BC$ in Fig. 4.3b as the recompression index, $C_r$, which we can express as

$$C_r = -\frac{e_2 - e_1}{\log \dfrac{(\sigma_z')_2}{(\sigma_z')_1}} = \frac{|\Delta e_{zr}|}{\log \dfrac{(\sigma_z')_2}{(\sigma_z')_1}} \tag{4.9}$$

where the subscripts 1 and 2 denote two arbitrarily selected points on the URL.

The slope $BC$ in Fig. 4.3c is called the modulus of volume recompressibility, $m_{vr}$, and is expressed as

$$m_{vr} = -\frac{(\varepsilon_z)_2 - (\varepsilon_z)_1}{(\sigma_z')_2 - (\sigma_z')_1} = \frac{|\Delta \varepsilon_{zr}|}{\Delta \sigma_z'} \left(\frac{\text{m}^2}{\text{kN}}\right) \tag{4.10}$$

where the subscripts 1 and 2 denote two arbitrarily selected points on the URL.

From Hooke's law, we know that Young's modulus of elasticity is

$$E_c' = \frac{\Delta \sigma_z'}{\Delta \varepsilon_z} \tag{4.11}$$

where the subscript $c$ denotes constrained because we are constraining the soil to settle only in one direction (one-dimensional consolidation). We can rewrite Eq. (4.11) as

$$E_c' = \frac{1}{m_{vr}} \tag{4.12}$$

The slopes $C_c$, $C_r$, $m_v$, and $m_{vr}$ are taken as positive values to satisfy our sign convention of compression or recompression as positive.

### 4.3.10 Effects of Loading History

In our experiment, we found that during reloading the soil follows the normal consolidation line once the past maximum vertical effective stress is exceeded. The history of loading of a soil is locked in its fabric and the soil maintains a memory of the past maximum effective stress. To understand how the soil will respond to loads, we have to unlock its memory. If a soil were to be consolidated to stresses below its past maximum vertical effective stress, then settlement would be small because the soil fabric was permanently changed by a higher stress in the past. However, if the soil were to be consolidated beyond its past maximum effective stress, settlement would be large for stresses beyond its past maximum effective stress because the soil fabric would now undergo further change from a current loading that is higher than its past maximum effective stress.

The practical significance of this soil behavior is that if the loading imposed on the soil by a structure is such that the vertical effective stress in the soil does not exceed its past maximum vertical effective stress, the settlement of the structure would be small, otherwise significant permanent settlement would occur. The preconsolidation stress defines the limit of elastic behavior. For stresses that are lower than the preconsolidation stress, the soil will follow the URL and we can reasonably assume that the soil will behave like an elastic material. For stresses greater than the preconsolidation stress the soil would behave like an elastoplastic material.

### 4.3.11 Overconsolidation Ratio

We will create a demarcation for soils based on their consolidation history. We will label a soil whose current vertical effective stress or overburden effective stress, $\sigma'_{zo}$, is less than its past maximum vertical effective stress or preconsolidation stress, $\sigma'_{zc}$, as an overconsolidated soil. An overconsolidated soil will follow a void ratio versus vertical effective stress path similar to $CDE$ (Fig. 4.3) during loading. The degree of overconsolidation, called overconsolidation ratio, OCR, is defined as

$$\text{OCR} = \frac{\sigma'_{zc}}{\sigma'_{zo}}$$ (4.13)

If OCR = 1, the soil is normally consolidated soil. Normally consolidated soils follow paths similar to $ABE$ (Fig. 4.3).

### 4.3.12 Possible and Impossible Consolidation Soil States

The normal consolidation line delineates possible from impossible soil states. Unloading of a soil or reloading it cannot bring it to soil states right of the normal consolidation line, which we will call impossible soil states (Fig. 4.3). Possible soil states only occur on or to the left of the normal consolidation line.

*The essential points are:*

1. *Path AB (Fig. 4.3), called the normal consolidation line (NCL), describes the response of a normally consolidated soil—a soil that has never experienced a vertical effective stress greater than its current vertical effective stress. The NCL is approximately a straight line in a plot of e versus log $\sigma'_z$ and is defined by a slope, $C_c$, called the compression index.*

2. *A normally consolidated soil would behave like an elastoplastic material. That is, part of the settlement under the load is recoverable while the other part is permanent.*

> **3.** *An overconsolidated soil has experienced vertical effective stresses greater than its current vertical effective stress.*
>
> **4.** *An overconsolidated soil will follow paths such as* **CDE** *(Fig. 4.3). For stresses below the preconsolidation stress, an overconsolidated soil would approximately behave like an elastic material and settlement would be small. However, for stresses greater than the preconsolidation stress, an overconsolidated soil will behave like an elastoplastic material, similar to a normally consolidated soil.*

*What's next* . . . Next, we will consider how to use the basic concepts to calculate one-dimensional settlement.

## 4.4  CALCULATION OF PRIMARY CONSOLIDATION SETTLEMENT

### 4.4.1 Effects of Unloading/Reloading of a Soil Sample Taken from the Field

Let us consider a soil sample that we wish to take from the field at a depth $z$ (Fig. 4.4a). We will assume that the groundwater level is at the surface. The current vertical effective stress or overburden effective stress is

$$\sigma'_{zo} = (\gamma_{sat} - \gamma_w)z = \gamma' z$$

and the current void ratio can be found from $\gamma_{sat}$ using Eq. (2.11). On a plot of $e$ versus $\log \sigma'_z$, the current vertical effective stress can be represented as $A$ as depicted in Fig. 4.4b.

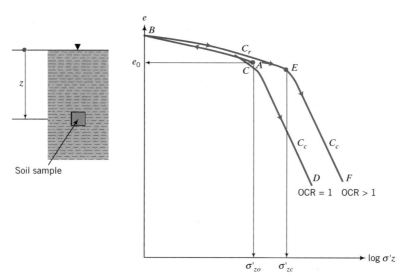

**FIGURE 4.4**   (a) Soil sample at a depth $z$ below ground surface. (b) Expected one-dimensional response.

To obtain a sample, we would have to make a borehole and remove the soil above it. The act of removing the soil and extracting the sample reduces the total stress to zero; that is, we have fully unloaded the soil. From the principle of effective stress [Eq. (3.39)], $\sigma' = -\Delta u$. Since $\sigma'$ cannot be negative—that is, soil cannot sustain tension—the pore water pressure must be negative. As the pore water pressure dissipates with time, volume changes (swelling) occur. Using the basic concepts of consolidation described in Section 4.3, the sample will follow an unloading path $AB$ (Fig. 4.4b). The point $B$ does not correspond to zero effective stress because we cannot represent zero on a logarithm scale. However, the effective stress level at the start of the logarithmic scale is assumed to be small ($\approx 0$). If we were to reload our soil sample, the reloading path followed depends on the OCR. If OCR $= 1$ (normally consolidated soil), the path followed during reloading would be $BCD$ (Fig. 4.4b). The average slope of $ABC$ is $C_r$. Once $\sigma'_{zo}$ is exceeded, the soil will follow the normal consolidation line, $CD$, of slope $C_c$. If the soil were overconsolidated, OCR $> 1$, the reloading path followed would be $BEF$ because we have to reload the soil beyond its preconsolidation stress before it behaves like a normally consolidated soil. The average slope of $ABE$ is $C_r$ and the slope of $EF$ is $C_c$. The point $E$ marks the preconsolidation stress. Later in this chapter, we will determine the position of $E$ from laboratory tests (Section 4.7).

### 4.4.2 Primary Consolidation Settlement of Normally Consolidated Fine-Grained Soils

Let us consider a site consisting of a normally consolidated soil on which we wish to construct a building. We will assume that the increase in vertical stress due to the building at depth $z$, where we took our soil sample, is $\Delta\sigma_z$. (Recall that you can find $\Delta\sigma_z$ using the methods described in Section 3.11.) The final vertical stress is

$$\sigma'_{\text{fin}} = \sigma'_{zo} + \Delta\sigma_z$$

The increase in vertical stress will cause the soil to settle following the NCL and the primary consolidation settlement is

$$\rho_{\text{pc}} = H_o \frac{\Delta e}{1 + e_o} = \frac{H_o}{1 + e_o} C_c \log \frac{\sigma'_{\text{fin}}}{\sigma'_{zo}}; \quad \text{OCR} = 1 \tag{4.14}$$

where $\Delta e = C_c \log(\sigma'_{\text{fin}}/\sigma'_{zo})$.

### 4.4.3 Primary Consolidation Settlement of Overconsolidated Fine-Grained Soils

If the soil is overconsolidated, we have to consider two cases depending on the magnitude of $\Delta\sigma_z$. We will approximate the curve in the log $\sigma'_z$ versus $e$ space as two straight lines, as shown in Fig. 4.5. In Case 1, the increase in $\Delta\sigma_z$ is such that $\sigma'_{\text{fin}} = \sigma'_{zo} + \Delta\sigma_z$ is less than $\sigma'_{zc}$ (Fig. 4.5a). In this case, consolidation occurs along the URL and

$$\rho_{\text{pc}} = \frac{H_o}{1 + e_o} C_r \log \frac{\sigma'_{\text{fin}}}{\sigma'_{zo}}; \quad \sigma'_{\text{fin}} < \sigma'_{zc} \tag{4.15}$$

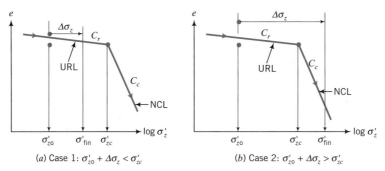

(a) Case 1: $\sigma'_{zo} + \Delta\sigma_z < \sigma'_{zc}$     (b) Case 2: $\sigma'_{zo} + \Delta\sigma_z > \sigma'_{zc}$

**FIGURE 4.5** Two cases to consider for calculating settlement of overconsolidated fine-grained soils.

In Case 2, the increase in $\Delta\sigma_z$ is such that $\sigma'_{fin} = \sigma'_{zo} + \Delta\sigma_z$ is greater than $\sigma'_{zc}$ (Fig. 4.5b). In this case, we have to consider two components of settlement—one along the URL and the other along the NCL. The equation to use in Case 2 is

$$\rho_{pc} = \frac{H_o}{1 + e_o}\left(C_r \log\frac{\sigma'_{zc}}{\sigma'_{zo}} + C_c \log\frac{\sigma'_{fin}}{\sigma'_{zc}}\right); \quad \sigma'_{fin} > \sigma'_{zc} \tag{4.16}$$

or

$$\rho_{pc} = \frac{H_o}{1 + e_o}\left[C_r \log(OCR) + C_c \log\frac{\sigma'_{fin}}{\sigma'_{zc}}\right]; \quad \sigma'_{fin} > \sigma'_{zc} \tag{4.17}$$

### 4.4.4 Procedure to Calculate Primary Consolidation Settlement

The procedure to calculate primary consolidation settlement is as follows:

1. Calculate the current vertical effective stress ($\sigma'_{zo}$) and the current void ratio ($e_o$) at the center of the soil layer for which settlement is required.
2. Calculate the applied vertical stress increase ($\Delta\sigma_z$) at the center of the soil layer using the appropriate method in Section 3.11.
3. Calculate the final vertical effective stress $\sigma'_{fin} = \sigma'_{zo} + \Delta\sigma_z$.
4. Calculate the primary consolidation settlement.
   a. If the soil is normally consolidated (OCR = 1), the primary consolidation settlement is

   $$\rho_{pc} = \frac{H_o}{1 + e_o}C_c \log\frac{\sigma'_{fin}}{\sigma'_{zo}}$$

   b. If the soil is overconsolidated and $\sigma'_{fin} < \sigma'_{zc}$, the primary consolidation settlement is

   $$\rho_{pc} = \frac{H_o}{1 + e_o}C_r \log\frac{\sigma'_{fin}}{\sigma'_{zo}}$$

**c.** If the soil is overconsolidated and $\sigma'_{fin} > \sigma'_{zc}$, the primary consolidation settlement is

$$\rho_{pc} = \frac{H_o}{1 + e_o} \left( C_r \log(\text{OCR}) + C_c \log \frac{\sigma'_{fin}}{\sigma'_{zc}} \right)$$

where $H_o$ is the thickness of the soil layer.

You can also calculate the primary consolidation settlement using $m_v$. However, unlike $C_c$, which is constant, $m_v$ varies with stress levels. You should compute an average value of $m_v$ over the stress range $\sigma'_{zo}$ to $\sigma'_{fin}$. The primary consolidation settlement, using $m_v$, is

$$\rho_{pc} = H_o m_v \Delta\sigma_z \tag{4.18}$$

The advantage of using Eq. (4.18) is that $m_v$ is readily determined from displacement data in consolidation tests; you do not have to calculate void ratio changes from the test data as required to determine $C_c$.

### 4.4.5 Thick Soil Layers

For better accuracy, when dealing with thick layers ($H_o > 2$ m), you should divide the soil layer into sublayers (about two to five sublayers) and find the settlement for each sublayer. Add up the settlement of each sublayer to find the total primary consolidation settlement. You must remember that the value of $H_o$ in the primary consolidation equations is the thickness of the sublayer. An alternative method is to use a harmonic mean value of the vertical stress increase for the sublayers in the equations for primary consolidation settlement. The harmonic mean stress increase is

$$\Delta\sigma_z = \frac{n(\Delta\sigma_z)_1 + (n-1)(\Delta\sigma_z)_2 + (n-2)(\Delta\sigma_z)_3 + \cdots + (\Delta\sigma_z)_n}{n + (n-1) + (n-2) + \cdots + 1} \tag{4.19}$$

where $n$ is the number of sublayers and the subscripts 1, 2, etc., mean the first (top) layer, the second layer, and so on. The advantage of using the harmonic mean is that the settlement is skewed in favor of the upper part of the soil layer. You should recall from Chapter 3 that the increase in vertical stress decreases with depth. Therefore, the primary consolidation settlement of the upper portion of the soil layer can be expected to be more than the lower portion because the upper portion of the soil layer is subjected to higher vertical stress increases.

### EXAMPLE 4.1

The soil profile at a site for a proposed office building consists of a layer of fine sand 10.4 m thick above a layer of soft normally consolidated clay 2 m thick. Below the soft clay is a deposit of coarse sand. The groundwater table was observed at 3 m below ground level. The void ratio of the sand is 0.76 and the water content of the clay is 43%. The building will impose a vertical stress increase of 140 kPa at the middle of the clay layer. Estimate the primary consolidation settlement of the clay. Assume the soil above the water table to be saturated, $C_c = 0.3$ and $G_s = 2.7$.

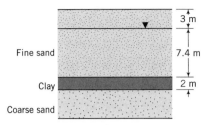

**FIGURE E4.1**

***Strategy*** You should write down what is given or known and draw a diagram of the soil profile (see Fig. E4.1). In this problem, you are given the stratigraphy, groundwater level, vertical stress increase, and the following soil parameters and soil condition:

$e_o$ (for sand) = 0.76; $w$ (for clay) = 43%

$H_o$ = 2 m, $\Delta\sigma_z$ = 140 kPa, $C_c$ = 0.3, $G_s$ = 2.7

Since you are given a normally consolidated clay, the primary consolidation settlement response of the soil will follow path *ABE* (Fig. 4.3). The appropriate equation to use is Eq. (4.14).

## Solution 4.1

**Step 1:** Calculate $\sigma'_{zo}$ and $e_o$ at the center of the clay layer.

$$\text{Sand:} \quad \gamma_{\text{sat}} = \left(\frac{G_s + e}{1 + e}\right)\gamma_w = \left(\frac{2.7 + 0.76}{1 + 0.76}\right)9.8 = 19.3 \text{ kN/m}^3$$

$$\gamma' = \left(\frac{G_s - 1}{1 + e}\right)\gamma_w = \left(\frac{2.7 - 1}{1 + 0.76}\right)9.8 = 9.5 \text{ kN/m}^3$$

$$\text{or} \quad \gamma' = \gamma_{\text{sat}} - \gamma_w = 19.3 - 9.8 = 9.5 \text{ kN/m}^3$$

$$\text{Clay:} \quad e_o = wG_s = 2.7 \times 0.43 = 1.16;$$

$$\gamma' = \left(\frac{G_s - 1}{1 + e}\right)\gamma_w = \left(\frac{2.7 - 1}{1 + 1.16}\right)9.8 = 7.7 \text{ kN/m}^3$$

The vertical effective stress at the mid-depth of the clay layer is

$$\sigma'_{zo} = (19.3 \times 3) + (9.5 \times 7.4) + (7.7 \times 1) = 135.9 \text{ kPa}$$

**Step 2:** Calculate the increase of stress at the mid-depth of the clay layer. You do not need to calculate $\Delta\sigma_z$ for this problem. It is given as $\Delta\sigma_z$ = 140 kPa.

**Step 3:** Calculate $\sigma'_{\text{fin}}$.

$$\sigma'_{\text{fin}} = \sigma'_{zo} + \Delta\sigma_z = 135.9 + 140 = 275.9 \text{ kPa}$$

**Step 4:** Calculate the primary consolidation settlement.

$$\rho_{\text{pc}} = \frac{H_o}{1 + e_o} C_c \log\frac{\sigma'_{\text{fin}}}{\sigma'_{zo}} = \frac{2}{1 + 1.16} \times 0.3 \log\frac{275.9}{135.9} = 0.085 \text{ m} = 85 \text{ mm}$$

∎

## EXAMPLE 4.2

Assume the same soil stratigraphy as in Example 4.1. But now the clay is over-consolidated with an OCR = 2.5, $w$ = 38%, and $C_r$ = 0.05. All other soil values given in Example 4.1 remain unchanged. Determine the primary consolidation settlement of the clay.

***Strategy*** Since the soil is overconsolidated, you will have to check whether the preconsolidation stress is less than or greater than the sum of the current vertical effective stress and the applied vertical stress at the center of the clay. This check will determine the appropriate equation to use. In this problem, the unit weight of the sand is unchanged but the clay has changed.

## Solution 4.2

**Step 1:** Calculate $\sigma'_{zo}$ and $e_o$ at mid-depth of the clay layer.
You should note that this settlement is small compared with the settlement obtained in Example 4.1.

Clay: $e_o = wG_s = 0.38 \times 2.7 = 1.03$

$$\gamma' = \left(\frac{G_s - 1}{1 + e}\right)\gamma_w = \left(\frac{2.7 - 1}{1 + 1.03}\right)9.8 = 8.2 \text{ kN/m}^3$$

$\sigma'_{zo} = (19.3 \times 3) + (9.5 \times 7.4) + (8.2 \times 1) = 136.4 \text{ kPa}$ (note that the increase in vertical effective stress from the unit weight change in this overconsolidated clay is very small)

**Step 2:** Calculate the preconsolidation stress.

$$\sigma'_{zc} = 136.4 \times 2.5 = 341 \text{ kPa}$$

**Step 3:** Calculate $\sigma'_{fin}$

$$\sigma'_{fin} = \sigma'_{zo} + \Delta\sigma_z = 136.4 + 140 = 276.4 \text{ kPa}$$

**Step 4:** Check if $\sigma'_{fin}$ is less than or greater than $\sigma'_{zc}$.

$$(\sigma'_{fin} = 276.4 \text{ kPa}) < (\sigma'_{zc} = 341 \text{ kPa})$$

Therefore, use Eq. (4.15).

**Step 5:** Calculate the total primary consolidation settlement.

$$\rho_{pc} = \frac{H_o}{1 + e_o} C_r \log \frac{\sigma'_{fin}}{\sigma'_{zo}} = \frac{2}{1 + 1.03} \times 0.05 \log \frac{276.4}{136.4} = 0.015 \text{ m} = 15 \text{ mm}$$

■

## EXAMPLE 4.3

Assume the same soil stratigraphy and soil parameters as in Example 4.2 except that the clay has an overconsolidation ratio of 1.5. Determine the primary consolidation settlement of the clay.

***Strategy*** Since the soil is overconsolidated, you will have to check whether the preconsolidation stress is less than or greater than the sum of the current vertical effective stress and the applied vertical stress at the center of the clay. This check will determine the appropriate equation to use.

## Solution 4.3

**Step 1:**  Calculate $\sigma'_{zo}$ and $e_o$.
From Example 4.2, $\sigma'_{zo} = 136.4$ kPa.

**Step 2:**  Calculate the preconsolidation stress.

$$\sigma'_{zc} = 136.4 \times 1.5 = 204.6 \text{ kPa}$$

**Step 3:**  Calculate $\sigma'_{\text{fin}}$.

$$\sigma'_{\text{fin}} = \sigma'_{zo} + \Delta\sigma_z = 136.4 + 140 = 276.4 \text{ kPa}$$

**Step 4:**  Check if $\sigma'_{\text{fin}}$ is less than or greater than $\sigma'_{zc}$.

$$(\sigma'_{\text{fin}} = 276.4 \text{ kPa}) > (\sigma'_{zc} = 204.6 \text{ kPa})$$

Therefore, use either Eq. (4.16) or (4.17).

**Step 5:**  Calculate the total primary consolidation settlement.

$$\rho_{pc} = \frac{H_o}{1 + e_o}\left\{ C_r \log \frac{\sigma'_{zc}}{\sigma'_{zo}} + C_c \log \frac{\sigma'_{\text{fin}}}{\sigma'_{zc}} \right\} = \frac{2}{1 + 1.03}$$

$$\times \left( 0.05 \log \frac{204.6}{136.4} + 0.3 \log \frac{276.4}{204.6} \right) = 0.047 \text{ m} = 47 \text{ mm}$$

or

$$\rho_{pc} = \frac{H_o}{1 + e_o}\left\{ C_r \log(\text{OCR}) + C_c \log \frac{\sigma'_{\text{fin}}}{\sigma'_{zc}} \right\} = \frac{1}{1 + 1.03}$$

$$\times \left( 0.05 \log 1.5 + 0.3 \log \frac{276.4}{204.6} \right) = 0.047 \text{ m} = 47 \text{ mm} \quad \blacksquare$$

## EXAMPLE 4.4

A vertical section through a building foundation at a site is shown in Fig. E4.4. The average modulus of volume compressibility of the clay is $m_v = 5 \times 10^{-5}$ m²/kN. Determine the primary consolidation settlement.

***Strategy***  To find the primary consolidation settlement, you need to know the vertical stress increase in the clay layer from the building load. Since the clay layer is finite, we will have to use the vertical stress influence values in Appendix B. If we assume a rough base, we can use the influence values specified by Milovic and Tournier (1971) or if we assume a smooth base we can use the values specified by Sovinc (1961). The clay layer is 10 m thick, so it is best to subdivide the clay layer into sublayers ≤2 m thick.

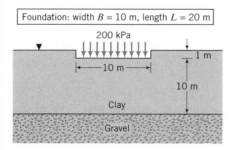

**FIGURE E4.4**

## Solution 4.4

**Step 1:**  Find the vertical stress increase at the center of the clay layer below the foundation.

Divide the clay layer into five sublayers, each of thickness 2 m—that is, $H_o = 2$ m. Find the vertical stress increase at the middle of each sublayer under the center of the rectangular foundation. Assume a rough base and use Table B1 (Appendix B).

$$B = 10 \text{ m}, L = 20 \text{ m}, \frac{L}{B} = 2, \quad q_s = 200 \text{ kPa}$$

| Layer | $z$ (m) | $\dfrac{z}{B}$ | $I_{zp}$ | $\Delta\sigma_z = I_{zp}q_s$ (kPa) |
|-------|---------|----------------|----------|-------------------------------------|
| 1 | 1 | 0.1 | 0.992 | 198.4 |
| 2 | 3 | 0.3 | 0.951 | 190.2 |
| 3 | 5 | 0.5 | 0.876 | 175.2 |
| 4 | 7 | 0.7 | 0.781 | 156.2 |
| 5 | 9 | 0.9 | 0.686 | 137.2 |

**Step 2:**  Calculate the primary consolidation settlement. Use Eq. (4.18).

$$\rho_{pc} = \sum_{i=1}^{n} (H_o m_v \, \Delta\sigma_z)_i = 2 \times 5 \times 10^{-5} \times (198.4 + 190.2 + 175.2$$
$$+ 156.2 + 137.2) = 0.086 \text{ m} = 86 \text{ mm}$$

*Alternatively:* Use the harmonic mean value of $\Delta\sigma_z$ with $n = 5$; that is, Eq. (4.19)

$$\Delta\sigma_z = \frac{5(198.4) + 4(190.2) + 3(175.2) + 2(156.2) + 1(137.2)}{5 + 4 + 3 + 2 + 1} = 181.9 \text{ kPa}$$

$$\rho_{pc} = 10 \times 5 \times 10^{-5} \times 181.9 = 0.091 \text{ m} = 91 \text{ mm}$$

The greater settlement in this method results from the bias toward the top layer. ∎

## EXAMPLE 4.5

A laboratory test on a saturated clay taken at a depth of 10 m below the ground surface gave the following results: $C_c = 0.3$, $C_r = 0.08$, OCR = 5, $w = 23\%$, and $G_s = 2.7$. The groundwater level is at the surface. Determine and plot the variation of water content and overconsolidation ratio with depth up to 50 m.

*Strategy*  The overconsolidation state lies on the unloading/reloading line (Fig. 4.3), so you need to find an equation for this line using the data given. Identify what given data is relevant to finding the equation for the unloading/reloading line. Here you are given the slope, $C_r$, so you need to use the other data to find the complete question. You can find the coordinate of one point on the unloading/reloading line from the water content and the depth as shown in Step 1.

## Solution 4.5

**Step 1:**   Determine $e_o$ and $\sigma'_{zo}$.

$$\gamma' = \left(\frac{G_s - 1}{1 + e}\right)\gamma_w = \left(\frac{2.7 - 1}{1 + 0.621}\right)9.8 = 10.3 \text{ kN/m}^3$$

$$e_o = G_s w = 2.7 \times 0.23 = 0.621$$

$$\sigma'_{zo} = \gamma' z = 10.3 \times 10 = 103 \text{ kPa}$$

**Step 2:**   Determine the preconsolidation stress.

$$\sigma'_{zc} = \sigma'_{zo} \times \text{OCR} = 103 \times 5 = 515 \text{ kPa}$$

**Step 3:**   Find the equation for the URL (slope $BC$ in Fig. E4.5a).

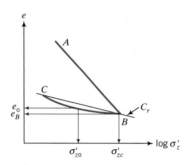

**FIGURE E4.5a**

$$e_B = e_o - C_r \log \frac{\sigma'_{zc}}{\sigma'_{zo}}$$

Therefore,

$$e_B = 0.621 - 0.08 \log(5) = 0.565$$

Hence, the equation for the unloading/reloading line is

$$e = 0.565 + 0.08 \log(\text{OCR}) \tag{1}$$

Substituting $e = wG_s$ ($G_s = 2.7$) and OCR $= 515/\gamma'z$ ($\gamma' = 10.3$ kN/m$^3$, $z$ is depth) in Eq. (1) gives

$$w = 0.209 + 0.03 \log\left(\frac{50}{z}\right)$$

You can now substitute values of $z$ from 1 to 50 and find $w$ and substitute $e = wG_s$ in Eq. (1) to find the OCR. The table below shows the calculated values and the results, which are plotted in Fig. E4.5b.

| Depth | w (%) | OCR |
|-------|-------|-----|
| 1  | 26.0 | 51.4 |
| 5  | 23.9 | 10   |
| 10 | 23.0 | 5    |
| 20 | 22.1 | 2.5  |
| 30 | 21.6 | 1.7  |
| 40 | 21.2 | 1.2  |
| 50 | 20.9 | 1.0  |

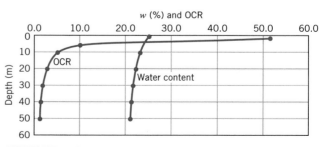

**FIGURE E4.5b**

You should note that the soil becomes normally consolidated as the depth increases. This is a characteristic of real soils.    ∎

*What's next* . . .So far, we have only considered how to determine the final primary consolidation settlement. This settlement might take months or years to occur, depending essentially on the permeability of the soil, the soil thickness, drainage conditions, and the magnitude of the applied stress. Geotechnical engineers have to know the magnitude of the final primary consolidation settlement and also the rate of settlement so that the settlement at any given time can be evaluated.

The next section deals with a theory to determine the settlement at any time. Several assumptions are made in developing this theory. However, you will see that many of the observations we made in Section 4.3 are well described by this theory.

## 4.5 ONE-DIMENSIONAL CONSOLIDATION THEORY

### 4.5.1 Derivation of Governing Equation

We now return to our experiment described in Section 4.3 to derive the theory for time rate of settlement using an element of the soil sample of thickness $dz$ and cross-sectional area $dA = dx\, dy$ (Fig. 4.6). We will assume the following:

1. The soil is saturated, isotropic, and homogeneous.
2. Darcy's law is valid.
3. Flow only occurs vertically.
4. The strains are small.

We will use the following observations made in Section 4.3:

1. The change in volume of the soil ($\Delta V$) is equal to the change in volume of pore water expelled ($\Delta V_w$), which is equal to the change in the volume of the voids ($\Delta V_v$). Since the area of the soil is constant (the soil is laterally constrained), the change in volume is directly proportional to the change in height.
2. At any depth, the change in vertical effective stress is equal to the change in excess pore water pressure at that depth. That is, $\partial\sigma_z' = \partial u$.

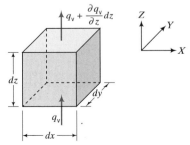

**FIGURE 4.6** One-dimensional flow through a two-dimensional soil element.

For our soil element in Fig. 4.6, the inflow of water is $q_v \, dA$ and the outflow over the elemental thickness $dz$ is $q_v + (\partial q_v/\partial z) \, dz \, dA$. The change in flow is then $(\partial q_v/\partial z) \, dz \, dA$. The rate of change in volume of water expelled, which is equal to the rate of change of volume of the soil, must equal the change in flow. That is,

$$\frac{\partial V}{\partial t} = \frac{\partial q_v}{\partial z} \, dz \, dA \tag{4.20}$$

Recall [Eq. (4.2)] that the volumetric strain $\varepsilon_p = \partial V/V = \partial e/(1 + e_o)$, and therefore

$$\partial V = \frac{\partial e}{1 + e_o} \, dz \, dA = m_v \, \partial \sigma_z' \, dz \, dA = m_v \, \partial u \, dz \, dA \tag{4.21}$$

Substituting Eq. (4.21) into Eq. (4.20) and simplifying, we obtain

$$\frac{\partial q_v}{\partial z} = \frac{\partial u}{\partial t} \, m_v \tag{4.22}$$

The one-dimensional flow of water from Darcy's law is

$$q_v = Ak_z i = Ak_z \frac{\partial h}{\partial z} \tag{4.23}$$

where $k_z$ is the coefficient of permeability in the vertical direction.
Partial differentiation of Eq. (4.23) with respect to $z$ gives

$$\frac{\partial q_v}{\partial z} = k_z \frac{\partial^2 h}{\partial z^2} \tag{4.24}$$

The pore water pressure at any time from our experiment in Section 4.3 is

$$u = h\gamma_w \tag{4.25}$$

where $h$ is the height of water in the burette.
Partial differentiation of Eq. (4.25) with respect to $z$ gives

$$\frac{\partial^2 h}{\partial z^2} = \frac{1}{\gamma_w} \frac{\partial^2 u}{\partial z^2} \tag{4.26}$$

By substitution of Eq. (4.26) into Eq. (4.24), we get

$$\frac{\partial q_v}{\partial z} = \frac{k_z}{\gamma_w} \frac{\partial^2 u}{\partial z^2} \tag{4.27}$$

Equating Eq. (4.22) and Eq. (4.27), we obtain

$$\frac{\partial u}{\partial t} = \frac{k_z}{m_v \gamma_w} \frac{\partial^2 u}{\partial z^2} \tag{4.28}$$

We can replace

$$\boxed{\frac{k_z}{m_v \gamma_w} \text{ by a coefficient } C_v \text{ called the coefficient of consolidation}}$$

The units for $C_v$ are length$^2$/time, for example, cm$^2$/min. Rewriting Eq. (4.28) by substituting $C_v$, we get the general equation for one-dimensional consolidation as

$$\boxed{\frac{\partial u}{\partial t} = C_v \frac{\partial^2 u}{\partial z^2}} \tag{4.29}$$

This equation describes the spatial variation of excess pore water pressure ($\Delta u$) with time ($t$) and depth ($z$). It is a common equation in many branches of engineering. For example, the heat diffusion equation commonly used in mechanical engineering is similar to Eq. (4.29) except that temperature, $T$, replaces $u$ and heat factor, $K$, replaces $C_v$. Equation (4.29) is sometimes called the Terzaghi one-dimensional consolidation equation because Terzaghi (1925) developed it.

In the derivation of Eq. (4.29), we tacitly assumed that $k_z$ and $m_v$ are constants. This is usually not the case because as the soil consolidates the void spaces are reduced and $k_z$ decreases. Also, $m_v$ is not linearly related to $\sigma'_z$ (Fig. 4.3c). The consequence of $k_z$ and $m_v$ not being constants is that $C_v$ is not a constant. In practice, $C_v$ is assumed to be a constant and this assumption is reasonable only if the stress changes are small enough such that $k_z$ and $m_v$ do not change significantly.

> *The* essential point *is:*
>
> 1. *The one-dimensional consolidation equation allows us to predict the changes in excess pore water pressure at various depths within the soil with time. We need to know the excess pore water pressure at a desired time because we have to determine the vertical effective stress to calculate the primary consolidation settlement.*

*What's next . . .*In the next section, the solution to the one-dimensional consolidation equation is found for the case where the soil can drain from the top and bottom boundaries using two methods. One method is based on the Fourier series and the other method is based on the finite difference numerical scheme. The latter is simpler in programming and in spreadsheet applications for any boundary condition.

### 4.5.2 Solution of Governing Consolidation Equation Using Fourier Series

The solution of any differential equation requires a knowledge of the boundary conditions. By specification of the initial distribution of excess pore water pressures at the boundaries, we can obtain solutions for the spatial variation of excess pore water pressure with time and depth. Various distributions of pore water pressures within a soil layer are possible. Two of these are shown in Fig. 4.7. One of these is a uniform distribution of initial excess pore water pressure with depth (Fig. 4.7a). This may occur in a thin layer of fine-grained soils. The other (Fig. 4.7b) is a triangular distribution. This may occur in a thick layer of fine-grained soils.

The boundary conditions for a uniform distribution of initial excess pore water pressure in which double drainage occurs are

When $t = 0$, $\Delta u = \Delta u_\mathrm{o} = \Delta \sigma_z$.

At the top boundary, $z = 0$, $\Delta u = 0$.

At the bottom boundary, $z = 2H_\mathrm{dr}$, $\Delta u = 0$, where $H_\mathrm{dr}$ is the length of the drainage path.

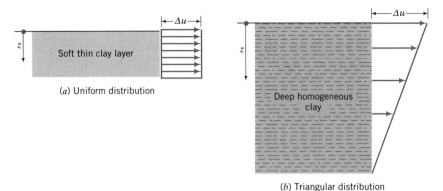

(a) Uniform distribution

(b) Triangular distribution

**FIGURE 4.7**   Two types of excess pore water pressure distribution with depth: (a) uniform distribution with depth in a thin layer and (b) triangular distribution with depth in a thick layer.

A solution for the governing consolidation equation (4.29), which satisfies these boundary conditions, is obtained using the Fourier series,

$$\Delta u(z,\,t) = \sum_{m=0}^{\infty} \frac{2\Delta u_{\mathrm{o}}}{M} \sin\left(\frac{Mz}{H_{\mathrm{dr}}}\right) \exp(-M^2 T_v) \tag{4.30}$$

where $M = (\pi/2)(2m + 1)$ and $m$ is a positive integer with values from 0 to $\infty$ and

$$T_v = \frac{C_v t}{H_{\mathrm{dr}}^2} \tag{4.31}$$

where $T_v$ is known as the time factor; it is a dimensionless term.

A plot of Eq. (4.30) gives the variation of excess pore water pressure with depth at different times. Let us examine Eq. (4.30) for an arbitrarily selected isochrone at any time $t$ or time factor $T_v$ as shown in Fig. 4.8. At time $t = 0$ ($T_v = 0$), the initial excess pore water pressure, $\Delta u_{\mathrm{o}}$, is equal to the applied vertical stress throughout the soil layer. As soon as drainage occurs, the initial excess pore water pressure will immediately fall to zero at the permeable boundaries. The maximum excess pore water pressure occurs at the center of the soil layer because the drainage path there is the longest, as obtained earlier in our experiment in Section 4.3.

At time $t > 0$, the total applied vertical stress increment $\Delta\sigma_z$ at a depth $z$ is equal to the sum of the vertical effective stress increment $\Delta\sigma_z'$ and the excess pore water pressure $\Delta u_z$. After considerable time ($t \to \infty$), the excess pore water pressure decreases to zero and the vertical effective stress increment becomes equal to the vertical total stress increment.

We now define a parameter, $U_z$, called the degree of consolidation or consolidation ratio, which gives us the amount of consolidation completed at a particular time and depth. This parameter can be expressed mathematically as

$$U_z = 1 - \frac{\Delta u_z}{\Delta u_{\mathrm{o}}} = 1 - \sum_{m=0}^{\infty} \frac{2}{M} \sin\left(\frac{Mz}{H_{\mathrm{dr}}}\right) \exp(-M^2 T_v) \tag{4.32}$$

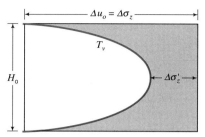

**FIGURE 4.8** An isochrone illustrating the theoretical excess pore water pressure distribution with depth.

The consolidation ratio is equal to zero everywhere at the beginning of the consolidation ($\Delta u_z = \Delta u_o$) but increases to unity as the initial excess pore water pressure dissipates.

A geotechnical engineer is often concerned with the average degree of consolidation, $U$, of a whole layer at a particular time rather than the consolidation at a particular depth. The shaded area in Fig. 4.8 represents the amount of consolidation of a soil layer at any given time. The average degree of consolidation can be expressed mathematically from the solution of the one-dimensional consolidation equation as

$$U = 1 - \sum_{m=0}^{\infty} \frac{2}{M^2} \exp(-M^2 T_v) \tag{4.33}$$

Figure 4.9 shows the variation of the average degree of consolidation with time factor $T_v$ for a uniform and a triangular distribution of excess pore water pressure.

A convenient set of equations for double drainage, found by curve fitting Fig. 4.9, is

$$T_v = \frac{\pi}{4}\left(\frac{U}{100}\right)^2 \quad \text{for } U < 60\% \tag{4.34}$$

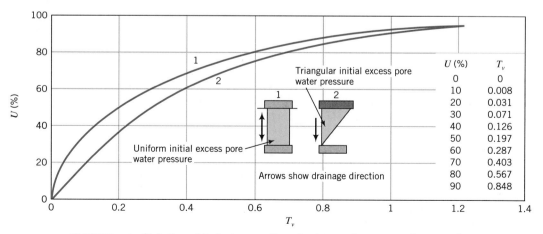

**FIGURE 4.9** Relationship between time factor and average degree of consolidation for a uniform distribution and a triangular distribution of initial excess pore water pressure.

and

$$T_v = 1.781 - 0.933 \log(100 - U) \quad \text{for } U \geq 60\% \tag{4.35}$$

The time factor corresponding to every 10% of average degree of consolidation for double drainage conditions is shown in the inset table in Fig. 4.9. The time factors corresponding to 50% and 90% consolidation are often used in interpreting consolidation test results. You should remember that $T_v = 0.848$ for 90% consolidation, and $T_v = 0.197$ for 50% consolidation.

### 4.5.3 Finite Difference Solution of the Governing Consolidation Equation

Numerical methods (finite difference, finite element, and boundary element) provide approximate solutions to differential and integral equations for boundary conditions in which closed form solutions (analytical solutions) are not possible. We will use the finite difference method here to find a solution to the consolidation equation because it involves only the expansion of the differential equation using Taylor's theorem and can easily be adopted for spreadsheet applications.

Using Taylor's theorem,

$$\frac{\partial u}{\partial t} = \frac{1}{\Delta t}(u_{i,j+1} - u_{i,j}) \tag{4.36}$$

and

$$\frac{\partial^2 u}{\partial z^2} = \frac{1}{(\Delta z)^2}(u_{i-1,j} - 2u_{i,j} + u_{i+1,j}) \tag{4.37}$$

where $(i, j)$ denotes a nodal position at the intersection of row $i$ and column $j$. Columns represent time divisions and rows represent soil depth divisions. The assumption implicit in Eq. (4.36) is that the excess pore water pressure between two adjacent nodes changes linearly with time. This assumption is reasonable if the distance between the two nodes is small. Substituting Eqs. (4.36) and (4.37) in the governing consolidation equation (4.29) and rearranging, we get

$$u_{i,j+1} = u_{i,j} + \frac{C_v \, \Delta t}{(\Delta z)^2}(u_{i-1,j} - 2u_{i,j} + u_{i+1,j}) \tag{4.38}$$

Equation (4.38) is valid for nodes that are not boundary nodes. There are special conditions that apply to boundary nodes. For example, at an impermeable boundary, no flow across it can occur and, consequently, $\partial u / \partial z = 0$ for which the finite difference equation is

$$\frac{\partial u}{\partial z} = 0 = \frac{1}{2\Delta z}(u_{i-1,j} - u_{i+1,j}) = 0 \tag{4.39}$$

and the governing consolidation equation becomes

$$u_{i,j+1} = u_{i,j} + \frac{C_v \, \Delta t}{(\Delta z)^2}(2u_{i-1,j} - 2u_{i,j}) \tag{4.40}$$

To determine how the pore water pressure is distributed within a soil at a given time, we have to establish the initial excess pore water pressure at the boundaries. Once we do this, we have to estimate the variation of the initial excess pore water pressure within the soil. We may, for example, assume a linear distribution of initial excess pore water pressure with depth if the soil layer is thin or a triangular distribution for a thick soil layer. If you cannot estimate the initial excess pore water pressure, then you can guess reasonable values for the interior of the soil or use linear interpolation. Then you successively apply Eq. (4.38) to each interior nodal point and replace the old value by the newly calculated value until the old value and the new value differ by a small tolerance. At impermeable boundaries, you have to apply Eq. (4.40).

The procedure to apply the finite difference form of the governing consolidation equation to determine the variation of excess pore water pressure with time and depth is as follows:

1. Divide the soil layer into a depth–time grid (Fig. 4.10). Rows represent subdivisions of the depth, columns represent subdivisions of time. Let's say we divide the depth into $m$ rows and the time into $n$ columns; then $\Delta z = H_o/m$ and $\Delta t = t/n$, where $H_o$ is the thickness of the soil layer and $t$ is the total time. A nodal point represents the $i$th depth position and the $j$th elapsed time. To avoid convergence problems, researchers have found that $\alpha = C_v \Delta t/(\Delta z)^2$ must be less than $\frac{1}{2}$. This places a limit on the number of subdivisions in the grid. Often, the depth is subdivided arbitrarily and the time step $\Delta t$ is selected so that $\alpha < \frac{1}{2}$. In many practical situations, $\alpha = 0.25$ usually ensures convergence.

2. Identify the boundary conditions. For example, if the top boundary is a drainage boundary then the excess pore water pressure there is zero. If, however, the top boundary is an impermeable boundary, then no flow can occur across it and Eq. (4.40) applies.

3. Estimate the distribution of initial excess pore water pressure and determine the nodal initial excess pore water pressures.

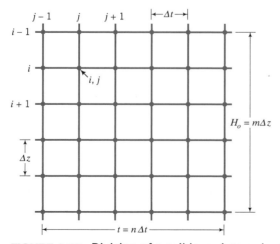

**FIGURE 4.10** Division of a soil layer into a depth (row)–time (column) grid.

4. Calculate the excess pore water pressure at interior nodes using Eq. (4.38) and at impermeable boundary nodes using Eq. (4.40). If the boundary is permeable, then the excess pore water pressure is zero at all nodes on this boundary.

> *The* **essential points** *are:*
>
> 1. *An ideal soil (isotropic, homogeneous, saturated) is assumed in developing the governing one-dimensional consolidation equation.*
> 2. *Strains are assumed to be small.*
> 3. *Excess pore water pressure dissipation depends on the time, soil thickness, drainage conditions, and permeability of the soil.*
> 4. *The decrease in initial excess pore water pressure causes an equivalent increase in vertical effective stress and settlement increases.*
> 5. *The average degree of consolidation is conventionally used to find the time rate of settlement.*

### EXAMPLE 4.6

A soft clay layer 1.5 m thick is sandwiched between layers of sand. The initial vertical total stress at the center of the clay layer is 200 kPa and the pore water pressure is 100 kPa. The increase in vertical stress at the center of the clay layer from a building foundation is 100 kPa. What is the vertical effective stress and excess pore water pressure at the center of the clay layer when 60% consolidation occurs?

**Strategy**  You are given the increment in applied stress and the degree of consolidation. You know that the initial change in excess pore water pressure is equal to the change in applied vertical stress. From the data given decide on the appropriate equation, which in this case is Eq. (4.32).

### Solution 4.6

**Step 1:**  Calculate the initial excess pore water pressure.

$$\Delta u_o = \Delta \sigma_z = 100 \text{ kPa}$$

**Step 2:**  Calculate the current excess pore water pressure at 60% consolidation. From Eq. (4.32),

$$\Delta u_z = \Delta u_o (1 - U_z) = 100(1 - 0.6) = 40 \text{ kPa}$$

**Step 3:**  Calculate the vertical total stress and total excess pore water pressure.

$$\text{Vertical total stress:} \quad \Delta \sigma_z = 200 + 100 = 300 \text{ kPa}$$
$$\text{Total pore water pressure:} \quad 100 + 40 = 140 \text{ kPa}$$

**Step 4:**  Calculate the current vertical effective stress.

$$\sigma_z' = \sigma_z - \Delta u_z = 300 - 140 = 160 \text{ kPa}$$

*Alternatively:*

**Step 1:** Calculate the initial vertical effective stress.

$$\text{Initial vertical effective stress} = 200 - 100 = 100 \text{ kPa}$$

**Step 2:** Same as Step 2 above.

**Step 3:** Calculate the increase in vertical effective stress at 60% consolidation.

$$\Delta\sigma_z' = 100 - 40 = 60 \text{ kPa}$$

**Step 4:** Calculate the current vertical effective stress.

$$\sigma_z' = 100 + 60 = 160 \text{ kPa} \qquad\blacksquare$$

## EXAMPLE 4.7

A layer of soft clay, 5 m thick, is drained at the top surface only. The initial excess pore water pressure from an applied load at time $t = 0$ is distributed according to $\Delta u_o = 80 - 2z^2$, where $z$ is the depth measured from the top boundary. Determine the distribution of excess pore water pressure with depth after 6 months using the finite difference method. The coefficient of consolidation, $C_v$, is $8 \times 10^{-4}$ cm$^2$/s.

**Strategy** Divide the clay layer into, say, five equal layers of 1 m thickness and find the value of the initial excess pore water pressure at each node at time $t = 0$ using $\Delta u_o = 80 - 2z^2$. Then find a time step $\Delta t$ that will lead to $\alpha < \frac{1}{2}$. The top boundary is a drainage boundary. Therefore, the excess pore water pressure is zero for all times at all nodes along this boundary. Use a spreadsheet or write a short program to do the iterations to solve the governing consolidation equation. You must note that the bottom boundary is not a drainage boundary and the relevant equation to use for the nodes at this boundary is Eq. (4.40).

## Solution 4.7

**Step 1:** Divide the soil layer into a grid.
Divide the depth into five layers:

$$\Delta z = \frac{5}{5} = 1 \text{ m}$$

$$C_v = 8 \times 10^{-4} \text{ cm}^2/\text{s} = 8 \times 10^{-8} \text{ m}^2/\text{s} = 2.52 \text{ m}^2/\text{yr}$$

Assume $\Delta t = 0.1$ yr.

$$\alpha = \frac{C_v \, \Delta t}{\Delta z^2} = \frac{2.52 \times 0.1}{1^2} = 0.25 \ < 0.5$$

**Step 2:** Identify boundary conditions.
The bottom boundary is impermeable, therefore Eq. (4.40) applies to the nodes along this boundary. The top boundary is pervious, therefore the excess pore water pressure is zero at all times greater than zero.

**Step 3:** Determine the distribution of initial excess pore water pressure. You are given the distribution of initial excess pore water pressure as $\Delta u_o = 80 - 2z^2$. At time $t = 0$ (column 1), insert the nodal values of initial excess for water pressure. For example, at row 2, column 1 (node 2), $\Delta u_o = 80 - 2 \times 1^2 = 78$ kPa. The initial excess pore water pressures are listed in column 1; see the table below.

**Step 4:** Calculate the excess pore water pressure at each node of the grid. The governing equation, except at the impermeable boundary, is

$$u_{i,j+1} = u_{i,j} + 0.25(u_{i-1,j} + u_{i+1,j})$$

Let us calculate the excess pore water pressure after 0.1 yr at the node located at row 2, column 2.

$$u_{2,1+1} = 78 + 0.25(0 - 2 \times 78 + 72) = 57 \text{ kPa}$$

At the bottom impermeable boundary (row 6, column 2), we have to apply Eq. (4.40), which is

$$u_{i,j+1} = u_{i,j} + 0.25(2u_{i-1,j} - 2u_{i,j})$$

Therefore,

$$u_{2,1+1} = 30 + 0.25(2 \times 48 - 2 \times 30) = 39 \text{ kPa}$$

The complete results, after 6 months, using a spreadsheet program are shown in the table below. The results are plotted in Fig. E4.7.

| Column | 1 | 2 | 3 | 4 | 5 |
|---|---|---|---|---|---|
| | | | Time (yr) | | |
| Depth (m) | 0.00 | 0.10 | 0.20 | 0.30 | 0.50 |
| 0.0 | 0.0 | 0.0 | 0.0 | 0.0 | 0.0 |
| 1.0 | 78.0 | 57 | 46.3 | 39.4 | 34.5 |
| 2.0 | 72.0 | 71 | 65.0 | 59.1 | 54.0 |
| 3.0 | 62.0 | 61 | 60.0 | 58.4 | 56.5 |
| 4.0 | 48.0 | 47 | 48.5 | 50.0 | 51.0 |
| 5.0 | 30.0 | 39 | 43.0 | 45.8 | 47.9 |

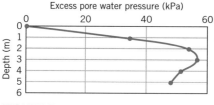

**FIGURE E4.7**

*What's next . . .* We have only described primary consolidation settlement. The other part of the total consolidation settlement is secondary compression, which will be discussed next.

## 4.6   SECONDARY COMPRESSION SETTLEMENT

You will recall from our experiment in Section 4.3 that consolidation settlement consisted of two parts. The first part is primary consolidation, which occurs at early times. The second part is secondary compression, or creep, which takes place under a constant vertical effective stress. The physical reasons for secondary compression in soils are not fully understood. One plausible explanation is the expulsion of water from micropores; another is viscous deformation of the soil structure.

We can make a plot of void ratio versus the logarithm of time from our experimental data in Section 4.3, as shown in Fig. 4.11. Primary consolidation is assumed to end at the intersection of the projection of the two straight parts of the curve.

The secondary compression index is

$$C_\alpha = -\frac{(e_t - e_p)}{\log(t/t_p)} = \frac{|\Delta e|}{\log(t/t_p)}; t > t_p \qquad (4.41)$$

where $(t_p, e_p)$ is the coordinate at the intersection of the tangents to the primary consolidation and secondary compression parts of the void ratio versus logarithm of time curve and $(t, e_t)$ is the coordinate of any point on the secondary compression curve as shown in Fig. 4.11. The secondary consolidation settlement is

$$\rho_{sc} = \frac{H_o}{(1 + e_p)} C_\alpha \log\left(\frac{t}{t_p}\right) \qquad (4.42)$$

Overconsolidated soils do not creep significantly but creep settlements in normally consolidated soils can be very significant.

*What's next . . .*You will recall that there are several unknown parameters in the theoretical solution of the one-dimensional consolidation equation. We need to know these parameters to calculate consolidation settlement and rate of settlement for fine-grained soils. The one-dimensional consolidation test is used to find these parameters. This test is usually one of the tests that you will perform in the laboratory component of your course. In the next section, the test procedures will be briefly discussed followed by the methods used to determine the various soil consolidation parameters.

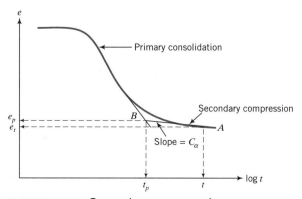

**FIGURE 4.11**   Secondary compression.

## 4.7   ONE-DIMENSIONAL CONSOLIDATION LABORATORY TEST

### 4.7.1 Oedometer Test

The one-dimensional consolidation test, called the oedometer test, is used to find $C_c$, $C_r$, $C_\alpha$, $C_v$, $m_v$, and $\sigma'_{zc}$. The coefficient of permeability, $k_z$, can also be calculated from the test data. The experimental arrangement we used in Section 4.3 is similar to an oedometer test setup. The details of the test apparatus and the testing procedures are described in ASTM D2435. A disk of soil is enclosed in a stiff metal ring and placed between two porous stones in a cylindrical container filled with water, as shown in Fig. 4.12. A metal load platen mounted on top of the upper porous stone transmits the applied vertical stress (vertical total stress) to the soil sample. Both the metal platen and the upper porous stone can move vertically inside the ring as the soil settles under the applied vertical stress. The ring containing the soil sample can be fixed to the container by a collar (fixed ring cell, Fig. 4.12b) or is unrestrained (floating ring cell, Fig. 4.12c).

Incremental loads, including unloading sequences, are applied to the platen and the settlement of the soil at various fixed times, under each load increment, is measured by a displacement gauge. Each loading increment is allowed to remain on the soil until the change in settlement is negligible and the excess pore water pressure developed under the current load increment has dissipated. For many soils, this usually occurs within 24 hours, but longer monitoring times may be required for exceptional soil types, for example, montmorillonite. Each loading increment is doubled. The ratio of the load increment to the previous load is called the load increment ratio (LIR); conventionally, LIR = 1. To determine $C_r$, the soil sample is unloaded using a load decrement ratio—load decrement divided by current load—of 2.

At the end of the oedometer test, the apparatus is dismantled and the water content of the sample is determined. It is best to unload the soil sample to a small pressure before dismantling the apparatus, because if you remove the final

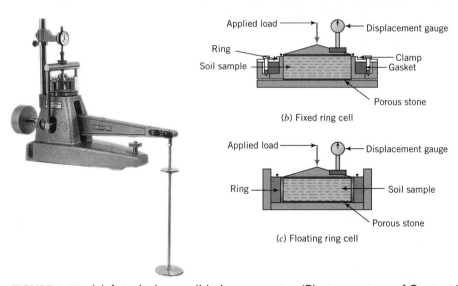

**FIGURE 4.12**   (a) A typical consolidation apparatus (Photo courtesy of Geotest.) (b) a fixed ring cell and (c) a floating ring cell.

consolidation load completely, a negative excess pore water pressure that equals the final consolidation pressure would develop. This negative excess pore water pressure can cause water to flow into the soil and increase the soil's water content. Consequently, the final void ratio calculated from the final water content would be erroneous.

The data obtained from the one-dimensional consolidation test are as follows:

1. Initial height of the soil, $H_o$, which is fixed by the height of the ring.
2. Current height of the soil at various time intervals under each load (time–settlement data).
3. Water content at the beginning and at the end of the test, and the dry weight of the soil at the end of the test.

You now have to use these data to determine $C_c$, $C_r$, $C_\alpha$, $C_v$, $m_v$, and $\sigma'_{zc}$. We will start with finding $C_v$.

### 4.7.2 Determination of the Coefficient of Consolidation

There are two popular methods that can be used to calculate $C_v$. Taylor (1942) proposed one method called the root time method. Casagrande and Fadum (1940) proposed the other method called the log time method. The root time method utilizes the early time response, which theoretically should appear as a straight line in a plot of square root of time versus displacement gauge reading.

*4.7.2.1 Root Time Method* Let us arbitrarily choose a point, $C$, on the displacement versus square root of time factor gauge reading as shown in Fig. 4.13. We will assume that this point corresponds to 90% consolidation ($U = 90\%$)

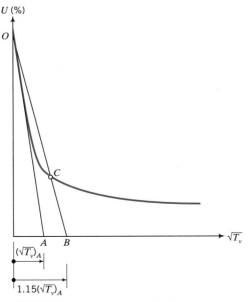

**FIGURE 4.13** Correction of laboratory early time response to determine $C_v$.

for which $T_v = 0.848$ (Fig. 4.9). If point $C$ were to lie on a straight line, the theoretical relationship between $U$ and $T_v$ would be $U = 0.98\sqrt{T_v}$; that is, if you substitute $T_v = 0.848$, you get $U = 90\%$.

At early times, the theoretical relationship between $U$ and $T_v$ is given by Eq. (4.34); that is,

$$U = \sqrt{\frac{4}{\pi} T_v} = 1.13\sqrt{T_v}; \quad U < 0.6$$

The laboratory early time response is represented by the straight line $OA$ in Fig. 4.13. You should note that $O$ is below the initial displacement gauge reading because there is an initial compression of the soil before consolidation begins. The ratio of the gradient of $OA$ and the gradient of the theoretical early time response, line $OCB$, is

$$\frac{1.13\sqrt{T_v}}{0.98\sqrt{T_v}} = 1.15$$

We can use this ratio to establish the time when 90% consolidation is achieved in the one-dimensional consolidation test.

The procedure, with reference to Fig. 4.14, is as follows:

1. Plot the displacement gauge readings versus square root of times.
2. Draw the best straight line through the initial part of the curve intersecting the ordinate (displacement reading) at $O$ and the abscissa ($\sqrt{\text{time}}$) at $A$.
3. Note the time at point $A$; let us say it is $\sqrt{t_A}$.
4. Locate a point $B$, $1.15\sqrt{t_A}$, on the abscissa.
5. Join $OB$.

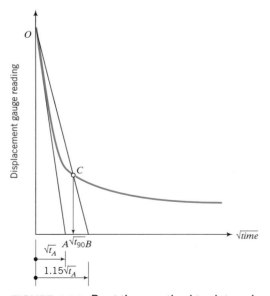

**FIGURE 4.14** Root time method to determine $C_v$.

6. The intersection of the line *OB* with the curve, point *C*, gives the displacement gauge reading and the time for 90% consolidation ($t_{90}$). You should note that the value read off the abscissa is $\sqrt{t_{90}}$. Now when $U = 90\%$, $T_v = 0.848$ (Fig. 4.9) and from Eq. (4.31) we obtain

$$C_v = \frac{0.848 H_{dr}^2}{t_{90}} \qquad (4.43)$$

where $H_{dr}$ is the length of the drainage path.

### 4.7.2.2 Log Time Method
In the log time method, the displacement gauge readings are plotted against the logarithm of times. A typical curve obtained is shown in Fig. 4.15. The theoretical early time settlement response in a plot of logarithm of times versus displacement gauge readings is a parabola (Section 4.3). The experimental early time curve is not normally a parabola and a correction is often required.

The procedure, with reference to Fig. 4.15, is as follows:

1. Project the straight portions of the primary consolidation and secondary compression to intersect at *A*. The ordinate of *A*, $d_{100}$, is the displacement gauge reading for 100% primary consolidation.
2. Correct the initial portion of the curve to make it a parabola. Select a time $t_1$, point *B*, near the head of the initial portion of the curve ($U < 60\%$) and then another time $t_2$, point *C*, such that $t_2 = 4t_1$.
3. Calculate the difference in displacement reading, $\Delta d = d_2 - d_1$, between $t_2$ and $t_1$. Plot a point *D* at a vertical distance $\Delta d$ from *B*. The ordinate of point *D* is the corrected initial displacement gauge reading, $d_o$, at the beginning of primary consolidation.
4. Calculate the ordinate for 50% consolidation as $d_{50} = (d_{100} + d_o)/2$. Draw a horizontal line through this point to intersect the curve at *E*. The abscissa of point *E* is the time for 50% consolidation, $t_{50}$.

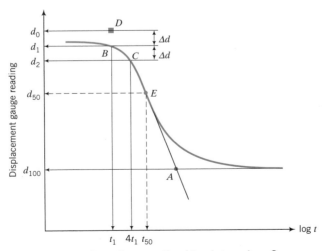

**FIGURE 4.15**   Log time method to determine $C_v$.

**5.** You will recall (Fig. 4.9) that the time factor for 50% consolidation is 0.197 and from Eq. (4.31) we obtain

$$C_v = \frac{0.197 H_{dr}^2}{t_{50}}$$

(4.44)

The log time method makes use of the early (primary consolidation) and later time responses (secondary compression) while the root time method only utilizes the early time response, which is expected to be a straight line. In theory, the root time method should give good results except when nonlinearities arising from secondary compression cause substantial deviations from the expected straight line. These deviations are most pronounced in fine-grained soils with organic materials.

### 4.7.3 Determination of Void Ratio at the End of a Loading Step

To determine $\sigma'_{zc}$, $C_c$, $C_r$, and $C_\alpha$, we need to know the void ratio for each loading step. Recall that at the end of the consolidation test, we determined the water content ($w$) of the soil sample. Using this data, initial height ($H_o$), and the specific gravity ($G_s$) of the soil sample, you can calculate the void ratio for each loading step as follows:

**1.** Calculate the final void ratio, $e_{fin} = wG_s$.
**2.** Calculate the total consolidation settlement of the soil sample during the test, $(\Delta z)_{fin} = d_{fin} - d_i$, where $d_{fin}$ is the final displacement gauge reading and $d_i$ is the displacement gauge reading at the start of the test.
**3.** Back-calculate the initial void ratio, $e_o = e_{fin} + \Delta e$, where $\Delta e$ is found from Eq. (4.4) as

$$\Delta e = \frac{(\Delta z)_{fin}}{H_o}(1 + e_o)$$

Therefore,

$$e_o = \frac{e_{fin} + \dfrac{(\Delta z)_{fin}}{H_o}}{1 - \dfrac{(\Delta z)_{fin}}{H_o}}$$

**4.** Calculate $e$ for each loading step using Eq. (4.6).

### 4.7.4 Determination of the Preconsolidation Stress

Now that we have calculated $e$ for each loading step, we can plot a graph of the void ratio versus the logarithm of vertical effective stress as shown in Fig. 4.16. We will call Fig. 4.16 the $e$ versus $\log \sigma'_z$ curve. You will now determine the preconsolidation stress using a method proposed by Casagrande (1936).

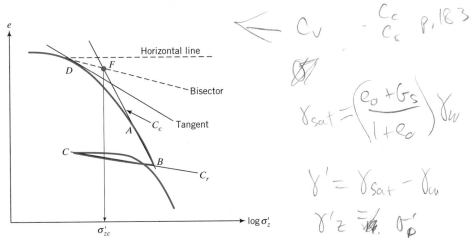

**FIGURE 4.16** Determination of the preconsolidation stress using Casagrande's method.

The procedure, with reference to Fig. 4.16, is as follows:

1. Identify the point of maximum curvature, point $D$, on the initial part of the curve.
2. Draw a horizontal line through $D$.
3. Draw a tangent to the curve at $D$.
4. Bisect the angle formed by the tangent and the horizontal line at $D$.
5. Extend backward the straight portion of the curve (the normal consolidation line), $BA$, to intersect the bisector line at $F$.
6. The abscissa of $F$ is the preconsolidation effective stress, $\sigma'_{zc}$.

A simpler method that is also used in practice is to project the straight portion of the initial recompression curve to intersect the backward projection of the normal consolidation line at $F$ as shown in Fig. 4.17. The abscissa of $F$ is $\sigma'_{zc}$.

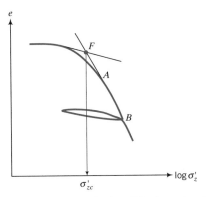

**FIGURE 4.17** A simplified method of determining the preconsolidation stress.

Both of these methods are based on individual judgment. The actual value of $\sigma'_{zc}$ for real soils is more difficult to ascertain than described above. Degradation of the soil from its intact condition caused by sampling, transportation, handling, and sample preparation usually does not produce the ideal curve shown in Fig. 4.16.

### 4.7.5 Determination of Compression and Recompression Indices

The slope of the normal consolidation line, $BA$, gives the compression index, $C_c$. To determine the recompression index, $C_r$, draw a line $(BC)$ approximately midway between the unloading and reloading curves (Fig. 4.16). The slope of this line is the recompression index $(C_r)$.

Field observations indicate that, in many instances, the predictions of the total settlement and the rate of settlement using oedometer test results do not match recorded settlement from actual structures. Disturbances from sampling and sample preparation tend to decrease $C_c$ and $C_v$. Schmertmann (1953) suggested a correction to the laboratory curve to obtain a more representative in situ value of $C_c$. His method is as follows. Locate a point $A$ at coordinate $(\sigma'_{zo}, e_o)$ and a point $B$ at ordinate $0.42e_o$ on the laboratory $e$ versus log $\sigma'_z$ curve, as shown in Fig. 4.18. The slope of the line $AB$ is the corrected value for $C_c$.

### 4.7.6 Determination of the Modulus of Volume Change

The modulus of volume compressibility, $m_v$, is found from plotting a curve similar to Fig. 4.3c and determining the slope as shown in this figure. You do not need to calculate void ratio to determine $m_v$. You need the final change in height at the end of each loading $(\Delta z)$ and then you calculate the vertical strain, $\varepsilon_z = \Delta z/H_o$, where $H_o$ is the initial height. The modulus of volume compressibility is not constant but depends on the range of vertical effective stress that is used in the calculation. A representative value for $m_v$ can be obtained by finding the slope between the current vertical effective stress and the final vertical effective stress $(\sigma'_{zo} + \Delta\sigma_z)$ at the center of the soil layer in the field or 100 kPa, whichever is less.

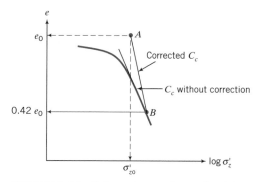

**FIGURE 4.18**  Schmertmann's method to correct $C_v$ for soil disturbances.

## 4.7.7 Determination of the Secondary Compression Index

The secondary compression index, $C_\alpha$, can be found by making a plot similar to Fig. 4.11. You should note that Fig. 4.11 is for a single load. The value of $C_\alpha$ usually varies with the magnitude of the applied loads and other factors such as the LIR.

*What's next* . . .Three examples and their solutions are presented next to show you how to find various consolidation soil parameters as discussed above. The first two examples are intended to illustrate the determination of the compression indices and how to use them to make predictions. The third example illustrates how to find $C_v$ using the root time method.

### EXAMPLE 4.8

At a vertical stress of 200 kPa, the void ratio of a saturated soil sample tested in an oedometer is 1.52 and lies on the normal consolidation line. An increment of vertical stress of 150 kPa compresses the sample to a void ratio of 1.43.

**(1)** Determine the compression index $C_c$ of the soil.

**(2)** The sample was unloaded to a vertical stress of 200 kPa and the void ratio increased to 1.45. Determine the slope of the recompression index, $C_r$.

**(3)** What is the overconsolidation ratio of the soil at stage (2)?

**(4)** If the soil were reloaded to a vertical stress of 500 kPa, what void ratio would be attained?

**Strategy**    Draw a sketch of the soil response on an $e$ versus $\log \sigma'_z$ curve. Use this sketch to answer the various questions.

### Solution 4.8

**Step 1:**    Determine $C_c$.

$C_c$ is the slope $AB$ shown in Fig. E4.8.

$$C_c = \frac{-(1.43 - 1.52)}{\log(350/200)} = 0.37$$

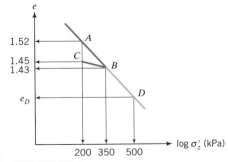

**FIGURE E4.8**

**Step 2:**   Determine $C_r$.
$C_r$ is the slope of $BC$ in Fig. E4.8.

$$C_r = \frac{-(1.43 - 1.45)}{\log(350/200)} = 0.08$$

**Step 3:**   Determine the overconsolidation ratio.

Preconsolidation effective stress:   $\sigma'_{zc} = 350$ kPa

Current vertical effective stress:   $\sigma'_z = 200$ kPa

$$\text{OCR} = \frac{\sigma'_{zc}}{\sigma'_z} = \frac{350}{200} = 1.75$$

**Step 4:**   Calculate the void ratio at 500 kPa.
The void ratio at 500 kPa is the void ratio at $D$ on the normal consolidation line (Fig. E4.8).

$$e_D = e_B - C_c \log\left(\frac{500}{350}\right) = 1.43 - 0.37 \log 1.43 = 1.37$$   ∎

## EXAMPLE 4.9

During a one-dimensional consolidation test, the height of a normally consolidated soil sample at a vertical effective stress of 200 kPa was 18 mm. The vertical effective stress was reduced to 50 kPa, causing the soil to swell by 0.5 mm. Determine $m_{vr}$ and $E'_c$.

**Strategy**   From the data, you can find the vertical strain. You know the increase in vertical effective stress, so the appropriate equations to use to calculate $m_{vr}$ and $E'_c$ are Eqs. (4.10) and (4.11).

## Solution 4.9

**Step 1:**   Calculate the vertical strain.

$$\Delta\varepsilon_z = \frac{\Delta z}{H_o} = \frac{0.5}{18} = 0.028$$

**Step 2:**   Calculate the modulus of volume recompressibility.

$$m_{vr} = \frac{\Delta\varepsilon_z}{\Delta\sigma'_z} = \frac{0.028}{150} = 1.9 \times 10^{-4} \text{ m}^2/\text{kN}$$

**Step 3:**   Calculate the constrained elastic modulus.

$$E'_c = \frac{1}{m_{vr}} = \frac{1}{1.9 \times 10^{-4}} = 5263 \text{ kPa}$$   ∎

## EXAMPLE 4.10

The following readings were taken for an increment of vertical stress of 20 kPa in an oedometer test on a saturated clay sample, 75 mm in diameter and 20 mm thick. Drainage was permitted from the top and bottom boundaries.

| Time (min) | 0.25 | 1 | 2.25 | 4 | 9 | 16 | 25 | 36 | 24 hours |
|---|---|---|---|---|---|---|---|---|---|
| $\Delta H$ (mm) | 0.12 | 0.23 | 0.33 | 0.43 | 0.59 | 0.68 | 0.74 | 0.76 | 0.89 |

Determine the coefficient of consolidation using the root time method.

**Strategy**  Plot the data in a graph of displacement reading versus $\sqrt{\text{time}}$ and follow the procedures in Section 4.7.2.1.

### Solution 4.10

**Step 1:**  Make a plot of settlement (decrease in thickness) versus $\sqrt{\text{time}}$ as shown in Fig. E4.10.

**Step 2:**  Follow the procedures outlined in Section 4.7.2.1 to find $t_{90}$. From Fig. E4.10,

$$\sqrt{t_{90}} = 3.22 \text{ min}^{1/2}; \quad t_{90} = 10.4 \text{ min}$$

**Step 3:**  Calculate $C_v$ from Eq. (4.43).

$$C_v = \frac{0.848 H_{\text{dr}}^2}{t_{90}}$$

where $H_{\text{dr}}$ is the length of the drainage path. The current height is $20 - 0.89 = 19.1$ mm. From Eq. (4.1),

$$H_{\text{dr}} = \frac{H_o + H_f}{4} = \frac{20 + 19.1}{4} = 9.8 \text{ mm}$$

$$\therefore C_v = \frac{0.848 \times 9.8^2}{10.4} = 7.8 \text{ mm}^2/\text{min}$$

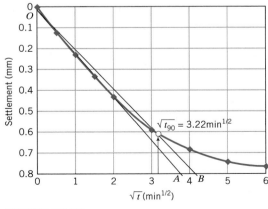

**FIGURE E4.10**

*What's next* . . .We have described the consolidation test of a small sample of soil and the soil consolidation parameters that can be obtained. What is the relationship between this small test sample and the soil in the field? Can you readily calculate the settlement of the soil in the field based on the results of your consolidation test? The next section provides the relationship between the small test sample and the soil in the field.

## 4.8 RELATIONSHIP BETWEEN LABORATORY AND FIELD CONSOLIDATION

The time factor $(T_v)$ provides a useful expression to estimate the settlement in the field from the results of a laboratory consolidation test. If two layers of the same clay have the same degree of consolidation, then their time factors and coefficients of consolidation are the same. Hence,

$$T_v = \frac{(C_v t)_{lab}}{(H_{dr}^2)_{lab}} = \frac{(C_v t)_{field}}{(H_{dr}^2)_{field}} \tag{4.45}$$

and, by simplification,

$$\boxed{\frac{t_{field}}{t_{lab}} = \frac{(H_{dr}^2)_{field}}{(H_{dr}^2)_{lab}}} \tag{4.46}$$

*What's next* . . .The following example shows you how to find the expected field settlement from consolidation test results for a particular degree of consolidation.

### EXAMPLE 4.11

A sample, 75 mm in diameter and 20 mm high, taken from a clay layer 10 m thick, was tested in an oedometer with drainage at the upper and lower boundaries. It took the laboratory sample 15 minutes to reach 50% consolidation.

**(1)** If the clay layer in the field has the same drainage condition as the laboratory sample, calculate how long it will take the 10 m clay layer to achieve 50% and 90% consolidation.

**(2)** How much more time would it take the 10 m clay layer to achieve 50% consolidation if drainage existed only on one boundary?

**Strategy** You are given all the data to directly use Eq. (4.46). For part (1) there is double drainage in the field and the lab, so the drainage path is one-half the soil thickness. For part (2), there is single drainage in the field, so the drainage path is equal to the soil thickness.

### Solution 4.11

**(1)** We proceed as follows:

**Step 1:** Calculate the drainage path.

$$(H_{dr})_{lab} = \frac{20}{2} = 10 \text{ mm} = 0.01 \text{ m}; \quad (H_{dr})_{field} = \frac{10}{2} = 5 \text{ m}$$

**Step 2:** Calculate the field time using Eq. (4.46).

$$t_{field} = \frac{t_{lab}(H_{dr}^2)_{field}}{(H_{dr}^2)_{lab}} = \frac{15 \times 5^2}{0.01^2} = 375 \times 10^4 \text{ min} = 7.13 \text{ years}$$

**(2)** We proceed as follows:

**Step 1:** Calculate the drainage path.

$$(H_{dr})_{lab} = \frac{20}{2} = 10 \text{ mm} = 0.01 \text{ m}; \quad (H_{dr})_{field} = 10 \text{ m}$$

**Step 2:** Calculate field time using Eq. (4.46).

$$t_{field} = \frac{t_{lab}(H_{dr}^2)_{field}}{(H_{dr}^2)_{lab}} = \frac{15 \times 10^2}{0.01} = 15 \times 10^6 \text{ min} = 28.54 \text{ years}$$

You should take note that if drainage exists only on one boundary rather than both boundaries of the clay layer, the time taken for a given percent consolidation in the field is four times longer. ∎

*What's next*...Several empirical equations are available linking consolidation parameters to simple, less time consuming, soil tests such as the Atterberg limits and water content. In the next section, some of these relationships are presented.

## 4.9 TYPICAL VALUES OF CONSOLIDATION SETTLEMENT PARAMETERS AND EMPIRICAL RELATIONSHIPS

Some relationships between simple soil tests and consolidation settlement parameters are given below. You should be cautious in using these relationships because they may not be applicable to your soil type.

**Typical Values**

$C_c = 0.1$ to $0.8$

$C_r = C_c/5$ to $C_c/10$

$C_\alpha/C_c = 0.01$ to $0.07$

**Empirical Relationships**

| | |
|---|---|
| $C_c = 0.009(w_{LL} - 10)$ | (Skempton, 1944) |
| $C_c = 0.40(e_o - 0.25)$ | (Azzouz et al., 1976) |
| $C_c = 0.01(w - 5)$ | (Azzouz et al., 1976) |
| $C_c = 0.37(e_o + 0.003w_{LL} - 0.34)$ | (Azzouz et al., 1976) |
| $C_c = 0.00234w_{LL}G_s$ | (Nagaraj and Murthy, 1986) |
| $C_r = 0.15(e_o + 0.007)$ | (Azzouz et al., 1976) |
| $C_r = 0.003(w + 7)$ | (Azzouz et al., 1976) |
| $C_r = 0.126(e_o + 0.003w_{LL} - 0.06)$ | (Azzouz et al., 1976) |
| $C_r = 0.000463w_{LL}G_s$ | (Nagaraj and Murthy, 1985) |

$w$ is the natural water content (%), $w_{LL}$ is the liquid limit (%), $e_o$ is the initial void ratio

*What's next* . . .Sometimes, we may have to build structures on a site for which the calculated settlement of the soil is intolerable. One popular method to reduce the consolidation settlement to tolerable limits is to preload the soil and use sand drains to speed up the drainage of the excess pore water pressure. Next, we will discuss sand drains.

## 4.10  SAND DRAINS

The purpose of sand drains is to accelerate the consolidation settlement of soft saturated clays by reducing the drainage path. A sand drain is constructed by making a borehole through the soil ending on an impervious boundary, for example, rock, shale, or a stiff clay. The borehole is then back-filled with coarse-grained soils. A sand drain in which one end is on an impervious boundary is called a half-closed drain (Fig. 4.19). Sometimes, the drain may penetrate into a pervious layer below an impervious layer, allowing the pore water to be expelled from the top and bottom of the drain. Such a drain provides two-way drainage and would accelerate the consolidation of the soil. Recall that for one dimen-

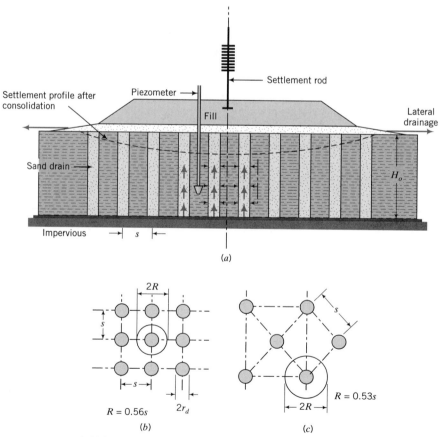

**FIGURE 4.19**  (a) Vertical section of a half-closed sand drain. (b) Plan of a square grid sand drain. (c) Plan of a triangular grid sand drain.

sional consolidation two-way drainage reduces the time for a given degree of consolidation by four times compared with single drainage.

The diameter of sand drains ranges from about 200 to 600 mm. The diameter required must only be large enough to drain the pore water and prevent premature clogging from fines in the soil to be drained. Filter fabrics are now commonly used at the interface of the natural soil and the back fill to prevent clogging. The spacing of sand drains depends on the permeability of the soil and the desired time to achieve the required degree of consolidation. The spacing of the drains must be less than the thickness of the soil layer. Typical spacing ranges from 2 m to about 5 m in either a square or triangular grid (Fig. 4.19). The excess pore water from an applied loading drains radially toward the sand drains. The radius of influence of each drain is $R \approx 0.56$ s for a square grid and $R \approx 0.53$ s for a triangular grid; s is spacing of drain. The consolidation of the soil has two components—one component is due to vertical drainage and the other component is due to radial drainage.

The governing equation for axisymmetric radial drainage is

$$\frac{\partial u}{\partial t} = C_h \left( \frac{\partial^2 u}{\partial r^2} + \frac{1}{r} \frac{\partial u}{\partial r} \right) \tag{4.47}$$

where $r$ is the radial distance from the center of the drain and $C_h$ is the coefficient of consolidation in the horizontal or radial direction. The boundary conditions to solve Eq. (4.47) are:

$$\text{At } r = r_d: \quad u = 0 \quad \text{when } t > 0$$

$$\text{At } r = R: \quad \frac{\partial u}{\partial r} = 0$$

and the initial condition is $t = 0$, $u = u_o$, where $r_d$ is the radius of the drains, $t$ is time, and $R$ is the radius of the cylindrical influence zone (Fig. 4.19). Richart (1959) reported solutions to Eq. (4.47) for two cases—free strain and equal strain. Free strain occurs when the surface load is uniformly distributed (flexible foundation) and the resulting surface settlement is uneven, as shown in Fig. 4.19. Equal strain occurs when the surface settlement is forced to be uniform (rigid foundation) and the resulting surface load is not uniformly distributed. Richart (1959) showed that the differences in the two cases are small and the solution for equal strain is often used in practice and shown in Fig. 4.20.

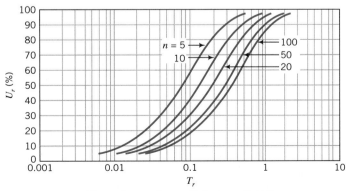

**FIGURE 4.20** Time factor for radial consolidation.

The time factor for consolidation in the vertical direction is given by Eq. (4.31) while the time factor for consolidation in the radial direction $(T_r)$ is

$$T_r = \frac{C_h t}{4R^2}$$ 
(4.48)

The permeability of the soil in the horizontal or radial direction is sometimes much greater (2 to 10 times for many soils) than in the vertical direction (Chapter 2) and, consequently, $C_h$ is greater than $C_v$, usually $C_h/C_v \simeq 1.2$ to 2. During drilling of the borehole and installation of the drain, a thin layer of soil at the interface of the drain is often remolded. This thin layer of remolded soil is called a smear zone. The values of $C_v$ and $C_h$ are often much lower in the smear zone than in the natural soil. It is customary to use reduced values of $C_v$ and $C_h$ to account for the smear zone.

The average degree of consolidation for vertical and radial dissipation of pore water pressure $(U_{vr})$ is

$$U_{vr} = 1 - (1 - U)(1 - U_r)$$ 
(4.49)

where $U$ is the average degree of consolidation for vertical drainage [Eq. (4.33)] and $U_r$ is the average degree of consolidation for radial drainage.

The analysis of sand drains is often centered on the spacing required to produce a desired settlement in a desired time. The procedure to find the spacing is as follows:

1. Obtain values of $C_v$, $C_h$, $C_c$, and $C_r$ or $m_v$ from laboratory tests and other pertinent data, for example, thickness of layer and structural load.

2. Calculate the primary consolidation settlement for the applied vertical stress.

3. Determine the degree of consolidation for the desired amount of primary consolidation. This is your desired value of $U_{vr}$.

4. Determine $T_v$ from Eq. (4.31).

5. Determine $U$ from Fig. 4.9 or either Eq. (4.34) or (4.35).

6. Substitute $U_{vr}$ and $U$ in Eq. (4.49) and find $U_r$; $U_r = 1 - (1 - U_{vr})/(1 - U)$.

7. Find $T_r$ in terms of $n$ by solving $T_r = C_h t/4R^2 = C_h t/4(nr_d)^2$, where $r_d$ is the radius of the drain and $n = R/r_d$.

8. You now have to find a value of $T_r$ such that the degree of consolidation matches $U_r$ calculated in Step 6. You do this by iteration since you do not know either $T_r$ or $n$. Select a value of $n$ from the values shown in Fig. 4.20 and find $T_r$. Determine $U_r$ from the curve corresponding to the selected value of $n$. Compare this value of $U_r$ with $U_r$ calculated in Step 6. If they are not approximately equal (<10% difference) then reiterate until you get a satisfactory solution. You may interpolate from Fig. 4.20 for intermediate values of $n$ from 5 to 100. You can also plot a graph of $n$ versus $U_r$ and find the $n$ value corresponding to the desired $U_r$ (Step 6).

9. Calculate the spacing, $s$, depending on the grid you desire. For a square grid $s = 1.8R = 1.8nr_d$ and for a triangular grid $s = 1.9R = 1.9nr_d$.

## EXAMPLE 4.12

A foundation for a structure is to be constructed on a soft deposit of clay. Below the soft clay is a stiff overconsolidated clay 20 m thick. The calculated settlement cannot be tolerated and it was decided that the soil should be preconsolidated by an embankment equivalent to the building load. The data available are:

$$C_v = 6 \text{ m}^2/\text{yr}, \quad C_h = 10 \text{ m}^2/\text{yr}, \quad m_v = 0.2 \text{ m}^2/\text{MN}$$

The foundation size is 10 m × 10 m and $q_s = 400$ kPa. The drain diameter is 300 mm. Determine the spacing of a square grid of the sand drains to achieve 90% consolidation in 4 months.

**Strategy**   Follow the procedure outlined in Section 4.10.

## Solution 4.12

**Step 1:**   The given values are:

$$C_v = 6 \text{ m}^2/\text{yr}, \quad C_h = 10 \text{ m}^2/\text{yr}, \quad m_v = 0.2 \text{ m}^2/\text{MN},$$
$$H_o = 20 \text{ m}, \quad q_s = 400 \text{ kPa}$$

**Step 2:**   Calculate the primary consolidation settlement. Determine the average increase in applied vertical stress at mid-depth of soil under the center of the foundation.

$$\frac{B}{2} = \frac{10}{2} = 5 \text{ m}, \quad \frac{L}{2} = \frac{10}{2} = 5 \text{ m}, \quad z = \frac{20}{2} = 10 \text{ m}$$

$$m = \frac{(B/2)}{z} = \frac{5}{10} = 0.5, \quad n = \frac{(L/2)}{z} = \frac{5}{10} = 0.5$$

From Fig. 3.25, $I_z = 0.085$.

$$\Delta\sigma_z = 4 \times 0.085 \times 400 = 136 \text{ kPa}$$

$$\rho_{pc} = m_v H_o \Delta\sigma_z = \frac{0.2}{10^3} \times 20 \times 136 = 0.544 \text{ m} = 544 \text{ mm}$$

**Step 3:**   Determine the required degree of consolidation.
You are given $U_{vr} = 90\% = 0.9$; that is, the settlement desired is 0.9 × 544 = 489.6 mm.

**Step 4:**   Determine $T_v$.
The drainage system is half-closed (stiff clay layer at bottom is assumed impervious). Therefore, $H_{dr} = H_o = 20$ m.

$$T_v = \frac{C_v t}{H_{dr}^2} = \frac{6 \times \dfrac{4}{12}}{(20)^2} = 0.005$$

**Step 5:**   Determine $U$.
From Fig. 4.9, $U = 8\%$ for $T_v = 0.005$.

**Step 6:**   Find $U_r$.
From Eq. (4.49),

$$U_r = 1 - \frac{1 - U_{vr}}{1 - U} = 1 - \frac{1 - 0.9}{1 - 0.08} = 0.89$$

**Step 7:** Determine $T_r$.

$$T_r = \frac{C_h t}{4(nr_d)^2} = \frac{10 \times \frac{4}{12}}{4 \times n^2 \times \left(\frac{0.3}{2}\right)^2} = \frac{37}{n^2}$$

**Step 8:** Determine $n$.

$$\text{Try } n = 5: \quad T_r = \frac{37}{5^2} = 1.48$$

From Fig. 4.20 with $T_r = 1.48$, $U_r \approx 1 > 0.89$.

$$\text{Try } n = 10: \quad T_r = \frac{37}{100} = 0.37$$

From Fig. 4.20 with $T_r = 0.37$, $U_r = 0.85 < 0.89$.

$$\therefore \text{Use } n = 10$$

**Step 9:** Calculate the spacing. For a square grid,

$$s = 1.8nr_d = 1.8 \times 10 \times 0.15 = 2.7 \text{ m} \quad \blacksquare$$

## 4.11 LATERAL EARTH PRESSURE AT REST DUE TO OVERCONSOLIDATION

The lateral earth pressure coefficient at rest, $K_o = \sigma_3'/\sigma_1'$, was presented in Section 3.10. For normally consolidated soil, $K_o = K_o^{nc}$ is reasonably predicted by an equation suggested by Jaky (1944) as

$$\boxed{K_o^{nc} \approx 1 - \sin \phi_{cs}'} \tag{4.50}$$

where $\phi_{cs}'$ is a fundamental soil constant that will be discussed in Chapter 5.

The value of $K_o^{nc}$ is constant. During unloading or reloading, the soil stresses must adjust to be in equilibrium with the applied stress. This means that stress changes take place not only vertically but also horizontally. From Chapter 3, you know that, for a given surface stress, the changes in horizontal stresses and vertical stresses are different. Therefore, $K_o$ for overconsolidated soils, denoted by $K_o^{oc}$, would not be a constant. Various equations have been suggested linking $K_o^{oc}$ to $K_o^{nc}$. One equation that is popular and found to match test data reasonably well is an equation proposed by Meyerhoff (1976) as

$$\boxed{K_o^{oc} = K_o^{nc}(OCR)^{1/2} = (1 - \sin \phi_{cs}')(OCR)^{1/2}} \tag{4.51}$$

## 4.12 SUMMARY

Consolidation settlement of a soil is a time-dependent process that depends on the soil's permeability, thickness and the drainage conditions. When an increment of vertical stress is applied to a soil, the instantaneous (initial) excess pore water pressure is equal to the vertical stress increment. With time, the initial excess pore water pressure decreases, the vertical effective stress increases by the amount of decrease of the initial excess pore water pressure, and settlement increases. The consolidation settlement is made up of two parts—the early time response called primary consolidation and a later time response called secondary compression.

Soils retain a memory of the past maximum effective stress, which may be erased by loading to a higher stress level. If the current vertical effective stress on a soil was never exceeded in the past (a normally consolidated soil), it would behave elastoplastically when stressed. If the current vertical effective stress on a soil was exceeded in the past (an overconsolidated soil), it would behave elastically (approximately) for stresses less than its past maximum effective stress.

## *Practical Examples*

### EXAMPLE 4.13

A foundation for an oil tank is proposed for a site with a soil profile as shown in Fig. E4.13a. A specimen of the fine-grained soil, 75 mm in diameter and 20 mm thick, was tested in an oedometer in a laboratory. The initial water content was 62% and $G_s = 2.7$. The vertical stresses were applied incrementally—each increment remaining on the specimen until the pore water pressure change was negligible. The cumulative settlement values at the end of each loading step are as follows:

| Vertical stress (kPa) | 15 | 30 | 60 | 120 | 240 | 480 |
|---|---|---|---|---|---|---|
| Settlement (mm) | 0.10 | 0.11 | 0.21 | 1.13 | 2.17 | 3.15 |

The time–settlement data when the vertical stress was 200 kPa are:

| Time (min) | 0 | 0.25 | 1 | 4 | 9 | 16 | 36 | 64 | 100 |
|---|---|---|---|---|---|---|---|---|---|
| Settlement (mm) | 0 | 0.22 | 0.42 | 0.6 | 0.71 | 0.79 | 0.86 | 0.91 | 0.93 |

The tank, when full, will impose vertical stresses of 90 kPa and 75 kPa at the top and bottom of the fine-grained soil layer, respectively. You may assume that the vertical stress is linearly distributed in this layer.

**(a)** Determine the primary consolidation settlement of the fine-grained soil layer when the tank is full.

**(b)** Calculate and plot the settlement–time curve.

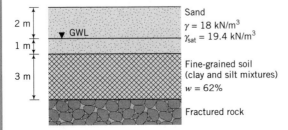

**FIGURE E4.13a**

*Strategy* To calculate the primary consolidation settlement you need to know $C_c$ and $C_r$ or $m_v$, and $\sigma'_{zo}$, $\Delta\sigma$, and $\sigma'_{zc}$. Use the data given to find the values of these parameters. To find time for a given degree of consolidation, you need to find $C_v$ from the data.

## Solution 4.13

**Step 1:** Find $C_v$ using the root time method.

Use the data from the 240 kPa load step to plot a settlement versus $\sqrt{\text{time}}$ curve as depicted in Fig. E4.13b. Follow the procedures set out in section 4.7 to find $C_v$. From the curve, $t_{90} = 1.2$ min.

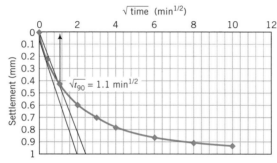

**FIGURE E4.13b**

Height of sample at beginning of loading $= 20 - 1.2 = 18.8$ mm

Height of sample at end of loading $= 20 - 2.17 = 17.83$ mm

Equation (4.1): $H_{dr} = \dfrac{H_o + H_f}{4} = \dfrac{18.8 + 17.83}{4} = 9.16$ mm

Equation (4.31): $C_v = \dfrac{T_v H_{dr}^2}{t_{90}} = \dfrac{0.848 \times (9.16)^2}{1.2} = 59.3$ mm$^2$/min

$$= 59.3 \times 10^{-6} \text{ m}^2/\text{min}$$

**Step 2:** Determine the void ratio at the end of each load step.

Initial void ratio: $e_o = wG_s = 0.62 \times 2.7 = 1.67$

Equation (4.6): $e = e_o - \dfrac{\Delta z}{H_o}(1 + e_o) = 1.67 - \dfrac{\Delta z}{20}(1 + 1.67)$

$$= 1.67 - 13.35 \times 10^{-2} \Delta z$$

The void ratio for each load step is shown in the table below.

| $\sigma'_z$ (kPa) | 15 | 30 | 60 | 120 | 240 | 480 |
|---|---|---|---|---|---|---|
| Void ratio | 1.66 | 1.66 | 1.64 | 1.52 | 1.38 | 1.25 |

A plot of $e - \log \sigma'_z$ versus $e$ is shown in Fig. E4.13c.

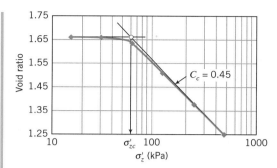

**FIGURE E4.13c**

**Step 3:** Determine $\sigma'_{zc}$ and $C_c$.
Follow the procedures in Section 4.7 to find $\sigma'_{zc}$.

$$\sigma'_{zc} = 60 \text{ kPa}$$

$$C_c = \frac{1.52 - 1.25}{\log\left(\dfrac{480}{120}\right)} = 0.45$$

**Step 4:** Calculate $\sigma'_{zo}$.

Clay:  $\gamma_{\text{sat}} = \dfrac{G_s + e_o}{1 + e_o} \gamma_w = \left(\dfrac{2.7 + 1.67}{1 + 1.67}\right) 9.8 = 16 \text{ kN/m}^3$

$\sigma'_{zo} = (18 \times 2) + (19.4 - 9.8)1 + (16 - 9.8)1.5 = 54.9 \text{ kPa}$

**Step 5:** Calculate settlement.

$$\text{OCR} = \frac{\sigma'_{zc}}{\sigma'_{zo}} = \frac{60}{54.9} = 1.1$$

For practical purpose, the OCR is very close to 1; that is, $\sigma'_z \approx \sigma'_{zc}$. Therefore, the soil is normally consolidated. Also, inspection of the $e$ versus $\log \sigma'_z$ curve shows that $C_r$ is approximately zero, which lends further support to the assumption that the soil is normally consolidated.

$$\rho_{\text{pc}} = \frac{H_o}{1 + 1.67} 0.45 \left(\frac{\sigma'_{zo} + \Delta\sigma_z}{\sigma'_{zo}}\right) = 0.17 H_o \log\left(\frac{\sigma'_{zo} + \Delta\sigma_z}{\sigma'_{zo}}\right)$$

Divide the clay layer into three sublayers of 1.0 m thick and compute the settlement for each sublayer. The primary consolidation settlement is the sum of the settlement of each sublayer. The vertical stress increase in the fine-grained soil layer is

$$90 - \left(\frac{90 - 75}{3}\right)z = 90 - 5z$$

where $z$ is the depth below the top of the layer. Calculate the vertical stress increase at the center of each sublayer and then the settlement from the above equation. The table below summarizes the computation.

| Layer | $z$ (m) | $\sigma'_{zo}$ at center of sublayer (kPa) | $\Delta\sigma_z$ | $\sigma'_{zo} + \Delta\sigma_z$ (kPa) | $\rho_{pc}$ (mm) |
|-------|---------|---------------------------------------------|------------------|---------------------------------------|------------------|
| 1 | 0.5 | 48.7 | 87.5 | 136.2 | 75.9 |
| 2 | 1.5 | 54.9 | 82.5 | 137.4 | 67.7 |
| 3 | 2.5 | 61.1 | 77.5 | 138.6 | 60.5 |
|   |     |      |      | Total | 204.1 |

Alternatively, by considering the whole fine-grained soil layer and taking the average vertical stress increment, we obtain

$$\rho_{pc} = \frac{3000}{1 + 1.67} \, 0.45 \, \log\left(\frac{137.4}{54.9}\right) = 201.4 \text{ mm}$$

In general, the former approach is more accurate for thick layers.

**Step 6:**   Calculate settlement–time values.

$$C_v = 59.3 \times 10^{-6} \times 60 \times 24 = 85392 \times 10^{-6} \text{ m}^2/\text{day}$$

$$t = \frac{T_v H_{dr}^2}{C_v} = \frac{T_v \times \left(\dfrac{3}{2}\right)^2}{85392 \times 10^{-6}} = 26.3 T_v \text{ days}$$

The calculation of settlement at discrete times is shown in the table below and the data are plotted in Fig. E4.13d.

| $U$ (%) | $T_v$ | Settlement (mm) $\rho_{pc} \times \dfrac{U}{100}$ | $t = 26.3 T_v$ (days) |
|---------|-------|---------------------------------------------------|------------------------|
| 10 | 0.008 | 20.4 | 0.2 |
| 20 | 0.031 | 20.8 | 0.8 |
| 30 | 0.071 | 61.3 | 1.9 |
| 40 | 0.126 | 81.6 | 3.3 |
| 50 | 0.197 | 102.1 | 5.2 |
| 60 | 0.287 | 122.5 | 7.6 |
| 70 | 0.403 | 142.9 | 10.6 |
| 80 | 0.567 | 163.3 | 14.9 |
| 90 | 0.848 | 183.7 | 22.3 |

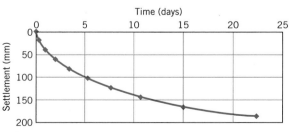

**FIGURE E4.13d**

### EXAMPLE 4.14

A geotechnical engineer made a preliminary settlement analysis for a foundation of an office building, which is to be constructed at a location where the soil strata contain a compressible clay layer. She calculated 50 mm of primary consolidation settlement. The building will impose an average vertical stress of 150 kPa in the clay layer. As often happens in design practice, design changes are required. In this case, the actual thickness of the clay is 30% more than the original soil profile indicated and, during construction, the groundwater table has to be lowered by 2 m. Estimate the new primary consolidation settlement.

*Strategy* From Section 4.3, the primary consolidation settlement is proportional to the thickness of the soil layer and also to the increase in vertical stress [see Eq. (4.18)]. Use proportionality to find the new primary consolidation settlement.

### Solution 4.14

**Step 1:** Estimate the new primary consolidation settlement due to the increase in thickness.

$$\frac{(\rho_{pc})_n}{(\rho_{pc})_o} = \frac{H_n}{H_o}$$

where subscripts $o$ and $n$ denote original and new, respectively.

$$(\rho_{pc})_n = 50 \times \frac{1.3 \, H_o}{H_o} = 65 \text{ mm}$$

**Step 2:** Estimate primary consolidation settlement from vertical stress increase due to lowering of the groundwater level.

Increase in vertical effective stress due to lowering of water table

$$= 2 \times 9.8 = 19.6 \text{ kPa}$$

Primary consolidation settlement is also proportional to the vertical effective stress:

$$\frac{(\rho_{pc})_n}{(\rho_{pc})_o} = \frac{(\sigma'_z)_n}{(\sigma'_z)_o}$$

$$\therefore (\rho_{pc})_n = 65 \times \frac{(150 + 19.6)}{150} = 73.5 \text{ mm} \qquad \blacksquare$$

### EXAMPLE 4.15

The foundations supporting two columns of a building are shown in Fig. E4.15. An extensive soil investigation was not carried out and it was assumed in the design of the footing that the clay layer has a uniform thickness of 1.2 m. Two years after construction, the building settled with a differential settlement of 10 mm. Walls of the building began to crack. The doors have not jammed but by measuring the out-of-vertical distance of the doors, it is estimated that they would become jammed if the differential settlement exceeded 24 mm. A subsequent soil investigation showed that the thickness of the clay layer was not uniform but

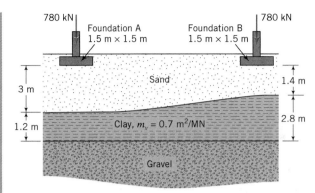

**FIGURE E4.15**

varies as shown in Fig. E4.15. The owners would like to get an estimate of the expected total differential settlement and how long it would take before the doors become jammed.

**Strategy**   Determine the settlement under each foundation and then calculate the differential settlement. Since you know the differential settlement that occurs over a two year period after construction, you can find the degree of consolidation and then use this information to calculate the expected time for the doors to become jammed.

## Solution 4.15

**Step 1:**   Calculate the vertical stress increase at the center of the clay layer under each foundation.
You can use Eq. (3.75) or Eq. (3.79) or, for a quick estimate, Eq. (3.85) to determine the stress increase under the foundations. Let's use the approximate method, Eq. (3.85).

$$\Delta\sigma_z = \frac{P}{(B + z)(L + z)}$$

$$(\Delta\sigma_z)_A = \frac{780}{(1.5 + 3.6)(1.5 + 3.6)} = 30 \text{ kPa}$$

$$(\Delta\sigma_z)_B = \frac{780}{(1.5 + 2.8)(1.5 + 2.8)} = 42.2 \text{ kPa}$$

Note: For a more accurate value of $\Delta\sigma_z$ you should use the vertical stress increase due to surface loads on multilayered soils (Poulos and Davis, 1974).

**Step 2:**   Calculate the primary consolidation settlement.
Use Eq. (4.18), $\rho_{pc} = H_o m_v \Delta\sigma$, to calculate the primary consolidation settlement.

$$(\rho_{pc})_A = 1.2 \times 0.7 \times 10^{-3} \times 30 = 25.2 \times 10^{-3} \text{ m} = 25.2 \text{ mm}$$
$$(\rho_{pc})_B = 2.8 \times 0.7 \times 10^{-3} \times 42.2 = 82.7 \times 10^{-3} \text{ m} = 82.7 \text{ mm}$$

**Step 3:** Calculate the differential settlement.

Differential settlement: $\delta = 82.7 - 25.2 = 57.5$ mm

**Step 4:** Calculate the time for 24 mm differential settlement to occur.

Current differential settlement: $\delta_c = 10$ mm

Degree of consolidation: $U = \dfrac{\delta_c}{\delta} = \dfrac{10}{57.5} = 0.17$

From Eq. (4.34): $T_v = \dfrac{4}{\pi} U^2 = \dfrac{4}{\pi} \times 0.17^2 = 0.037$

From Eq. (4.31): $C_v = \dfrac{T_v H_{dr}^2}{t} = \dfrac{0.037 \times (2.8/2)^2}{2} = 0.036$ m²/yr

For 24 mm differential settlement: $U = \dfrac{24}{57.5} = 0.42$, $T_v = \dfrac{4}{\pi} \times 0.42^2 = 0.225$

From Eq. (4.31): $t = \dfrac{T_v H_{dr}^2}{C_v} = \dfrac{0.225 \times (2.8/2)^2}{0.036} = 12.25$ years

Therefore, in the next 10.25 years, the total differential settlement would be 24 mm. ∎

## EXERCISES

For all problems, assume $G_s = 2.7$ unless otherwise stated.

### Theory

4.1 A clay soil of thickness $H$ is allowed to drain on the top boundary through a thin sand layer. A vertical stress of $\sigma$ was applied to the clay. The excess pore water pressure distribution was linear in the soil layer with a value of $u_t$ at the top boundary and $u_b$ ($u_b > u_t$) at the bottom boundary. The excess pore water pressure at the top boundary was not zero because the sand layer was partially blocked. Derive an equation for the excess pore water pressure distribution with soil thickness and time.

4.2 A soil layer of thickness $H_o$ has only single drainage through the top boundary. The excess pore water pressure distribution when a vertical stress, $\sigma$, is applied varies parabolically with a value of zero at the top boundary and $u_b$ at the bottom boundary. Show that

$$C_v = \frac{H_o^2}{2u_b} \frac{d\sigma'}{dt} \quad \text{and} \quad k_z = \frac{\gamma_w H_o}{2u_b} \frac{dH_o}{dt}$$

4.3 Show that, for a linear elastic soil,

$$m_v = \frac{(1 + v')(1 - 2v')}{E'(1 - v')}$$

4.4 Show that, if an overconsolidated soil behaves like a linear elastic material,

$$K_o^{oc} = (OCR)K_o^{nc} - \frac{v'}{1 - v'}(OCR - 1)$$

**4.5**    The excess pore water pressure distribution in a 10 m thick clay varies linearly from 100 kPa at the top to 10 kPa at the bottom of the layer when a vertical stress was applied. Assuming drainage only at the top of the clay layer, determine the excess pore water pressure in 1 year's time using the finite difference method if $C_v = 1.5$ m²/yr.

**4.6**    At a depth of 4 m in a clay deposit, the overconsolidation ratio is 3.0. Plot the variation of overconsolidation ratio and water content with depth for this deposit up to a depth of 15 m. The recompression index is $C_r = 0.05$ and the water content at 4 m is 32%.

**4.7**    The overconsolidation ratio of a saturated clay at a depth of 5 m is 6.0 and its water content is 38%. It is believed that the clay has become overconsolidated as a result of erosion. Calculate the thickness of the soil layer that was eroded. Assume that the present and past groundwater level is at the ground surface.

## Problem Solving

**4.8**    An oedometer test on a clay soil gave the following results: $C_c = 0.4$, $C_r = 0.08$, OCR = 4.5. The existing vertical effective stress in the field is 130 kPa. A building foundation will increase the vertical stress at the center of the clay by 150 kPa. The thickness of the clay layer is 2 m and its water content is 28%. Calculate the primary consolidation settlement. What would be the difference in settlement if the OCR were 1.5 instead of 4.5?

**4.9**    A building is expected to increase the vertical stress at the center of a 2 m thick clay layer by 100 kPa. If $m_v$ is $4 \times 10^{-4}$ m²/kN, calculate the primary consolidation settlement.

**4.10**    Two adjacent bridge piers rest on clay layers of different thickness but with the same properties. Pier #1 imposes a stress increment of 100 kPa to a 3 m thick layer while Pier #2 imposes a stress increment of 150 kPa to a 5 m thick layer. What is the differential settlement between the two piers if $m_v = 3 \times 10^{-4}$ m²/kN?

**4.11**    The table below shows data recorded during an oedometer test on a soil sample for an increment of vertical stress of 200 kPa.

| Time (min) | 0 | 0.25 | 1 | 4 | 9 | 16 | 36 | 64 | 100 |
|---|---|---|---|---|---|---|---|---|---|
| Settlement (mm) | 0 | 0.30 | 0.35 | 0.49 | 0.61 | 0.73 | 0.90 | 0.95 | 0.97 |

After 24 hours the settlement was negligible and the void ratio was 1.20, corresponding to a sample height of 18.2 mm. Determine $C_v$ using the root time and the log time methods.

**4.12**    A sample of saturated clay of height 20 mm and water content of 30% was tested in an oedometer. Loading and unloading of the sample were carried out. The thickness $H_f$ of the sample at the end of each stress increment/decrement is shown in the table below.

| $\sigma_z'$ (kPa) | 100 | 200 | 400 | 200 | 100 |
|---|---|---|---|---|---|
| $H_f$ (mm) | 20 | 19.31 | 18.62 | 18.68 | 18.75 |

**(a)** Plot the results as void ratio versus log $\sigma_z'$.

**(b)** Determine $C_c$ and $C_r$.

**(c)** Determine $m_v$ between $\sigma_z' = 200$ kPa and $\sigma_z' = 300$ kPa.

4.13 A sample of saturated clay, taken from a depth of 5 m, was tested in a conventional oedometer. The table below gives the vertical stress and the corresponding thickness recorded during the test.

| $\sigma_z'$ (kPa) | 100 | 200 | 400 | 800 | 1600 | 800 | 400 | 100 |
|---|---|---|---|---|---|---|---|---|
| $h$(mm) | 19.2 | 19.0 | 17.0 | 14.8 | 12.6 | 13.1 | 14.3 | 15.9 |

The water content at the end of the test was 40% and the initial height was 20 mm.
(a) Plot the graph of void ratio versus log $\sigma_z'$.
(b) Determine $C_c$ and $C_r$.
(c) Determine $m_v$ between $\sigma_z' = 400$ kPa and $\sigma_z' = 500$ kPa.
(d) Determine the relationship between $e$ (void ratio) and $h$ (thickness).
(e) Determine $\sigma_{zc}'$ using Casagrande's method.

4.14 The following observations were recorded in an oedometer test on a clay sample 100 mm in diameter and 30 mm high.

| Load (N) | 0 | 50 | 100 | 200 | 400 | 800 | 0 |
|---|---|---|---|---|---|---|---|
| Displacement gauge reading (mm) | 0 | 0.48 | 0.67 | 0.98 | 1.24 | 1.62 | 1.4 |

At the end of the test, the wet mass of the sample was 507.3 grams and, after oven drying, its dry mass was 412.5 grams. The specific gravity was 2.65.
(a) Calculate the void ratio at the end of the test.
(b) Calculate the void ratio at the end of each loading step.
(c) Calculate the initial thickness of the soil sample from the initial void ratio and compare this with the initial thickness.
(d) Determine $m_v$ between $\sigma_z' = 50$ kPa and $\sigma_z' = 150$ kPa.

4.15 A laboratory consolidation test on a 20 mm thick sample of soil shows that 90% consolidation occurs in 30 minutes. Plot a settlement–time curve for a 10 m layer of this clay in the field for (a) single drainage and (b) double drainage.

4.16 A clay layer below a building foundation settles 15 mm in 200 days after the building was completed. According to the oedometer results, this settlement corresponds to an average degree of consolidation of 25%. Plot the settlement–time curve for a 10 year period, assuming double drainage.

4.17 An oil tank is to be sited on a soft alluvial deposit of clay. Below the soft clay is a thick layer of stiff clay. It was decided that a circular embankment, 10 m diameter, with sand drains inserted into the clay would be constructed to preconsolidate the soil. The height of the embankment is 6 m and the saturated unit weight of the soil comprising the embankment is 18 kN/m³. The following data are available: thickness of clay = 7 m, $m_v$ = 0.2 m²/MN, $C_v$ = 3.5 m²/yr, $C_h$ = 6.2 m²/yr, diameter of drain = 300 mm. The desired degree of consolidation is 90% in 6 months. Determine the spacing of a square grid of the sand drains such that when the tank is constructed the maximum primary consolidation should not exceed 20 mm.

## Practical

4.18 Fig. P4.18 shows the soil profile at a site for a proposed office building. It is expected that the vertical stress at the top of the clay will increase by 150 kPa and at the bottom by 90

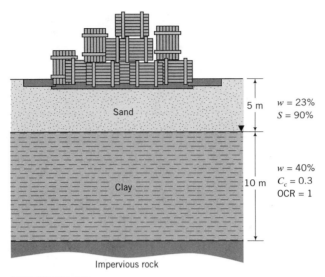

**FIGURE P4.18**

kPa. Assuming a linear stress distribution within the clay, calculate the consolidation settlement. [*Hint:* You should divide the clay into five equal layers, compute the settlement for each layer, and then find the total settlement.] An oedometer test on a sample of the clay revealed that the time for 90% consolidation on a 20 mm thick sample occurred in 40 minutes. The sample was drained on the upper and lower boundaries. How long will it take for 50% consolidation to occur in the field?

4.19    A borehole at a site for a proposed building reveals the following soil profile:

| 0–5 m | Dense sand, $\gamma = 18$ kN/m³, $\gamma_{sat} = 19$ kN/m³ |
|---|---|
| At 4 m | Groundwater level |
| 5–10 m | Soft normally consolidated clay, $\gamma_{sat} = 17.5$ kN/m³ |
| Below 10 m | Impervious rock |

A building is to be constructed on this site with its foundation at 2 m below ground level. The building load is 30 MN and the foundation is rectangular with a width of 10 m and length of 15 m. A sample of the clay was tested in an oedometer and the following results were obtained:

| Vertical stress (kPa) | 50 | 100 | 200 | 400 | 800 |
|---|---|---|---|---|---|
| Void ratio | 0.945 | 0.895 | 0.815 | 0.750 | 0.705 |

Calculate the primary consolidation settlement. Assuming that the primary consolidation took 5 years to achieve in the field, calculate the secondary compression for a period of 10 years beyond primary consolidation. The secondary compression index is $C_c/6$. [*Hint:* Determine $e_p$ for your $\sigma'_{fin}$ from a plot of $e$ versus log $\sigma'_z$.]

## CHAPTER 5

# *SHEAR STRENGTH OF SOILS*

| ABET | ES | ED |
|------|-----|-----|
|      | 85 | 15 |

## 5.0 INTRODUCTION

The safety of any geotechnical structure is dependent on the strength of the soil. If the soil fails, a structure founded on it can collapse, endangering lives and causing economic damage. The strength of soils is therefore of paramount importance to geotechnical engineers. The word strength is used loosely to mean shear strength, which is the internal frictional resistance of a soil to shearing forces. Shear strength is required to make estimates of the load bearing capacity of soils, the stability of geotechnical structures, and in analyzing the stress–strain characteristics of soils.

In this chapter we will define, describe, and determine the shear strength of soils. When you complete this chapter, you should be able to:

- Determine the shear strength of soils
- Understand the differences between drained and undrained shear strength
- Determine the type of shear test that best simulates field conditions
- Interpret laboratory and field test results to obtain shear strength parameters

You will use the following principles learned from previous chapters and other courses.

- Stresses, strains, Mohr's circle of stresses and strains, and stress paths (Chapter 3)
- Friction (statics and/or physics)

*Sample Practical Situation* You are the geotechnical engineer in charge of a soil exploration program for a dam and housing project. You are expected to specify laboratory and field tests to determine the shear strength of the soil and to recommend soil strength parameters for the design of the dam.

In Fig. 5.1 a house is shown in a precarious position because the shear strength of the soil within the slope near the house was exceeded. Would you like this to be your house? The content of this chapter will help you to understand the shear behavior of soils so that you can prevent catastrophes like that shown in Fig. 5.1.

**FIGURE 5.1**    Shear failure of soil under a house. (© Leverett Bradley/Tony Stone Images/New York, Inc.)

## 5.1  DEFINITIONS OF KEY TERMS

*Shear strength* of a soil ($\tau_f$) is the maximum internal resistance to applied shearing forces.

*Effective friction angle* ($\phi'$) is a measure of the shear strength of soils.

*Cohesion* ($c_o$) is a measure of the forces that cement particles of soil.

*Undrained shear strength* ($s_u$) is the shear strength of a soil when sheared at constant volume.

*Critical state* is a stress state reached in a soil when continuous shearing occurs at constant shear stress and constant volume.

*Dilation* is a measure of the change in volume of a soil when it is distorted by shearing.

## 5.2  QUESTIONS TO GUIDE YOUR READING

1. What is meant by the shear strength of soils?
2. What factors affect the shear strength?
3. How is shear strength determined?
4. What are the assumptions in the Mohr–Coloumb failure criterion?
5. Do soils fail on a plane?
6. What are the differences between peak, critical, and residual effective friction angles?

7. What are peak shear strength, critical shear strength, and residual shear strength?

8. Are there differences between the shear strengths of dense and loose sands or normally and overconsolidated clays?

9. What are the differences between drained and undrained shear strengths?

10. Under what conditions should the drained shear strength or the undrained shear strength parameters be used?

11. What laboratory and field tests are used to determine shear strength?

12. What are the differences between the results of various laboratory and field tests?

13. How do I know what laboratory test to specify for a project?

## 5.3 TYPICAL RESPONSE OF SOILS TO SHEARING FORCES

We are going to describe the behavior of two groups of soils when they are subjected to shearing forces. One group, called uncemented soils, has very weak interparticle bonds and comprises most soils. The other group, called cemented soils, has strong interparticle bonds through ion exchange or substitution. The particles of cemented soils are chemically bonded or cemented together. An example of a cemented soil is caliche, which is a mixture of clay, sand, and gravel cemented by calcium carbonate.

Let us incrementally deform two samples of soil by applying simple shear deformation (Fig. 5.2) to each of them. One sample, which we call Type I, represents mostly loose sands and normally consolidated and lightly overconsolidated clays (OCR $\leq$ 2). The other, which we call Type II, represents mostly dense sands and overconsolidated clays (OCR $>$ 2). In classical mechanics, simple shear deformation refers to shearing under constant volume. In soil mechanics, we relax this restriction (constant volume) because we wish to know the volume change characteristics of soils under simple shear (Fig. 5.2). The implication is that the mathematical interpretation of simple shear tests becomes complicated because we now have to account for the influence of volumetric strains

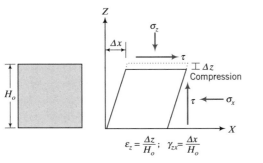

(a) Original soil sample  (b) Simple shear deformation of Type I soils  (c) Simple shear deformation of Type II soils

**FIGURE 5.2** Simple shear deformation of Type I and Type II soils.

on soil behavior. Since in simple shear $\varepsilon_x = \varepsilon_y = 0$, the volumetric strain is equal to the vertical strain, $\varepsilon_z = \Delta z/H_o$, where $\Delta z$ is the vertical displacement (positive for compression) and $H_o$ is the initial sample height. The shear strain is the small angular distortion expressed as $\gamma_{zx} = \Delta x/H_o$, where $\Delta x$ is the horizontal displacement.

We are going to summarize the important features of the responses of these two groups of soils when subjected to a constant vertical (normal) effective stress and increasing shear strain. We will consider the shear stress versus the shear strain, the volumetric strain versus the shear strain, and the void ratio versus the shear strain responses, as illustrated in Fig. 5.3.

When a shear band(s) develops in some types of overconsolidated clays, the particles become oriented parallel to the direction of the shear band, causing the final shear stress of these clays to decrease below the critical state shear stress. We will call this type of soil, Type II-A, and the final shear stress attained the residual shear stress, $\tau_r$. Type I soils at very low normal effective stress can also exhibit a peak shear stress during shearing.

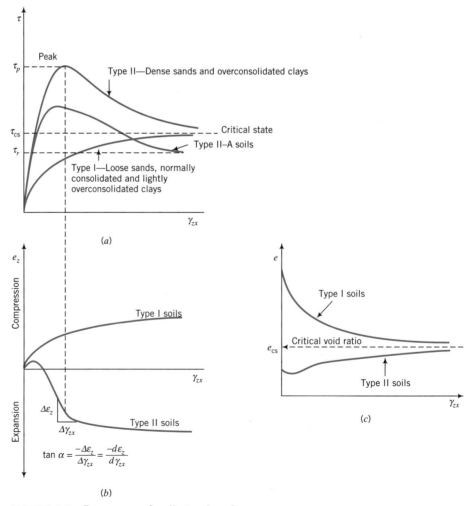

**FIGURE 5.3** Response of soils to shearing.

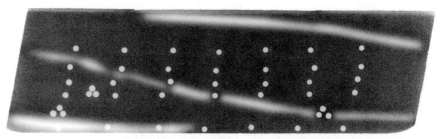

**FIGURE 5.4** Radiographs of shear bands in a dense fine sand (the white circles are lead shot used to trace internal displacements; white lines are shear bands). (After Budhu, 1979.)

**Type I soils—loose sands, normally consolidated and lightly overconsolidated clays (OCR $\leq$ 2)—are observed to:**

- Show gradual increase in shear stresses as the shear strain increases (strain hardens) until an approximately constant shear stress, which we will call the critical state shear stress, $\tau_{cs}$, is attained (Fig. 5.3a).
- Compress, that is, they become denser (Fig. 5.2b and Figs. 5.3b,c) until a constant void ratio, which we will call the critical ratio, $e_{cs}$, is reached (Fig. 5.3c).

**Type II soils—dense sands and heavily overconsolidated clays (OCR > 2)—are observed to:**

- Show a rapid increase in shear stress reaching a peak value, $\tau_p$, at low shear strains (compared to Type I soils) and then show a decrease in shear stress with increasing shear strain (strain softens) ultimately attaining a critical state shear stress (Fig. 5.3a). The strain softening response generally results from localized failure zones called shear bands (Fig. 5.4). The thickness of shear bands was observed from laboratory tests to be about 10 to 15 grain diameters. The soil within a shear band undergoes intense shearing while the soil masses above and below it behave as rigid bodies. The development of shear bands depends on the boundary conditions imposed on the soil samples.
- Compress initially (attributed to particle adjustment) and then expand, that is, they become looser (Fig. 5.2c and Figs. 5.3b,c) until a critical void ratio (the same void ratio as Type I soils) is attained.

The critical state shear stress is reached for all soils when no further volume change occurs under continued shearing. We will use the term critical state to define the stress state reached by a soil when no further change in shear stress and volume occurs under continuous shearing.

## 5.3.1 Effects of Increasing the Normal Effective Stress

So far, we have only used a single normal effective stress in our presentation of the responses of Type I and Type II soils. What is the effect of increasing the normal effective stress? For Type I soils, the amount of compression and the magnitude of the critical state shear stress will increase (Figs. 5.5a,b). For Type II soils, the peak shear stress tends to disappear, the critical shear stress increases and the change in volume expansion decreases (Figs. 5.5a,b).

If we were to plot the peak shear stress and the critical state shear stress for each constant normal effective stress for Type I and II soils, we would get:

1. An approximate straight line (OA, Fig. 5.5c) that links all the critical state shear stress values of Type I and Type II soils. We will call the angle between OA and the $\sigma'_n$ axis, the critical state friction angle, $\phi'_{cs}$. The line OA will be called the failure envelope because any shear stress that lies on it is a critical state shear stress.

2. A curve (OBCA, Fig. 5.5c) that links all peak shear stress values for Type II soils. We will call OBC (the curved part of OBCA), the peak shear stress envelope because any shear stress that lies on it is a peak shear stress.

At large normal effective stresses, the peak shear stress for Type II soils is suppressed and only a critical state shear stress is observed and appears as a point (point 9) located on OA (Fig. 5.5c). For Type II-A soils, the residual shear stresses would lie on a line OD below OA. We will call the angle between OD and the $\sigma'_n$ axis, the residual friction angle, $\phi'_r$. As the normal effective stress increases, the critical void ratio decreases (Fig. 5.5d). Thus, the critical void ratio is dependent on the magnitude of the normal effective stress.

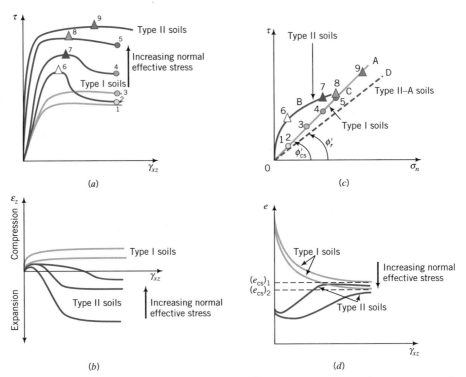

**FIGURE 5.5** Effects of increasing normal effective stresses on the response of soils.

## 5.3.2 Effects of Overconsolidation Ratio

The initial state of the soil dictates the response of the soil to shearing forces. For example, two overconsolidated soils with different overconsolidation ratios but the same mineralogical composition would exhibit different peak shear stresses and volume expansion as shown in Fig. 5.6. The higher overconsolidated soil gives a higher peak shear strength and greater volume expansion.

*The* **essential points** *are:*

1. *Type I soils—loose sands and normally consolidated and lightly over-consolidated clays—strain harden to a critical state shear stress and compress toward a critical void ratio.*

2. *Type II soils—dense sands and overconsolidated clays—reach a peak shear stress, strain soften to a critical state shear stress and expand toward a critical void ratio after an initial compression at low shear strains.*

3. *The peak shear stress of Type II soils is suppressed and the volume expansion decreases when the normal effective stress is large.*

4. *All soils reach a critical state, irrespective of their initial state, at which continuous shearing occurs without changes in shear stress and volume.*

5. *At large strains, the particles of some overconsolidated clays become oriented parallel to the direction of shear bands and the final shear stress attained is lower than the critical state shear stress.*

6. *The critical state shear stress and the critical void ratio depend on the normal effective stress. Higher normal effective stresses result in higher critical state shear stresses and lower critical void ratios.*

7. *Higher overconsolidation ratios result in higher peak shear stresses and greater volume expansion.*

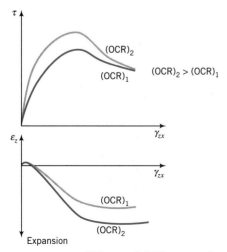

**FIGURE 5.6**   Effects of OCR on peak strength and volume expansion.

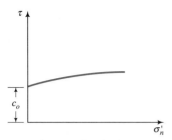

**FIGURE 5.7**    Peak shear stress envelope for cemented soils.

### 5.3.3 Cemented Soils

The particles of some soils are cemented (chemically bonded). These soils possess shear strength even when the normal effective stress is zero. Cemented soils behave much like Type II soils except they have an initial shear strength under zero normal effective stress. In a plot of peak shear stress versus normal effective stress, an intercept shear stress, $c_o$, called cohesion is observed (Fig. 5.7). You should not rely on $c_o$ in design because at large shear strains any shear strength due to cementation in the soil is destroyed.

*What's next . . .*You should now have a general idea on the responses of soils to shearing forces. Can we interpret these responses using a simple model? In the next section, we will develop a simple model to gain an insight into the behavior of soils that will later help us to interpret the shear strength of soils.

## 5.4    SIMPLE MODEL FOR THE SHEAR STRENGTH OF SOILS USING COULOMB'S LAW

You may recall Coulomb's frictional law from your courses in statics or physics. If a wooden block is pushed horizontally across a table (Fig. 5.8a), the horizontal force ($H$) required to initiate movement is

$$H = \mu W \tag{5.1}$$

where $\mu$ is the coefficient of static friction between the block and the table and $W$ is the weight of the block. The angle between the resultant force and the normal force is called the friction angle, $\phi' = \tan^{-1} \mu$.

In terms of stresses, Coulomb's law is expressed as

$$\tau_f = (\sigma'_n)_f \tan \phi' \tag{5.2}$$

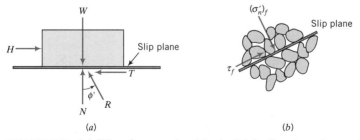

**FIGURE 5.8**    (a) Slip of a wooden block. (b) A slip plane in a soil mass.

where $\tau_f$ ($= T/A$, where $T$ is the shear force at impending slip and $A$ is the area of the plane parallel to $T$) is the shear stress when slip is initiated, $(\sigma_n')_f$ is the normal effective stress on the plane on which slip is initiated. The subscript $f$ denotes failure, which according to Coulomb's law occurs when rigid body movement of one body relative to another is initiated. Failure does not necessarily mean collapse but the initiation of movement of one rigid body relative to another.

Coulomb's law requires the existence or the development of a critical sliding plane, also called slip plane or failure plane. In the case of the wooden block on the table, the slip plane is the horizontal plane at the interface between the wooden block and the table. Unlike the wooden block, we do not know where the sliding plane is located for soils.

If you plot Coulomb's equation (5.2) on a graph of shear stress, $\tau_f$, versus normal effective stress, $(\sigma_n')_f$, you get a straight line similar to $OA$ (Fig. 5.5) if $\phi' = \phi_{cs}'$. Thus, Coulomb's law can be used to model soil behavior at the critical state. But what about modeling the peak behavior that is characteristic of Type II soils?

You should recall from Chapter 2 that soils can have different unit weights depending on the arrangement of the particles. Let us simulate two extreme arrangements of soil particles for coarse-grained soils—one loose, the other dense. We will assume that the soil particles are spheres. In two dimensions, arrays of spheres become arrays of disks. The loose array is obtained by stacking the spheres one on top of another while the dense packing is obtained by staggering the rows as illustrated in Fig. 5.9.

For simplicity, let us consider the first two rows. If we push (shear) row 2 relative to row 1 in the loose assembly, sliding would be initiated on the horizontal plane, $a$–$a$, consistent with Coulomb's frictional law [Eq. (5.2)]. Once motion is initiated the particles in the loose assembly would tend to move into the void spaces. The direction of movement would have a downward component, that is, compression.

In the dense packing, relative sliding of row 2 with respect to row 1 is restrained by the interlocking of the disks. Sliding, for the dense assembly, would be initiated on an inclined plane rather than on a horizontal plane. For the dense assembly, the particles must ride up over each other or be pushed aside or both. The direction of movement of the particles would have an upward component, that is, expansion.

We are going to use our knowledge of statics to investigate impending sliding of particles up or down a plane to assist us in interpreting the shearing behavior of soils using Coulomb's frictional law. The shearing of the loose array can be idealized by analogy with the sliding of our wooden block on the horizontal plane. At failure (impending motion),

$$\frac{\tau_f}{(\sigma_n')_f} = \frac{H}{W} = \tan \phi' \tag{5.3}$$

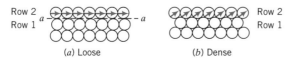

(a) Loose    (b) Dense

**FIGURE 5.9**  Packing of disks representing loose and dense sand.

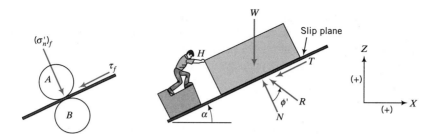

(a) Stresses on failure plane        (b) Simulated shearing of a dense array of particles

**FIGURE 5.10**  Simulation of failure in dense sand.

Consider two particles A and B in the dense assembly and draw the free-body diagram of the stresses at the sliding contact between A and B as depicted in Fig. 5.10. We now appeal to our wooden block for an analogy to describe the shearing behavior of the dense array. For the dense array, the wooden block is placed on a plane oriented at an angle $\alpha$ to the horizontal (Fig. 5.10b). Our goal is to find the horizontal force to initiate movement of the block up the incline. You may have solved this problem in statics. Anyway, we are going to solve it again. At impending motion, $T = \mu N$ where $N$ is the normal force. Using the force equilibrium equations in the $X$ and $Z$ directions, we get

$$\Sigma F_x = 0: \quad H - N \sin \alpha - \mu N \cos \alpha = 0 \tag{5.4}$$

$$\Sigma F_z = 0: \quad N \cos \alpha - \mu N \sin \alpha - W = 0 \tag{5.5}$$

Solving for $H$ and $W$, we obtain

$$H = N(\sin \alpha + \mu \cos \alpha) \tag{5.6}$$

$$W = N(\cos \alpha - \mu \sin \alpha) \tag{5.7}$$

Dividing Eq. (5.6) by Eq. (5.7) and simplifying, we obtain

$$\frac{H}{W} = \frac{\mu + \tan \alpha}{1 - \mu \tan \alpha} = \frac{\tan \phi' + \tan \alpha}{1 - \tan \phi' \tan \alpha}$$

By analogy with the loose assembly, we can replace $H$ by $\tau_f$ and $W$ by $(\sigma_n')_f$, resulting in

$$\boxed{\tau_f = (\sigma_n')_f \frac{\tan \phi' + \tan \alpha}{1 - \tan \phi' \tan \alpha} = (\sigma_n')_f \tan(\phi' + \alpha)} \tag{5.8}$$

Let us investigate the implications of Eq. (5.8). If $\alpha = 0$, Eq. (5.8) reduces to Coulomb's frictional equation (5.2). If $\alpha$ increases, the shear strength, $\tau_f$, gets larger. For instance, assume $\phi' = 30°$ and $(\sigma_n')_f$ is constant; then for $\alpha = 0$ we get $\tau_f = 0.58(\sigma_n')_f$, but if $\alpha = 10°$ we get $\tau_f = 0.84(\sigma_n')_f$, that is, an increase of 45% in shear strength for a 10% increase in $\alpha$.

If the normal effective stress increases on our dense disk assembly, the amount of "riding up" of the disks will decrease. In fact, we can impose a sufficiently high normal effective stress to suppress the "riding up" tendencies of the dense disk assembly. Therefore, the ability of the dense disk assembly to expand depends on the magnitude of the normal effective stress. The lower the normal effective stress, the greater the value of $\alpha$. The net effect of $\alpha$ due to normal effective stress increases is that the failure envelope becomes curved as illustrated

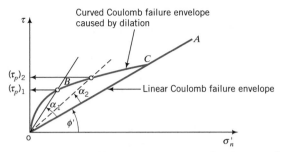

**FIGURE 5.11** Effects of dilation on Coulomb's failure envelope.

by OBC in Fig. 5.11, which is similar to the expected peak shear stress response of Type II soils (Fig. 5.5c).

The geometry of soil grains and the structural arrangement of soil fabrics are much more complex than our loose and dense assembly of disks. However, the model using disks is applicable to soils if we wish to interpret their (soils) shear strength using Coulomb's frictional law. In real soils, the particles are randomly distributed and often irregular. Shearing of a given volume of soil would cause impending slip of some particles to occur up the plane while others occur down the plane. The general form of Eq. (5.8) is then

$$\tau_f = (\sigma_n')_f \tan(\phi' \pm \alpha) \tag{5.9}$$

where the positive sign refers to soils in which the net movement of the particles is initiated up the plane and the negative sign refers to net particle movement down the plane.

We will call the angle $\alpha$ the dilation angle. It is a measure of the change in volumetric strain with respect to the change in shear strain. Soils that have positive values of $\alpha$ expand during shearing while soils with negative values of $\alpha$ contract during shearing. In Mohr's circle of strain (Fig. 5.12), the dilation angle is

$$\alpha = \sin^{-1}\left(-\frac{\Delta\varepsilon_1 + \Delta\varepsilon_3}{\Delta\varepsilon_1 - \Delta\varepsilon_3}\right) = \sin^{-1}\left(-\frac{\Delta\varepsilon_1 + \Delta\varepsilon_3}{(\Delta\gamma_{zx})_{max}}\right) \tag{5.10}$$

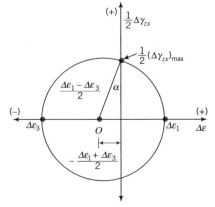

**FIGURE 5.12** Mohr's circle of strain and angle of dilation.

where $\Delta$ denotes change. The negative sign is used because we want $\alpha$ to be positive when the soil is expanding. You should recall that compression is taken as positive in soil mechanics. The angle $\alpha$ is also the tangent to the curve in a plot of volumetric strain versus shear strain as illustrated for simple shear in Fig. 5.3b.

If a soil mass is constrained in the lateral directions, the dilation angle is represented as

$$\alpha = \tan^{-1}\left(\frac{-\Delta z}{\Delta x}\right) \tag{5.11}$$

Dilation is not a peculiarity of soils but occurs in many other materials, for example, rice and wheat. The ancient traders of grains were well aware of the phenomenon of volume expansion of grains. However, it was Osborne Reynolds (1885) who described the phenomenon of dilatancy and brought it to the attention of the scientific community.

For cemented soils, Coulomb's frictional law can be written as

$$\tau_f = c_o + (\sigma_n')_f \tan(\phi' + \alpha) \tag{5.12}$$

where $c_o$ is called cohesion (Fig. 5.7).

> *The* **essential points** *are:*
>
> 1. *Shear failure of soils can be modeled using Coulomb's frictional law, $\tau_f = (\sigma_n')_f \tan(\phi' \pm \alpha)$ where $\tau_f$ is the shear stress when slip is initiated, $(\sigma_n')_f$ is the normal effective stress on the slip plane, $\phi'$ is the friction angle, and $\alpha$ is the dilation angle.*
> 2. *The effect of dilation is to increase the shear strength of the soil and cause the Coulomb's failure envelope to be curved.*
> 3. *Large normal effective stresses tend to suppress dilation.*
> 4. *At the critical state, the dilation angle is zero.*
> 5. *For cemented soils, Coulomb's frictional law is $\tau_f = c_o + (\sigma_n')_f \tan(\phi' + \alpha)$ where $c_o$ is called cohesion.*

*What's next . . .*In the next section, we will define and describe various parameters to interpret the shear strength of soils. It is an important section, which you should read carefully, because it is an important juncture in our understanding of shear strength of soils for soil stability analyses and design considerations.

## 5.5  INTERPRETATION OF THE SHEAR STRENGTH OF SOILS

The shear strength of a soil is its resistance to shearing stresses. In this book, we will interpret the shear strength of soils based on their capacity to dilate. Dense sands and overconsolidated clays (OCR > 2) tend to show peak shear stresses and expand (positive dilation angle), while loose sands and normally consolidated and lightly overconsolidated clays do not show peak shear stresses except at very low normal effective stresses and tend to compress (negative dilation angle). In our interpretation of shear strength, we will describe soils as dilating soils when they exhibit peak shear stresses at $\alpha > 0$ and nondilating soils when they exhibit no peak shear stress and attain a maximum shear stress at $\alpha = 0$. However, a nondilating soil does not mean that it does not change volume (expand or contract) during shearing. The terms dilating and nondilating only refer to particular stress states (peak and critical) during soil deformation.

We will refer to key soil shear strength parameters using the following notation. The peak shear strength, $\tau_p$, is the peak shear stress attained by a dilating soil (Fig. 5.3). The dilation angle at peak shear stress will be denoted as $\alpha_p$. The shear stress attained by all soils at large shear strains ($\gamma_{zx} > 10\%$), when the dilation angle is zero, is the critical state shear strength denoted by $\tau_{cs}$. The void ratio corresponding to the critical state shear strength is the critical void ratio denoted by $e_{cs}$. The effective friction angle corresponding to the critical state shear strength and critical void ratio is $\phi'_{cs}$.

The peak effective friction angle for a dilating soil is

$$\phi'_p = \phi'_{cs} + \alpha_p \tag{5.13}$$

Test results (Bolton, 1986) show that for plane strain tests

$$\phi'_p = \phi'_{cs} + 0.8\alpha_p \tag{5.14}$$

We will continue to use Eq. (5.13) but in practice you can make the adjustment [Eq. (5.14)] suggested by Bolton (1986).

Typical values of $\phi'_{cs}$, $\phi'_p$ and $\phi'_r$ for soils are shown in Table 5.1.

We will drop the term effective in describing friction angle and accept it by default such that effective critical state friction angle becomes critical state friction angle, $\phi'_{cs}$, and effective peak friction angle becomes peak friction angle, $\phi'_p$.

The Coulomb equations for the shear strength of soils are:

All soils, critical state shear strength: $\boxed{\tau_{cs} = (\sigma'_n)_f \tan \phi'_{cs}} \tag{5.15}$

Dilating soils, peak shear strength: $\boxed{\tau_p = (\sigma'_n)_f \tan(\phi'_{cs} + \alpha_p) = (\sigma'_n)_f \tan \phi'_p} \tag{5.16}$

### TABLE 5.1 Ranges of Friction Angles for Soils (degrees)

| Soil Type | $\phi'_{cs}$ | $\phi'_p$ | $\phi'_r$ |
|---|---|---|---|
| Gravel | 30–35 | 35–50 | |
| Mixtures of gravel and sand with fine-grained soils | 28–33 | 30–40 | |
| Sand | 27–37* | 32–50 | |
| Silt or silty sand | 24–32 | 27–35 | |
| Clays | 15–30 | 20–30 | 5–15 |

*a*Higher values (32°–37°) in the range are for sands with significant amount of feldspar (Bolton, 1986). Lower values (27°–32°) in the range are for quartz sands.

The Coulomb equation for soils that exhibit residual shear strength is

$$\tau_r = (\sigma_n')_f \tan \phi_r' \qquad (5.17)$$

where $\phi_r'$ is the residual friction angle. The residual shear strength is very important in the analysis and design of slopes in overconsolidated clays and previously failed slopes.

In designing geotechnical structures you should not rely on $\phi_p'$. In a soil mass, not all planes will mobilize the peak shear strength. Regions of a soil mass that reach the peak shear strength condition would likely fail suddenly. Geotechnical engineers prefer to design geotechnical systems on the basis that if the failure state were to occur, the soil will not collapse suddenly but will continuously deform under constant load. You should use the critical state value, $\phi_{cs}'$, in most design problems since at this state continuous shearing occurs at constant shear stress and volume. Also, $\phi_{cs}'$ is a constant and a fundamental soil parameter, while $\phi_p'$ is not a fundamental soil parameter but depends on the dilation capacity of the soil, which may be suppressed at large normal effective stresses.

In certain problems, dealing with the determination of the peak load for dilating soils, sometimes called collapse load, the use of $\phi_{cs}'$ underpredicts the collapse load. For these types of calculations, it is advisable to use $\phi_p'$. You must, however, obtain $\phi_p'$ under the expected normal effective stresses that the soil mass will be subjected to in the field.

The Coulomb equation gives no information on the shear strains required to initiate slip. Strains (shear and volumetric) are important in the evaluation of shear strength and deformation of soils for design of safe foundations, slopes, and other geotechnical systems. In Chapter 6, we will develop a simple model in which we will consider strains at which soil failure occurs.

If the shear stress ($\tau$) induced in a soil is less than the peak or critical shear strength, then the soil has reserved shear strength and we can characterize this reserved shear strength by a factor of safety (FS).

For peak condition in dilating soils: $\quad FS = \dfrac{\tau_p}{\tau} \qquad (5.18)$

For critical state condition in all soils: $\quad FS = \dfrac{\tau_{cs}}{\tau} \qquad (5.19)$

> *The essential points are:*
>
> 1. *The friction angle at the critical state, $\phi_{cs}'$, is a fundamental soil parameter.*
>
> 2. *The friction angle at peak shear stress for dilating soils, $\phi_p'$, is not a fundamental soil parameter but depends on the capacity of the soil to dilate.*
>
> 3. *Coulomb equation only gives information of the soil shear strength when slip is initiated. It does not give any information on the strains at which soil failure occurs.*

*What's next . . .* Coulomb's frictional law for finding the shear strength of soils requires that we know the friction angle and the normal effective stress on the slip plane. Both of these are not readily known for soils because soils are usually subjected to a variety of stresses. You should recall from Chapter 3 that Mohr's circle can be used to determine the stress within a soil mass. By combining Mohr's circle for finding stress states with Coulomb's frictional law we can develop a generalized failure criterion.

## 5.6 MOHR–COULOMB FAILURE CRITERION

Let us draw a Coulomb frictional failure line as illustrated by $AB$ in Fig. 5.13 and subject a cylindrical sample of soil to principal stresses so that Mohr's circle touches the Coulomb failure line.

Of course, several circles can share $AB$ as the common tangent but we will show only one for simplicity. The point of tangency is at $B$ $[\tau_f, (\sigma'_n)_f]$ and the center of the circle is at $O$. We are going to discuss the top half of the circle; the bottom half is a reflection of the top half. The major and minor principal effective stresses at failure are $(\sigma'_1)_f$ and $(\sigma'_3)_f$. Our objective is to find a relationship between the principal effective stresses and $\phi'$ at failure. We will discuss the appropriate $\phi'$ later in this section.

From the geometry of Mohr's circle,

$$\sin \phi' = \frac{OB}{OA} = \frac{\dfrac{(\sigma'_1)_f - (\sigma'_3)_f}{2}}{\dfrac{(\sigma'_1)_f + (\sigma'_3)_f}{2}}$$

which reduces to

$$\sin \phi' = \frac{(\sigma'_1)_f - (\sigma'_3)_f}{(\sigma'_1)_f + (\sigma'_3)_f} \tag{5.20}$$

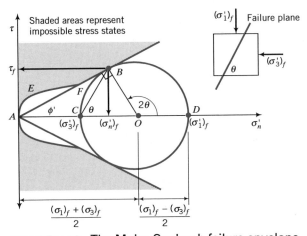

**FIGURE 5.13**  The Mohr–Coulomb failure envelope.

Rearranging Eq. (5.20) gives

$$\frac{(\sigma_1')_f}{(\sigma_3')_f} = \frac{1 + \sin \phi'}{1 - \sin \phi'} = K_p \qquad (5.21)$$

or

$$\frac{(\sigma_3')_f}{(\sigma_1')_f} = \frac{1 - \sin \phi'}{1 + \sin \phi'} = K_a \qquad (5.22)$$

where $K_p$ and $K_a$ are called the passive and active earth pressure coefficients. In Chapter 10, we will discuss $K_p$ and $K_a$ and use them in connection with the analysis of earth retaining walls. The angle $BOD = 2\theta$ represents the inclination of the failure plane or slip plane to the plane on which the major principal effective stress acts in Mohr's circle. Let us find a relationship between $\theta$ and $\phi'$. From the geometry of Mohr's circle (Fig. 5.13),

$$\angle BOC = 90 - \phi' \quad \text{and} \quad \angle BOD = 2\theta = 90° + \phi'$$

$$\therefore \quad \theta = 45° + \frac{\phi'}{2} = \frac{\pi}{4} + \frac{\phi'}{2} \qquad (5.23)$$

The Mohr–Coulomb failure criterion requires that stresses on the soil mass cannot lie within the shaded region shown in Fig. 5.13. That is, the soil cannot have stress states greater than the failure stress state. The shaded areas are called the regions of impossible stress states. The bounding curve for possible stress states is the failure envelope, AEFB, for dilating soils. For nondilating soils, the bounding curve is the linear line AFB. The Mohr–Coulomb failure criterion derived here is independent of the intermediate principal effective stress $\sigma_2'$ and does not consider the strains at which failure occurs.

Traditionally, failure criteria are defined in terms of stresses. Strains are considered at working stresses (stresses below the failure stresses) using stress–strain relationships (also called constitutive relationships) such as Hooke's law. Strains are important because although the stress or load imposed on a soil may not cause it to fail, the resulting strains or displacements may be intolerable. We will describe a simple model in Chapter 6 that considers the effects of the intermediate principal effective stress and strains on soil behavior.

If we normalize (make the quantity a number, i.e., no units) Eq. (5.20) by dividing the nominator and denominator by $\sigma_3'$, we get

$$\sin \phi' = \frac{\dfrac{(\sigma_1')_f}{(\sigma_3')_f} - 1}{\dfrac{(\sigma_1')_f}{(\sigma_3')_f} + 1}$$

The implication of this equation is that the Mohr–Coulomb failure criterion defines failure when the maximum principal effective stress ratio, called maximum effective stress obliquity, $\dfrac{(\sigma_1')_f}{(\sigma_3')_f}$, is achieved and not when the maximum shear stress, $[(\sigma_1' - \sigma_3')/2]_{\max}$, is achieved. The failure shear stress is then less than the maximum shear stress.

You should use the appropriate value of $\phi'$ in the Mohr–Coulomb equation [Eq. (5.20)]. For nondilating soils, $\phi' = \phi_{cs}'$; while for dilating soils, $\phi' = \phi_p'$.

*The* **essential points** *are:*

1. *Coupling Mohr's circle with Coulomb's frictional law allows us to define shear failure based on the stress state of the soil.*

2. *The Mohr–Coulomb failure criterion is*

$$\sin \phi' = \frac{(\sigma_1')_f - (\sigma_3')_f}{(\sigma_1')_f + (\sigma_3')_f}$$

*or*

$$\frac{(\sigma_1')_f}{(\sigma_3')_f} = \frac{1 + \sin \phi'}{1 - \sin \phi'}$$

*or*

$$\frac{(\sigma_3')_f}{(\sigma_1')_f} = \frac{1 - \sin \phi'}{1 + \sin \phi'}$$

3. *Failure occurs, according to the Mohr–Coulomb failure criterion, when the soil reaches the maximum effective stress obliquity, that is,*
$$\left\{ \frac{(\sigma_1')_f}{(\sigma_3')_f} \right\}_{max}.$$

4. *The failure plane or slip plane is inclined at an angle $\theta = \pi/4 + \phi'/2$ to the plane on which the major principal effective stress acts.*

5. *The maximum shear stress, $\tau_{max} = [(\sigma_1' - \sigma_3')/2]_{max}$, is not the failure shear stress.*

## EXAMPLE 5.1

A cylindrical soil sample was subjected to axial principal stresses ($\sigma_1'$) and radial principal stresses ($\sigma_3'$). The soil could not support additional stresses when $\sigma_1' = 300$ kPa and $\sigma_3' = 100$ kPa. Determine the friction angle and the inclination of the slip plane to the horizontal. Assume no significant dilational effects.

**Strategy** This example is a straightforward application of Eqs. (5.20) and (5.23). Since dilation is neglected $\phi' = \phi_{cs}'$.

## Solution 5.1

**Step 1:** Find $\phi_{cs}'$.
From Eq. (5.20),

$$\sin \phi_{cs}' = \frac{(\sigma_1')_f - (\sigma_3')_f}{(\sigma_1')_f + (\sigma_3')_f} = \frac{300 - 100}{300 + 100} = \frac{2}{4} = \frac{1}{2}$$

$$\therefore \phi_{cs}' = 30°$$

**Step 2:** Find $\theta$.
From Eq. (5.23),

$$\theta = 45° + \frac{\phi_{cs}'}{2} = 45° + \frac{30°}{2} = 60°$$

■

## EXAMPLE 5.2

Figure E5.2 shows the soil profile at a site for a proposed building. Determine the increase in vertical effective stress at which a soil element at a depth of 3 m, under the center of the building, will fail if the increase in lateral effective stress is 40% of the increase in vertical effective stress. The coefficient of lateral earth pressure at rest, $K_o$, is 0.5.

**Strategy**   You are given a uniform deposit of sand and its properties. Use the data given to find the initial stresses and then use the Mohr–Coulomb equation to solve the problem. Since the soil element is under the center of the building, axisymmetric conditions prevail. Also, you are given that $\Delta\sigma_3' = 0.4\Delta\sigma_1'$. Therefore, all you need to do is to find $\Delta\sigma_1'$.

## Solution 5.2

**Step 1:**   Find the initial effective stresses.
Assume the top 1 m of soil to be saturated.

$$\sigma_{zo}' = (\sigma_1')_o = (18 \times 3) - 9.8 \times 2 = 34.4 \text{ kPa}$$

The subscript $o$ denotes original or initial.
The lateral earth pressure is

$$(\sigma_x')_o = (\sigma_3')_o = K_o(\sigma_z')_o = 0.5 \times 34.4 = 17.2 \text{ kPa}$$

**Step 2:**   Find $\Delta\sigma_1'$.

$$\text{At failure:} \quad \frac{(\sigma_1')_f}{(\sigma_3')_f} = \frac{1 + \sin\phi_{cs}'}{1 - \sin\phi_{cs}'} = \frac{1 + \sin 30°}{1 - \sin 30°} = 3$$

But

$$(\sigma_1')_f = (\sigma_1')_o + \Delta\sigma_1' \quad \text{and} \quad (\sigma_3')_f = (\sigma_3')_o + 0.4\Delta\sigma_1'$$

where $\Delta\sigma_1'$ is the additional vertical effective stress to bring the soil to failure.

$$\therefore \frac{(\sigma_1')_o + \Delta\sigma_1'}{(\Delta\sigma_3)_o + 0.4\Delta\sigma_1'} = \frac{34.4 + \Delta\sigma_1'}{17.2 + 0.4\Delta\sigma_1'} = 3$$

The solution is $\Delta\sigma_1' = 86$ kPa.

Ground surface

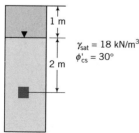

$\gamma_{sat} = 18$ kN/m$^3$
$\phi_{cs}' = 30°$

**FIGURE E5.2**

*What's next . .* .In the next section, we will consider two rather extreme conditions—drained and undrained conditions—under which soil is loaded and the effects these loading conditions have on the shear strength. Drained and undrained conditions are the bounds to evaluate soil stability.

## 5.7 UNDRAINED AND DRAINED SHEAR STRENGTH

You were introduced to drained and undrained conditions when we discussed stress paths in Chapter 3. Drained condition occurs when the excess pore water pressure developed during loading of a soil dissipates, i.e., $\Delta u = 0$. Undrained condition occurs when the excess pore water pressure cannot drain, at least quickly, from the soil; that is, $\Delta u \neq 0$. The existence of either condition—drained or undrained—depends on the soil type, geological formation (fissures, sand layers in clays, etc.), and the rate of loading.

The rate of loading under the undrained condition is often much faster than the rate of dissipation of the excess pore water pressure and the volume change tendency of the soil is suppressed. The result of this suppression is a change in excess pore water pressure during shearing. A soil with a tendency to compress during drained loading will exhibit an increase in excess pore water pressure (positive excess pore water pressure, Fig. 5.14) under undrained condition resulting in a decrease in effective stress. A soil that expands during drained loading will exhibit a decrease in excess pore water pressure (negative excess pore water pressure, Fig. 5.14) under undrained condition resulting in an increase in effective stress. These changes in excess pore water pressure occur because the void ratio does not change during undrained loading; that is, the volume of the soil remains constant.

During the life of a geotechnical structure, called the long-term condition, the excess pore water pressure developed by a loading dissipates and drained condition applies. Clays usually take many years to dissipate the excess pore water pressures. During construction and shortly after, called the short-term condition, soils with low permeability (fine-grained soils) do not have sufficient time

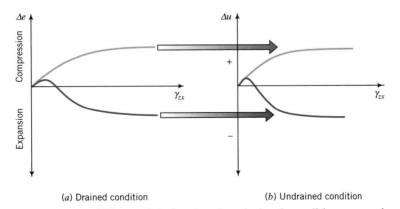

(*a*) Drained condition                    (*b*) Undrained condition

**FIGURE 5.14**  Effects of drained and undrained conditions on volume changes.

for the excess pore water pressure to dissipate and undrained condition applies. The permeability of coarse-grained soils is sufficiently large that under static loading conditions the excess pore water pressure dissipates quickly. Consequently, undrained condition does not apply to clean coarse-grained soils under static loading but only to fine-grained soils and to mixtures of coarse and fine grained soils. Dynamic loading, such as during an earthquake, is imposed so quickly that even coarse-grained soils do not have sufficient time to dissipate the excess pore water pressure and undrained condition applies.

The shear strength of a fine-grained soil under undrained condition is called the undrained shear strength, $s_u$. The undrained shear strength $s_u$, is the radius of the Mohr total stress circle; that is,

$$s_u = \frac{(\sigma_1)_f - (\sigma_3)_f}{2} = \frac{(\sigma'_1)_f - (\sigma'_3)_f}{2} \tag{5.24}$$

as shown in Fig. 5.15a. The shear strength under undrained loading depends only on the initial void ratio or the initial water content. An increase in initial normal stress, sometimes called confining pressure, causes a decrease in initial void ratio and a larger change in excess pore water pressure when a soil is sheared under undrained condition. The result is that the Mohr's circle of total stress expands and the undrained shear strength increases (Fig. 5.15b). Thus, $s_u$ is not a fundamental soil shear strength property. The value of $s_u$ depends on the magnitude of the confining pressure. Analyses of soil strength and soil stability problems using $s_u$ are called total stress analyses (TSA).

When designing geotechnical systems, geotechnical engineers must consider both drained and undrained conditions to determine which of these conditions is critical. The decision on what shear strength parameters to use depends on whether you are considering the short-term (undrained) or the long-term (drained) conditions. In the case of analyses for drained condition, called effective stress analyses (ESA), the shear strength parameters are $\phi'_p$ and $\phi'_{cs}$. The value of $\phi'_{cs}$ is constant for a soil regardless of its initial condition and the magnitude of the normal effective stress. But the value of $\phi'_p$ depends on the normal effective stress. In the case of fine-grained soils, the shear strength parameter for short-term loading is $s_u$.

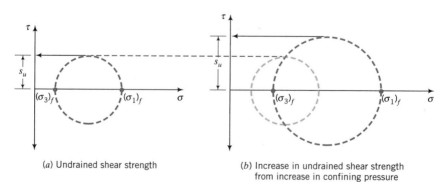

(a) Undrained shear strength

(b) Increase in undrained shear strength from increase in confining pressure

**FIGURE 5.15**   Mohr's circles for undrained conditions.

*The* **essential points** *are:*

1. *Volume changes that occur under drained condition are suppressed under undrained condition. The result of this suppression is that a soil with a compression tendency under drained condition will respond with positive excess pore water pressures during undrained condition, and a soil with an expansion tendency during drained condition will respond with negative excess pore water pressures during undrained condition.*

2. *For an effective stress analysis, the shear strength parameters are $\phi'_{cs}$ and $\phi'_{p}$.*

3. *For a total stress analysis, which applies to fine-grained soils, the shear strength parameter is the undrained shear strength, $s_u$.*

4. *The undrained shear strength depends on the initial void ratio. It is not a fundamental soil shear strength parameter.*

*What's next* . . .We have identified the shear strength parameters ($\phi'_{cs}$, $\phi'_p$ and $s_u$) that are important for analyses and design of geotechnical systems. A variety of laboratory tests and field tests are used to determine these parameters. We will describe many of these tests and the interpretation of the results. You may have to perform some of these tests in the laboratory section of your course.

## 5.8 LABORATORY TESTS TO DETERMINE SHEAR STRENGTH PARAMETERS

### 5.8.1 Shear Box or Direct Shear Test

A popular apparatus to determine the shear strength parameters is the shear box. This test is useful when a soil mass is likely to fail along a thin zone under plane strain conditions. The shear box (Fig. 5.16) consists of a horizontally split, open metal box. Soil is placed in the box, and one-half of the box is moved relative to the other half. Failure is thereby constrained along a thin zone of soil on the horizontal plane (*AB*). Serrated or grooved metal plates are placed at the top and bottom faces of the soil to generate the shearing force.

Vertical forces are applied through a metal platen resting on the top serrated plate. Horizontal forces are applied through a motor for displacement con-

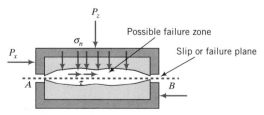

**FIGURE 5.16** Shear box.

trol or by weights through a pulley system for load control. Most shear box tests are conducted using displacement control because we can get both the peak shear force and the critical shear force. In load control tests, you cannot get data beyond the maximum or peak shear force.

The horizontal displacement, $\Delta x$, the vertical displacement, $\Delta z$, the vertical loads, $P_z$, and the horizontal loads, $P_x$, are measured. Usually, three or more tests are carried out on a soil sample using three different constant vertical forces. Failure is determined when the soil cannot resist any further increment of horizontal force. The stresses and strains in the shear box test are difficult to calculate from the forces and displacements measured. The stresses in the thin (dimension unknown) constrained failure zone (Fig. 5.16) are not uniformly distributed and strains cannot be determined.

The shear box apparatus cannot prevent drainage, but one can get an estimate of the undrained shear strength of clays by running the shear box test at a fast rate of loading so that the test is completed quickly. However, the shear box test should not be used for an accurate determination of the undrained shear strength of soils.

In summary, drained tests are generally conducted in a shear box test. Three or more tests are performed on a soil. The soil sample in each test is sheared under a constant vertical force, which is different in each test. The data recorded for each test are the horizontal displacements, the horizontal forces, the vertical displacements, and the constant vertical force under which the test is conducted. From the recorded data, you can find the following parameters: $\tau_p$, $\tau_{cs}$, $\phi'_p$, $\phi'_{cs}$, $\alpha$ (and $s_u$, if fine-grained soils are tested quickly). These parameters are generally determined from plotting the data, as illustrated in Fig. 5.17 for sand.

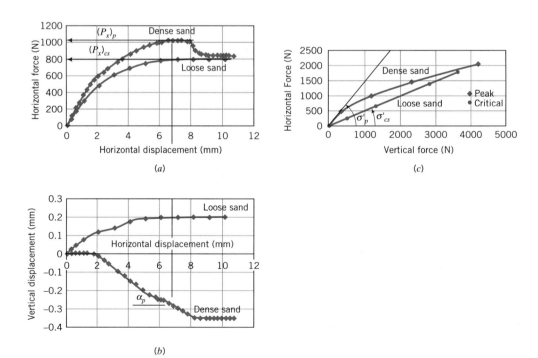

**FIGURE 5.17** Results from a shear box test on a dense and a loose sand.

Only the results of one test at a constant value of $P_z$ are shown in Figs. 5.17a,b. The results of $(P_x)_p$ and $(P_x)_{cs}$ plotted against $P_z$ for all tests are shown in Fig. 5.17c. If the soil is dilatant, it would exhibit a peak shear force (Fig. 5.17a, dense sand) and expand (Fig. 5.17b, dense sand), and the failure envelope would be curved (Fig. 5.17c, dense sand). The peak shear stress is the peak shear force divided by the cross-sectional area $(A)$ of the test sample; that is,

$$\tau_p = \frac{(P_x)_p}{A} \tag{5.25}$$

The critical shear stress is

$$\tau_{cs} = \frac{(P_x)_{cs}}{A} \tag{5.25}$$

In a plot of vertical forces versus horizontal forces (Fig. 5.17c), the points representing the critical horizontal forces should ideally lie along a straight line through the origin. Experimental results usually show small deviations from this straight line and a "best-fit" straight line is conventionally drawn. The angle subtended by this straight line and the horizontal axis is $\phi'_{cs}$. Alternatively,

$$\phi'_{cs} = \tan^{-1}\frac{(P_x)_{cs}}{P_z} \tag{5.27}$$

For dilatant soils, each point representing the peak horizontal force in Fig. 5.17c that does not lie on the "best-fit" straight line represents a value of $\phi'_p$ at the corresponding vertical force. Recall from Section 5.5 that $\phi'_p$ is not constant but varies with the magnitude of the normal effective stress $(P_z/A)$. Usually, the normal effective stress at which $\phi'_p$ is determined should correspond to the maximum anticipated normal effective stress in the field. The value of $\phi'_p$ is largest at the lowest value of the applied normal effective stress, as illustrated in Fig. 5.17c. You would determine $\phi'_p$ by drawing a line from the origin to the point representing the peak horizontal force at the desired normal force, and measuring the angle subtended by this line and the horizontal axis. Alternatively,

$$\phi'_p = \tan^{-1}\frac{(P_x)_p}{P_z} \tag{5.28}$$

You can also determine the peak dilation angle directly for each test from a plot of horizontal displacement versus vertical displacement, as illustrated in Fig. 5.17b. The peak dilation angle is

$$\boxed{\alpha_p = \tan^{-1}\left(\frac{-\Delta z}{\Delta x}\right)} \tag{5.29}$$

We can find $\alpha_p$ from

$$\alpha_p = \phi'_p - \phi'_{cs} \tag{5.30}$$

## EXAMPLE 5.3

The shear box test results of two samples of the same soil but with different initial unit weights are shown in the table below. Sample A did not show any peak values but Sample B did.

| Soil | Test number | Vertical force (N) | Horizontal force (N) |
|------|-------------|--------------------|-----------------------|
| A | Test 1 | 250 | 150 |
|   | Test 2 | 500 | 269 |
|   | Test 3 | 750 | 433 |
| B | Test 1 | 100 | 98 |
|   | Test 2 | 200 | 175 |
|   | Test 3 | 300 | 210 |
|   | Test 4 | 400 | 248 |

Determine the following:

**(a)** $\phi'_{cs}$

**(b)** $\phi'_p$ at vertical forces of 200 N and 400 N for sample B

**(c)** The dilation angle at vertical forces of 200 N and 400 N for sample B

***Strategy***    To obtain the desired values, it is best to plot a graph of vertical force versus horizontal force.

## Solution 5.3

**Step 1:**    Plot a graph of the vertical forces versus failure horizontal forces for each sample. See Fig. E5.3.

**Step 2:**    Extract $\phi'_{cs}$.
All the plotted points for sample A fall on a straight line through the origin. Sample A is a nondilatant soil, possibly a loose sand or a normally consolidated clay. The effective friction angle is $\phi'_{cs} = 30°$.

**Step 3:**    Determine $\phi'_p$.
The horizontal forces at 200 N and 400 N for sample B do not lie on the straight line corresponding to $\phi'_{cs}$. Therefore, each of these forces has a $\phi'_p$ associated with it.

$$(\phi'_p)_{200\ N} = \tan^{-1}\left(\frac{175}{200}\right) = 41.2°$$

$$(\phi'_p)_{400\ N} = \tan^{-1}\left(\frac{248}{400}\right) = 31.8°$$

**Step 4:**    Determine $\alpha_p$.

$$\alpha_p = \phi'_p - \phi'_{cs}$$
$$(\alpha)_{200\ N} = 41.2 - 30 = 11.2°$$
$$(\alpha)_{400\ N} = 31.8 - 30 = 1.8°$$

Note that as the normal force increases $\alpha_p$ decreases.

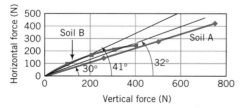

**FIGURE E5.3**

## EXAMPLE 5.4

The critical state friction angle of a soil is 28°. Determine the critical state shear stress if the normal effective stress is 200 kPa.

**Strategy** This is a straightforward application of the Coulomb's frictional equation.

## Solution 5.4

**Step 1:** Determine the failure shear stress.

$$\tau_f = \tau_{cs} = (\sigma_n')_f \tan \phi_{cs}'$$
$$\tau_{cs} = 200 \tan 28° = 106.3 \text{ kPa} \qquad \blacksquare$$

## EXAMPLE 5.5

The data recorded during a shear box test on a sand sample, 10 cm × 10 cm × 3 cm, at a constant vertical force of 1200 N are shown in the table below. A negative sign denotes vertical expansion.

**(a)** Plot graphs of (1) horizontal forces versus horizontal displacements and (2) vertical displacements versus horizontal displacements.

**(b)** Would you characterize the behavior of this sand as that of a dense or a loose sand? Explain your answer.

**(c)** Determine (1) the maximum or peak shear stress, (2) the critical state shear stress, (3) the peak dilation angle, (4) $\phi_p'$, and (5) $\phi_{cs}'$.

| Horizontal displacement (mm) | Horizontal force (N) | Vertical displacement (mm) | Horizontal displacement (mm) | Horizontal force (N) | Vertical displacement (mm) |
|---|---|---|---|---|---|
| 0.00 | 0.00 | 0.00 | 6.10 | 988.29 | −0.40 |
| 0.25 | 82.40 | 0.00 | 6.22 | 988.29 | −0.41 |
| 0.51 | 157.67 | 0.00 | 6.48 | 993.68 | −0.45 |
| 0.76 | 249.94 | 0.00 | 6.60 | 998.86 | −0.46 |
| 1.02 | 354.31 | 0.00 | 6.86 | 991.52 | −0.49 |
| 1.27 | 425.72 | 0.01 | 7.11 | 999.76 | −0.51 |
| 1.52 | 488.90 | 0.00 | 7.37 | 1005.26 | −0.53 |
| 1.78 | 538.33 | 0.00 | 7.75 | 1002.51 | −0.57 |
| 2.03 | 571.29 | −0.01 | 7.87 | 994.27 | −0.57 |
| 2.41 | 631.62 | −0.03 | 8.13 | 944.83 | −0.58 |
| 2.67 | 663.54 | −0.05 | 8.26 | 878.91 | −0.58 |
| 3.30 | 759.29 | −0.09 | 8.51 | 807.50 | −0.58 |
| 3.68 | 807.17 | −0.12 | 8.64 | 791.02 | −0.59 |
| 4.06 | 844.47 | −0.16 | 8.89 | 774.54 | −0.59 |
| 4.45 | 884.41 | −0.21 | 9.14 | 766.30 | −0.60 |
| 4.97 | 928.35 | −0.28 | 9.40 | 760.81 | −0.59 |
| 5.25 | 939.34 | −0.31 | 9.65 | 760.81 | −0.59 |
| 5.58 | 950.32 | −0.34 | 9.91 | 758.06 | −0.60 |
| 5.72 | 977.72 | −0.37 | 10.16 | 758.06 | −0.59 |
| 5.84 | 982.91 | −0.37 | 10.41 | 758.06 | −0.59 |
| 5.97 | 988.29 | −0.40 | 10.67 | 755.32 | −0.59 |

***Strategy***   After you plot the graphs, you can get an idea as to whether you have a loose or dense sand. A dense sand may show a peak horizontal force in the plot of horizontal force versus horizontal displacement and would expand.

## Solution 5.5

**Step 1:**   Plot graphs.
See Fig. E5.5.

**Step 2:**   Determine whether the sand is dense or loose.
The sand appears to be dense—it shows a peak horizontal force and dilated.

**Step 3:**   Extract the required values.

Cross-sectional area of sample:   $A = 10 \times 10 = 100 \text{ cm}^2 = 10^{-2} \text{ m}^2$

**(c1)** $\tau_p = \dfrac{(P_x)_p}{A} = \dfrac{1005 \text{ N}}{10^{-2}} \times 10^{-3} = 100.5 \text{ kPa}$

**(c2)** $\tau_{cs} = \dfrac{(P_x)_{cs}}{A} = \dfrac{758 \text{ N}}{10^{-2}} \times 10^{-3} = 75.8 \text{ kPa}$

**(c3)** $\alpha_p = \tan^{-1}\left(\dfrac{-\Delta z}{\Delta x}\right) = \tan^{-1}\left(\dfrac{0.1}{0.8}\right) = 7.1°$

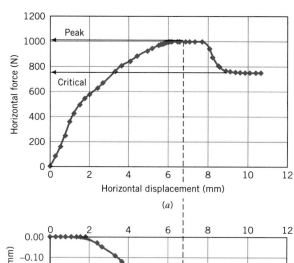

(a)

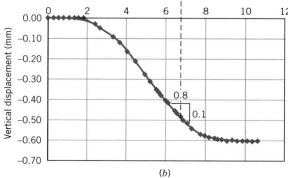

(b)

**FIGURE E5.5**

**(c4)**    Normal stress:  $\sigma_n' = \left(\dfrac{1200\ \text{N}}{10^{-2}}\right) \times 10^{-3} = 120\ \text{kPa}$

$$\phi_p' = \tan^{-1}\!\left(\dfrac{\tau_p}{\sigma_n}\right) = \tan^{-1}\!\left(\dfrac{100.5}{120}\right) = 39.9°$$

$$\phi_{cs}' = \tan^{-1}\!\left(\dfrac{\tau_{cs}}{\sigma_n'}\right) = \tan^{-1}\!\left(\dfrac{75.8}{120}\right) = 32.3°$$

Also,

$$\alpha_p = \sigma_p' - \sigma_{cs}' = 39.9 - 32.3 = 7.6° \qquad\blacksquare$$

### 5.8.2 Conventional Triaxial Apparatus

A widely used apparatus to determine the shear strength parameters and the stress-strain behavior of soils is the triaxial apparatus. The name is a misnomer since two, not three, stresses can be controlled. In the triaxial test, a cylindrical sample of soil, usually with a length to diameter ratio of 2, is subjected to either controlled increases in axial stresses or axial displacements and radial stresses. The sample is laterally confined by a membrane and radial stresses are applied by pressuring water in a chamber (Fig. 5.18). The axial stresses are applied by loading a plunger. If the axial stress is greater than the radial stress, the soil is compressed vertically and the test is called triaxial compression. If the radial stress is greater than the axial stress, the soil is compressed laterally and the test is called triaxial extension.

The applied stresses (axial and radial) are principal stresses and the loading

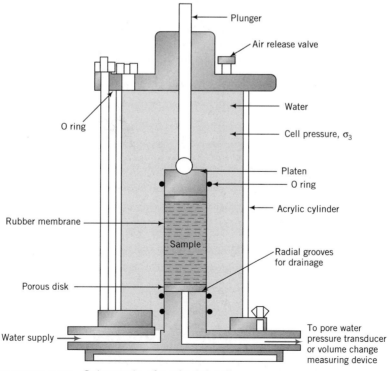

**FIGURE 5.18**   Schematic of a triaxial cell.

condition is axisymmetric. For compression tests, we will denote the radial stresses $\sigma_r$ as $\sigma_3$ and the axial stresses $\sigma_z$ as $\sigma_1$. For extension tests, we will denote the radial stresses $\sigma_r$ as $\sigma_1$ and the axial stresses $\sigma_z$ as $\sigma_3$.

The average stresses and strains on a soil sample in the triaxial apparatus for compression tests are as follows:

$$\text{Axial total stress:} \quad \boxed{\sigma_1 = \frac{P_z}{A} + \sigma_3} \tag{5.31}$$

$$\text{Deviatoric stress:} \quad \boxed{\sigma_1 - \sigma_3 = \frac{P_z}{A}} \tag{5.32}$$

$$\text{Axial strain:} \quad \boxed{\varepsilon_1 = \frac{\Delta z}{H_o}} \tag{5.33}$$

$$\text{Radial strain:} \quad \boxed{\varepsilon_3 = \frac{\Delta r}{r_o}} \tag{5.34}$$

$$\text{Volumetric strain:} \quad \boxed{\varepsilon_p = \frac{\Delta V}{V_o} = \varepsilon_1 + 2\varepsilon_3} \tag{5.35}$$

$$\text{Deviatoric strain:} \quad \boxed{\varepsilon_q = \tfrac{2}{3}(\varepsilon_1 - \varepsilon_3)} \tag{5.36}$$

where $P_z$ is the load on the plunger, $A$ is the cross-sectional area of the soil sample, $r_o$ is the initial radius of the soil sample, $\Delta r$ is the change in radius, $V_o$ is the initial volume, $\Delta V$ is the change in volume, $H_o$ is the initial height, and $\Delta z$ is the change in height. We will call the plunger load, the deviatoric load, and the corresponding stress, the deviatoric stress, $q = (\sigma_1 - \sigma_3)$. The shear stress is $\tau = q/2$.

The area of the sample changes during loading and at any given instance the area is

$$\boxed{A = \frac{V}{H} = \frac{V_o - \Delta V}{H_o - \Delta z} = \frac{V_o \left(1 - \dfrac{\Delta V}{V_o}\right)}{H_o \left(1 - \dfrac{\Delta z}{H_o}\right)} = \frac{A_o(1 - \varepsilon_p)}{1 - \varepsilon_1}} \tag{5.37}$$

where $A_o \ (= \pi r_o^2)$ is the initial cross-sectional area and $H$ is the current height of the sample. The dilation angle for a triaxial test is given by Eq. (5.10).

The triaxial apparatus is versatile because we can (1) independently control the applied axial and radial stresses, (2) conduct tests under drained and undrained conditions, and (3) control the applied displacements or stresses.

A variety of stress paths can be applied to soil samples in the triaxial apparatus. However, only a few stress paths are used in practice to mimic typical geotechnical problems. We will discuss the tests most often used, why they are used, and typical results obtained.

### 5.8.3 Unconfined Compression (UC) Test

The purpose of this test is to determine the undrained shear strength of saturated clays quickly. In the UC test, no radial stress is applied to the sample ($\sigma_3 = 0$). The plunger load, $P_z$, is increased rapidly until the soil sample fails, that is, cannot

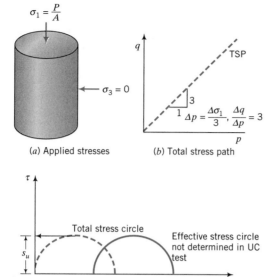

(a) Applied stresses   (b) Total stress path

(c) Mohr's circles

**FIGURE 5.19** Stresses, stress paths, and Mohr's circle for UC test.

support any additional load. The loading is applied quickly so that the pore water cannot drain from the soil; the sample is sheared at constant volume.

The stresses applied on the soil sample and the total stress path followed are shown in Figs. 5.19a,b. The effective stress path is unknown since pore water pressure changes are not normally measured. Mohr's circle using total stresses is depicted in Fig. 5.19c. If the excess pore water pressures were to be measured, they would be negative. The theoretical reason for negative excess pore water pressures is as follows. Since $\sigma_3 = 0$, then from the principle of effective stresses, $\sigma_3' = \sigma_3 - \Delta u = 0 - \Delta u = -\Delta u$. The effective radial stress, $\sigma_3'$, cannot be negative because soils cannot sustain tension. Therefore, the excess pore water pressure must be negative so that $\sigma_3'$ is positive. Mohr's circle of effective stresses would be to the right of the total stress circle as shown in Fig. 5.19c.

The undrained shear strength is

$$s_u = \frac{P_z}{2A} = \frac{1}{2}\sigma_1 \tag{5.38}$$

where, from Eq. (5.37), $A = A_o/(1 - \varepsilon_1)$ (no volume change, i.e., $\varepsilon_p = 0$). The undrained elastic modulus, $E_u$, is determined from a plot of $\varepsilon_1$ versus $\sigma_1$.

The results from UC tests are used to:

- Estimate the short-term bearing capacity of fine-grained soils for foundations
- Estimate the short-term stability of slopes
- Compare the shear strengths of soils from a site to establish soil strength variability quickly and cost-effectively (the UC test is cheaper to perform than other triaxial tests)
- Determine the stress–strain characteristics under fast (undrained) loading conditions

### EXAMPLE 5.6

An unconfined compression test was carried out on a saturated clay sample. The maximum load the clay sustained was 127 N and the vertical displacement was 0.8 mm. The size of the sample was 38 mm diameter × 76 mm long. Determine the undrained shear strength. Draw Mohr's circle of stress for the test and locate $s_u$.

**Strategy**   Since the test is a UC test, $\sigma_3 = 0$ and $(\sigma_1)_f$ is the failure axial stress. You can find $s_u$ by calculating one-half the failure axial stress.

### Solution 5.6

**Step 1:**   Determine the sample area at failure.
Diameter $D_o = 38$ mm, length, $H_o = 76$ mm.

$$A_o = \frac{\pi \times D_o^2}{4} = \frac{\pi \times 0.038^2}{4} = 11.3 \times 10^{-4}\ \text{m}^2, \quad \varepsilon_1 = \frac{\Delta z}{H_o} = \frac{0.8}{76} = 0.01$$

$$A = \frac{A_o}{1 - \varepsilon_1} = \frac{11.3 \times 10^{-4}}{1 - 0.01} = 11.4 \times 10^{-4}\ \text{m}^2$$

**Step 2:**   Determine the major principal stress at failure.

$$(\sigma_1)_f = \frac{P_z}{A} = \frac{127 \times 10^{-3}}{11.4 \times 10^{-4}} = 111.4\ \text{kPa}$$

**Step 3:**   Calculate $s_u$.

$$s_u = \frac{(\sigma_1)_f - (\sigma_3)_f}{2} = \frac{111.4 - 0}{2} = 55.7\ \text{kPa}$$

**Step 4:**   Draw Mohr's circle.
See Fig. E5.6. The values extracted from the graphs are

$$(\sigma_3)_f = 0, \quad (\sigma_1)_f = 111\ \text{kPa}, \quad s_u = 56\ \text{kPa}$$

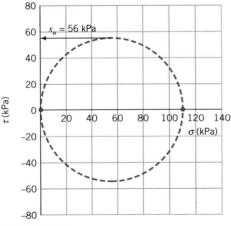

**FIGURE E5.6**

### 5.8.4 Consolidated Drained (CD) Compression Test

The purpose of a CD test is to determine the drained shear strength parameters, $\phi'_{cs}$ and $\phi'_p$, to analyze long-term loading of a soil mass. The effective elastic moduli for drained condition $E'$ and $E'_s$ are also obtained from this test. A consolidated drained compression test is performed in two stages. The first stage is consolidating the soil to a desired effective stress level by pressurizing the water in the cell and allowing the soil sample to drain until the excess pore water pressure dissipates. In the second stage, the pressure in the cell (cell pressure or confining pressure) is kept constant and additional axial loads or displacements are added very slowly until the soil sample fails. The displacement rate (or strain rate) used must be slow enough to allow the excess pore water pressure to dissipate. Because the permeability of fine-grained soils is much lower than coarse-grained soils, the displacement rate for testing fine-grained soils is much lower than for coarse-grained soils. Drainage of the excess pore water is permitted and the amount of water expelled is measured. It is customary to perform a minimum of three tests at different cell pressures. The stresses on the soil sample for the two stages of a CD test are as follows:

**Stage 1: Isotropic consolidation phase**

$$\Delta\sigma_1 = \Delta\sigma_3 = \Delta\sigma'_1 = \Delta\sigma'_3; \quad \Delta\sigma_1 > 0, \quad \Delta u = 0$$

When the load is applied, the initial excess pore water pressure for a saturated soil is equal to the cell pressure (Chapter 4); that is, $\Delta u = \Delta\sigma_3$. At the end of the consolidation phase, the excess pore water pressure dissipates; that is, $\Delta u = 0$.

$$\Delta p' = \Delta p = \frac{\Delta\sigma_1 + 2\Delta\sigma_1}{3} = \Delta\sigma_1; \quad \Delta q = \Delta\sigma_1 - \Delta\sigma_3 = 0, \quad \text{and} \quad \frac{\Delta q}{\Delta p'} = \frac{\Delta q}{\Delta p} = 0$$

**Stage 2: Shearing phase**

$$\Delta\sigma_1 = \Delta\sigma'_1 > 0; \quad \Delta\sigma_3 = \Delta\sigma'_3 = 0; \quad \Delta u = 0$$

Therefore,

$$\Delta p' = \Delta p = \frac{\Delta\sigma'_1}{3}; \quad \Delta q = \sigma_1; \quad \frac{\Delta q}{\Delta p'} = \frac{\Delta q}{\Delta p} = 3$$

The stresses and stress path applied are illustrated in Fig. 5.20. The change in volume of the soil is measured by continuously recording the volume of water expelled. The volumetric strain is

$$\boxed{\varepsilon_p = \frac{\Delta V}{V_o} = \varepsilon_1 + 2\varepsilon_3} \tag{5.39}$$

where $\Delta V$ is the change in volume and $V_o$ is the original volume of the soil. Also, the axial displacements are recorded and the axial strain is calculated as $\varepsilon_1 = \Delta z/H_o$. The radial strains are calculated by rearranging Eq. (5.39) to yield

$$\boxed{\varepsilon_3 = \tfrac{1}{2}(\varepsilon_p - \varepsilon_1)} \tag{5.40}$$

The maximum shear strain is

$$\boxed{(\gamma_{zx})_{max} = (\varepsilon_1 - \varepsilon_3)_{max}}$$

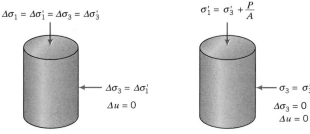

$\Delta\sigma_1 = \Delta\sigma_1' = \Delta\sigma_3 = \Delta\sigma_3'$

$\sigma_1' = \sigma_3' + \dfrac{P}{A}$

$\Delta\sigma_3 = \Delta\sigma_1'$

$\Delta u = 0$

$\sigma_3 = \sigma_3'$

$\Delta\sigma_3 = 0$

$\Delta u = 0$

(a) Stage 1: Consolidation phase

(b) Stage 2: Shearing phase

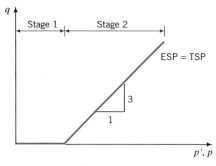

(c) Stress path

**FIGURE 5.20**   Stresses and stress paths during a CD test.

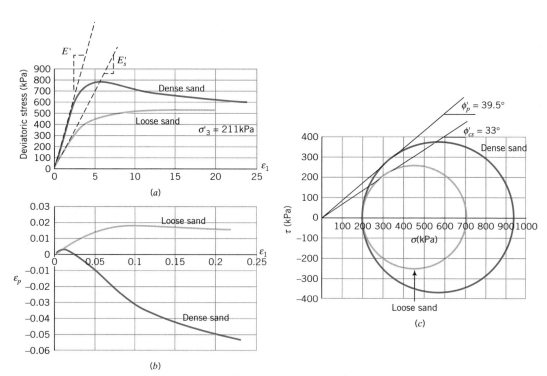

(a)

(b)

$\phi_p' = 39.5°$

$\phi_{cs}' = 33°$

(c)

**FIGURE 5.21**   Results from CD tests on dense and loose sand. (From 'The Measurement of Soil Properties in the Triaxial Test,' by Bishop and Henkel, Edward Arnold 1962.)

which, by substitution of Eq. (5.39), gives

$$(\gamma_{zx})_{max} = \tfrac{1}{2}(3\varepsilon_1 - \varepsilon_p)$$
(5.41)

Since the CD test is a drained test, a single test can take several days if the permeability of the soil is low (e.g., fine-grained soils). Typical results of consolidated drained tests on a sand are shown in Fig. 5.21.

The elastic moduli for drained conditions, $E'$ and $E'_s$, are obtained from the CD test from the plot of deviatoric stress, $(\sigma'_1 - \sigma'_3)$, as ordinate and $\varepsilon_1$ as abscissa as shown in Fig. 5.21a. The results of CD tests are used to determine the long-term stability of slopes, foundations, retaining walls, excavations and other earthworks.

## EXAMPLE 5.7

The results of three CD tests on a soil at failure are as follows:

| Test number | $\sigma'_3$ (kPa) | Deviatoric stress (kPa) |
|---|---|---|
| 1 | 100 | 250 (peak) |
| 2 | 180 | 362 (peak) |
| 3 | 300 | 564 (no peak observed) |

The detailed results for Test 1 are as follows. The negative sign indicates expansion.

| $\Delta z$ (mm) | $\Delta V$ (cm³) | Plunger load— $P_z$ (N) |
|---|---|---|
| 0 | 0.00 | 0.0 |
| 0.152 | 0.02 | 61.1 |
| 0.228 | 0.03 | 94.3 |
| 0.38 | −0.09 | 124.0 |
| 0.76 | −0.50 | 201.5 |
| 1.52 | −1.29 | 257.5 |
| 2.28 | −1.98 | 292.9 |
| 2.66 | −2.24 | 298.9 |
| 3.04 | −2.41 | 298.0 |
| 3.8 | −2.55 | 279.2 |
| 4.56 | −2.59 | 268.4 |
| 5.32 | −2.67 | 252.5 |
| 6.08 | −2.62 | 238.0 |
| 6.84 | −2.64 | 229.5 |
| 7.6 | −2.66 | 223.2 |
| 8.36 | −2.63 | 224.3 |

The initial size of the sample is 38 mm in diameter and 76 mm in length.

(a) Determine the friction angle for each test.

(b) Determine $\tau_p$, $\tau_{cs}$, $E'$, and $E'_s$ at peak shear stress for Test 1.

**(c)** Determine $\phi'_{cs}$.

**(d)** Determine $\alpha_p$ for Test 1.

***Strategy*** From a plot of deviatoric stress versus axial strain for Test 1, you will get $\tau_p$, $\tau_{cs}$, $E'$ and $E'_s$. The friction angles can be calculated or found from Mohr's circle.

### Solution 5.7

**Step 1:** Determine the friction angles. Use a table to do the calculations.

| Test No. | $\sigma'_3$ kPa | $\sigma'_1 - \sigma'_3$ kPa | $\sigma'_1$ kPa | $\sigma'_1 + \sigma'_3$ | $\phi'_p = \sin^{-1} \dfrac{\sigma'_1 - \sigma'_3}{\sigma'_1 + \sigma'_3}$ |
|---|---|---|---|---|---|
| Test 1 | 100 | 250 | 350 | 450 | 33.7° |
| Test 2 | 180 | 362 | 542 | 722 | 30.1° |
| Test 3 | 300 | 564 | 864 | 1164 | 29° |

Alternatively, plot Mohr's circles and determine the friction angles as shown for Test 1 and Test 2 in Fig. E5.7a.

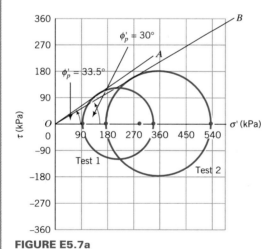

**FIGURE E5.7a**

**Step 2:** Determine $\tau_p$ and $\tau_{cs}$ from a plot of deviatoric stress versus axial strain response for Test 1.

$$\text{The initial area is: } A_o = \frac{\pi D_o^2}{4} = \frac{\pi \times 38^2}{4} = 1134 \text{ mm}^2$$

$$V_o = \frac{\pi D_o H_o}{4} = \frac{\pi \times 38^2 \times 76}{4} = 86193 \text{ mm}^3$$

$$A = \frac{A_o(1 - \varepsilon_p)}{1 - \varepsilon_1}$$

See the table for calculations and Fig. E5.7b for a plot of the results.

| | | $A_o = 1134$ mm² | | | |
| $\Delta z$ (mm) | $\varepsilon_1$<br>$\Delta z/H_o$ | $\Delta V$ (cm³) | $\varepsilon_p$<br>($\Delta V/V_o$) | $A$<br>(mm²) | $q = P_z/A$<br>(kPa) |
|---|---|---|---|---|---|
| 0.00 | 0.00 | 0.00 | 0.00 | 1134 | 0 |
| 0.15 | 0.20 | 0.02 | 0.02 | 1136 | 53.8 |
| 0.23 | 0.30 | 0.03 | 0.03 | 1137 | 83 |
| 0.38 | 0.50 | −0.09 | −0.10 | 1140 | 108.8 |
| 0.76 | 1.00 | −0.50 | −0.58 | 1150 | 175.3 |
| 1.52 | 2.00 | −1.29 | −1.50 | 1169 | 220.3 |
| 2.28 | 3.00 | −1.98 | −2.30 | 1187 | 246.7 |
| 2.66 | 3.50 | −2.24 | −2.60 | 1196 | 250 |
| 3.04 | 4.00 | −2.41 | −2.80 | 1203 | 247.8 |
| 3.80 | 5.00 | −2.55 | −2.97 | 1214 | 230 |
| 4.56 | 6.00 | −2.59 | −3.01 | 1224 | 219.2 |
| 5.32 | 7.00 | −2.67 | −3.10 | 1235 | 204.4 |
| 6.08 | 8.00 | −2.62 | −3.05 | 1245 | 191.2 |
| 6.84 | 9.00 | −2.64 | −3.07 | 1255 | 182.9 |
| 7.60 | 10.00 | −2.66 | −3.09 | 1265 | 176.4 |
| 8.36 | 11.00 | −2.63 | −3.06 | 1276 | 175.8 |

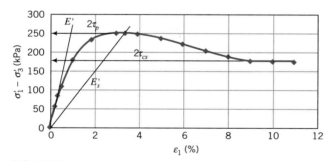

**FIGURE E5.7b**

Extract $\tau_p$ and $\tau_{cs}$.

$$\tau_p = \frac{(\sigma_1' - \sigma_3')_p}{2} = \frac{250}{2} = 125 \text{ kPa}, \quad \tau_{cs} = \frac{(\sigma_1' - \sigma_3')_{cs}}{2} = \frac{175.8}{2} = 87.9 \text{ kPa}$$

**Step 3:** Determine $E'$ and $E_s$.
The initial slope of Fig. E5.7b gives $E'$ and the slope of the line from the origin to $2\tau_p$ gives $E_s'$.

$$E' = \frac{54}{0.002} = 27{,}000 \text{ kPa}$$

$$E_s' = \frac{250}{0.035} = 7143 \text{ kPa}$$

**Step 4:** Determine $\phi'_{cs}$.

The deviatoric stress and the volumetric change appear to be constant from about $\varepsilon_1 \approx 10\%$. We can use the result at $\varepsilon_1 = 11\%$ to determine $\phi'_{cs}$. $(\sigma'_3)_{cs} = 100$ kPa, $(\sigma'_1)_{cs} = 175.8 + 100 = 275.8$ kPa.

$$\phi'_{cs} = \sin^{-1}\left(\frac{\sigma'_1 - \sigma'_3}{\sigma'_1 + \sigma'_3}\right)_{cs} = \sin^{-1}\frac{175.8}{275.8 + 100} = 27.9°$$

**Step 5:** Determine $\alpha_p$

$$\alpha_p = \phi'_p - \phi'_{cs} = 33.7 - 27.9 = 5.8° \qquad \blacksquare$$

### 5.8.5 Consolidated Undrained Compression (CU) Test

The purpose of a CU test is to determine the undrained and drained shear strength parameters $(s_u, \phi'_{cs}, \phi'_p)$. The CU test is conducted in a similar manner to the CD test except that after isotropic consolidation, the axial load is increased under undrained condition and the excess pore water pressure is measured. The applied stresses are as follows:

**Stage 1: Isotropic consolidation phase**

$$\Delta\sigma_1 = \Delta\sigma_3 = \Delta\sigma'_1 = \Delta\sigma'_3; \quad \Delta\sigma_1 > 0, \quad \Delta u = 0$$

Therefore,

$$\Delta p' = \Delta p = \Delta\sigma_1; \quad \Delta q = 0, \quad \frac{\Delta q}{\Delta p'} = \frac{\Delta q}{\Delta p} = 0$$

**Stage 2: Shearing phase**

$$\Delta\sigma_1 > 0, \quad \Delta\sigma_3 = 0; \quad \Delta\sigma'_1 = \Delta\sigma_1 - \Delta u, \quad \Delta\sigma'_3 = -\Delta u$$

$$\Delta p = \frac{\Delta\sigma_1}{3}; \quad \Delta q = \Delta\sigma_1, \quad \frac{\Delta q}{\Delta p} = 3$$

$$\Delta p' = \Delta p - \Delta u = \frac{\Delta\sigma_1}{3} - \Delta u$$

$$\Delta q = \Delta\sigma_1, \quad \frac{\Delta q}{\Delta p'} = \frac{\Delta\sigma_1}{\frac{\Delta\sigma_1}{3} - \Delta u} = \frac{3}{1 - \frac{3\Delta u}{\Delta\sigma_1}}$$

While the total stress path is determinate, the effective stress path can be determined only if we measure the changes in excess pore water pressures. The stresses on the soil samples and the stress paths in a CU test are illustrated in Fig. 5.22. The effective stress path is nonlinear because when the soil yields, the excess pore water pressures increase nonlinearly causing the ESP to bend.

In a CU test, the volume of the soil is constant during the shear phase. Therefore,

$$\varepsilon_p = \varepsilon_1 + 2\varepsilon_3 = 0$$

which leads to

$$\boxed{\varepsilon_3 = -\frac{\varepsilon_1}{2}} \qquad (5.42)$$

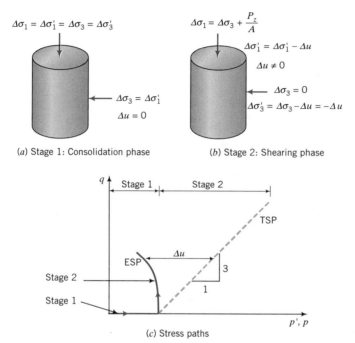

**FIGURE 5.22** Stresses and stress paths for triaxial CU test.

The axial displacement is measured and $\varepsilon_1 = \Delta z/H_o$ is calculated. The maximum shear strain is

$$(\gamma_{zx})_{\max} = \varepsilon_1 - \varepsilon_3 = \varepsilon_1 - \left(\frac{-\varepsilon_1}{2}\right) = 1.5\varepsilon_1 \qquad (5.43)$$

The elastic moduli $E_u$ and $(E_u)_s$ are determined from a plot of $(\sigma_1 - \sigma_3)$ as ordinate and $\varepsilon_1$ as abscissa.

Typical results from a CU test are shown in Fig. 5.23. Two sets of Mohr's circles at failure can be drawn, as depicted in Fig. 5.23. One represents total stress condition and the other, effective stress condition. For each test, Mohr's circle representing the total stresses has the same size as Mohr's circle representing the effective stresses but they are separated horizontally by the excess pore water pressure. Mohr's circle of effective stresses is shifted to the right if the excess pore water pressure at failure is negative and to the left if the excess pore water pressure is positive.

Each Mohr's circle of total stress is associated with a particular value of $s_u$ because each test has a different initial water content resulting from the different confining pressure, or applied consolidating stresses. The value of $s_u$ is obtained by drawing a horizontal line from the top of the desired Mohr's circle of total stress to intersect the vertical shear axis. The intercept is $s_u$. The value of $s_u$ at a cell pressure of about 830 kPa is 234 kPa, as shown in Fig. 5.23. Alternatively, you can calculate $s_u$ from

$$s_u = \frac{(\sigma_1 - \sigma_3)_f}{2} = \frac{(P_z)_{\max}}{2A}$$

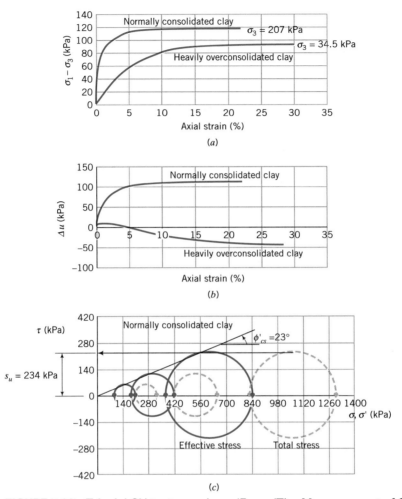

**FIGURE 5.23**    Triaxial CU tests on clays. (From 'The Measurement of Soil Properties in the Triaxial Test,' by Bishop and Henkel, Edward Arnold 1962.)

The drained shear strength parameters ($\phi'_{cs}$ and $\phi'_p$) are found as described in the previous section dealing with the CD test. You would normally determine $s_u$ at the maximum anticipated stress level in the field.

The CU test is the most popular triaxial test because you can obtain not only $s_u$ but $\phi'_{cs}$ and $\phi'_p$, and most tests can be completed within a few minutes after consolidation compared with more than a day for a CD test. Fine-grained soils with low $k$ values must be sheared slowly to allow the excess pore water pressure to equilibrate throughout the test sample. The results from CU tests are used to analyze the stability of slopes, foundations, retaining walls, excavations, and other earthworks.

## EXAMPLE 5.8

A CU test was conducted on a saturated clay soil by isotropically consolidating the soil using a cell pressure of 150 kPa and then incrementally applying loads on the plunger while keeping the cell pressure constant. Failure was observed when the stress exerted by the plunger was 160 kPa and the pore water pressure recorded was 54 kPa. Determine (a) $s_u$ and (b) $\phi'_{cs}$. Illustrate your answer by plotting Mohr's circle for total and effective stresses.

***Strategy*** You can calculate the effective strength parameters by using the Mohr–Coulomb failure criterion [Eq. (5.20)] or you can determine them from plotting Mohr's circle. Remember that the stress imposed by the plunger is not the major principal stress $\sigma_1$ but $(\sigma_1 - \sigma_3) = (\sigma'_1 - \sigma'_3)$.

## Solution 5.8

**Step 1:** Calculate the stresses at failure.

$$\frac{P_z}{A} = (\sigma_1)_f - (\sigma_3)_f = 160 \text{ kPa}$$

$$(\sigma_1)_f = \frac{P_z}{A} + \sigma_3 = 160 + 150 = 310 \text{ kPa}$$

$$(\sigma'_1)_f = (\sigma_1)_f - \Delta u_f = 310 - 54 = 256 \text{ kPa}$$

$$(\sigma_3)_f = 150 \text{ kPa}, \quad (\sigma'_3)_f = (\sigma_3)_f - \Delta u_f = 150 - 54 = 96 \text{ kPa}$$

**Step 2:** Determine the undrained shear strength.

$$s_u = \frac{(\sigma_1)_f - (\sigma_3)_f}{2} = \frac{160}{2} = 80 \text{ kPa}$$

**Step 3:** Determine $\phi'_{cs}$.

$$\sin \phi'_{cs} = \frac{(\sigma_1)_f - (\sigma_3)_f}{(\sigma_1)_f + (\sigma_3)_f} = \frac{160}{256 + 96} = 0.45$$

$$\phi'_{cs} = 26.7°$$

**Step 4:** Draw Mohr's circle.
See Fig. E5.8.

$$\phi'_{cs} = 27°$$

$$s_u = 80 \text{ kPa}$$

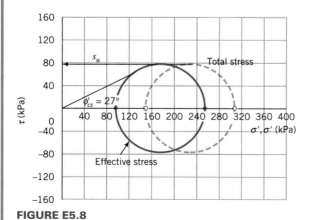

**FIGURE E5.8**

### 5.8.6 Unconsolidated Undrained (UU) Test

The purpose of a UU test is to determine the undrained shear strength of a saturated soil. The UU test consists of applying a cell pressure to the soil sample without drainage of pore water followed by increments of axial stress. The cell pressure is kept constant and the test is completed very quickly because in neither of the two stages—consolidation and shearing—is the excess pore water pressure allowed to drain. The stresses applied are:

**Stage 1: Isotropic compression (not consolidation) phase**

$$\Delta\sigma_1 = \Delta\sigma_3, \quad \Delta u \neq 0$$

$$\Delta p = \Delta\sigma_1, \quad \Delta q = 0, \quad \frac{\Delta q}{\Delta p} = 0$$

**Stage 2: Shearing phase**

$$\Delta\sigma_1 > 0, \quad \Delta\sigma_3 = 0$$

$$\Delta p = \frac{\Delta\sigma_1}{3}, \quad \Delta q = \Delta\sigma_1, \quad \frac{\Delta q}{\Delta p} = 3$$

Two or more samples of the same soil and the same initial void ratio are normally tested at different cell pressures. Each Mohr's circle is the same size but the circles are translated horizontally by the difference in the magnitude of the cell pressures. Mohr's circles, stresses, and stress paths for a UU test are shown in Fig. 5.24. Only the total stress path is known since the pore water pressures are not measured to enable the calculation of the effective stresses.

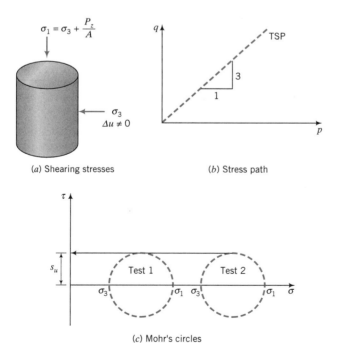

(a) Shearing stresses

(b) Stress path

(c) Mohr's circles

**FIGURE 5.24** Stresses, stress path, and Mohr's circles for UU tests.

The undrained shear strength, $s_u$, and the undrained elastic moduli, $E_u$ and $(E_u)_s$, are obtained from a UU test. The UU tests, like the UC tests, are quick and inexpensive compared with CD and CU tests. The advantage that the UU test has over the UC test is that the soil sample is stressed in the lateral direction to simulate the field condition. Both the UU and UC tests are useful in preliminary analyses for the design of slopes, foundations, retaining walls, excavations, and other earthworks.

## EXAMPLE 5.9

A UU test was conducted on saturated clay. The cell pressure was 200 kPa and failure occurred under a deviatoric stress of 220 kPa. Determine the undrained shear strength.

*Strategy* You can calculate the radius of the total stress circle to give $s_u$ but a plot of Mohr's circle is instructive.

## Solution 5.9

**Step 1:** Draw Mohr's circle.
See Fig. E5.9

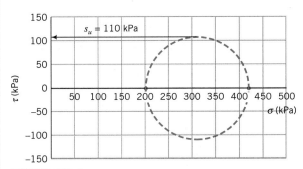

**FIGURE E5.9**

**Step 2:** Determine the undrained shear strength.
Draw a horizontal line to the top of Mohr's circle. The intersection of the horizontal line with the ordinate is the undrained shear strength.

$$s_u = 110 \text{ kPa}$$

By calculation:    $s_u = \dfrac{(\sigma_1)_f - (\sigma_3)_f}{2} = \dfrac{220}{2} = 110 \text{ kPa}$    ■

*What's next* . . .In the UU test and sometimes in the CU test, the excess pore water pressures are not measured. However, we need to know the magnitude of the excess pore water pressures to calculate effective stresses. Next, we will present a method to predict the excess pore water pressure from axisymmetric tests.

## 5.9 PORE WATER PRESSURE UNDER AXISYMMETRIC UNDRAINED LOADING

Pore water pressure changes in soils are due to the change in mean total and deviatoric stresses. Skempton (1954) proposed the following equation to determine the pore water pressure under axisymmetric conditions:

$$\Delta u = B[\Delta\sigma_3 + A(\Delta\sigma_1 - \Delta\sigma_3)] \tag{5.44}$$

where $\Delta\sigma_3$ is the increase in lateral principal stress, $\Delta\sigma_1 - \Delta\sigma_3$ is the deviatoric stress increase, $B$ is a coefficient indicating the level of saturation, and $A$ is an excess pore water pressure coefficient. The coefficient $B$ is 1 for saturated soils and 0 for dry soils. However, $B$ is not directly correlated with saturation except at high values of saturation ($S > 90\%$). At failure,

$$A = A_f = \left(\frac{\Delta u_d}{\Delta\sigma_1 - \Delta\sigma_3}\right)_f \tag{5.45}$$

where $\Delta u_d$ is the change in excess pore water pressure resulting from changes in deviatoric (shear) stresses. Experimental results of $A_f$ presented by Skempton (1954) are shown in Table 5.2. The coefficient $A$ was found to be dependent on the overconsolidation ratio (OCR). A typical variation of $A_f$ with OCR is shown in Fig. 5.25.

Equation (5.44) is very useful in determining whether a soil is saturated in an axisymmetric test. Let us manipulate Eq. (5.44) by dividing both sides by $\Delta\sigma_3$, resulting in

$$\frac{\Delta u}{\Delta\sigma_3} = B\left[1 + A\left(\frac{\Delta\sigma_1}{\Delta\sigma_3} - 1\right)\right] \tag{5.46}$$

During isotropic consolidation $\Delta\sigma_3 = \Delta\sigma_1$ and Eq. (5.46) becomes

$$\frac{\Delta u}{\Delta\sigma_3} = B \tag{5.47}$$

If a soil is saturated, then $B = 1$ and $\Delta u = \Delta\sigma_3$. That is, if we were to increase the consolidation stress or confining pressure by $\Delta\sigma_3$, the instantaneous excess

**TABLE 5.2** $A_f$ **Values**

| Type of clay | $A_f$ |
| --- | --- |
| Normally consolidated | 0.5 to 1 |
| Compacted sandy clay | 0.25 to 0.75 |
| Lightly overconsolidated clays | 0 to 0.5 |
| Compacted clay-gravel | −0.25 to 0.25 |
| Heavily overconsolidated clays | −0.5 to 0 |

SOURCE: From 'The Measurement of Soil Properties in the Triaxial Test,' by Bishop and Henkel, Edward Arnold 1962.

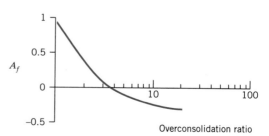

**FIGURE 5.25** Variation of OCR with $A_f$.

pore water pressure change must be equal to the increase of confining pressure. Equation (5.47) then provides a basis to evaluate the level of saturation of a soil sample in an axisymmetric test. The coefficients $A$ and $B$ are referred to as Skempton's pore water pressure coefficients.

> **The essential points are:**
> 1. **Under an axisymmetric loading condition, the pore water pressure can be predicted using Skempton's pore water pressure coefficients, A and B.**
> 2. **For a saturated soil, B = 1.**

*What's next . . .* In the next section, we will briefly describe a few other laboratory apparatuses to determine the shear strength of soils. These apparatuses are more complex and were developed to more closely represent the field stresses on the laboratory soil sample than the triaxial and direct shear apparatuses.

## 5.10  OTHER LABORATORY DEVICES TO MEASURE SHEAR STRENGTH

There are several other types of apparatus that are used to determine the shear strength of soils in the laboratory. These apparatuses are, in general, more sophisticated than the shear box and the triaxial apparatus.

### 5.10.1 Simple Shear Apparatuses

The purpose of a simple shear test is to determine shear strength parameters and the stress-strain behavior of soils under loading conditions that closely simulate plane strain and allow for the principal axes of stresses and strains to rotate. Principal stress rotations also occur in the direct shear test but are indeterminate. The stress states in soils for many geotechnical structures are akin to simple shear.

There are two types of commercially available simple shear devices. One deforms an initial cuboidal sample under plane strain conditions into a parallelepiped (Fig. 5.26a). The sample is contained in a box made by stacking square hollow plates between two platens. The top platen can be maintained at a fixed height for constant-volume tests or allowed to move vertically to permit volume

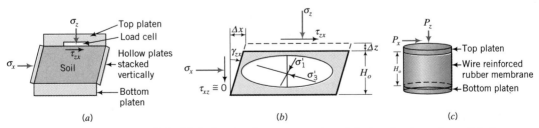

**FIGURE 5.26** Cuboidal simple shear apparatus: (a) simple shear box, (b) stresses imposed on samples, and (c) direct simple shear.

change to occur. By displacing the bottom of the box relative to the top, the soil is transformed from a cube to a parallelepiped.

A load cell mounted on the top platen measures the excess pore water pressures. The lateral stresses are deduced from one of the hollow plates outfitted with strain gauges. The stresses and strains deduced from measurements in the cuboidal simple shear apparatus are shown in Fig. 5.26b. If the excess pore water pressures are measured in undrained (constant-volume) tests, then the effective stresses can be determined.

The other apparatus tests a cylindrical sample whose vertical side is enclosed by a wire reinforced rubber membrane (Fig. 5.26c). Rigid rough metal plates are placed at the top and bottom of the sample. Displacing the top of the sample relative to the bottom deforms the sample. The vertical and horizontal loads (usually on the top boundary) as well as displacements on the boundaries are measured and thus the average normal and shear stresses and boundary strains can be deduced. In the cylindrical apparatus, the stresses measured are $\sigma_z$ and $\tau_{zx}$, and the test is referred to as direct simple shear.

Simple shear apparatuses do not subject the sample as a whole to uniform stresses and strains. However, the stresses and strains in the central region of the sample are uniform. In simple shear, the strains are $\varepsilon_x = \varepsilon_y = 0$, $\varepsilon_z = \Delta z / H_o$, and $\gamma_{zx} = \Delta x / H_o$. A plot of shear stress $\tau_{zx}$ versus $\gamma_{zx}$ is used to determine $G$.

The shear displacement $\Delta x$ must be applied in small increments to comply with the above definition. The principal strains from Eqs. (3.35) and (3.36) are

$$\varepsilon_1 = \tfrac{1}{2}(\varepsilon_z + \sqrt{\varepsilon_z^2 + \gamma_{zx}^2}) \tag{5.48}$$

$$\varepsilon_3 = \tfrac{1}{2}(\varepsilon_z - \sqrt{\varepsilon_z^2 + \gamma_{zx}^2}) \tag{5.49}$$

and

$$(\gamma_{zx})_{max} = \varepsilon_1 - \varepsilon_3 = \sqrt{\varepsilon_z^2 + \gamma_{zx}^2} \tag{5.50}$$

The dilation angle is determined from Eq. (5.10).

## EXAMPLE 5.10

A cuboidal soil sample, with 50 mm sides, fails in a simple shear constant-volume test under a vertical load ($P_z$) of 500 N, a horizontal load ($P_x$) of 375 N, and a shear load ($T$) of 150 N. The excess pore water pressure developed is 60 kPa.

(a) Plot Mohr's circles for total and effective stresses.

(b) Determine the friction angle and the undrained shear strength, assuming the soil is nondilating.

(c) Determine the failure stresses.

(d) Find the magnitudes of the principal effective stresses and the inclination of the major principal axis of stress to the horizontal.

(e) Determine the shear and normal stresses on a plane oriented at 20° clockwise to the horizontal.

**Strategy** Draw a diagram of the forces on the soil sample, calculate the stresses, and plot Mohr's circle. You can find all the required values from Mohr's circle or you can calculate them. You must use effective stresses to calculate the friction angle.

## Solution 5.10

**Step 1:** Determine the total and effective stresses.

$$\sigma_z = \frac{P_z}{A} = \frac{500 \times 10^{-3}}{(0.05)^2} = 200 \text{ kPa}$$

$$\sigma_x = \frac{P_x}{A} = \frac{375 \times 10^{-3}}{(0.05)^2} = 150 \text{ kPa}$$

$$\tau_{zx} = \frac{T}{A} = \frac{150 \times 10^{-3}}{(0.05)^2} = 60 \text{ kPa}$$

$$\sigma_z' = \sigma_z - \Delta u = 200 - 60 = 140 \text{ kPa}$$

$$\sigma_x' = \sigma_x - \Delta u = 150 - 60 = 90 \text{ kPa}$$

**Step 2:** Draw Mohr's circle of total and effective stresses. See Fig. E5.10.

**Step 3:** Determine $\phi_{cs}'$ and $s_u$.
Draw a tangent to Mohr's circle of effective stress from the origin of the axes.

From Mohr's circle:   $\phi_{cs}' = 34.5°$

The undrained shear strength is found by drawing a horizontal line from the top of Mohr's circle of total stresses to intersect the ordinate.

$$s_u = 65 \text{ kPa}$$

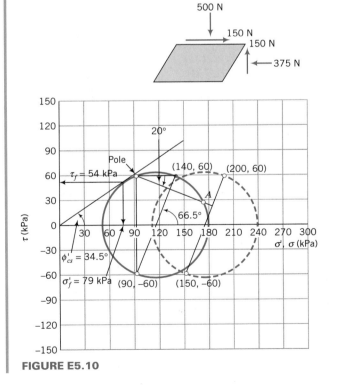

**FIGURE E5.10**

**Step 4:**    Determine the failure stresses.

At the point of tangency of the failure envelope and Mohr's circle of effective stress, we get

$$\tau_f = 54 \text{ kPa}, \quad (\phi'_n)_f = 79 \text{ kPa}$$

**Step 5:**    Determine $\sigma'_1$, $\sigma'_3$, and $\psi$.

From Mohr's circle of effective stresses, we get

$$\sigma'_1 = 180 \text{ kPa} \quad \text{and} \quad \sigma'_3 = 50 \text{ kPa}; \quad 2\psi = 66.5° \quad \text{and} \quad \psi = 33.3°$$

**Step 6:**    Determine the stresses on a plane oriented at 20°.

Identify the pole as shown in Fig. E5.10. Draw a line inclined at 20° to the horizontal from the pole as shown in Fig. E5.10. Remember that we are using clockwise shear as positive. Point A gives the stresses on a plane oriented at 20° to the horizontal.

$$\tau_{20°} = 30 \text{ kPa}; \quad \sigma'_{20°} = 173 \text{ kPa}$$

Alternatively, by calculation,

Equation (3.27):    $\sigma'_1 = \dfrac{140 + 90}{2} + \sqrt{\left(\dfrac{140 - 90}{2}\right)^2 + 60^2} = 180 \text{ kPa}$

Equation (3.28):    $\sigma'_3 = \dfrac{140 + 90}{2} - \sqrt{\left(\dfrac{140 - 90}{2}\right)^2 + 60^2} = 50 \text{ kPa}$

$s_u = \dfrac{(\sigma'_1 - \sigma'_3)_f}{2} = \dfrac{180 - 50}{2} = 65 \text{ kPa}$

$\phi'_{cs} = \sin^{-1}\left(\dfrac{\sigma'_1 - \sigma'_3}{\sigma' + \sigma'_3}\right) = \sin^{-1}\left(\dfrac{180 - 50}{180 + 50}\right) = 34.4°$

Equation (3.29):    $\tan \psi = \dfrac{\tau_{zx}}{\sigma'_1 - \sigma'_x} = \dfrac{60}{180 - 90} = 0.67; \quad \psi = 33.7°$

Equation (3.32):    $\sigma'_{20°} = \dfrac{140 + 90}{2} + \dfrac{140 - 90}{2} \cos 40° + 60 \sin 40°$

$$= 172.7 \text{ kPa}$$

Equation (3.33):    $\tau_{20°} = 60 \cos 40° - \dfrac{140 - 90}{2} \sin 40° = 29.9 \text{ kPa}$

∎

## EXAMPLE 5.11

A cuboidal soil sample, with 50 mm sides, was tested under drained conditions and the maximum shear stress occurred when the shear displacement was 1 mm and the vertical movement was −0.05 mm.

(a) Plot Mohr's circle of strain.

(b) Determine the principal strains.

(c) Determine the maximum shear strain.

(d) Determine α.

**Strategy**    Calculate the vertical and shear strains and then plot Mohr's circle of strain. You can determine all the required values from Mohr's circle or you can calculate them.

**Solution 5.11**

**Step 1:** Determine the strains.

$$\varepsilon_z = \frac{\Delta z}{H_o} = -\frac{0.05}{50} = -0.001$$

(negative sign because the sample expands; compression is positive)

$$\varepsilon_x = \varepsilon_y = 0$$

$$\gamma_{zx} = \frac{\Delta x}{H_o} = \frac{1}{50} = 0.02$$

**Step 2:** Plot Mohr's circle of strain and determine $\varepsilon_1$, $\varepsilon_3$, $(\gamma_{zx})_{max}$, and $\alpha$. See Mohr's circle of strain, Fig. E5.11.

$$\varepsilon_1 = 9.5 \times 10^{-3}, \quad \varepsilon_3 = -10.5 \times 10^{-3}$$

$$\alpha = 3°$$

By calculation,

Equation (3.35):  $\varepsilon_1 = \frac{1}{2}(-0.001 + \sqrt{(-0.001)^2 + (0.02)^2}) = 9.5 \times 10^{-3}$

Equation (3.36):  $\varepsilon_3 = \frac{1}{2}(-0.001 - \sqrt{(-0.001)^2 + (0.02)^2}) = -10.5 \times 10^{-3}$

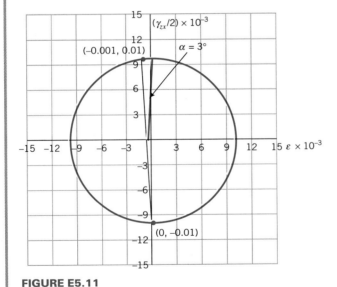

**FIGURE E5.11**    ■

## 5.10.2 True Triaxial Apparatus

The purpose of a true triaxial test is to determine soil behavior and properties by loading the soil in three dimensions. In a true triaxial test, a cuboidal sample is subjected to independent displacements or stresses on three Cartesian axes. Displacements are applied through a system of rigid metal plates moving perpendicularly and tangentially to each face, as shown by the arrows in Fig. 5.27a. Pressure transducers are fixed to the inside of the faces to measure the three principal stresses. Like the conventional triaxial apparatus, the directions of principal stresses are prescribed and can only be changed instantaneously through

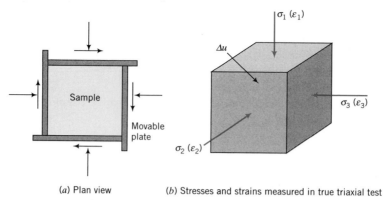

(a) Plan view      (b) Stresses and strains measured in true triaxial test

**FIGURE 5.27** Schematic of true triaxial cell and stresses imposed on a sample of soil.

an angle of 90°. The stresses and strains that can be measured in a true triaxial test are shown in Fig. 5.27b.

### 5.10.3 Hollow Cylinder Apparatus

The purpose of a hollow cylinder test is to determine soil properties from a variety of plane strain stress paths. In the hollow cylinder apparatus (Fig. 5.28a), a hollow thin-walled cylindrical sample is enclosed in a pressure chamber and can be subjected to vertical loads or displacements, radial stresses on the inner and outer cylindrical surfaces, and a torque as shown in Fig. 5.28b.

The vertical stress acting on a typical element of soil in this device is

$$\sigma_z = \frac{P_z}{A}$$

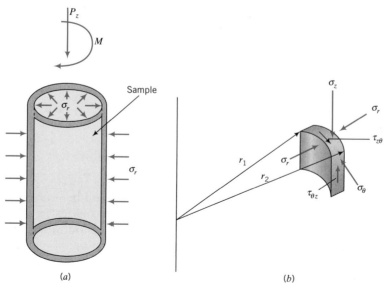

(a)      (b)

**FIGURE 5.28** (a) Hollow cylinder cell and (b) stresses on an element of soil.

If the internal and external radial pressures are equal, then

$$\sigma_r = \sigma_\theta$$

The shearing stress applied is

$$\tau_{z\theta} = \frac{3M}{2\pi(r_2^3 - r_1^3)} \tag{5.51}$$

where $M$ is the applied torque and $r_1$ and $r_3$ are the inner and outer radii. We can obtain $\phi_p'$, $\phi_{cs}'$, $s_u$, and $G$ from the hollow cylinder test.

*What's next* . . .Sampling disturbances and sample preparation for laboratory tests may significantly impair the shear strength parameters. Consequently, a variety of field tests have been developed to obtain more reliable soil shear strength parameters by testing soils in situ. We will briefly describe some of the popular field tests. Each test has its own advantages and disadvantages. Field test results are often related to laboratory test results using empirical factors.

## 5.11  FIELD TESTS

### 5.11.1 Shear Vane

The shear vane device consists of four thin metal blades welded orthogonally (90°) to a rod (Fig. 5.29). The vane is pushed, usually from the bottom of a borehole, to the desired depth. A torque is applied at a rate of 6° per minute by a torque head device located above the soil surface and attached to the shear vane rod. The undrained shear strength is calculated from

$$s_u = \frac{2T}{\pi d^3(h/d + \frac{1}{3})} \tag{5.52}$$

where $T$ is the maximum torque, $h$ is the height, and $d$ is the diameter of the vane.

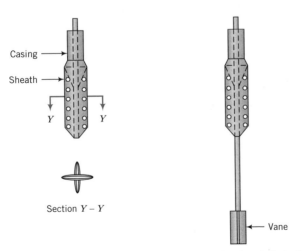

Casing

Sheath

Y          Y

Section Y – Y

Vane probe in protective sheath        Vane extended and ready for testing

Vane

**FIGURE 5.29**  Shear vane tester.

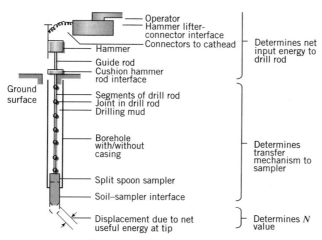

**FIGURE 5.30** Standard penetration test.

## 5.11.2 The Standard Penetration Test

The standard penetration test (SPT) was developed circa 1927 and it is perhaps the most popular field test. The SPT is performed by driving a standard split spoon sampler into the ground by blows from a drop hammer of mass 64 kg falling 760 mm (Fig. 5.30). The sampler is driven 150 mm into the soil at the bottom of a borehole, and the number of blows ($N$) required to drive it an additional 300 mm is counted. The number of blows ($N$) is called the standard penetration number.

The word "standard" is a misnomer for the standard penetration test. Several methods are used in different parts of the world to release the hammer. Also, different types of anvil, rods, and rod lengths are prevalent. Various corrections are applied to the $N$ values to account for energy losses, overburden pressure, rod length, and so on. It is customary to correct the $N$ values to a rod energy ratio of 60%. The rod energy ratio is the ratio of the energy delivered to the split spoon sampler to the free-falling energy of the hammer. The corrected $N$ values are denoted as $N_{60}$. The $N$ value is used to estimate the relative density, friction angle, and settlement in coarse-grained soils. The test is very simple, but the results are difficult to interpret.

Typical correlation among $N$ values, relative density, and $\phi'$ are given in Tables 5.3 and 5.4. You should be cautious in using the correlation in Tables 5.3

**TABLE 5.3 Correlation of $N$, $N_{60}$, $\gamma$, $D_r$, and $\phi'$ for Coarse-Grained Soils**

| $N$ | $N_{60}$ | Description | $\gamma$ (kN/m³) | $D_r$ (%) | $\phi'$ (degrees) |
|------|----------|-------------|------------------|-----------|-------------------|
| 0–5 | 0–3 | Very loose | 11–13 | 0–15 | 26–28 |
| 5–10 | 3–9 | Loose | 14–16 | 16–35 | 29–34 |
| 10–30 | 9–25 | Medium | 17–19 | 36–65 | 35–40[a] |
| 30–50 | 25–45 | Dense | 20–21 | 66–85 | 38–45[a] |
| >50 | >45 | Very dense | >21 | >86 | >45[a] |

[a]These values correspond to $\phi'_p$.

**TABLE 5.4 Correlation of $N_{60}$ and $s_u$ for Saturated Fine-Grained Soils**

| $N_{60}$ | Description | $s_u$ (kPa) |
|---|---|---|
| 0–2 | Very soft | <10 |
| 3–5 | Soft | 10–25 |
| 6–9 | Medium | 25–50 |
| 10–15 | Stiff | 50–100 |
| 15–30 | Very stiff | 100–200 |
| >30 | Extremely stiff | >200 |

and 5.4 to determine the properties of soils and to design foundations because the scatter in data is generally large; that is, the correlation coefficients are low. Experience and judgment are required to successfully use Tables 5.3 and 5.4.

The SPT is very useful for determining changes in stratigraphy and locating bedrock. Also, you can inspect the soil in the split spoon sampler to describe the soil profile and extract disturbed samples for laboratory tests.

### 5.11.3 The "Dutch" Cone Penetrometer and Piezocone

The "Dutch" cone penetrometer is a cone with a maximum area of 10 cm² (Fig. 5.31a) that is attached to a rod. An outer sleeve encloses the rod. The thrusts required to drive the cone and the sleeve into the ground are measured independently so that the end resistance or cone resistance and side friction or sleeve resistance may be estimated separately. Although originally developed for the design of piles, the cone penetrometer has also been used to estimate the bearing capacity and settlement of foundations.

The piezocone is a "Dutch" cone penetrometer that has porous elements inserted into the cone or sleeve to allow for pore water pressure measurements (Fig. 5.31b). The measured pore water pressure depends on the location of the porous elements. A load cell is often used to measure the force of penetration. The piezocone is a very useful tool for soil profiling. Researchers have claimed that the piezocone provides useful data to estimate the shear strength, bearing capacity, and consolidation characteristics of soils. Typical results from a piezocone are shown in Fig. 5.31c.

The cone resistance $q_c$ is normally correlated with the undrained shear strength. Several adjustments are made to $q_c$. One correlation equation is

$$s_u = \frac{q_c - \sigma_z}{N_k} \tag{5.52}$$

where $N_k$ is a cone factor that depends on the geometry of the cone and the rate of penetration. Average values of $N_k$ as a function of plasticity index can be estimated from

$$N_k = 19 - \frac{I_p - 10}{5}; \quad I_p > 10 \tag{5.54}$$

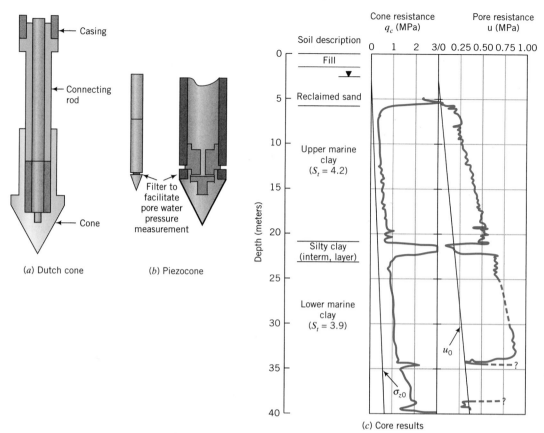

**FIGURE 5.31**    (a) Dutch cone and (b) piezocone. (c) Cone results. (From Chang, 1988.)

Results of cone penetrometer tests have been correlated with the peak friction angle. A number of correlations exist. Based on published data for sand (Robertson and Campanella, 1983), you can estimate $\phi'_p$ using

$$\phi'_p = 35° + 11.5 \log\left(\frac{q_c}{30\sigma'_{zo}}\right); \quad 25° < \phi'_p < 50° \tag{5.55}$$

### 5.11.4 Pressure Meters

The Menard pressure meter (Fig. 5.32a) is a probe that is placed at the desired depth in an unlined borehole and pressure is applied to a measuring cell of the probe. The stresses near the probe are shown in Fig. 5.32b. The pressure applied is analogous to the expansion of a cylindrical cavity. The pressure is raised in stages at constant time intervals and volume changes are recorded at each stage. A pressure–volume change curve is then drawn from which the elastic modulus, shear modulus, and the undrained shear strength may be estimated.

One of the disadvantages of the Menard pressure meter is that it has to be inserted into a predrilled hole and consequently the soil is disturbed. The Cam-

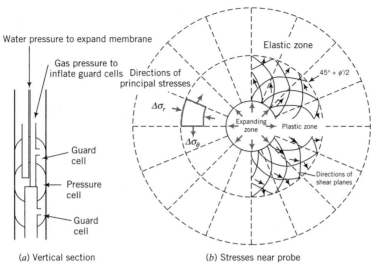

**FIGURE 5.32**  (a) Menard pressure meter and (b) stresses near the probe.

bridge Camkometer (Fig. 5.33) is a self-boring pressure meter, which minimizes soil disturbances. A pressure is applied to radially expand a rubber membrane, which is built into the side wall of the Camkometer, and a feeler gauge measures the radial displacement. Thus, the stress–strain response of the soil can be obtained. Interpretation of the pressure meter test is beyond the scope of this book.

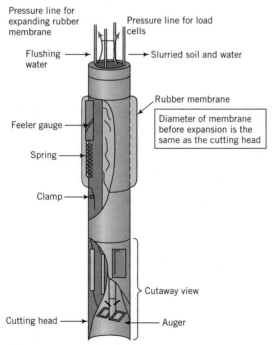

**FIGURE 5.33**  Schematic of Cambridge Camkometer. (Redrawn from Hughes and Wroth, 1972.)

An excellent source on the interpretation of the pressure meter test is Wroth (1984).

> *The* **essential points** *are:*
> 1. *Various field tests are used to determine soil strength parameters.*
> 2. *The most simple field test, and the most popular, is the standard penetration test (SPT).*

*What's next* . . .Several empirical relationships have been proposed to obtain soil strength parameters from laboratory tests, for example, the Atterberg limits, or from statistical analyses of field and laboratory test results. Some of these relationships are presented in the next section.

## 5.12  EMPIRICAL RELATIONSHIPS FOR SHEAR STRENGTH PARAMETERS

Some suggested empirical relationships for the shear strength of soils are shown in Table 5.5. These relationships should only be used as a guide and in preliminary design calculations.

**TABLE 5.5  Empirical Soil Strength Relationships**

| Soil type | Equation | Reference |
|---|---|---|
| Normally consolidated clays | $\left(\dfrac{s_u}{\sigma'_z}\right)_{nc} = 0.11 + 0.0037 I_p$ | Skempton (1957) |
| Overconsolidated clays | $\dfrac{(s_u/\sigma'_z)_{oc}}{(s_u/\sigma'_z)_{nc}} = (OCR)^{0.8}$ | Ladd et al. (1977) |
|  | $\dfrac{s_u}{\sigma'_z} = (0.23 \pm 0.04)OCR^{0.8}$ | Jamiolkowski et al. (1985) |
| All clays | $\dfrac{s_u}{\sigma'_{zc}} = 0.22$ | Mesri (1975) |
| Clean quartz sand | $\phi'_p = \phi'_{cs} + 3D_r(10 - \ln p'_f) - 3$, where $p'_f$ is the mean effective stress at failure (in kPa) and $D_r$ is relative density. This equation should only be used if $12 > (\phi'_p - \phi'_{cs}) > 0$. | Bolton (1986) |

## 5.13  SUMMARY

The strength of soils is interpreted using Coulomb's frictional law. All soils regardless of their initial state of stress will reach a critical state characterized by continuous shearing at constant shear stress and constant volume. The initial void ratio of a soil and the normal effective stresses determine whether the soil will dilate or not. Dilating soils often exhibit (1) a peak shear stress and then strain soften to a constant shear stress, and (2) initial contraction followed by expansion toward a critical void ratio. Nondilating soils (1) show a gradual increase of shear stress, ultimately reaching a constant shear stress, and (2) contract toward a critical void ratio. The shear strength parameters are the friction angles ($\phi'_p$ and $\phi'_{cs}$) for drained conditions and $s_u$ for undrained conditions. Only $\phi'_{cs}$ is a fundamental soil strength parameter.

A number of laboratory and field tests are available to determine the shear strength parameters. All these tests have shortcomings and you should use careful judgment in deciding on the test to be used for a particular project and in the interpretation of the results.

## *Practical Examples*

### EXAMPLE 5.12

A rectangular foundation 4 m × 5 m transmits a total load of 5 MN to a deep, uniform deposit of stiff overconsolidated clay with an OCR = 4 and $\gamma_{sat}$ = 18 kN/m³ (Fig. E5.12). Groundwater level is at 1 m below the ground surface. A CU test was conducted on a soil sample taken at a depth 5 m below the center of the foundation. The results obtained are $s_u$ = 40 kPa, $\phi'_p$ = 27°, and $\phi'_{cs}$ = 24°. Determine if the soil will reach the failure state for short-term and long-term loading conditions. If the soil does not reach the failure state, what are the factors of safety? Assume the soil above the groundwater level is saturated.

**Strategy**   The key is to find the stresses imposed on the soil at a depth of 5 m and check whether the imposed shear stress exceeds the critical shear stress. For short-term condition, the critical shear strength is $s_u$ while for long-term conditions the critical shear strength is $(\sigma'_n)_f \tan \phi'_{cs}$.

### Solution 5.12

**Step 1:**   Determine the initial stresses.

$$\sigma'_z = \sigma'_1 = \gamma_{sat}z_1 + (\gamma_{sat} - \gamma_w)z_2 = (18 \times 1) + (18 - 9.8)4 = 50.8 \text{ kPa}$$
$$K_o^{oc} = K_o^{nc}(OCR)^{0.5} = (1 - \sin \phi'_{cs})(OCR)^{0.5} = (1 - \sin 18°)(4)^{0.5} = 1.4$$
$$\sigma'_3 = K_o^{oc}\sigma'_z = 1.4 \times 50.8 = 71.1 \text{ kPa}$$
$$\sigma_1 = \sigma'_z + z_2\gamma_w = 50.8 + 4 \times 9.8 = 90 \text{ kPa}$$
$$\sigma_3 = \sigma'_x + z_2\gamma_w = 71.1 + 4 \times 9.8 = 110.3 \text{ kPa}$$

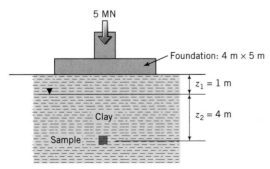

**FIGURE E5.12**

**Step 2:**   Determine the vertical stress increases at $z = 5$ m under the center of the rectangular foundation.
Use Eq. (3.79) or the computer program on the CD.

$$\Delta\sigma_z = 71.1 \text{ kPa}, \quad \Delta\sigma_x = 5.1 \text{ kPa}$$

Neglect the effect of shear stresses $(\Delta\tau_{zx})$.

**Step 3:**   Determine imposed shear stress for short-term loading.

Current vertical total stress:   $(\sigma_1)_T = \sigma_1 + \Delta\sigma_z = 90 + 71.1 = 161.1$ kPa

Current horizontal total stress:   $(\sigma_3)_T = \sigma_3 + \Delta\sigma_x = 110.3 + 5.1 = 115.4$ kPa

Current shear stress:   $\tau_u = \dfrac{(\sigma_1)_T - (\sigma_3)_T}{2} = \dfrac{161.1 - 110.4}{2} = 25.4$ kPa $< 40$ kPa

The soil will not reach the failure state.

$$\text{Factor of safety:}\quad \text{FS} = \frac{s_u}{\tau_u} = \frac{40}{25.4} = 1.6$$

**Step 4:**   Determine failure shear stress for long-term loading.
For long-term loading, we will assume that the excess pore water pressure dissipated.

Final effective stresses:   $(\sigma_1')_F = \sigma_1' + \Delta\sigma_z = 50.8 + 71.1 = 121.9$ kPa

$(\sigma_3')_F = \sigma_3' + \Delta\sigma_x = 71.1 + 5.1 = 76.2$ kPa

Angle of friction mobilized:   $\phi' = \sin^{-1}\left(\dfrac{\sigma_1' - \sigma_3'}{\sigma_1' + \sigma_3'}\right) = \sin^{-1}\left(\dfrac{121.9 - 76.2}{121.9 + 76.2}\right) = 13.3°$

The critical state angle of friction is 24°, which is greater than $\phi'$. Therefore, failure would not occur.

$$\text{Factor of safety:}\quad \text{FS} = \frac{\tan 24°}{\tan 13.3°} = 1.9 \qquad \blacksquare$$

**EXAMPLE 5.13**

An earth dam is proposed for a site consisting of a homogeneous stiff clay, as shown in Fig. E5.13a. You, the geotechnical engineer, are required to specify soil strength tests to determine the stability of the dam. One of the critical situations is the possible failure of the dam by a rotational slip plane in the stiff clay. What laboratory strength tests would you specify and why?

*Strategy*   The key is to determine the stress states for short-term and long-term conditions along the failure surface.

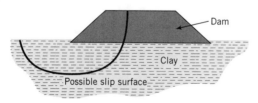

**FIGURE E5.13a**

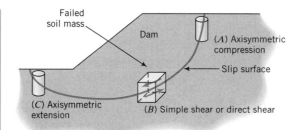

**FIGURE E5.13b**

## Solution 5.13

**Step 1:** Determine stresses along the failure surface.
Let us select three points—$A$, $B$, and $C$—on the possible failure surface. The rotational slip surface will introduce compression on element $A$, shear on element $B$, and extension on element $C$ Fig. E5.13b. The stresses on element $A$ are analogous to a triaxial compression test. Element $B$ will deform in a manner compatible with simple shear, while element $C$ will suffer from an upward thrust that can be simulated by a triaxial extension test.

**Step 2:** Make recommendations.
The following strength tests are recommended.

**(a)** Triaxial CU compression tests with pore water pressure measurements. Parameters required are $\phi'_p$, $\phi'_{cs}$ and $s_u$.

**(b)** Simple shear constant-volume tests.
Parameters required are $\phi'_p$, $\phi'_{cs}$, and $s_u$. If a simple shear apparatus is unavailable, then direct shear (shear box) tests should be substituted.

Undrained tests should be carried out at the maximum anticipated stress on the soil. You can determine the stress increases from the dam using the methods described in Chapter 3. ∎

## EXAMPLE 5.14

You have contracted a laboratory to conduct soil tests for a site, which consists of a layer of sand, 6 m thick, with $\gamma_{sat}$ = 18 kN/m³. Below the sand is a deep, soft, bluish clay with $\gamma_{sat}$ = 20 kN/m³ (Fig. E5.14). The site is in a remote area. Groundwater level was located at 2.5 m below the surface. You specified a consolidation test and a triaxial consolidated undrained test for samples of the soil

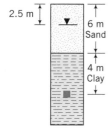

**FIGURE E5.14**

taken at 10 m below ground surface. The consolidation test shows that the clay is lightly overconsolidated with an OCR = 1.3. The undrained shear strength at a cell pressure approximately equal to the initial vertical stress is 72 kPa. Do you think the undrained shear strength value is reasonable, assuming the OCR of the soil is accurate? Show calculations to support your thinking.

***Strategy***   Because the site is in a remote area, it is likely that you may not find existing soil results from neighboring constructions. In such a case, you can use empirical relationships as guides but you are warned that soils are notorious for having variable strengths.

## Solution 5.14

**Step 1:**   Determine the initial effective stresses. Assume that the sand above the groundwater level is saturated.

$$(\sigma_z')_o = (18 \times 2.5) + (18 - 9.8) \times 3.5 + (20 - 9.8) \times 4 = 114.5 \text{ kPa}$$
$$\sigma_{zc}' = (\sigma_z')_o \times \text{OCR} = 114.5 \times 1.3 = 148.9 \text{ kPa}$$

**Step 2:**   Determine $s_u/(\sigma_z')_o$ and $s_u/\sigma_{zc}'$.

$$\frac{s_u}{(\sigma_z')_o} = \frac{72}{114.5} = 0.63$$

$$\frac{s_u}{\sigma_{zc}'} = \frac{72}{148.9} = 0.48$$

**Step 3:**   Use empirical equations. See Table 5.5.

$$\text{Jamiolkowski et al. (1985): } \frac{s_u}{(\sigma_z')_o} = (0.23 \pm 0.04)\text{OCR}^{0.8}$$

$$\text{Range of } \frac{s_u}{(\sigma_z')}: \quad 0.27(\text{OCR})^{0.8} \text{ to } 0.19(\text{OCR})^{0.8}$$

$$= 0.27(1.3)^{0.8} \text{ to } 0.19(1.3)^{0.8}$$

$$= 0.33 \text{ to } 0.23 < 0.63$$

$$\text{Mesri (1975): } \frac{s_u}{\sigma_{zc}'} = 0.22 < 0.48$$

The differences between the reported results and the empirical relationships are substantial. The undrained shear strength is therefore suspicious. You should request a repeat of the test.   ■

# EXERCISES

## Theory

**5.1**   A CD triaxial test was conducted on a loose sand. The axial stress was held constant and the radial stress was increased until failure occurred. At failure the radial stress was greater than the axial stress. Show, using Mohr's circle and geometry, that the slip plane is inclined at $\pi/4 - \phi_{cs}'/2$ to the plane on which the principal stress acts.

**5.2**   The initial stresses on a soil are $\sigma_1' = \sigma_z'$ and $\sigma_3' = K_o\sigma_z'$, where $K_o$ is the lateral earth pressure coefficient at rest (see Chapter 3). The soil was then brought to failure by reducing $\sigma_3'$ while keeping $\sigma_1'$ constant. At failure, $\sigma_3' = K\sigma_1'$, where $K$ is a lateral earth pressure coefficient. Show that

$$\frac{\tau_{cs}}{\tau_{zx}} = \frac{1 + K}{1 - K_o} \sin \phi_{cs}'$$

where $\tau_{zx}$ is the maximum shear stress under the initial stresses and $\tau_{cs}$ is the failure shear stress.

5.3   Sand is placed on a clay slope as shown in Fig. P5.3. (a) Show that sand will be unstable (i.e., fail by sliding) if $\theta > \phi'$. (b) Does the thickness of the clay layer influence impending failure? (c) If $\phi' = 25°$ and $\theta = 23°$, determine the factor of safety.

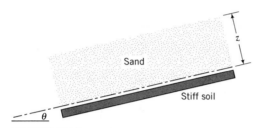

**FIGURE P5.3**

## Problem Solving

5.4   A structure will impose a normal effective stress of 100 kPa and a shear stress of 30 kPa on a plane inclined at 58° to the horizontal. The friction angle of the soil is 25°. Will the soil fail? If not, what is the factor of safety?

5.5   Figure P5.5 shows the stress–strain behavior of a soil. Determine the peak and critical shear stresses. Estimate the peak and critical state friction angle and the dilation angle. The normal effective stress is 160 kPa.

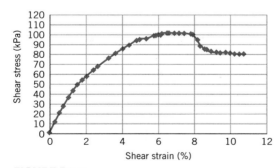

**FIGURE P5.5**

5.6   The following results were obtained from three shear box tests on a sample of sandy clay. The cross section of the shear box is 6 cm × 6 cm.

| Normal force (N) | 1250 | 1000 | 500 | 250 |
|---|---|---|---|---|
| Shearing force (N) | 506 | 405 | 325 | 255 |

(a) Determine the critical state friction angle.
(b) If the soil is dilatant, determine $\phi_p'$ at a normal force of 250 N.

**5.7**    The results of a direct shear test on a dense sand are shown in the table below. Determine $\phi'_p$, $\phi'_{cs}$ and $\alpha_p$. The vertical force is 200 N and the sample area is 100 mm × 100 mm.

| Horizontal | 0.00 | 0.25 | 0.38 | 0.64 | 0.76 | 1.02 | 1.52 | 1.78 | 2.67 | 2.92 |
|---|---|---|---|---|---|---|---|---|---|---|
| displacement | 3.18 | 3.68 | 4.06 | 4.57 | 5.08 | 5.59 | 6.10 | 6.48 | 6.60 | 6.86 |
| (mm) | 7.11 | 7.62 | 8.13 | 8.64 | 9.14 | 9.65 | 10.16 | | | |
| Vertical | 0.00 | 0.00 | −0.02 | −0.03 | −0.04 | −0.04 | −0.03 | −0.01 | 0.04 | 0.06 |
| displacement | 0.09 | 0.14 | 0.15 | 0.19 | 0.22 | 0.24 | 0.26 | 0.27 | 0.28 | 0.28 |
| (mm) | 0.28 | 0.28 | 0.28 | 0.28 | 0.28 | 0.28 | 0.28 | | | |
| Horizontal | 0.0 | 17.7 | 19.3 | 33.4 | 40.3 | 54.9 | 83.3 | 94.9 | 127.2 | 133.8 |
| force (N) | 137.4 | 148.8 | 155.9 | 161.7 | 169.6 | 174.4 | 176.8 | 177.7 | 177.7 | 176.4 |
| | 172.9 | 163.5 | 161.2 | 160.8 | 159.8 | 158.9 | 158.8 | | | |

**5.8**    A cylindrical sample of soil 50 mm in diameter × 100 mm long was subjected to axial and radial effective stresses. When the vertical displacement was 2 mm, the soil failed. The stresses at failure are $(\sigma'_1)_f = 280$ kPa and $(\sigma'_3)_f = 100$ kPa. The change in soil volume at failure was 800 mm³. Determine (a) the axial strain at failure, (b) the volumetric strain at failure, (c) the friction angle, and (d) the inclination of the slip plane to the horizontal.

**5.9**    A CD test was conducted on a sample of dense sand, 38 mm in diameter × 76 mm long. The cell pressure was 200 kPa. The results are shown in the table below.

(a) Determine the $\phi'_p$, $\phi'_{cs}$, $\alpha_p$, $\tau_p$, $E'$, and $E'_s$.
(b) If $\nu' = 0.3$, calculate $G$.

| Axial displacement (mm) | Axial load (N) | Change in volume (cm³) | Axial displacement (mm) | Axial load (N) | Change in volume (cm³) |
|---|---|---|---|---|---|
| 0 | 0 | 0 | 3.2 | 604.3 | −2.7 |
| 0.3 | 128.1 | 0.1 | 5.2 | 593.0 | −3.2 |
| 0.6 | 225.9 | 0.1 | 6.4 | 576.0 | −3.4 |
| 1.0 | 338.9 | 0.1 | 7.1 | 564.8 | −3.5 |
| 1.3 | 451.8 | −0.25 | 9.7 | 525.2 | −3.6 |
| 1.6 | 508.3 | −0.7 | 12.9 | 497.0 | −3.7 |
| 1.9 | 564.8 | −1.9 | 15.5 | 480.0 | −3.7 |

**5.10**    CU tests were carried out on two samples of a clay. Each sample was isotropically consolidated before the axial stress was increased. The following results were obtained.

| Sample No. | $(\sigma_3)_f$ (kPa) | $(\sigma_1 - \sigma_3)_f$ (kPa) | $\Delta u_f$ (kPa) |
|---|---|---|---|
| I | 420 | 320 | 205 |
| II | 690 | 365 | 350 |

(a) Draw Mohr's circles (total and effective stresses) for each test on the same graph.
(b) Why are the total stress circles not the same size?
(c) Determine the friction angle for each test.
(d) Determine the undrained shear strength at a cell pressure, $(\sigma_3)_f$ of, 690 kPa.
(e) Determine the shear stress on the failure plane for each sample.

5.11 CU tests were conducted on a compacted clay. Each sample was brought to a saturated state before shearing. The results, when no further change in excess pore water pressure or deviatoric stress occurred, are shown in the table below. Determine (a) $\phi'_{cs}$ and (b) $s_u$ at a cell pressure of 420 kPa.

| $\sigma_3$ (kPa) | $(\sigma_1 - \sigma_3)$ (kPa) | $\Delta u$ (kPa) |
|---|---|---|
| 140 | 636 | −71 |
| 280 | 1008 | −51.2 |
| 420 | 1323 | −19.4 |

5.12 Three samples of a loose sand were tested under CU conditions. The failure stresses and excess pore water pressures for each sample are given below.

| Sample No. | $(\sigma_3)_f$ (kPa) | $(\sigma_1 - \sigma_3)_f$ (kPa) | $\Delta u_f$ (kPa) |
|---|---|---|---|
| 1 | 210 | 123 | 112 |
| 2 | 360 | 252 | 162 |
| 3 | 685 | 448 | 323 |

(a) Plot Mohr's circle of effective stress from these data.

(b) Determine the friction angle for each test.

(c) If the sand were to be subjected to a vertical effective stress of 300 kPa, what magnitude of horizontal effective stress will cause failure?

(d) Determine the orientations of (1) the failure plane and (2) the plane of maximum shear stress to the horizontal.

(e) Is the failure stress the maximum shear stress? Give reasons.

5.13 A CU triaxial test was carried out on a silty clay that was isotropically consolidated with a cell pressure of 125 kPa. The following data were obtained:

| Stress imposed by plunger (kPa) | Axial strain, $\varepsilon_1$ (%) | $\Delta u$ (kPa) | Stress imposed by plunger (kPa) | Axial strain, $\varepsilon_1$ (%) | $\Delta u$ (kPa) |
|---|---|---|---|---|---|
| 0 | 0 | 0 | 35.0 | 0.56 | 34.8 |
| 5.5 | 0.05 | 4.0 | 50.5 | 1.08 | 41.0 |
| 11.0 | 0.12 | 8.6 | 85.0 | 2.43 | 49.7 |
| 24.5 | 0.29 | 19.1 | 105.0 | 4.02 | 55.8 |
| 28.5 | 0.38 | 29.3 | 120.8 | 9.15 | 59.0 |

(a) Plot the deviatoric stress against axial strain and excess pore water pressure against axial strain.

(b) Determine the undrained shear strength and the friction angle.

(c) Determine $E'$ and $E'_s$.

(d) Determine Skempton's $A_f$.

**5.14**    Three CU tests with pore water pressure measurements were made on a clay soil. The results at failure are shown below.

| $\sigma_3$ (kPa) | $\sigma_1 - \sigma_3$ (kPa) | $\Delta u$ (kPa) |
|---|---|---|
| 150 | 87.5 | 80 |
| 275 | 207.5 | 115 |
| 500 | 345 | 230 |

(a) Determine the friction angle for each test.

(b) Determine the undrained shear strength at a cell pressure of 275 kPa.

(c) Determine the maximum shear stress and the failure shear stress for each test.

**5.15**    The following data at failure were obtained from two CU tests on a soil:

| Test | $\sigma_3$ (kPa) | $\sigma_1 - \sigma_3$ (kPa) | $\Delta u$ (kPa) |
|---|---|---|---|
| 1 | 100 | 240 | −60 |
| 2 | 150 | 270 | −90 |

(a) Determine the friction angle for each test.

(b) Determine the change in dilation angle between the two tests, if a CD test were carried out.

(c) Determine the undrained shear strength for each test.

**5.16**    A CU test on a stiff, overconsolidated clay at a cell pressure at 150 kPa fails at a deviatoric stress of 448 kPa. The excess pore water pressure at failure is −60 kPa. Determine the friction angle.

**5.17**    The results of UU and CU tests on samples of soil are shown below. Determine (a) the friction angle and (b) $s_u$ at a cell pressure of 210 kPa in the CU tests. Predict the excess pore water pressure at failure for each of the UU test samples.

| Type of Test | $\sigma_3$ (kPa) | $\sigma_1 - \sigma_3$ (kPa) | $\Delta u$ (kPa) |
|---|---|---|---|
| UU | 57 | 105 | To be determined |
|  | 210 | 246 | To be determined |
| CU | 57 | 74.2 | 11.9 |
|  | 133 | 137.9 | 49.0 |
|  | 210 | 203 | 86.1 |

**5.18**    The failure stresses in a simple shear constant-volume test are shown in the table below.

| | |
|---|---|
| Total normal stress on the horizontal plane | 300 kPa |
| Total normal stress on the vertical plane | 200 kPa |
| Total shear stress on the horizontal and vertical planes | 100 kPa |
| Pore water pressure | 50 kPa |

(a) Draw Mohr's circles of total and effective stresses and determine the magnitude of the principal effective stresses and direction of the major principal effective stresses.

(b) Determine the friction angle assuming the soil is nondilational.

(c) Determine the undrained shear strength.

## Practical

5.19    You are in charge of designing a retaining wall. What laboratory tests would you specify for the backfill soil? Give reasons.

5.20    The following results were obtained from CU tests on a clay soil that is the foundation material for an embankment:

| $\sigma_3$ (kPa) | $\sigma_1 - \sigma_3$ (kPa) | $\Delta u$ (kPa) |
|---|---|---|
| 300 | 331 | 111 |
| 400 | 420 | 160 |
| 600 | 487 | 322 |

Recommend the shear strength parameters to be used for short-term and long-term analyses. The maximum confining pressure (cell pressure) at the depth of interest is 300 kPa.

## CHAPTER 6

# A CRITICAL STATE MODEL TO INTERPRET SOIL BEHAVIOR

| ABET | ES | ED |
|------|----|----|
|      | 90 | 10 |

## 6.0 INTRODUCTION

So far, we have painted individual pictures of soil behavior. We looked at the physical characteristics of soil in Chapter 2, effective stresses and stress paths in Chapter 3, one-dimensional consolidation in Chapter 4, and shear strength in Chapter 5. You know that if you consolidate a soil to a higher stress state than its current one, the shear strength of the soil will increase. But the amount of increase depends on the soil type, the loading conditions (drained or undrained condition), and the stress paths. Therefore, the individual pictures should all be linked together: But how?

In this chapter, we are going to take the individual pictures and build a mosaic that will provide a base for us to interpret and anticipate soil behavior. Our mosaic is mainly intended to unite consolidation and shear strength. Real soils, of course, require a complex mosaic not only because soils are natural, complex materials but also because the loads and loading paths cannot be anticipated accurately.

We are going to build a mosaic to provide a simple framework to describe, interpret, and anticipate soil responses to various loadings. The framework is essentially a theoretical model based on critical state soil mechanics—critical state model (Schofield and Wroth, 1968). Laboratory and field data, especially results from soft normally consolidated clays, lend support to the underlying concepts embodied in the development of the critical state model. The emphasis in this chapter will be on using the critical state model to provide a generalized understanding of soil behavior rather than on the mathematical formulation.

The critical state model (CSM) we are going to study is a simplification and an idealization of soil behavior. However, the CSM captures the behavior of soils that are of greatest importance to geotechnical engineers. The central idea in the CSM is that all soils will fail on a unique failure surface in $(q, p', e)$ space. Thus, the CSM incorporates volume changes in its failure criterion unlike the Mohr–Coulomb failure criterion, which defines failure only as the attainment of the maximum stress obliquity. According to the CSM, the failure stress state is insufficient to guarantee failure; the soil structure must also be loose enough.

The CSM is a tool to make estimates of soil responses when you cannot conduct sufficient soil tests to completely characterize a soil at a site or when you have to predict the soil's response from changes in loading during and after construction. Although there is a debate on the application of the CSM to real soils, the ideas behind the CSM are simple. It is a very powerful tool to get insights into soil behavior, especially in the case of the "what-if" situation. There is also a plethora of soil models in the literature that have critical state as their core. By studying the CSM, albeit a simplified version in this chapter, you will be able to better understand these other soil models.

When you have studied this chapter, you should be able to:

- Estimate failure stresses for soil
- Estimate strains at failure
- Predict stress–strain characteristics of soils from a few parameters obtained from simple soil tests
- Evaluate possible soil stress states and failure if the loading on a geotechnical system were to change

You will make use of all the materials you studied in Chapters 2 to 5 but particularly:

- Index properties (Chapter 2)
- Effective stresses, stress invariants, and stress paths (Chapter 3)
- Primary consolidation (Chapter 4)
- Shear strength (Chapter 5)

*Sample Practical Situation*   An oil tank is to be constructed on a soft alluvial clay. It was decided that the clay would be preloaded with a circular embankment imposing a stress equal to, at least, the total applied stress of the tank when filled. Sand drains are to be used to accelerate the consolidation process. The foundation for the tank is a circular slab of concrete and the purpose of the preloading is to reduce the total settlement of the foundation. You are required to advise the owners on how the tank should be filled during preloading to prevent premature failure. After preloading, the owners decided to increase the height of the tank. You are requested to determine whether the soil has enough shear strength to support an additional increase in tank height, and if so the amount of settlement that can be expected. The owners do not want to finance any further preloading and soil testing.

## 6.1   DEFINITIONS OF KEY TERMS

*Overconsolidation ratio ($R_o$)* is the ratio by which the current mean effective stress in the soil was exceeded in the past ($R_o = p_c'/p_o'$ where $p_c'$ is the past maximum mean effective stress and $p_o'$ is the current mean effective stress).

*Compression index ($\lambda$)* is the slope of the normal consolidation line in a plot of void ratio versus the natural logarithm of mean effective stress.

*Unloading/reloading index or recompression index ($\kappa$)* is the average slope of the unloading/reloading curves in a plot of void ratio versus the natural logarithm of mean effective stress.

*Critical state line (CSL)* is a line that represents the failure state of soils. In ($q, p'$) space the critical state line has a slope $M$, which is related to the friction angle of the soil at the critical state. In ($e, \ln p'$) space, the critical state line has a slope $\lambda$, which is parallel to the normal consolidation line. In three-dimensional $(q, p', e)$ space, the critical state line becomes a critical state surface.

## 6.2 QUESTIONS TO GUIDE YOUR READING

1. What is soil yielding?
2. What is the difference between yielding and failure in soils?
3. What parameters affect the yielding and failure of soils?
4. Does the failure stress depend on the consolidation pressure?
5. What are the critical state parameters and how can you determine them from soil tests?
6. Are strains important in soil failure?
7. What are the differences in the stress–strain responses of soils due to different stress paths?

## 6.3 BASIC CONCEPTS

### 6.3.1 Parameter Mapping

In our development of the basic concepts on critical state, we are going to map certain plots we have studied in Chapters 4 and 5 using stress and strain invariants and concentrate on a saturated soil under axisymmetric loading. However, the concepts and method hold for any loading condition. Rather than plotting $\tau$ versus $\sigma'_n$ or $\sigma'_z$, we will plot the data as $q$ versus $p'$ (Fig. 6.1a). This means that you must know the principal stresses acting on the element. For axisymmetric (triaxial) condition, you only need to know two principal stresses.

The Mohr–Coulomb failure line in $(\tau, \sigma'_z)$ space of slope $\phi'_{cs} = \tan^{-1}[\tau_{cs}/(\sigma'_z)_f]$ is now mapped in $(q, p')$ space as a line of slope $M = q_f/p'_f$, where the subscript $f$ denotes failure. Instead of a plot of $e$ versus $\sigma'_z$, we will plot the data as $e$ versus $p'$ (Fig. 6.1b) and instead of $e$ versus $\log \sigma'_z$, we will plot $e$ versus $\ln p'$ (Fig. 6.1c). We will denote the slope of the normal consolidation line in the plot of $e$ versus $\ln p'$ as $\lambda$ and the unloading/reloading line as $\kappa$. There are now relationships between $\phi'_{cs}$ and $M$, $C_c$ and $\lambda$, and $C_r$ and $\kappa$. The relationships for the slopes of the normal consolidation line (NCL), $\lambda$, and the unloading/reloading line (URL), $\kappa$, are

$$\lambda = \frac{C_c}{\ln(10)} = \frac{C_c}{2.3} = 0.434 C_c \qquad (6.1)$$

$$\kappa = \frac{C_r}{\ln(10)} = \frac{C_r}{2.3} = 0.434 C_r \qquad (6.2)$$

Both $\lambda$ and $\kappa$ are positive for compression. For many soils, $\kappa/\lambda$ has values within the range $\frac{1}{10}$ to $\frac{1}{5}$. We will formulate the relationship between $\phi'_{cs}$ and $M$ later. The overconsolidation ratio using stress invariants is

$$R_o = \frac{p'_c}{p'_o} \qquad (6.3)$$

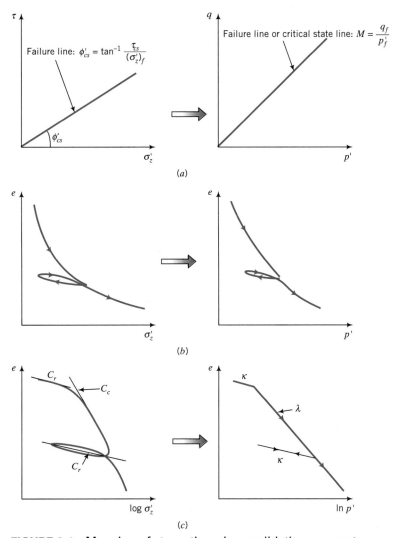

**FIGURE 6.1** Mapping of strength and consolidation parameters.

where $p_o'$ is the initial mean effective stress or overburden mean pressure and $p_c'$ is the preconsolidated mean effective stress. The overconsolidation ratio, $R_o$, defined by Eq. (6.3) is not equal to OCR [Eq. (4.13)]:

$$R_o = \frac{1 + 2K_o^{nc}}{1 + 2K_o^{oc}} \text{OCR}$$

(You will be required to prove this equation in Exercise 6.1.)

## 6.3.2 Failure Surface

The fundamental concept in CSM is that a unique failure surface exists in $(q, p', e)$ space, which defines failure of a soil irrespective of the history of loading or the stress paths followed. Failure and critical state are synonymous. We will refer to the failure line as the critical state line (CSL) in this chapter. You should

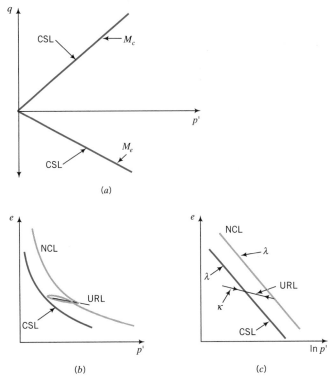

**FIGURE 6.2**  Critical state lines, normal compression, and unloading/reloading lines.

recall that critical state is a constant stress state characterized by continuous shear deformation at constant volume. In stress space $(q, p')$ the CSL is a straight line of slope $M = M_c$, for compression, and $M = M_e$, for extension (Fig. 6.2a). Extension does not mean tension but refers to the case where the lateral stress is greater than the vertical stress. There is a corresponding CSL in $(p', e)$ space (Fig. 6.2b) or $(e, \ln p')$ space (Fig. 6.2c) that is parallel to the normal consolidation line.

We can represent the CSL in a single three-dimensional plot with axes $q$, $p'$, $e$ (see book cover), but we will use the projections of the failure surface in the $(q, p')$ space and the $(e, p')$ space for simplicity.

### 6.3.3 Soil Yielding

You should recall from Chapter 3 (Fig. 3.8) that there is a yield surface in stress space that separates stress states that produce elastic responses from stress states that produce plastic responses. We are going to use the yield surface in $(q, p')$ space (Fig. 6.3) rather than $(\sigma_1, \sigma_3)$ space so that our interpretation of soil responses is independent of the axis system.

The yield surface is assumed to be an ellipse and its initial size or major axis is determined by the preconsolidation stress, $p_c'$. Experimental evidence (Wong and Mitchell, 1975) indicates that an elliptical yield surface is a reasonable approximation for soils. The higher the preconsolidation stress, the larger the

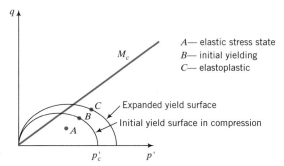

**FIGURE 6.3**  Expansion of the yield surface.

initial ellipse. We will consider the yield surface for compression but the ideas are the same for extension except that the minor axis of the elliptical yield surface in extension is smaller than in compression. All combinations of $q$ and $p'$ that lie within the yield surface, for example, point $A$ in Fig. 6.3, will cause the soil to respond elastically. If a combination of $q$ and $p'$ lies on the initial yield surface (point $B$, Fig. 6.3), the soil yields similar to the yielding of a steel bar. Any tendency of a stress combination to move outside the current yield surface is accompanied by an expansion of the current yield surface such that during plastic loading the stress point $(q, p')$ lies on the expanded yield surface and not outside, as depicted by $C$. Effective stress paths such as $BC$ (Fig. 6.3) cause the soil to behave elastoplastically. If the soil is unloaded from any stress state below failure, the soil will respond like an elastic material. As the yield surface expands, the elastic region gets larger.

### 6.3.4 Prediction of the Behavior of Normally Consolidated and Lightly Overconsolidated Soils Under Drained Conditions

Let us consider a hypothetical situation to illustrate the ideas presented so far. We are going to try to predict how a sample of soil of initial void ratio $e_o$ will respond when tested under drained condition in a triaxial apparatus, that is, a CD test. You should recall that the soil sample in a CD test is isotropically consolidated and then axial loads or displacements are applied, keeping the cell pressure constant. We are going to consolidate our soil sample up to a maximum mean effective stress $p_c'$, and then unload it to a mean effective stress $p_o'$ such that $R_o = p_c'/p_o' < 2$. We can sketch a curve of $e$ versus $p'$ ($AB$, Fig. 6.4b) during the consolidation phase. You should recall from Fig. 6.1 that the line $AB$ is the normal consolidation line of slope $\lambda$. Because we are applying isotropic loading, the line $AB$ (Fig. 6.4c) is called the isotropic consolidation line. The line $BC$ is the unloading/reloading line of slope $\kappa$.

The preconsolidated mean effective stress, $p_c'$, determines the size of the initial yield surface. A semi-ellipse is sketched in Fig. 6.4a to illustrate the initial yield surface for compression. We can draw a line, $OS$, from the origin to represent the critical state line in $(q, p')$ space as shown in Fig. 6.4a and a similar line in $(e, p')$ space as shown in Fig. 6.4b. Of course, we do not know, as yet, the slope $M$, or the equation to draw the initial yield surface. We have simply selected

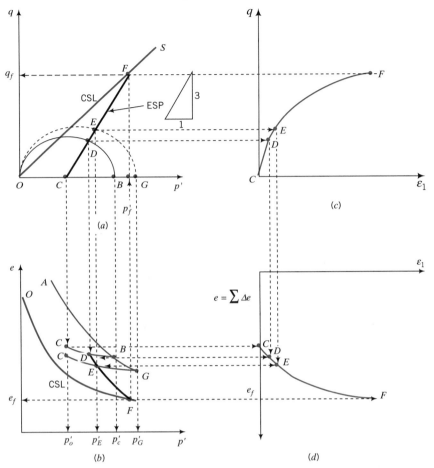

**FIGURE 6.4**    Illustrative predicted results from a CD test ($R_o \leq 2$) using CSM.

arbitrary values. Later, we are going to develop equations to define the slope $M$, the shape of the yield surface, and the critical state line in $(e, p')$ space or $(e, \ln p')$ space.

Let us now shear the soil sample at its current mean effective stress, $p'_o$, by increasing the axial stress, keeping the cell pressure, $\sigma_3$, constant and allowing the sample to drain. You should recall from Chapter 5 that the effective stress path for a CD test has a slope $q/p' = 3$. The effective stress path is shown by $CF$ in Fig. 6.4a. The effective stress path intersects the initial yield surface at $D$. All stress states from $C$ to $D$ lie within the initial yield surface and, therefore, from $C$ to $D$ on the ESP the soil behaves elastically. Assuming linear elastic response of the soil, we can draw a line $CD$ in $(q, \varepsilon_1)$ space (Fig. 6.4c) to represent the elastic stress–strain response. At this stage, we do not know the slope of $CD$ but later you will learn how to get this slope. Since the line $BC$ in $(e, p')$ space represents the unloading/reloading line (URL), the elastic response must lie along this line. The change in void ratio is $\Delta e = e_C - e_D$ (Fig. 6.4b) and we can plot the $e$ versus $\varepsilon_1$ response as shown by $CD$ in Fig. 6.4d.

Further loading from $D$ along the stress path $CF$ causes the soil to yield.

The initial yield surface expands (Fig. 6.4a) and the stress–strain response is a curved path (Fig. 6.4c) because the soil behaves elastoplastically (Chapter 3). At some arbitrarily chosen loading point, $E$, along the ESP, the size (major axis) of the yield surface is $p'_G$ corresponding to point $G$ in $(e, p')$ space.

The total change in void ratio as you load the sample from $D$ to $E$ is $DE$ (Fig. 6.4b). Since $E$ lies on the yield surface corresponding to a mean effective stress $p'_E$, then $E$ must be on the unloading line, $EC'$, as illustrated in Fig. 6.4b. If you unload the soil sample from $E$ back to $C$, the soil will follow an unloading path, $EC'$, parallel to $BC$ as shown in Fig. 6.4b.

We can continue to add increments of loading along the ESP until the soil fails. For each load increment, we can sketch the stress–strain curve and the path followed in $(e, p')$ space. Failure occurs when the ESP intersects the critical state line as indicated by $F$ in Fig. 6.4a. The failure stresses are $p'_f$ and $q_f$ (Fig. 6.4a) and the failure void ratio is $e_f$ (Fig. 6.4b). For each increment of loading, we can determine $\Delta e$ and plot $\varepsilon_1$ versus $\Sigma\Delta e$ [or $\varepsilon_p = (\Sigma\Delta e)/(1 + e_o)$] as shown in Fig. 6.4d.

Each point on one of the figures has a corresponding point on another figure in each of the quadrants shown in Fig. 6.4. Thus, each point on any figure can be obtained by projection as illustrated in Fig. 6.4. Of course, the scale of the axis on one figure must match the scale of the corresponding axis on the other figure.

## 6.3.5 Prediction of the Behavior of Normally Consolidated and Lightly Overconsolidated Soils Under Undrained Condition

Instead of a CD test we could have conducted a CU test after consolidating the sample. Let us examine what would have occurred according to our CSM. We know (Chapter 5) that for undrained condition the soil volume remains constant, that is, $\Delta e = 0$, and the ESP for stresses that produce an elastic response is vertical, that is, the change in mean effective stress, $\Delta p'$, is zero for linearly elastic soils. Because the change in volume is zero, the mean effective stress at failure can be represented by drawing a horizontal line from the initial void ratio to intersect the critical state line in $(e, p')$ space as illustrated by $CF$ in Fig. 6.5b. Projecting a vertical line from the mean effective stress at failure in $(e, p')$ space to intersect the critical state line in $(q, p')$ space gives the deviatoric stress at failure (Fig. 6.5a). Since the ESP is vertical within the initial yield surface ($CD$, Fig. 6.5a), the yield stress can readily be found from the intersection of the ESP and the initial yield surface. Points $C$ and $D$ are coincident in the $(e, p')$ plot as illustrated in Fig. 6.5b because $\Delta p' = 0$. For normally consolidated and lightly overconsolidated soils, the effective stress path after initial yielding (point $D$, Fig. 6.5a) curves toward the critical state line as the excess pore water pressure increases significantly after yielding occurs.

The TSP has a slope of 3 (Chapter 5) as illustrated by $CG$ in Fig. 6.5a. The difference in mean stress between the total stress path and the effective path gives the change in excess pore water pressure. The intersection of the TSP with the critical state line at $G$ is not the failure point because failure and deformation in a soil mass depend on effective not total stress. By projection, we can sketch the stress–strain response and the excess pore water pressure versus strain as illustrated in Figs. 6.5c,d.

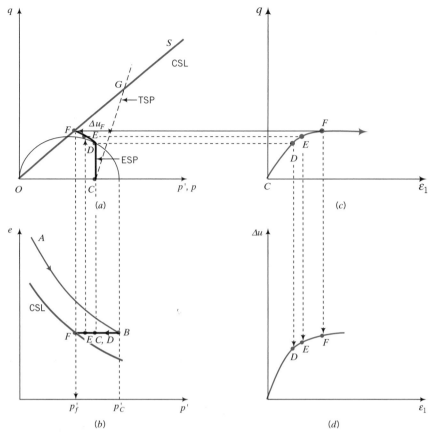

**FIGURE 6.5** Illustrative predicted results from a CU test using the CSM ($R_o \leq 2$).

### 6.3.6 Prediction of the Behavior of Heavily Overconsolidated Soils

So far we have considered a lightly overconsolidated soil ($R_o < 2$). What is the situation regarding heavily overconsolidated soils, that is, $R_o > 2$? We can model a heavily overconsolidated soil by unloading it so that $p'_c/p'_o > 2$ as shown by point $C$ in Figs. 6.6a,b. Heavily overconsolidated soils have initial stress states that lie to the left of the critical state line in the $e$ versus $p'$ plot. The ESP for a CD test has a slope of 3 and intersects the initial yield surface at $D$. Therefore, from $C$ to $D$ the soil behaves elastically as shown by $CD$ in Figs. 6.6b,c. The intersection of the ESP with the critical state line is at $F$ (Fig. 6.6a), so that the yield surface must contract as the soil is loaded to failure. The initial yield shear stress is analogous to the peak shear stress for dilating soils. From $D$, the soil expands (Figs. 6.6b,d) and strain softens (Fig. 6.6c) to failure at $F$.

The CSM simulates the mechanical behavior of heavily overconsolidated soils as elastic materials up to the peak shear stress and thereafter elastoplastically as the imposed loading causes the soil to strain soften toward the critical state line. In reality, heavily overconsolidated soils may behave elastoplastically before the peak shear stress is achieved but this behavior is not captured by the simple CSM described here.

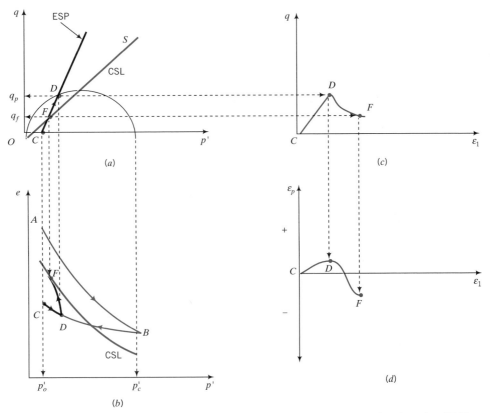

**FIGURE 6.6**  Illustrative predicted results from a CD test ($R_o > 2$) using the CSM.

In the case of a CU test on heavily overconsolidated soils, the path to failure in ($e, p'$) space is $CF$ as shown in Fig. 6.7b. Initial yielding is attained at $D$ and failure at $F$. The excess pore water pressures at initial yield, $\Delta u_y$, and at failure, $\Delta u_f$, are shown in the inset of Fig. 6.7a. The excess pore water pressure at failure is negative ($p'_f > p_f$).

### 6.3.7 Critical State Boundary

The CSL serves as a boundary separating normally consolidated and lightly over-consolidated soils and heavily overconsolidated soils. Stress states that lie to the right of the CSL will result in compression and strain hardening of the soil; stress states that lie to the left of the CSL will result in expansion and strain softening of the soil.

### 6.3.8 Volume Changes and Excess Pore Water Pressures

If you compare the responses of soils in drained and undrained tests as predicted by the CSM, you will notice that compression in drained tests translates as positive excess pore water pressures in undrained tests, and expansion in drained tests translates as negative excess pore water pressures in undrained tests. The

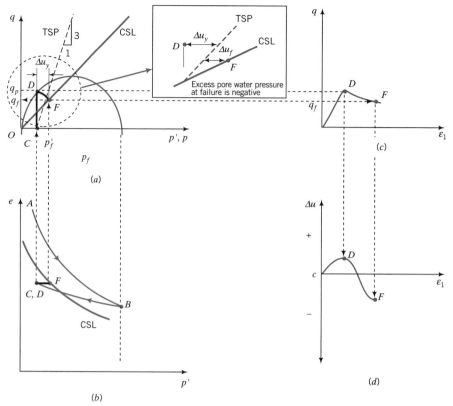

**FIGURE 6.7** Illustrative predicted results from a CU test ($R_o \geq 2$) using the CSM.

CSM also predicts that normally consolidated and lightly overconsolidated soils strain harden to failure, while heavily overconsolidated soils strain soften to failure. The predicted responses from the CSM then qualitatively match observed soil responses (Chapter 5).

## 6.3.9 Effects of Effective Stress Paths

The response of a soil depends on the ESP. Effective stress paths with slopes less than the CSL ($OA$, Fig. 6.8) will not produce shear failure in the soil because the ESP will never intersect the critical state line. You can load a normally consolidated or a lightly overconsolidated soil with an ESP that causes it to respond like

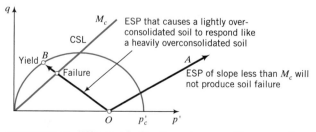

**FIGURE 6.8** Effects of effective stress paths on soil response.

an overconsolidated soil as shown by *OB* in Fig. 6.8. Effective stress paths similar to *OB* are possible in soil excavation. Remember that a soil must yield before it fails.

*The* essential points *are:*

1. *There is a unique critical state line in* (q, p′) *space and a corresponding critical state line in* (e, p′) *space for soils.*
2. *There is an initial yield surface for soils. The size of the initial yield surface depends on the preconsolidation mean effective stress.*
3. *The yield surface expands for* R$_o$ ≤ 2 *and contracts for* R$_o$ ≥ 2 *when the applied effective stresses exceed the initial yield stress.*
4. *The soil will behave elastically for stresses that are within the yield surface and elastoplastically for stresses outside the yield surface.*
5. *Every stress state must lie on an expanded or contracted yield surface and on a corresponding URL.*
6. *The critical state model qualitatively captures the essential features of soil responses under drained and undrained loading.*

*What's next . . .*You were given an illustration using projection geometry of the essential ingredients of the critical state model. There were many unknowns. For example, you did not know the slope of the critical state line and the equation of the yield surface. In the next section we will develop equations to find these unknowns. Remember that our intention is to build a simple mosaic coupling the essential features of consolidation and shear strength.

## 6.4 ELEMENTS OF THE CRITICAL STATE MODEL

### 6.4.1 Yield Surface

The equation for the yield surface is an ellipse given by

$$(p')^2 - p'p'_c + \frac{q^2}{M^2} = 0 \qquad (6.4)$$

The theoretical basis for the yield surface is presented by Schofield and Wroth (1968) and Roscoe and Burland (1968). You can draw the initial yield surface from the initial stresses on the soil if you know the value of *M*.

### 6.4.2 Critical State Parameters

*6.4.2.1 Failure Line in (q, p′) Space*   The Mohr–Coulomb failure criterion for soils as described in Chapter 5 can be written in terms of stress invariants as

$$q_f = Mp'_f \qquad (6.5)$$

where $q_f$ is the deviatoric stress at failure (similar to $\tau_f$), *M* is a friction constant (similar to tan $\phi'_{cs}$), and $p'_f$ is the mean effective stress at failure (similar to $\sigma'_n$).

For compression, $M = M_c$ and for extension $M = M_e$. The critical state line intersects the yield surface at $p_c'/2$.

Let us find a relationship between $M$ and $\phi_{cs}'$ for axisymmetric compression and axisymmetric extension.

### Axisymmetric Compression

$$M_c = \frac{q_f}{p_f'} = \frac{(\sigma_1' - \sigma_3')_f}{\left(\dfrac{\sigma_1' + 2\sigma_3'}{3}\right)_f} = \frac{3\left(\dfrac{\sigma_1'}{\sigma_3'} - 1\right)_f}{\left(\dfrac{\sigma_1'}{\sigma_3'} + 2\right)_f}$$

We know from Chapter 5 that

$$\left(\frac{\sigma_1'}{\sigma_3'}\right)_f = \frac{1 + \sin \phi_{cs}'}{1 - \sin \phi_{cs}'}$$

Therefore,

$$M_c = \frac{6 \sin \phi_{cs}'}{3 - \sin \phi_{cs}'} \tag{6.6}$$

or

$$\sin \phi_{cs}' = \frac{3M_c}{6 + M_c} \tag{6.7}$$

**Axisymmetric Extension**   In an axisymmetric extension test, the radial stress is the major principal stress. Since in axial symmetry the radial stress is equal to the circumferential stress, we get

$$p_f' = \left(\frac{2\sigma_1' + \sigma_3'}{3}\right)_f$$

$$q_f = (\sigma_1' - \sigma_3')_f$$

and

$$M_e = \frac{q_f}{p_f'} = \frac{\left(2\dfrac{\sigma_1'}{\sigma_3'} + 1\right)_f}{\left(\dfrac{\sigma_1'}{\sigma_3'} - 1\right)_f} = \frac{6 \sin \phi_{cs}'}{3 + \sin \phi_{cs}'} \tag{6.8}$$

or

$$\sin \phi_{cs}' = \frac{3M_e}{6 - M_e} \tag{6.9}$$

An important point to note is that while the friction angle, $\phi_{cs}'$, is the same for compression and extension, the slope of the critical state line in $(q, p')$ space is not the same. Therefore, the failure deviatoric stresses in compression and extension are different. Since $M_e < M_c$, the failure deviatoric stress of a soil in extension is lower than that for the same soil in compression.

**6.4.2.2 Failure Line in (e, p') Space**  Let us now find the equation for the critical state line in $(e, p')$ space. We will use the $(e, \ln p')$ plot as shown in Fig. 6.9c. The CSL is parallel to the normal consolidation line and is represented by

$$e_f = e_\Gamma - \lambda \ln p'_f \tag{6.10}$$

where $e_\Gamma$ is the void ratio on the critical state line when $\ln p' = 1$. The value of $e_\Gamma$ depends on the units chosen for the $p'$ scale. In this book, we will use kPa for the units of $p'$.

We will now determine $e_\Gamma$ from the initial state of the soil. Let us isotropically consolidate a soil to a mean effective stress $p'_c$ and then isotropically unload it to a mean effective stress $p'_o$ (Figs. 6.9a,b). Let $X$ be the intersection of the unloading/reloading line with the critical state line. The mean effective stress at $X$ is $p'_c/2$ and from the unloading/reloading line

$$e_X = e_o + \kappa \ln \frac{p'_o}{p'_c/2} \tag{6.11}$$

where $e_o$ is the initial void ratio. From the critical state line,

$$e_X = e_\Gamma - \lambda \ln \frac{p'_c}{2} \tag{6.12}$$

Therefore, equating Eqs. (6.11) and (6.12) we get

$$e_\Gamma = e_o + (\lambda - \kappa) \ln \frac{p'_c}{2} + \kappa \ln p'_o \tag{6.13}$$

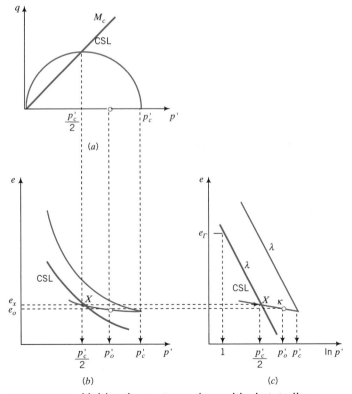

**FIGURE 6.9**  Void ratio, $e_\Gamma$, to anchor critical state line.

> *The* **essential critical state parameters** *are:*
>
> λ—*Compression index, which is obtained from an isotropic or a one-dimensional consolidation test.*
>
> κ—*Unloading/reloading index or recompression index, which is obtained from an isotropic or a one-dimensional consolidation test.*
>
> M—*Critical state frictional constant, which is a function of $\phi'_{cs}$ and is obtained from shear tests (direct shear, triaxial, simple shear, etc.).*

To use the critical state model, you must also know the initial stresses, for example, $p'_o$ and $p'_c$, and the initial void ratio, $e_o$.

## EXAMPLE 6.1

A CD test at a constant cell pressure, $\sigma_3 = \sigma'_3 = 120$ kPa, was conducted on a sample of normally consolidated clay. At failure, $q = \sigma'_1 - \sigma'_3 = 140$ kPa. What is the value of $M_c$? If an extension test were to be carried out, determine the mean effective and deviatoric stresses at failure.

**Strategy** You are given the final stresses, so you have to use these to compute $\phi'_{cs}$ and then use Eq. (6.6) to calculate $M_c$ and Eq. (6.8) to calculate $M_e$. You can then calculate $q'_f$ for the extension test by proportionality.

## Solution 6.1

**Step 1:** Find the major principal stress at failure.

$$(\sigma'_1)_f = 140 + 120 = 260 \text{ kPa}$$

**Step 2:** Find $\phi'_{cs}$.

$$\sin \phi'_{cs} = \frac{\sigma'_1 - \sigma'_3}{\sigma'_1 + \sigma'_3} = \frac{140}{260 + 120} = 0.37$$

$$\phi'_{cs} = 21.6°$$

**Step 3:** Find $M_c$ and $M_e$.

$$M_c = \frac{6 \sin \phi'_{cs}}{3 - \sin \phi'_{cs}} = \frac{6 \times 0.37}{3 - 0.37} = 0.84$$

$$M_e = \frac{6 \sin \phi'_{cs}}{3 + \sin \phi'_{cs}} = \frac{6 \times 0.37}{3 + 0.37} = 0.66$$

**Step 4:** Find $q_f$ for extension.

$$q_f = \frac{0.66}{0.84} \times 140 = 110 \text{ kPa}; \quad p'_f = \frac{q_f}{M_e} = \frac{110}{0.66} = 166.7 \text{ kPa} \quad \blacksquare$$

## EXAMPLE 6.2

A saturated soil sample was isotropically consolidated in a triaxial apparatus and a selected set of data is shown in the table. Determine λ, κ, and $e_\Gamma$.

| Condition | Cell pressure (kPa) | Final void ratio |
|---|---|---|
| Loading | 200 | 1.72 |
| | 1000 | 1.20 |
| Unloading | 500 | 1.25 |

***Strategy*** Make a sketch of the results in $(e, \ln p')$ space to provide a visual aid for solving this problem.

### Solution 6.2

**Step 1:** Make a plot of $\ln p'$ versus $e$.
See Fig. E6.2.

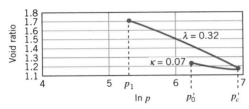

**FIGURE E6.2**

**Step 2:** Calculate $\lambda$.
From Fig. E6.2,

$$\lambda = \frac{|\Delta e|}{\ln(p'_c/p'_1)} = \frac{|1.20 - 1.72|}{\ln(\frac{1000}{200})} = 0.32$$

**Step 3:** Calculate $\kappa$.
From Fig. E6.2,

$$\kappa = \frac{|\Delta e|}{\ln(p'_c/p'_o)} = \frac{|1.20 - 1.25|}{\ln(\frac{1000}{500})} = 0.07$$

**Step 4:** Calculate $e_\Gamma$.

$$e_\Gamma = e_o + (\lambda - \kappa)\ln\frac{p'_c}{2} + \kappa \ln p'_o = 1.25 + (0.32 - 0.07)\ln\frac{1000}{2} + 0.07 \ln 500 = 3.24$$

■

***What's next . . .***We now know the key parameters to use in the CSM. Next, we will use the CSM to predict the shear strength of soils.

## 6.5  FAILURE STRESSES FROM THE CRITICAL STATE MODEL

### 6.5.1 Drained Triaxial Test

Let us consider a CD test in which we isotropically consolidate a soil to a mean effective stress $p'_c$ and unload it isotropically to a mean effective stress of $p'_o$ (Figs.

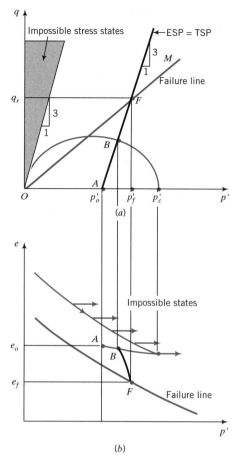

**FIGURE 6.10**  Failure in CD tests.

6.10a,b) such that $R_o \leq 2$. The slope of the ESP = TSP is 3:1 as shown by $AF$ in Fig. 6.10a. The ESP will intersect the critical state line at $F$. We need to find the stresses at $F$. The equation for the ESP is

$$q_f = 3(p'_f - p'_o) \tag{6.14}$$

The equation for the critical state line, using a generic $M$, which for compression is $M_c$ and for extension is $M_e$, is

$$q_f = Mp'_f \tag{6.15}$$

The intersection of these two lines is found by equating Eqs. (6.14) and (6.15), which leads to

$$p'_f = \frac{3p'_o}{3 - M} \tag{6.16}$$

and

$$q_f = Mp'_f = \frac{3Mp'_o}{3 - M} \tag{6.17}$$

Let us examine Eqs. (6.16) and (6.17). If $M = M_c = 3$, then $p'_f \to \infty$ and $q_f \to \infty$. Therefore, $M_c$ cannot have a value of 3 because soils cannot have infinite strength. If $M_c > 3$, then $p'_f$ is negative and $q_f$ is negative. Of course, $p'_f$ cannot be negative because soil cannot sustain tension. Therefore, we cannot have a value of $M_c$ greater than 3. Therefore, the region bounded by a slope $q/p' = 3$ originating from the origin and the deviatoric stress axis represents impossible soil states (Fig. 6.10a). For extension tests, the bounding slope is $q/p' = -3$. Also, you should recall from Chapter 4 that soil states to the right of the normal consolidation line are impossible (Fig. 6.10b).

We have now delineated regions in stress space $(q, p')$ and in void ratio space versus mean effective stress—that is, $(e, p')$ space, that are possible for soils. Soil states cannot exist outside these regions.

## 6.5.2 Undrained Triaxial Test

In an undrained test, no volume change occurs—that is, $\Delta V = 0$—which means that $\Delta \varepsilon_p = 0$ or $\Delta e = 0$ (Fig. 6.11) and, consequently.

$$e_f = e_o = e_\Gamma - \lambda \ln p'_f \tag{6.18}$$

By rearranging Eq. (6.18), we get

$$\boxed{p'_f = \exp\left(\frac{e_\Gamma - e_o}{\lambda}\right)} \tag{6.19}$$

Since $q_f = Mp'_f$, then

$$\boxed{q_f = M \exp\left(\frac{e_\Gamma - e_o}{\lambda}\right)} \tag{6.20}$$

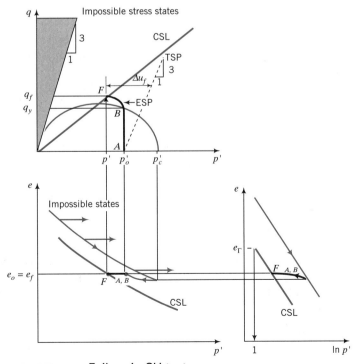

**FIGURE 6.11** Failure in CU tests.

For a CU test, the TSP has a slope of 3 (Fig. 6.11). For the elastic range of stress, the ESP is vertical ($\Delta p' = 0$) up to the yield stress and bends toward the critical state line as the pore water pressure increases considerably after yield.

The undrained shear strength, denoted by $s_u$, is defined as one-half the deviatoric stress at failure. That is,

$$s_u = \frac{M}{2} \exp\left(\frac{e_\Gamma - e_o}{\lambda}\right) \qquad (6.21)$$

For a given soil, $M$, $\lambda$, and $e_\Gamma$ are constants and the only variable in Eq. (6.21) is the initial void ratio. Therefore, the undrained shear strength of a particular saturated soil depends only on the initial void ratio or initial water content. You should recall that we discussed this in Chapter 5 but did not show any mathematical proof.

We can use Eq. (6.21) to compare the undrained shear strengths of two samples of the same soil tested at different void ratio or to predict the undrained shear strength of one sample if we know the undrained shear strength of the other. Consider two samples, A and B, of the same soil. The ratio of their undrained shear strength is

$$\frac{(s_u)_A}{(s_u)_B} = \frac{\left[\exp\left(\frac{e_\Gamma - e_o}{\lambda}\right)\right]_A}{\left[\exp\left(\frac{e_\Gamma - e_o}{\lambda}\right)\right]_B} = \exp\left(\frac{(e_o)_B - (e_o)_A}{\lambda}\right)$$

For a saturated soil, $e_o = wG_s$ and we can then rewrite the above equation as

$$\frac{(s_u)_A}{(s_u)_B} = \exp\left[\frac{G_s(w_B - w_A)}{\lambda}\right] \qquad (6.22)$$

Let us examine the difference in undrained shear strength for a 1% difference in water content between samples A and B. We will assume that the water content of sample B is greater than sample A, that is, $(w_B - w_A)$ is positive, $\lambda = 0.15$ (a typical value for a silty clay), and $G_s = 2.7$. Putting these values into Eq. (6.22), we get

$$\frac{(s_u)_A}{(s_u)_B} = 1.20$$

That is, a 1% increase in water content causes a reduction in undrained shear strength of 20% for this soil. The implication on soil testing is that you should preserve the water content of soil samples, especially samples taken from the field, because the undrained shear strength can be significantly altered by even small changes in water content.

For highly overconsolidated clays ($R_o > 2$) or dense sands, the peak shear stress ($q_p$) is equal to the initial yield stress (Fig. 6.7). Recall that the CSM predicts that soils with $R_o > 2$ will behave elastically up to the peak shear stress (initial yield stress). By substituting $p' = p'_o$ and $q = q_p$ in the equation for the yield surface [Eq. (6.4)], we obtain

$$(p'_o)^2 - p'_o p'_c + \frac{q_p^2}{M^2} = 0$$

which simplifies to

$$q_p = Mp'_o\sqrt{\frac{p'_c}{p'_o} - 1} = Mp'_o\sqrt{R_o - 1}; \quad R_o > 2 \tag{6.23}$$

and

$$s_u = \frac{M}{2}p'_o\sqrt{R_o - 1}; \quad R_o > 2 \tag{6.24}$$

The excess pore water pressure at failure is found from the difference between the mean total stress and the corresponding mean effective stress at failure; that is,

$$\Delta u_f = p_f - p'_f$$

From the TSP,

$$p_f = p'_o + \frac{q_f}{3}$$

Therefore,

$$\Delta u_f = p'_o + \left(\frac{M}{3} - 1\right)\exp\left(\frac{e_\Gamma - e_o}{\lambda}\right) \tag{6.25}$$

*The essential points are:*

1. *The intersection of the ESP and the critical state line gives the failure stresses.*
2. *The undrained shear strength depends only on the initial void ratio.*
3. *Small changes in water content can significantly alter the undrained shear strength.*

## EXAMPLE 6.3

Two specimens, A and B, of a clay were each isotropically consolidated under a cell pressure of 300 kPa and then unloaded isotropically to a mean effective stress of 200 kPa. A CD test is to be conducted on specimen A and a CU test is to be conducted on specimen B. Estimate, for each specimen, (a) the yield stresses, $p'_y$, $q_y$, $(\sigma'_1)_y$, and $(\sigma'_3)_y$; and (b) the failure stresses $p'_f$, $q_f$, $(\sigma'_1)_f$, and $(\sigma'_3)_f$. Estimate for sample B the excess pore water pressure at yield and at failure. The soil parameters are $\lambda = 0.3$, $\kappa = 0.05$, $e_o = 1.10$, and $\phi'_{cs} = 30°$. The cell pressure was kept constant at 200 kPa.

*Strategy* Both specimens have the same consolidation history but are tested under different drainage conditions. The yield stresses can be found from the intersection of the ESP and the initial yield surface. The initial yield surface is known since $p'_c = 300$ kPa, and $M$ can be found from $\phi'_{cs}$. The failure stresses can be obtained from the intersection of the ESP and the critical state line. It is always a good habit to sketch the $q$ versus $p'$ and the $e$ versus $p'$ graphs to help you solve problems using the critical state model. You can also find the yield and failure stresses using graphical methods as described in the alternative solution.

## Solution 6.3

**Step 1:**  Calculate $M_c$.

$$M_c = \frac{6 \sin 30°}{3 - \sin 30°} = 1.2$$

**Step 2:**  Calculate $e_\Gamma$.
From Eq. (6.13), with $p_o' = 200$ kPa and $p_c' = 300$ kPa,

$$e_\Gamma = e_o + (\lambda - \kappa) \ln \frac{p_c'}{2} + \kappa \ln p_o' = 1.10 + (0.3 - 0.05) \ln \frac{300}{2} + 0.05 \ln 200 = 2.62$$

**Step 3:**  Make a sketch or draw scaled plots of $q$ versus $p'$ and $e$ versus $p'$. See Figs. E6.3a,b.

**Step 4:**  Find the yield stresses.

**Drained Test**   Let $p_y'$ and $q_y$ be the yield stress (point $B$ in Fig. E6.3a). From the equation for the yield surface [Eq. (6.4)],

$$(p_y')^2 - 300p_y' + \frac{q_y^2}{(1.2)^2} = 0 \tag{1}$$

From the ESP,

$$q_y = 3(p_y' - p_o') = 3p_y' - 600 \tag{2}$$

Solving Eqs. (1) and (2) for $p_y'$ and $q_y$ gives two solutions: $p_y' = 140.1$ kPa, $q_y = -179.6$ kPa and $p_y' = 246.1$ kPa, $q_y = 138.2$ kPa. Of course, $q_y = -179.6$ kPa

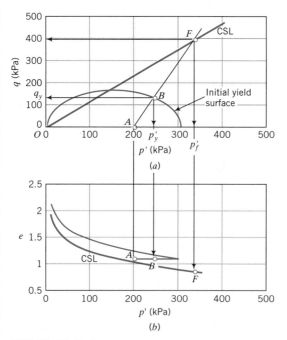

**FIGURE E6.3a,b**

is not possible because we are conducting a compression test. The yield stresses are then $p'_y = 246.1$ kPa, $q_y = 138.2$ kPa.

Now,

$$q_y = (\sigma'_1)_y - (\sigma'_3)_y = 138.2 \text{ kPa}; \quad (\sigma'_3)_f = 200 \text{ kPa}$$

Solving for $(\sigma'_1)_f$ gives

$$(\sigma'_1)_f = 138.2 + 200 = 338.2 \text{ kPa}$$

**Undrained Test**  The ESP for the undrained test is vertical for the region of stress paths below the yield stress, that is, $\Delta p' = 0$. From the yield surface [Eq. (6.4)] for $p' = p'_y = p'_o$, we get

$$200^2 - 200 \times 300 + \frac{q_y^2}{1.2^2} = 0$$

$$\therefore q_y^2 = 1.2^2 \times 200 \times 100$$

and

$$q_y = 169.7 \text{ kPa}$$

From the TSP, we can find $p_y$ ($B'$, Fig. E6.3c)

$$p_y = p'_o + \frac{q_y}{3} = 200 + \frac{169.7}{3} = 256.6 \text{ kPa}$$

The excess pore water pressure at yield is

$$\Delta u_y = p_y - p'_y = p_y - p'_o = 256.6 - 200 = 56.6 \text{ kPa}$$

Now

$$p'_y = p'_o = \frac{(\sigma'_1)_y + 2(\sigma'_3)_y}{3} = 200 \text{ kPa}$$

$$q_y = (\sigma'_1)_y - \sigma'_3 = 169.7 \text{ kPa}$$

Solving for $(\sigma'_1)_y$ and $(\sigma'_3)_y$ gives

$$(\sigma'_1)_y = 313.3 \text{ kPa}; \quad (\sigma'_3)_y = 143.4 \text{ kPa}$$

**Check**

$$(\sigma_3)_y = (\sigma'_3)_y + \Delta u_y = 143.4 + 56.6 = 200 \text{ kPa}$$

**Step 5:**  Find the failure stresses.

**Drained Test**

$$\text{Equation (6.16):} \quad p'_f = \frac{3 \times 200}{3 - 1.2} = 333.3 \text{ kPa}$$

$$\text{Equation (6.5):} \quad q_f = 1.2 \times 333.3 = 400 \text{ kPa}$$

Now,

$$q_F = (\sigma'_1)_f - (\sigma'_3)_f = 400 \text{ kPa} \quad \text{and} \quad (\sigma'_3)_f = 200 \text{ kPa}$$

Solving for $(\sigma'_1)_f$, we get

$$(\sigma'_1)_f = 400 + 200 = 600 \text{ kPa}$$

### Undrained Test

Equation (6.19):   $p_f' = \exp\left(\dfrac{2.62 - 1.10}{0.3}\right) = 158.6 \text{ kPa}$

Equation (6.5):   $q_f = 1.2 \times 158.6 = 190.3 \text{ kPa}$

Now,

$$p_f' = \frac{(\sigma_1')_f + 2(\sigma_3')_f}{3} = 158.6 \text{ kPa}$$

$$q_f = (\sigma_1')_f - (\sigma_3')_f = 190.4 \text{ kPa}$$

Solving for $(\sigma_1')_f$ and $(\sigma_3')_f$, we find

$$(\sigma_1')_f = 285.5 \text{ kPa} \quad \text{and} \quad (\sigma_3')_f = 95.1 \text{ kPa}$$

We can find the change in pore water pressure at failure from either Eq. (6.24)

$$\Delta u_f = 200 + \left(\frac{1.2}{3} - 1\right) \exp\left(\frac{2.62 - 1.10}{0.3}\right) = 104.9 \text{ kPa}$$

or

$$\Delta u_f = \sigma_3 - (\sigma_3')_f = 200 - 95.1 = 104.9 \text{ kPa}$$

**Graphical Method**   We need to find the equations for the normal consolidation line and the critical state lines.

### Normal Consolidation Line

Void ratio at preconsolidated mean effective stress:

$$e_c = e_o - \kappa \ln \frac{p_c'}{p_o'} = 1.10 - 0.05 \ln \frac{300}{200} = 1.08$$

Void ratio at $\ln p' = 1$ kPa on NCL:

$$e_n = e_c + \lambda \ln p_c' = 1.08 + 0.3 \ln 300 = 2.79$$

The equation for the normal consolidation line is then

$$e = 2.79 - 0.3 \ln p'$$

The equation for the unloading/reloading line is

$$e = 1.08 + 0.05 \ln \frac{p_c'}{p'}$$

The equation for the critical state line in $(e, p')$ space is

$$e = 2.62 - 0.3 \ln p'$$

Now you can plot the normal consolidation line, the unloading/reloading line, and the critical state line as shown in Fig. E6.3b.

**Plot Initial Yield Surface**   The yield surface is

$$(p')^2 - 300p' + \frac{q^2}{(1.2)^2} = 0$$

$$\therefore q = 1.2p' \sqrt{\frac{300}{p'} - 1}$$

For $p' = 0$ to 300, plot the initial yield surface as shown in Fig. E6.3a.

**Plot Critical State Line**   The critical state line is

$$q = 1.2p'$$

and is plotted as $OF$ in Fig. E6.3a.

**Drained Test**   The ESP for the drained test is

$$p' = 200 + \frac{q}{3}$$

and is plotted as $AF$ in Fig. E6.3a. The ESP intersects the initial yield surface at $B$ and the yield stresses are $p'_y = 240$ kPa and $q_y = 138$ kPa. The ESP intersects the critical state line at $F$ and the failure stresses are $p'_f = 333$ kPa and $q_f = 400$ kPa.

**Undrained Test**   For the undrained test, the initial void ratio and the final void ratio are equal. Draw a horizontal line from $A$ to intersect the critical state line in $(e, p')$ space at $F$ (Fig. E6.3d). Project a vertical line from $F$ to intersect the critical state line in $(q, p')$ space at $F$ (Fig. E6.3c). The failure stresses are $p'_f = 159$ kPa and $q_f = 190$ kPa. Draw the TSP as shown by $AS$ in Fig. E6.3c. The ESP within the elastic region is vertical as shown by $AB$. The yield stresses are $p'_y = 200$ kPa and $q_y = 170$ kPa. The excess pore water pressures are:

At yield—horizontal line $BB'$:   $\Delta u_y = 57$ kPa

At failure—horizontal line $FF'$:   $\Delta u_f = 105$ kPa

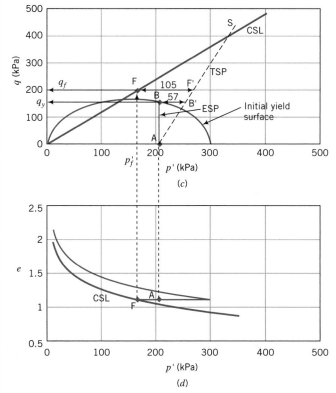

**FIGURE E6.3c,d**

## EXAMPLE 6.4

Determine the undrained shear strength in (a) a CU compression test and (b) a CU extension test for a soil with $R_o = 5$, $p'_o = 70$ kPa, and $\phi'_{cs} = 25°$.

***Strategy***   Since you are given $\phi'_{cs}$, you should use Eqs. (6.6) and (6.8) to find $M_c$ and $M_e$. Use Eq. (6.24) to solve the problem.

## Solution 6.4

**Step 1:**   Calculate $M_c$ and $M_e$.

$$M_c = \frac{6 \sin \phi'_{cs}}{3 - \sin \phi'_{cs}} = \frac{6 \sin 25°}{3 - \sin 25°} = 0.98$$

$$M_e = \frac{6 \sin \phi'_{cs}}{3 + \sin \phi'_{cs}} = 0.74$$

**Step 2:**   Calculate $s_u$.
Use Eq. (6.24).

$$\text{Compression:} \quad s_u = \frac{0.98}{2} \times 70\sqrt{5 - 1} = 68.6 \text{ kPa}$$

$$\text{Extension:} \quad s_u = \frac{0.74}{2} \times 70\sqrt{5 - 1} = 51.8 \text{ kPa}$$

Or, by proportion,

$$\text{Extension:} \quad s_u = \frac{0.74}{0.98} \times 68.6 = 51.8 \text{ kPa} \quad\blacksquare$$

## EXAMPLE 6.5

The in situ water content of a soil sample is 48%. The water content decreases to 44% due to transportation of the sample to the laboratory and during sample preparation. What difference in undrained shear strength could be expected if $\lambda = 0.13$ and $G_s = 2.7$?

***Strategy***   The solution to this problem is a straightforward application of Eq. (6.22).

## Solution 6.5

**Step 1:**   Determine the difference in $s_u$ values.
Use Eq. (6.22).

$$\frac{(s_u)_{\text{lab}}}{(s_u)_{\text{field}}} = \exp\left(\frac{2.7(0.48 - 0.44)}{0.13}\right) = 2.3$$

The laboratory undrained shear strength would probably show an increase over the in situ undrained shear strength by a factor greater than 2. $\quad\blacksquare$

---

***What's next . . .***We have discussed methods to calculate the failure stresses. But failure stresses are only one of the technical criteria in the analysis of soil behavior. We also need to know the deformations or strains. But before we can get the strains from the stresses we need to know the elastic, shear, and bulk moduli. In the next section, we will use the CSM to determine these moduli.

## 6.6  SOIL STIFFNESS

The elastic modulus, $E'$, or the shear modulus, $G$, and the bulk modulus, $K'$, characterize soil stiffness. In practice, $E'$ or $G$, and $K'$ are commonly obtained from triaxial or simple shear tests. We can obtain an estimate of $E'$ or $G$ and $K'$ using the critical state model and results from axisymmetric, isotropic consolidation tests. The void ratio during unloading/reloading is described by

$$e = e_\kappa - \kappa \ln p' \tag{6.26}$$

where $e_\kappa$ is the void ratio on the unloading/reloading line at $p' = 1$ unit of stress (Fig. 6.12). The unloading/reloading path $BC$ (Fig. 6.12) is reversible, which is a characteristic of elastic materials. Differentiating Eq. (6.26) gives

$$de = -\kappa \frac{dp'}{p'} \tag{6.27}$$

The elastic volumetric strain increment is

$$de_p^e = -\frac{de}{1 + e_o} = \frac{\kappa}{1 + e_o} \frac{dp'}{p'} \tag{6.28}$$

But, from Eq. (3.99),

$$de_p^e = \frac{dp'}{K'}$$

Therefore,

$$\frac{dp'}{K'} = \frac{\kappa}{1 + e_o} \frac{dp'}{p'}$$

Solving for $K'$, we obtain

$$K' = \frac{p'(1 + e_o)}{\kappa} \tag{6.29}$$

From Eq. (3.100),

$$E' = 3K'(1 - 2\nu')$$

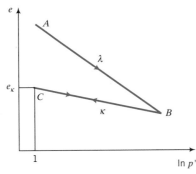

**FIGURE 6.12**  Loading and unloading/reloading (elastic) response of soils in $(e–p'$ ln) space.

Therefore,

$$E' = \frac{3p'(1 + e_o)(1 - 2v')}{\kappa} \qquad (6.30)$$

Also, from Eq. (3.102),

$$G = \frac{E'}{2(1 + v')}$$

Therefore,

$$G = \frac{3p'(1 + e_o)(1 - 2v')}{2\kappa(1 + v')} = \frac{1.5p'(1 + e_o)(1 - 2v')}{\kappa(1 + v')} \qquad (6.31)$$

Equations (6.30) and (6.31) indicate that the elastic constants, $E'$ and $G$, are proportional to the mean effective stress. This implies nonlinear elastic behavior and therefore calculations must be carried out incrementally. For overconsolidated soils, Eqs. (6.30) and (6.31) provide useful estimates of $E'$ and $G$ from conducting an isotropic consolidation test, which is a relatively simple soil test.

Soil stiffness is influenced by the amount of shear strains applied. Increases in shear strains tend to lead to decreases in $G$ and $E'$ while increases in volumetric strains lead to decreases in $K'$. The net effect is that the soil stiffness decreases with increasing strains.

It is customary to identify three regions of soil stiffness based on the level of applied shear strains. At small shear strains ($\gamma$ or $\varepsilon_q$ usually $< 0.001\%$), the soil stiffness is approximately constant (Fig. 6.13) and the soil behaves like a linearly elastic material. At intermediate shear strains between 0.001% and 1%, the soil stiffness decreases significantly and the soil behavior is elastoplastic (nonlinear). At large strains ($\gamma > 1\%$), the soil stiffness decreases slowly to an approximately constant value as the soil approaches critical state. At the critical state, the soil behaves like a viscous fluid.

In practical problems, the shear strains are in the intermediate range, typically $\gamma < 0.1\%$. However, the shear strain distribution within the soil is not uniform. The shear strains decrease with distance away from a structure and local shear strains near the edge of a foundation slab, for example, can be much greater than 0.1%. The implication of a nonuniform shear strain distribution is that the

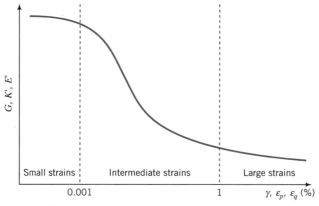

**FIGURE 6.13**   Schematic variation of shear, bulk, and Young's elastic moduli with strain levels.

soil stiffness varies within the loaded region of the soil. Consequently, large settlements and failures are usually initiated in the loaded soil region where the soil stiffness is the lowest.

In conventional laboratory tests, it is not practical to determine the soil stiffness at shear strains less than 0.001% because of inaccuracies in the measurement of the soil displacements due to displacements of the apparatuses themselves and to resolution and inaccuracies of measuring instruments. The soil stiffness at small strains is best determined in the field using wave propagation techniques. In one such technique, vibrations are created at the soil surface or at a prescribed depth in the soil, and the shear wave velocity ($v_{sh}$) is measured. The shear modulus at small strains is calculated from

$$G = \frac{\gamma(v_{sh})^2}{g} \qquad (6.32)$$

where $\gamma$ is the bulk unit weight of the soil, and $g$ is the acceleration due to gravity. In the laboratory, the shear modulus at small strains can be determined using a resonance column test (Drnevick, 1967). The resonance column test utilizes a hollow cylinder apparatus (Chapter 5) to induce resonance of the soil sample. Resonance column tests show that $G$ depends not only on the level of shear strain but also on void ratio, overconsolidation ratio, and mean effective stress. Various empirical relationships have been proposed linking $G$ to $e$, overconsolidation ratio, and $p'$. Two such relationships are presented below.

**Jamiolkowski et al. (1991) for clays**

$$G = \frac{198}{e^{1.3}} (R_o)^a \sqrt{p'} \text{ MPa} \qquad (6.33)$$

where $G$ is the initial shear modulus, $p'$ is the mean effective stress (MPa), and $a$ is a coefficient that depends on the plasticity index as follows:

| $I_p$ (%) | $a$ |
|---|---|
| 0 | 0 |
| 20 | 0.18 |
| 40 | 0.30 |
| 60 | 0.41 |
| 80 | 0.48 |
| $\geq 100$ | 0.50 |

**Seed and Idriss (1970) for sands**

$$G = k_1 \sqrt{p'} \text{ MPa}$$

| $e$ | $k_1$ | $D_r$ (%) | $k_1$ |
|---|---|---|---|
| 0.4 | 484 | 30 | 235 |
| 0.5 | 415 | 40 | 277 |
| 0.6 | 353 | 45 | 298 |
| 0.7 | 304 | 60 | 360 |
| 0.8 | 270 | 75 | 408 |
| 0.9 | 235 | 90 | 484 |

***What's next . . .*** Now that we know how to calculate the shear and bulk moduli, we can move on to determine strains, which we will consider next.

## 6.7 STRAINS FROM THE CRITICAL STATE MODEL

### 6.7.1 Volumetric Strains

The total change in volumetric strains consists of two parts: the recoverable part (elastic) and the unrecoverable part (plastic). We can write an expression for the total change in volumetric strain as

$$\Delta\varepsilon_p = \Delta\varepsilon_p^e + \Delta\varepsilon_p^p \tag{6.35}$$

where the superscripts $e$ and $p$ denote elastic and plastic, respectively. Let us consider a soil sample that is isotropically consolidated to a mean effective stress $p_c'$ and unloaded to a mean effective stress $p_o'$ as represented by $ABC$ in Figs. 6.14a,b. In a CD test, the soil will yield at $D$. Let us now consider a small increment of stress, $DE$, which causes the yield surface to expand as shown in Fig. 6.14a.

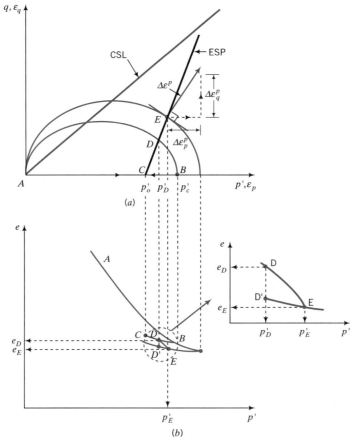

**FIGURE 6.14** Determination of plastic strains.

The change in void ratio for this stress increment is $\Delta e = |e_E - e_D|$ (Fig. 6.14b) and the corresponding total change in volumetric strain is

$$\Delta\varepsilon_p = \frac{\Delta e}{1 + e_o} = \left(\frac{|e_E - e_D|}{1 + e_o}\right) = \frac{\lambda}{1 + e_o} \ln \frac{p'_E}{p'_D} \tag{6.36}$$

The volumetric elastic strain component is represented by $ED'$. That is, if you were to unload the soil from $E$ back to its previous stress state at $D$, the rebound will occur along an unloading/reloading line associated with the maximum mean effective stress for the yield surface on which unloading starts. The elastic change in volumetric strain from $E$ to $D'$ is

$$\Delta\varepsilon_p^e = \frac{\Delta e}{1 + e_o} = \frac{|e_{D'} - e_E|}{1 + e_o} = \frac{\kappa}{1 + e_o} \ln \frac{p'_E}{p'_D} \tag{6.37}$$

We get a positive value of $\Delta\varepsilon_p^e$ because rather than computing the rebound (expansion) from $E$ to $D'$, we compute the compression from $D'$ to $E$.

The volumetric elastic strains can also be computed from Eq. (3.99); that is,

$$\Delta\varepsilon_p^e = \frac{\Delta p'}{K'} \tag{6.38}$$

The change in volumetric plastic strain is

$$\Delta\varepsilon_p^p = \Delta\varepsilon_p - \Delta\varepsilon_p^e = \left(\frac{\lambda - \kappa}{1 + e_0}\right) \ln \frac{p'_E}{p'_D} \tag{6.39}$$

Under undrained conditions, the total volumetric change is zero. Consequently, from Eq. (6.35),

$$\Delta\varepsilon_p^e = -\Delta\varepsilon_p^p \tag{6.40}$$

### 6.7.2 Shear Strains

Let the yield surface be represented by

$$F = (p')^2 - p'p'_c + \frac{q^2}{M^2} = 0 \tag{6.41}$$

To find the shear or deviatoric strains, we will assume that the resultant plastic strain increment, $\Delta\varepsilon^p$, for an increment of stress is normal to the yield surface (Fig. 6.14a). Normally, the plastic strain increment should be normal to a plastic potential function but we are assuming here that the plastic potential function and the yield surface (yield function, $F$) are the same. A plastic potential function is a scalar quantity that defines a vector in terms of its location in space. Classical plasticity demands that the surfaces defined by the yield and plastic potential coincide. If they do not, then basic work restrictions are violated. However, modern soil mechanics theories often use different surfaces for yield and potential functions to obtain more realistic stress–strain relationships. The resultant plastic strain increment has two components—a deviatoric or shear component, $\Delta\varepsilon_q^p$, and a volumetric component, $\Delta\varepsilon_p^p$, as shown in Fig. 6.14. We already found $\Delta\varepsilon_p^p$ in the previous section.

Since we know the equation for the yield surface [Eq. (6.41)], we can find the normal to it by differentiation of the yield function with respect to $p'$ and $q$. The tangent or slope of the yield surface is

$$dF = 2p'\, dp' - p'_c\, dp' + 2q\, \frac{dq}{M^2} = 0 \qquad (6.42)$$

Rearranging Eq. (6.42), we obtain the slope as

$$\frac{dq}{dp'} = \left( \frac{p'_c/2 - p'}{q/M^2} \right) \qquad (6.43)$$

The normal to the yield surface is

$$-\frac{1}{dq/dp'} = -\frac{dp'}{dq}$$

From Fig. 6.14a, the normal, in terms of plastic strains, is $d\varepsilon_q^p/d\varepsilon_p^p$. Therefore,

$$\frac{d\varepsilon_q^p}{d\varepsilon_p^p} = -\frac{dp'}{dq} = -\frac{q/M^2}{p'_c/2 - p'} \qquad (6.44)$$

which leads to

$$\boxed{d\varepsilon_q^p = d\varepsilon_p^p \frac{q}{M^2(p' - p'_c/2)}} \qquad (6.45)$$

The elastic shear strains can be obtained from Eq. (3.101); that is,

$$\boxed{\Delta\varepsilon_q^e = \frac{1}{3G}\Delta q} \qquad (6.46)$$

These equations for strains are valid only for small changes in stress. For example, you cannot use these equations to calculate the failure strains by simply substituting the failure stresses for $p'$ and $q$. You have to calculate the strains for small increments of stresses up to failure and then sum each component of strain separately. We need to do this because the critical state model considers soils as elastic–plastic materials and not linearly elastic materials.

## EXAMPLE 6.6

A sample of clay was isotropically consolidated to a mean effective stress of 225 kPa and was then unloaded to a mean effective stress of 150 kPa at which stress $e_o = 1.4$. A CD test is to be conducted. Calculate (a) the elastic strains at initial yield and (b) the total volumetric and deviatoric strains for an increase of deviatoric stress of 12 kPa after initial yield. For this clay, $\lambda = 0.16$, $\kappa = 0.05$, $\phi'_{cs} = 25.5°$, and $\nu' = 0.3$.

*Strategy*   It is best to sketch diagrams similar to Fig. 6.4 to help you visualize the solution to this problem. Remember that the strains within the yield surface are elastic.

## Solution 6.6

**Step 1:**  Calculate initial stresses and $M_c$.

$$p'_c = 225 \text{ kPa}, \, p'_o = 150 \text{ kPa}$$

$$R_o = \frac{225}{150} = 1.5$$

$$M_c = \frac{6 \sin \phi'_{cs}}{3 - \sin \phi'_{cs}} = \frac{6 \sin 25.5°}{3 - \sin 25.5°} = 1$$

**Step 2:**  Determine the initial yield stresses.
The yield stresses are the stresses at the intersection of the initial yield surface and the effective stress path.

Equation for the yield surface:  $(p')^2 - p'p'_c + \dfrac{q^2}{M_c^2} = 0$

Equation of the ESP:  $p' = p'_o + \dfrac{q}{3}$

At the initial yield point $D$ (Fig. 6.4):  $p'_y = p'_o + \dfrac{q_y}{3} = 150 + \dfrac{q_y}{3}$

Substituting $p' = p'_y$, $q = q_y$, and the values for $M_c$ and $p'_c$ into the equation for the initial yield surface [Eq. (6.4)] gives

$$\left(150 + \frac{q_y}{3}\right)^2 - \left(150 + \frac{q_y}{3}\right)225 + \frac{q_y^2}{1^2} = 0$$

Simplification results in

$$q_y^2 + 22.5q_y - 10125 = 0$$

The solution for $q_y$ is $q_y = 90 \text{ kPa}$ or $q_y = -112.5 \text{ kPa}$. The correct answer is $q_y = 90 \text{ kPa}$ since we are applying compression to the soil sample. Therefore,

$$p'_y = 150 + \frac{q'_y}{3} = 150 + \frac{90}{3} = 180 \text{ kPa}$$

**Step 3:**  Calculate the elastic strains at initial yield.

**Elastic volumetric strains**

Elastic volumetric strains: $\Delta\varepsilon_p^e = \dfrac{\kappa}{1 + e_o} \ln \dfrac{p'_y}{p'_o} = \dfrac{0.05}{1 + 1.4} \ln \dfrac{180}{150} = 38 \times 10^{-4}$

Alternatively, you can use Eq. (6.38). Take the average value of $p'$ from $p'_o$ to $p'_y$ to calculate $K'$.

$$p'_{av} = \frac{p'_o + p'_y}{2} = \frac{150 + 180}{2} = 165 \text{ kPa}$$

$$K' = \frac{3p'(1 + e_o)}{\kappa} = \frac{165(1 + 1.4)}{0.05} = 7920 \text{ kPa}$$

$$\Delta\varepsilon_p^e = \frac{\Delta p'}{K'} = \frac{180 - 150}{7920} = 38 \times 10^{-4}$$

**Elastic shear strains**

$$G = \frac{3p'(1 + e_o)(1 - 2v')}{2\kappa(1 + v')} = \frac{3 \times 165(1 + 1.4)(1 - 2 \times 0.3)}{2 \times 0.05(1 + 0.3)} = 3655 \text{ kPa}$$

$$\Delta\varepsilon_q^e = \frac{\Delta q}{3G} = \frac{90}{3 \times 3655} = 82 \times 10^{-4}$$

**Step 4:** Determine expanded yield surface.

$$\text{After initial yield: } \Delta q = 12 \text{ kPa}$$

$$\therefore \Delta p' = \frac{\Delta q}{3} = \frac{12}{3} = 4 \text{ kPa}$$

The stresses at $E$ (Fig. 6.4) are $p_E' = p_y' + \Delta p = 180 + 4 = 184$ kPa, and

$$q_E = q_y + \Delta q = 90 + 12 = 102 \text{ kPa}$$

The preconsolidated mean effective stress (major axis) of the expanded yield surface is obtained by substituting $p_E' = 184$ kPa and $q_E = 102$ kPa in the equation for the yield surface [Eq. (6.4)]:

$$(184)^2 - 184(p_c')_E + \frac{102^2}{1^2} = 0$$

$$\therefore (p_c')_E = 240.5 \text{ kPa}$$

**Step 5:** Calculate strain increments after yield.

Equation (6.36): $\Delta\varepsilon_p = \dfrac{\lambda}{1 + e_o} \ln \dfrac{p_E'}{p_y'} = \dfrac{0.16}{1 + 1.4} \ln \dfrac{184}{180} = 15 \times 10^{-4}$

Equation (6.39): $\Delta\varepsilon_p^p = \dfrac{\lambda - \kappa}{1 + e_o} \ln \dfrac{p_E'}{p_y'} = \dfrac{0.16 - 0.05}{1 + 1.4} \ln \dfrac{184}{180} = 10 \times 10^{-4}$

Equation (6.45): $\Delta\varepsilon_q^p = \Delta\varepsilon_p^p \dfrac{q_E}{M_c^2[p_E' - (p_c')_E/2]} = 10 \times 10^{-4} \dfrac{102}{1^2(184 - 240.5/2)}$

$$= 16 \times 10^{-4}$$

Assuming that $G$ remains constant, we can calculate the elastic shear strain from

Equation (6.46): $\Delta\varepsilon_q^e = \dfrac{\Delta q}{3G} = \dfrac{12}{3 \times 3655} = 11 \times 10^{-4}$

**Step 6:** Calculate total strains.

Total volumetric strains: $\varepsilon_p = \Delta\varepsilon_p^e + \Delta\varepsilon_p^p = (38 + 10)10^{-4} = 48 \times 10^{-4}$

Total shear strains: $\varepsilon_q = \Delta\varepsilon_q^e + \Delta\varepsilon_q^p = [(82 + 11) + 16]10^{-4} = 109 \times 10^{-4}$ ∎

## EXAMPLE 6.7

Show that the yield surface in an undrained test increases such that

$$p_c' = (p_c')_{\text{prev}} \left(\frac{p_{\text{prev}}'}{p'}\right)^{\kappa/(\lambda - \kappa)}$$

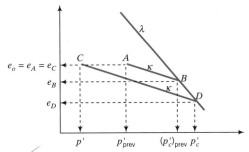

FIGURE E6.7

where $p'_c$ is the current value of the major axis of the yield surface, $(p'_c)_{\text{prev}}$ is the previous value of the major axis of the yield surface, $p'_{\text{prev}}$ is the previous value of mean effective stress, and $p'$ is the current value of mean effective stress.

**Strategy**  Sketch an $e$ versus $\ln p'$ diagram and then use it to prove the equation given.

### Solution 6.7

**Step 1:**  Sketch an $e$ versus $\ln p'$ diagram.
See Fig. E6.7.

**Step 2:**  Prove the equation.

**Line $AB$**

$$|e_B - e_A| = \kappa \ln\left(\frac{(p'_c)_{\text{prev}}}{p'_{\text{prev}}}\right) \tag{1}$$

**Line $CD$**

$$|e_D - e_C| = \kappa \ln\frac{p'_c}{p'} \tag{2}$$

Subtracting Eq. (2) from Eq. (1), noting that $e_A = e_C$, we obtain

$$|e_D - e_B| = \kappa \ln\left\{\frac{(p'_c)_{\text{prev}}}{p'_{\text{prev}}}\right\} - \kappa \ln\frac{p'_c}{p'} \tag{3}$$

But, from the normal consolidation line $BD$, we get

$$|e_D - e_B| = \lambda \ln\left\{\frac{p'_c}{(p'_c)_{\text{prev}}}\right\} \tag{4}$$

Substituting Eq. (4) into Eq. (3) and simplifying gives

$$p'_c = (p'_c)_{\text{prev}}\left(\frac{p'_{\text{prev}}}{p'}\right)^{\kappa/(\lambda-\kappa)}$$

∎

*What's next* . . We have calculated the yield stresses, the failure stresses, and strains for a given stress increment. In the next section, a procedure is outlined to calculate the stress–strain, volume change, and excess pore water pressure responses of a soil using the critical state model.

## 6.8   CALCULATED STRESS–STRAIN RESPONSE

You can predict the stress–strain response, volume changes, and excess pore water pressures from the initial stress state to the failure stress state using the methods described in the previous sections. The required soil parameters are $p'_o$, $e_o$, $p'_c$ or OCR, $\lambda$, $\kappa$, $\phi'_{cs}$, and $v'$. The procedures for a given stress path are as follows.

### 6.8.1 Drained Compression Tests

1. Determine the mean effective stress and the deviatoric stress at initial yield, that is, $p'_y$ and $q_y$, by finding the coordinates of the intersection of the initial yield surface with the effective stress path. For a CD test,

$$p'_y = \frac{(M^2 p'_c + 18 p'_o) + \sqrt{(M^2 p'_c + 18 p'_o)^2 - 36(M^2 + 9)(p'_o)^2}}{2(M^2 + 9)} \qquad (6.47)$$

$$q_y = 3(p'_y - p'_o) \qquad (6.48)$$

2. Calculate the mean effective stress and deviatoric stress at failure by finding the coordinate of the intersection of the critical state line and the effective stress path, that is, $p'_f$ and $q_f$. For a CD test, use Eqs. (6.16) and (6.17).

3. Calculate $G$ using Eq. (6.31) or empirical equations (6.33) and (6.34). Use an average value of $p'$ [$p' = (p'_o + p'_y)/2$] to calculate $G$.

4. Calculate the initial elastic volumetric strain using Eq. (6.37) and initial elastic deviatoric strain using Eq. (6.46).

5. Divide the ESP between the initial yield point and the failure point into a number of equal stress increments. Small increment sizes ($<5\%$ of the stress difference between $q_f$ and $q_y$) tend to give a more accurate solution than larger increment sizes.

For each mean effective stress increment up to failure:

6. Calculate the preconsolidation stress, $p'_c$, for each increment; that is, you are calculating the major axis of the ellipse using Eq. (6.4), which gives

$$p'_c = p' + \frac{q^2}{M^2 p'} \qquad (6.49)$$

where $p'$ is the current mean effective stress.

7. Calculate the total volumetric strain increment using Eq. (6.36).

8. Calculate the plastic volumetric strain using Eq. (6.39).

9. Calculate the plastic deviatoric strain increment using Eq. (6.45).

10. Calculate the elastic deviatoric strain increment using Eq. (6.46).

11. Add the plastic and elastic deviatoric strain increments to give the total deviatoric strain increment.

12. Sum the total volumetric strain increments ($\varepsilon_p$).
13. Sum the total deviatoric shear strain increments ($\varepsilon_q$).
14. Calculate

$$\varepsilon_1 = \frac{3\varepsilon_q + \varepsilon_p}{3} = \varepsilon_q + \frac{\varepsilon_p}{3} \tag{6.50}$$

15. If desired, you can calculate

$$\sigma_1' = \frac{2q}{3} + p' \quad \text{and} \quad \sigma_3' = p' - \frac{q}{3}$$

The last value of mean effective stress should be about $0.99p_f'$ to prevent instability in the solution.

## 6.8.2 Undrained Compression Tests

1. Determine the mean effective stress and the deviatoric stress at initial yield, that is, $p_y'$ and $q_y$. Remember that the effective stress path within the initial yield surface is vertical. Therefore, $p_o' = p_y$ and $q_y$ are found by determining the intersection of a vertical line originating at $p_o'$ with the initial yield surface. The equation to determine $q_y$ for an isotropically consolidated soil is

$$q_y = Mp_o'\sqrt{\frac{p_c'}{p_o'} - 1} \tag{6.51}$$

If the soil is heavily overconsolidated, then $q_y = q_p$.

2. Calculate the mean effective and deviatoric stress at failure from Eqs. (6.19) and (6.20).
3. Calculate $G$ using Eq. (6.31) or empirical equations (6.33) and (6.34).
4. Calculate the initial elastic deviatoric strain from Eq. (6.46).
5. Divide the horizontal distance between the initial mean effective stress, $p_o'$, and the failure mean effective stress, $p_f'$, in the $e$–$p'$ plot into a number of equal mean effective stress increments. You need to use small stress increment size, usually less than $0.05(p_o' - p_f')$.

For each increment of mean effective stress, calculate the following:

6. Determine the preconsolidation stress after each increment of mean effective stress from

$$p_c' = (p_c')_{\text{prev}} \left(\frac{p_{\text{prev}}'}{p'}\right)^{\kappa/(\lambda - \kappa)}$$

where the subscript "prev" denotes the previous increment, $p_c'$ is the current preconsolidation stress or the current size of the major axis of the yield surface, and $p'$ is the current mean effective stress.

7. Calculate $q$ at the end of each increment from

$$q = Mp'\sqrt{\frac{p_c'}{p'} - 1}$$

8. Calculate the volumetric elastic strain increment from Eq. (6.37).

9. Calculate the volumetric plastic strain increment. Since the total volumetric strain is zero, the volumetric plastic strain increment is equal to the negative of the volumetric elastic strain increment; that is, $\Delta\varepsilon_p^p = -\Delta\varepsilon_p^e$.

10. Calculate the deviatoric plastic strain increment from Eq. (6.45).

11. Calculate the deviatoric elastic strain increment from Eq. (6.46).

12. Add the deviatoric elastic and plastic strain increments to get the total deviatoric strain increment.

13. Sum the total deviatoric strain increments. For undrained conditions, $\varepsilon_1 = \varepsilon_d$.

14. Calculate the current mean total stress from the TSP. Remember you know the current value of $q$ from Step 7. For a CU test, $p = p_o' + q/3$.

15. Calculate the change in excess pore water pressure by subtracting the current mean effective stress from the current mean total stress.

**EXAMPLE 6.8**

Estimate and plot the stress–strain curve, volume changes (drained conditions), and excess pore water pressures (undrained conditions) for two samples of the same soil. The first sample, sample A, is to be subjected to conditions similar to a CD test and the second sample, sample B, is to be subjected to conditions similar to a CU test. The soil parameters are $\lambda = 0.25$, $\kappa = 0.05$, $\phi_{cs}' = 24°$, $\nu' = 0.3$, $e_o = 1.15$, $p_o' = 200$ kPa, and $p_c' = 250$ kPa.

**Strategy** Follow the procedures listed in Section 6.8. A spreadsheet can be prepared to do the calculations. However, you should manually check some of the spreadsheet results to be sure that you entered the correct formulation. A spreadsheet will be used here but we will calculate the results for one increment for each sample.

**Solution 6.8**

$$\text{Calculate } M_c: \quad M_c = \frac{6 \sin \phi_{cs}'}{3 - \sin \phi_{cs}'} = \frac{6 \sin 24°}{3 - \sin 24°} = 0.94$$

$$\text{Calculate } e_\Gamma: \quad e_\Gamma = e_o + (\lambda - \kappa) \ln \frac{p_c'}{2} + \kappa \ln p_o'$$

$$= 1.15 + (0.25 - 0.05) \ln \frac{250}{2} + 0.05 \ln 200 = 2.38$$

Each step corresponds to the procedures listed in Section 6.8.

**Sample A, Drained Test**

**Step 1:**

$$p_y' = \frac{(M^2 p_c') + 18 p_o') + \sqrt{(M^2 p_c' + 18 p_o')^2 - 36(M^2 + 9)(p_o')^2}}{2(M^2 + 9)}$$

$$= \frac{(0.94^2 \times 250 + 18 \times 200) + \sqrt{(0.94^2 \times 250 + 18 \times 200)^2 - 36(0.94^2 + 9)(200)^2}}{2(0.94^2 + 9)}$$

$$= 224 \text{ kPa}$$

$$q_y = 3(p_y' - p_o') = 3(224 - 200) = 72 \text{ kPa}$$

**Step 2:**

$$p'_f = \frac{3Mp'_o}{3 - M};$$

$$p'_f = \frac{3 \times 200}{3 - 0.94} = 291.3 \text{ kPa}, \quad q_f = Mp'_f = 0.94 \times 291.3 = 273.9 \text{ kPa}$$

**Step 3:**

$$p'_{av} = \frac{200 + 224}{2} = 212 \text{ kPa}$$

$$G = \frac{3p'(1 + e_o)(1 - 2v')}{2\kappa(1 + v')} = \frac{3 \times 212(1 + 1.15) \times (1 - 2 \times 0.3)}{2 \times 0.05(1 + 0.3)} = 4207 \text{ kPa}$$

**Step 4:**

$$(\Delta\varepsilon_q^e)_{\text{initial}} = \frac{\Delta q}{3G} = \frac{71.9}{3 \times 4207} = 5.7 \times 10^{-3},$$

$$(\Delta\varepsilon^e p)_{\text{initial}} = \frac{\kappa}{1 + e_o} \ln \frac{p'_y}{p'_o} = \frac{0.05}{1 + 1.15} \ln \frac{224}{200} = 2.6 \times 10^{-3}$$

**Step 5:**   Let $\Delta p' = 4$ kPa; then $\Delta q = 3 \times \Delta p' = 12$ kPa.

First stress increment after the initial yield follows.

**Step 6:**   $p' = 224 + 4 = 228$ kPa, $q = 71.9 + 12 = 83.9$ kPa,

$$p'_c = p' + \frac{q^2}{M^2 p'} = 228 + \frac{83.9^2}{0.94^2 \times 228} = 262.9 \text{ kPa}$$

**Step 7:**

$$\Delta\varepsilon_p = \frac{\lambda}{1 + e_o} \ln \frac{p'}{p'_y} = \frac{0.25}{1 + 1.15} \ln \frac{228}{224} = 2.1 \times 10^{-3}$$

**Step 8:**

$$\Delta\varepsilon_p^p = \frac{\lambda - \kappa}{1 + e_o} \ln \frac{p'}{p'_y} = \frac{(0.25 - 0.05)}{1 + 1.15} \ln \frac{228}{224} = 1.6 \times 10^{-3}$$

**Step 9:**

$$\Delta\varepsilon_q^p = \Delta\varepsilon_p^p \frac{q}{M^2(p' - p'_c/2)} = 1.6 \times 10^{-3} \frac{83.9}{0.94^2(228 - 262.9/2)} = 1.6 \times 10^{-3}$$

**Step 10:**

$$\Delta\varepsilon_q^e = \frac{\Delta q}{3G} = \frac{12}{3 \times 4207} = 1.0 \times 10^{-3}$$

**Step 11:**

$$\Delta\varepsilon_q = \Delta\varepsilon_q^e + \Delta\varepsilon_q^p = (1.0 + 1.6) \times 10^{-3} = 2.6 \times 10^{-3}$$

**Step 12:**

$$\varepsilon_p = (\Delta\varepsilon_p^e)_{\text{initial}} + \Delta\varepsilon_p = (2.6 + 2.1) \times 10^{-3} = 4.7 \times 10^{-3}$$

**Step 13:**

$$\varepsilon_q = (\Delta\varepsilon_q^e)_{\text{initial}} + \Delta\varepsilon_q = (5.7 + 2.6) \times 10^{-3} = 8.3 \times 10^{-3}$$

**Step 14:**

$$\varepsilon_1 = \varepsilon_q + \varepsilon_p/3 = (8.3 + 4.7/3) \times 10^{-3} = 9.8 \times 10^{-3}$$

The spreadsheet program and the stress–strain plots are shown in the table below and Figs. E6.8a,b. There are some slight differences between the calculated values shown above and the spreadsheet because of number rounding.

## Drained Case

| Given data | | | | Calculated values | |
|---|---|---|---|---|---|
| $\lambda$ | 0.25 | $M$ | 0.94 | $\Delta p'$ | 4 kPa |
| $\kappa$ | 0.05 | $R_o$ | 1.25 | $\Delta q$ | 12 kPa |
| $\phi_{cs}'$ | 24 | $e_{cs}$ | 2.38 | $G$ | 4207.0 kPa |
| $e_o$ | 1.15 | $p_f'$ | 291.4 kPa | $\Delta\varepsilon_p^e$ | 0.0026 |
| $p_o'$ | 200 kPa | $q_f'$ | 274.2 kPa | $\Delta\varepsilon_q^e$ | 0.0057 |
| $p_c'$ | 250 kPa | $p_y'$ | 224.0 kPa | | |
| $v'$ | 0.3 | $q_y$ | 71.9 kPa | | |

[a]Selected increment.

## Tabulation

| $p'$ (kPa) | $\Sigma\Delta q$ (kPa) | $q$ (kPa) | $p_c'$ (kPa) | $\Delta\varepsilon_p$ ($\times 10^{-3}$) | $\varepsilon_p = \Sigma\Delta\varepsilon_p$ ($\times 10^{-3}$) | $\Delta\varepsilon_p^p$ ($\times 10^{-3}$) | $\Delta\varepsilon_q^p$ ($\times 10^{-3}$) | $G$ (kPa) | $\Delta\varepsilon_q^e$ ($\times 10^{-3}$) | $\Delta\varepsilon_q$ ($\times 10^{-3}$) | $\varepsilon_q = \Sigma\Delta\varepsilon_q$ ($\times 10^{-3}$) | $\varepsilon_1$ ($\times 10^{-3}$) |
|---|---|---|---|---|---|---|---|---|---|---|---|---|
| 0 | 0 | 0 | 0 | 0.0 | 0.0 | 0.0 | 0.0 | | 0.0 | 0.0 | 0.0 | 0.0 |
| 224.0 | 0.0 | 71.9 | 250.0 | 2.6 | 2.6 | 0.0 | 0.0 | 4207.0 | 5.7 | 5.7 | 5.7 | 6.6 |
| 228.0 | 12.0 | 83.9 | 262.8 | 2.1 | 4.7 | 1.6 | 1.6 | 4484.4 | 0.9 | 2.5 | 8.2 | 9.8 |
| 232.0 | 24.0 | 95.9 | 276.7 | 2.0 | 6.7 | 1.6 | 1.9 | 4563.8 | 0.9 | 2.7 | 10.9 | 13.2 |
| 236.0 | 36.0 | 107.9 | 291.6 | 2.0 | 8.7 | 1.6 | 2.1 | 4643.1 | 0.9 | 3.0 | 14.0 | 16.9 |
| 240.0 | 48.0 | 119.9 | 307.6 | 2.0 | 10.7 | 1.6 | 2.5 | 4722.5 | 0.8 | 3.3 | 17.3 | 20.8 |
| 244.0 | 60.0 | 131.9 | 324.4 | 1.9 | 12.6 | 1.5 | 2.8 | 4801.9 | 0.8 | 3.6 | 20.9 | 25.1 |
| 248.0 | 72.0 | 143.9 | 342.2 | 1.9 | 14.5 | 1.5 | 3.2 | 4881.3 | 0.8 | 4.0 | 24.9 | 29.7 |
| 252.0 | 84.0 | 155.9 | 360.8 | 1.9 | 16.3 | 1.5 | 3.7 | 4960.7 | 0.8 | 4.5 | 29.4 | 34.8 |
| 256.0 | 96.0 | 167.9 | 380.3 | 1.8 | 18.2 | 1.5 | 4.2 | 5040.1 | 0.8 | 5.0 | 34.4 | 40.5 |
| 260.0 | 108.0 | 179.9 | 400.5 | 1.8 | 20.0 | 1.4 | 4.9 | 5119.4 | 0.8 | 5.7 | 40.1 | 46.7 |
| 264.0 | 120.0 | 191.9 | 421.4 | 1.8 | 21.7 | 1.4 | 5.8 | 5198.8 | 0.8 | 6.6 | 46.6 | 53.9 |
| 268.0 | 132.0 | 203.9 | 443.1 | 1.7 | 23.5 | 1.4 | 6.9 | 5278.2 | 0.8 | 7.7 | 54.3 | 62.2 |
| 272.0 | 144.0 | 215.9 | 465.4 | 1.7 | 25.2 | 1.4 | 8.6 | 5357.6 | 0.7 | 9.3 | 63.6 | 72.0 |
| 276.0 | 156.0 | 227.9 | 488.4 | 1.7 | 26.9 | 1.4 | 11.0 | 5437.0 | 0.7 | 11.7 | 75.4 | 84.4 |
| 280.0 | 168.0 | 239.9 | 512.0 | 1.7 | 28.6 | 1.3 | 15.1 | 5516.4 | 0.7 | 15.9 | 91.3 | 100.8 |
| 284.0 | 180.0 | 251.9 | 536.2 | 1.6 | 30.2 | 1.3 | 23.7 | 5595.8 | 0.7 | 24.4 | 115.7 | 125.7 |
| 288.0 | 192.0 | 263.9 | 561.0 | 1.6 | 31.9 | 1.3 | 52.0 | 5675.1 | 0.7 | 52.7 | 168.3 | 179.0 |
| 291.0 | 201.0 | 272.9 | 579.9 | 1.2 | 33.1 | 1.0 | 290.2 | 5745.0 | 0.7 | 290.9 | 459.3 | 470.3 |

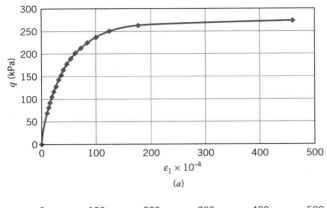

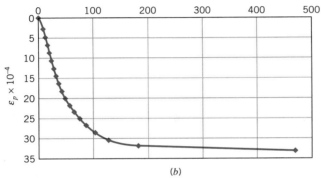

**FIGURE E6.8a,b**

### Sample B, Undrained Test

**Step 1:**

$$q_y = Mp_o'\sqrt{\frac{p_c'}{p_o'} - 1} = 0.94 \times 200 \sqrt{\frac{250}{200} - 1} = 94 \text{ kPa}$$

**Step 2:**

$$p_f' = \exp\left(\frac{e_\Gamma - e_o}{\lambda}\right) = \exp\left(\frac{2.38 - 1.15}{0.25}\right) = 137 \text{ kPa}$$
$$q_f = Mp_f' = 0.94 \times 137 = 128.8 \text{ kPa}$$

**Step 3:**

$$G = \frac{3p'(1 + e_o)(1 - 2v')}{2\kappa(1 + v')} = \frac{3 \times 200(1 + 1.15) \times (1 - 2 \times 0.3)}{2 \times 0.05(1 + 0.3)} = 3969.2 \text{ kPa}$$

**Step 4:**

$$(\Delta\varepsilon_q^e)_{\text{initial}} = \frac{\Delta q}{3G} = \frac{94}{3 \times 3939.2} = 7.9 \times 10^{-3}$$

**Step 5:**   Let $\Delta p' = 3$ kPa.

First stress increment after the initial yield follows.

**Step 6:**

$$p' = p'_o - \Delta p' = 200 - 3 = 197 \text{ kPa}$$

$$p'_c = (p'_c)_{\text{prev}}\left(\frac{p'_{\text{prev}}}{p'}\right)^{\kappa/(\lambda-\kappa)} = 250\left(\frac{200}{197}\right)^{0.05/(0.25-0.05)} = 250.9 \text{ kPa}$$

**Step 7:**

$$q = Mp'\sqrt{\frac{p'_c}{p'} - 1} = 0.94 \times 197\sqrt{\frac{250.9}{197} - 1} = 97 \text{ kPa}$$

**Step 8:**

$$\Delta\varepsilon^e_p = -\frac{\kappa}{1 + e_o}\ln\frac{p'_o}{p'} = -\frac{0.05}{1 + 1.15}\ln\frac{200}{197} = -0.35 \times 10^{-3}$$

**Step 9:**

$$\Delta\varepsilon^p_p = -\Delta\varepsilon^e_p = 0.35 \times 10^{-3}$$

**Step 10:**

$$\Delta\varepsilon^p_q = \Delta\varepsilon^p_p\frac{q}{M^2(p' - p'_c/2)} = 0.35 \times 10^{-3}\frac{97}{0.94^2(197 - 250.9/2)} = 0.54 \times 10^{-3}$$

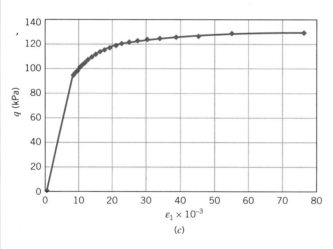

(c)

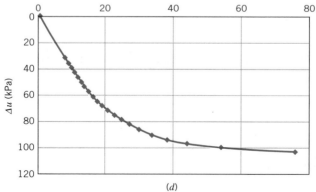

(d)

**FIGURE E6.8c,d**

**Step 11:**

$$\Delta\varepsilon_q^e = \frac{\Delta q}{3G} = \frac{97 - 94.1}{3 \times 3969.2} = 0.24 \times 10^{-3}$$

**Step 12:** $\Delta\varepsilon_q = \varepsilon_q^e + \Delta\varepsilon_q^p = (0.24 + 0.54) \times 10^{-3} = 0.78 \times 10^{-3}$

**Step 13:** $\varepsilon_q = \varepsilon_1 = (\Delta\varepsilon_q^e)_{\text{initial}} + \Delta\varepsilon_q = (7.9 + 0.78) \times 10^{-3} = 8.7 \times 10^{-3}$

**Step 14:** $p = p_o' + q/3 = 200 + \frac{97}{3} = 232.3$ kPa

**Step 15:** $\Delta u = p - p' = 232.3 - 197 = 35.3$ kPa

The spreadsheet program and the stress–strain plots are shown in the table below and Figs. E6.8c,d.

### Undrained Triaxial Test

| Given data | | Calculated values | | | |
|---|---|---|---|---|---|
| $\lambda$ | 0.25 | $M$ | 0.94 | $\Delta p$ | 3 kPa |
| $\kappa$ | 0.05 | $R_o$ | 1.25 | $\Delta q$ | 9 kPa |
| $\phi_{cs}'$ | 24 | $e_{cs}$ | 2.38 | $G$ | 3969.2 kPa |
| $e_o$ | 1.15 | $p_f'$ | 137.3 kPa | $\varepsilon_p^e$ | 0 |
| $p_o'$ | 200 kPa | $q_f'$ | 129.2 kPa | $\varepsilon_q^e$ | 0.0079 |
| $p_c'$ | 250 kPa | $p_y'$ | 200.0 kPa | | |
| $\nu'$ | 0.3 | $q_y$ | 94.1 kPa | | |
| | | $\Delta u_f$ | 105.8 kPa | | |

### Tabulation

| $p'$ (kPa) | $p_c'$ (kPa) | $q$ (kPa) | $\Delta\varepsilon_p^e$ ($\times\,10^{-3}$) | $\Delta\varepsilon_p^p$ ($\times\,10^{-3}$) | $\Delta\varepsilon_q^p$ ($\times\,10^{-3}$) | $G$ (kPa) | $\Delta\varepsilon_q^e$ ($\times\,10^{-3}$) | $\Delta\varepsilon_q$ ($\times\,10^{-3}$) | $\varepsilon_q = \Sigma\Delta\varepsilon q$ ($\times\,10^{-3}$) | $\varepsilon_1$ ($\times\,10^{-3}$) | $p$ (kPa) | $\Delta u$ (kPa) |
|---|---|---|---|---|---|---|---|---|---|---|---|---|
| 0 | 0 | 0 | | 0.0 | 0.0 | | 0.0 | 0.0 | 0.0 | 0.0 | 0 | 0 |
| 200.0 | 250.0 | 94.1 | 0.0 | 0.0 | 0.0 | 3969.2 | 7.9 | 7.9 | 7.9 | 7.9 | 231.4 | 31.4 |
| 197.0 | 250.9 | 97.0 | −0.4 | 0.4 | 0.5 | 3939.5 | 0.2 | 0.8 | 8.7 | 8.7 | 232.3 | 35.3 |
| 194.0 | 251.9 | 99.7 | −0.4 | 0.4 | 0.6 | 3879.9 | 0.2 | 0.8 | 9.5 | 9.5 | 233.2 | 39.2 |
| 191.0 | 252.9 | 102.3 | −0.4 | 0.4 | 0.6 | 3820.4 | 0.2 | 0.9 | 10.4 | 10.4 | 234.1 | 43.1 |
| 188.0 | 253.9 | 104.7 | −0.4 | 0.4 | 0.7 | 3760.8 | 0.2 | 0.9 | 11.3 | 11.3 | 234.9 | 46.9 |
| 185.0 | 254.9 | 107.0 | −0.4 | 0.4 | 0.8 | 3701.3 | 0.2 | 1.0 | 12.3 | 12.3 | 235.7 | 50.7 |
| 182.0 | 256.0 | 109.2 | −0.4 | 0.4 | 0.9 | 3641.8 | 0.2 | 1.1 | 13.4 | 13.4 | 236.4 | 54.4 |
| 179.0 | 257.0 | 111.2 | −0.4 | 0.4 | 1.0 | 3582.2 | 0.2 | 1.2 | 14.5 | 14.5 | 237.1 | 58.1 |
| 176.0 | 258.1 | 113.1 | −0.4 | 0.4 | 1.1 | 3522.7 | 0.2 | 1.3 | 15.8 | 15.8 | 237.7 | 61.7 |
| 173.0 | 259.2 | 114.9 | −0.4 | 0.4 | 1.2 | 3463.2 | 0.2 | 1.4 | 17.1 | 17.1 | 238.3 | 65.3 |
| 170.0 | 260.4 | 116.6 | −0.4 | 0.4 | 1.3 | 3403.6 | 0.2 | 1.5 | 18.7 | 18.7 | 238.9 | 68.9 |
| 167.0 | 261.5 | 118.2 | −0.4 | 0.4 | 1.5 | 3344.1 | 0.2 | 1.7 | 20.3 | 20.3 | 239.4 | 72.4 |
| 164.0 | 262.7 | 119.7 | −0.4 | 0.4 | 1.7 | 3284.5 | 0.2 | 1.9 | 22.2 | 22.2 | 239.9 | 75.9 |
| 161.0 | 263.9 | 121.1 | −0.4 | 0.4 | 2.0 | 3225.0 | 0.1 | 2.2 | 24.4 | 24.4 | 240.4 | 79.4 |
| 158.0 | 265.2 | 122.5 | −0.4 | 0.4 | 2.4 | 3165.5 | 0.1 | 2.5 | 26.9 | 26.9 | 240.8 | 82.8 |
| 155.0 | 266.4 | 123.7 | −0.4 | 0.4 | 2.9 | 3105.9 | 0.1 | 3.0 | 29.9 | 29.9 | 241.2 | 86.2 |
| 152.0 | 267.8 | 124.8 | −0.5 | 0.5 | 3.5 | 3046.4 | 0.1 | 3.7 | 33.6 | 33.6 | 241.6 | 89.6 |
| 149.0 | 269.1 | 125.9 | −0.5 | 0.5 | 4.6 | 2986.8 | 0.1 | 4.7 | 38.2 | 38.2 | 242.0 | 93.0 |
| 146.0 | 270.5 | 126.9 | −0.5 | 0.5 | 6.3 | 2927.3 | 0.1 | 6.4 | 44.7 | 44.7 | 242.3 | 96.3 |
| 143.0 | 271.9 | 127.8 | −0.5 | 0.5 | 9.9 | 2867.8 | 0.1 | 10.0 | 54.6 | 54.6 | 242.6 | 99.6 |
| 140.0 | 273.3 | 128.6 | −0.5 | 0.5 | 21.4 | 2808.2 | 0.1 | 21.5 | 76.1 | 76.1 | 242.9 | 102.9 |
| 137.4 | 274.6 | 129.2 | −0.4 | 0.4 | 636.6 | 2752.7 | 0.1 | 636.7 | 712.9 | 712.9 | 243.1 | 105.7 |

■

***What's next . . .***We have concentrated on isotropic consolidation of soils and axisym-metric conditions during shearing. The concepts and methodology developed are equally applicable to plane strain or other loading conditions. In nature, most soils are one-dimensionally consolidated, called $K_o$-consolidation. Next, we will consider $K_o$-consolidation using the critical state model.

## 6.9   $K_o$-CONSOLIDATED SOIL RESPONSE

When a soil is one-dimensionally consolidated, anisotropy is conferred on the soil structure. The soil properties are no longer the same in all directions. We can use our simple critical state model to provide insights into $K_o$-consolidated soils although the model, as described, cannot handle anisotropy. We will assume that the yield surface is unaltered, that is, remains an ellipse, for $K_o$-consolidated soils. The normal consolidation line for a $K_o$-consolidated soil is shifted to the left of the normal consolidation line of an isotropically consolidated soil (Fig. 6.15b) because $p'$ for a $K_o$-consolidated soil is

$$p' = \frac{1 + 2K_o}{3} \sigma_z'$$

compared with $p' = \sigma_z'$ for an isotropically consolidated soil. Recall that $K_o$ is the lateral earth pressure coefficient at rest.

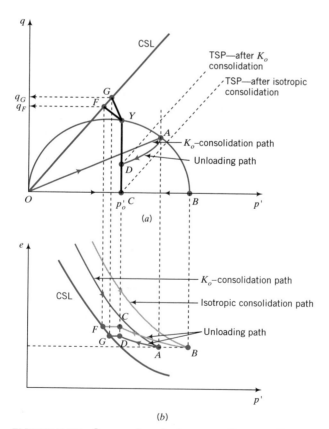

**FIGURE 6.15**   Comparison between a $K_o$-consolidated soil and an isotropically consolidated soil.

Let us compare the probable response of two samples, sample A and sample B, of a soil. Sample A is $K_o$-consolidated while sample B is isotropically consolidated. Both samples are normally consolidated to a void ratio $e$. The $K_o$-consolidated sample requires a lower mean effective stress to achieve the same void ratio as an isotropically consolidated sample (Fig. 6.15). The ESP from the isotropically consolidated sample is $OB$ and for the $K_o$-consolidated sample it is $OA$ (Fig. 6.15a). You should recall from Chapter 3 that the stress path for isotropic consolidation is $q/p' = 0$ and for $K_o$-consolidation is

$$\frac{q}{p'} = \frac{3(1 - K_o)}{1 + 2K_o}$$

Let us unload both samples to an effective stress $p'_o$ by reducing the vertical stress. The stress path during unloading of sample A will not follow the loading path because upon unloading $K_o$ increases nonlinearly with mean effective stress as the soil sample becomes overconsolidated (Chapter 4). The unloading effective stress path for sample A is $AD$ but for sample B it is $BC$ (Fig. 6.15a). The void ratio is now different—the initial void ratio for sample A is $e_D$ while for sample B it is $e_C$.

Let us now conduct a CU test on each sample. Because of the different initial void ratio of the two samples, prior to shearing, you should expect different undrained shear strength. The TSP for each sample has a slope of $3:1$ as depicted in Fig. 6.15a. The effective stress paths within the initial yield surface for both samples are vertical and intersect the initial yield surface at the same point, $Y$. Sample B requires a higher deviatoric stress to bring it to yield compared with sample A because the initial deviatoric stress on sample A is $q_o = (1 - K_o)\sigma'_z$ but is $q_o = 0$ for sample B. Therefore, sample A only requires a deviatoric stress increment of $\Delta q_o = q_y - (1 - K_o)\sigma'_z$ to bring it to yield compared with $q_y$ for sample B. Why do both samples have the same yield stress although each sample has a different consolidation stress history? Stress history has no effect on the elastic response: that is, the elastic response is independent of stress history.

Beyond $Y$, the yield surface expands, excess pore water pressures increase significantly, and the effective stress paths bend toward the critical state line (Fig. 6.15a). In the CU test, the volume of the soil remains constant, so the paths to failure in $(e, p')$ space for both samples are horizontal lines represented by $DG$ (sample A) and $CF$ (sample B). Sample A fails at $G$, which is at a lower deviatoric stress than at $F$, where sample B fails (Fig. 6.15a). The implication is that two samples of the same soil with different stress histories will have different shear strength even if the initial mean effective stresses before shearing and the slope of the stress path during shearing are the same.

Let us see whether we can develop an equation to estimate the undrained shear strength of a $K_o$-consolidated soil based on the ideas discussed in this chapter and using Skempton's pore water pressure coefficients (Chapter 5). Consider a saturated soil that has been $K_o$-consolidated and then subjected to total stresses $\Delta\sigma_1$ and $\Delta\sigma_3$ to bring it to failure. The initial stress conditions are $(\sigma'_1)_o > 0$ and $(\sigma'_3)_o = K_o(\sigma'_1)_o$. Upon application of the stresses, $\Delta\sigma_1$ and $\Delta\sigma_3$, the gross stresses on the soil are

$$\sigma_1 = (\sigma'_1)_o + \Delta\sigma_1 \tag{6.52}$$

$$\sigma'_1 = (\sigma'_1)_o + \Delta\sigma_1 - \Delta u \tag{6.53}$$

$$\sigma_3 = K_o(\sigma'_1)_o + \Delta\sigma_3 \tag{6.54}$$

$$\sigma'_3 = K_o(\sigma'_1)_o + \Delta\sigma_3 - \Delta u \tag{6.55}$$

For a saturated soil, Skempton's coefficient $B = 1$, and from Eq. (5.44)

$$\Delta u = \Delta \sigma_3 + A(\Delta \sigma_1 - \Delta \sigma_3) \tag{6.56}$$

Substituting Eq. (6.56) into Eq. (6.55) gives

$$\sigma_3' = K_o(\sigma_1')_o - A(\Delta \sigma_1 - \Delta \sigma_3)$$

Solving for $\Delta \sigma_1 - \Delta \sigma_3$, we obtain

$$\Delta \sigma_1 - \Delta \sigma_3 = \frac{K_o(\sigma_1')_o - \sigma_3'}{A} \tag{6.57}$$

At failure,

$$s_u = \left(\frac{\sigma_1 - \sigma_3}{2}\right)_f = \frac{1}{2}\{[(\sigma_1')_o + \Delta \sigma_1] - [K_o(\sigma_1')_o + \Delta \sigma_3]\}$$

$$= \frac{1}{2}[(\Delta \sigma_1 - \Delta \sigma_3) + (1 - K_o)(\sigma_1')_o] \tag{6.58}$$

Substituting Eq. (6.57) into Eq. (6.58) gives

$$\boxed{s_u = \frac{1}{2}\left[\frac{K_o(\sigma_1')_o - \sigma_3'}{A} + (1 - K_o)(\sigma_1')_o\right]} \tag{6.59}$$

At failure,

$$\frac{\sigma_1'}{\sigma_3'} = \frac{1 + \sin \phi_{cs}'}{1 - \sin \phi_{cs}'}$$

which by substitution into Eq. (6.59) leads to

$$\boxed{\frac{s_u}{\sigma_1'} = \frac{s_u}{\sigma_z'} = \frac{\sin \phi_{cs}'[K_o + A(1 - K_o)]}{1 + (2A - 1)\sin \phi_{cs}'}} \tag{6.60}$$

*The* essential points *are:*

1. *A $K_o$-consolidated sample of a soil is likely to have a different undrained shear strength than an isotropically consolidated sample of the same soil even if the initial confining pressures before shearing are the same and the slopes of the stress paths are also the same.*
2. *Failure stresses in soils are dependent on the stress history of the soil.*
3. *Stress history does not influence the elastic response of soils.*

*What's next . . .*We have established the main ideas behind the critical state model and used the model to estimate the response of soils to loading. The CSM can also be used with results from simple soil tests (e.g., Atterberg limits) to make estimates of the soil strengths. In the next section, we will employ the CSM to build some relationships among results from simple soil tests, critical state parameters, and soil strengths.

## 6.10 RELATIONSHIPS BETWEEN SIMPLE SOIL TESTS, CRITICAL STATE PARAMETERS, AND SOIL STRENGTHS

Wood and Wroth (1978) and Wood (1990) used the critical state model to correlate results from Atterberg limit tests with various engineering properties of fine-grained soils. We are going to present some of these correlations. These correlations are very useful when limited test data are available during the preliminary design of geotechnical systems or when you need to evaluate the quality of test results. The correlations utilized water content, which at best is accurate to 0.1%. Most often water content results are reported to the nearest whole number and consequently significant differences can occur between the actual test results and the correlations, especially those involving exponentials. Since we are using CSM and index properties, the relationships only pertain to remolded or disturbed soils.

### 6.10.1 Undrained Shear Strength of Clays at the Liquid and Plastic Limits

Wood (1990), using test results reported by Youssef et al. (1965) and Dumbleton and West (1970), showed that

$$\frac{(s_u)_{\mathrm{PL}}}{(s_u)_{\mathrm{LL}}} = R \tag{6.61}$$

where $R$ depends on activity (Chapter 2) and varies between 30 and 100, and the subscripts PL and LL denote plastic limit and liquid limit, respectively. Wood and Wroth (1978) recommend a value of $R = 100$ as reasonable for most soils. The recommended value of $(s_u)_{\mathrm{LL}}$, culled from the published data, is 2 kPa (the test data showed variations between 0.9 and 8 kPa) and that for $(s_u)_{\mathrm{PL}}$ is 200 kPa. Since most soils are within the plastic range these recommended values place lower (2 kPa) and upper (200 kPa) limits on the undrained shear strength of disturbed or remolded clays.

### 6.10.2 Vertical Effective Stresses at the Liquid and Plastic Limits

Wood (1990) used results from Skempton (1970) and recommended that

$$(\sigma'_z)_{\mathrm{LL}} = 8 \text{ kPa} \tag{6.62}$$

The test results showed that $(\sigma'_z)_{\mathrm{LL}}$ varies from 6 to 58 kPa. Laboratory and field data also showed that the undrained shear strength is proportional to the vertical effective stress. Therefore

$$\boxed{(\sigma'_z)_{\mathrm{PL}} = R(\sigma'_z)_{\mathrm{LL}} \approx 800 \text{ kPa}} \tag{6.63}$$

### 6.10.3 Undrained Shear Strength–Vertical Effective Stress Relationship

Normalizing the undrained shear strength with respect to the vertical effective stress we get a ratio of

$$\frac{s_u}{\sigma'_z} = \frac{2}{8} \text{ or } \frac{200}{800} = 0.25 \tag{6.64}$$

Mesri (1975) reported, based on soil test results, that $s_u/\sigma'_{zc} = 0.22$, which is in good agreement with Eq. (6.64) for normally consolidated soils.

### 6.10.4 Compressibility Indices ($\lambda$ and $C_c$) and Plasticity Index

The compressibility index $C_c$ or $\lambda$ is usually obtained from a consolidation test. In the absence of consolidation test results, we can estimate $C_c$ or $\lambda$ from the plasticity index. With reference to Fig. 6.16,

$$-(e_{PL} - e_{LL}) = \lambda \ln \frac{(\sigma'_z)_{PL}}{(\sigma'_z)_{LL}} = \lambda \ln R$$

Now, $e_{LL} = w_{LL}G_s$, $e_{PL} = w_{PL}G_s$, and $G_s = 2.7$. Therefore, for $R = 100$,

$$w_{LL} - w_{PL} = \frac{\lambda}{2.7} \ln R \approx 1.7\lambda$$

and

$$\lambda \approx 0.6I_p \tag{6.65}$$

or

$$C_c = 2.3\lambda \approx 1.38I_p \tag{6.66}$$

Equation (6.65) indicates that the compression index increases with plasticity index.

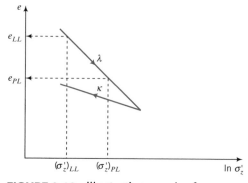

**FIGURE 6.16**    Illustrative graph of $e$ versus $\ln\sigma'_z$.

### 6.10.5 Undrained Shear Strength, Liquidity Index, and Sensitivity

Let us build a relationship between liquidity index and undrained shear strength. The undrained shear strength of a soil at a water content $w$, with reference to its undrained shear strength at the plastic limit, is obtained from Eq. (6.22) as

$$\frac{(s_u)_w}{(s_u)_{PL}} = \exp\left(G_s \frac{(w_{PL} - w)}{\lambda}\right)$$

Putting $G_s = 2.7$, $\lambda = 0.6 I_p$ in the above equation and recalling that

$$I_L = \frac{w - w_{PL}}{I_p}$$

we get

$$(s_u)_w = (s_u)_{PL} \exp(-4.6 I_L) \approx 200 \exp(-4.6 I_L) \tag{6.67}$$

Clays laid down in saltwater environments and having flocculated structure (Chapter 2) often have in situ (natural) water contents higher than their liquid limit but do not behave like a viscous liquid in their natural state. The flocculated structure becomes unstable when fresh water leaches out the salt. The undistributed or intact undrained shear strengths of these clays are significantly greater than their disturbed or remolded undrained shear strengths. The term sensitivity, $S_t$, is used to define the ratio of the intact undrained shear strength to the remolded undrained shear strength:

$$S_t = \frac{(s_u)_i}{(s_u)_r} \tag{6.68}$$

where $i$ denotes intact and $r$ denotes remolded. From Eq. (6.67) we can write

$$(s_u)_r \approx 200 \exp(-4.6 I_L) \tag{6.69}$$

For values of $S_t > 8$, the clay is called a quick clay. Quick clay, when disturbed, can flow like a viscous liquid ($I_L > 1$). Bjerrum (1954) reported test data on quick clays in Scandinavia, which yield an empirical relationship between $S_t$ and $I_L$ as

$$I_L = 1.2 \log_{10} S_t \tag{6.70}$$

## 6.11 SUMMARY

In this chapter, a simple critical state model (CSM) was used to provide some insight into soil behavior. The model replicates the essential features of soil behavior but the quantitative predictions of the model may not match real soil values. The key feature of the critical state model is that every soil fails on a unique surface in $(q, p', e)$ space. According to the CSM, the failure stress state is insufficient to guarantee failure; the soil must also be loose enough (reaches the critical void ratio). Every sample of the same soil will fail on a stress state that lies on the critical state line regardless of any differences in the initial stress state, stress history, and stress path among samples.

The model makes use of an elliptical yield surface that expands to simulate hardening or contracts to simulate softening during loading. Expansion and contraction of the yield surface are related to the normal consolidation line of the soil. Imposed stress states that lie within the initial yield surface will cause the soil to behave elastically. Imposed stress states that lie outside the initial yield surface will cause the soil to yield and to behave elastoplastically. Each imposed stress state that causes the soil to yield must lie on a yield surface and on an unloading/reloading line corresponding to the preconsolidation mean effective stress associated with the current yield surface.

The CSM is not intended to replicate all the details of the behavior of real soils but to serve as a simple framework from which we can interpret and understand the important features of soil behavior.

## Practical Examples

### EXAMPLE 6.9

An oil tank foundation is to be located on a very soft clay, 6 m thick, underlain by a deep deposit of stiff clay. Soil tests at a depth of 3 m gave the following results: $\lambda = 0.32$, $\kappa = 0.06$, $\sigma'_{cs} = 26°$, OCR = 1.2, and $w = 55\%$. The tank has a diameter of 8 m and is 5 m high. The dead load of the tank and its foundation is 350 kN. Because of the expected large settlement, it was decided to preconsolidate the soil by quickly filling the tank with water and then allowing consolidation to take place. To reduce the time to achieve the desired level of consolidation, sand drains were installed at the site. Determine whether the soil will fail if the tank is rapidly filled to capacity. What levels of water will cause the soil to yield and to fail? At the end of the consolidation, the owners propose to increase the tank capacity by welding a section on top of the existing tank. However, the owners do not want further preconsolidation or soil tests. What is the maximum increase in the tank height you would recommend so that the soil does not fail and settlement does not exceed 75 mm? The dead load per meter height of the proposed additional section is 40 kN. The unit weight of the oil is 8.5 kN/m³.

**Strategy**   The soil is one-dimensionally consolidated before the tank is placed on it. The loads from the tank will force the soil to consolidate along a path that depends on the applied stress increments. A soil element under the center of the tank will be subjected to axisymmetric loading conditions. If the tank is loaded quickly, then undrained conditions apply and the task is to predict the failure stresses and then use them to calculate the surface stresses that would cause failure. After consolidation, the undrained shear strength will increase and you would have to find the new failure stresses.

### Solution 6.9

**Step 1:**   Calculate initial values.

$$e_o = wG_s = 0.55 \times 2.7 = 1.49$$
$$K_o^{nc} = 1 - \sin \phi'_{cs} = 1 - \sin 26° = 0.56$$
$$K_o^{oc} = K_o^{oc}(OCR)^{1/2} = 0.56 \times (1.2)^{1/2} = 0.61$$

$$\gamma' = \frac{G_s - 1}{1 + e_o} \gamma_w = \frac{2.7 - 1}{1 + 1.49} \times 9.8 = 6.69 \text{ kN/m}^3$$

$$\sigma'_{zo} = \gamma' z = 6.69 \times 3 = 20.1 \text{ kPa}$$

$$\sigma'_{xo} = K_o^{oc} \sigma'_{zo} = 0.61 \times 20.1 = 12.3 \text{ kPa}$$

$$\sigma'_{zc} = \text{OCR} \times \sigma'_{zo} = 1.2 \times 20.1 = 24.1 \text{ kPa}$$

$$p'_o = \frac{1 + 2K_o^{oc}}{3} \sigma'_{zo} = \frac{1 + 2 \times 0.61}{3} \times 20.1 = 14.9 \text{ kPa} \approx 15 \text{ kPa}$$

$$q_o = (1 - K_o^{oc}) \sigma'_{zo} = (1 - 0.61) \times 20.1 = 7.8 \text{ kPa}$$

The stresses on the initial yield surface are:

$$(p'_c)_o = \frac{1 + 2K_o^{nc}}{3} \sigma'_{zc} = \frac{1 + 2 \times 0.56}{3} \times 24.1 = 17 \text{ kPa}$$

$$(q_c)_o = (1 - K_o^{nc}) \sigma'_{zc} = (1 - 0.56) \times 24.1 = 10.6 \text{ kPa}$$

$$M_c = \frac{6 \sin \phi'_{cs}}{3 - \sin \phi'_{cs}} = \frac{6 \sin 26°}{3 - \sin 26°} = 1.03$$

$$e_\Gamma = e_o + (\lambda - \kappa) \ln \frac{p'_c}{2} + \kappa \ln p'_o = 1.49 + (0.26 - 0.06) \ln \frac{17}{2} + 0.06 \ln 15 = 2.08$$

**Step 2:** Calculate the stress increase from the tank and also the consolidation stress path.

Area of tank: $A = \dfrac{\pi D^2}{4} = \dfrac{\pi \times 8^2}{4} = 50.27 \text{ m}^2$

Vertical surface stress from water: $\gamma_w h = 9.8 \times 5 = 49 \text{ kPa}$

Vertical surface stress from dead load: $\dfrac{350}{50.27} = 7 \text{ kPa}$

Total vertical surface stress: $q_s = 49 + 7 = 56 \text{ kPa}$

Vertical stress increase: $\Delta\sigma_z = q_s \left[ 1 - \left( \dfrac{1}{1 - (r/z)^2} \right)^{3/2} \right]$

$$= q_s \left[ 1 - \left( \frac{1}{1 + (4/3)^2} \right)^{3/2} \right] = 0.78 q_s$$

Radial stress increase: $\Delta\sigma_r = \dfrac{q_s}{2} \left( (1 + 2v) - \dfrac{2(1 + v)}{[1 + (r/z)^2]^{1/2}} + \dfrac{1}{[1 + (r/z)^2]^{3/2}} \right)$

$$= \frac{q_s}{2} \left( (1 + 2 \times 0.5) - \frac{2(1 + 0.5)}{[1 + (\frac{4}{3})^2]^{1/2}} + \frac{1}{[1 + (\frac{4}{3})^2]^{3/2}} \right)$$

$$= 0.21 q_s$$

$$\frac{\Delta\sigma_r}{\Delta\sigma_z} = \frac{0.21}{0.78} = 0.27$$

$$\Delta\sigma_z = 0.78 \times 56 = 43.7 \text{ kPa}, \quad \Delta\sigma_r = 0.21 \times 56 = 11.8 \text{ kPa},$$

$$\Delta p = \frac{43.7 + 2 \times 11.8}{3} = 22.4 \text{ kPa}$$

$$\Delta q = 43.7 - 11.8 = 31.9 \text{ kPa}$$

Slope of TSP = ESP during consolidation: $\dfrac{\Delta q}{\Delta p} = \dfrac{31.9}{22.4} = 1.42$

**Step 3:** Calculate the initial yield stresses and excess pore water pressure at yield.

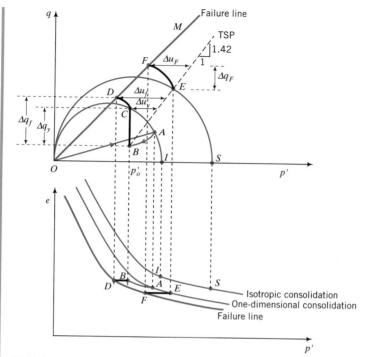

**FIGURE E6.9**

You need to calculate the preconsolidated mean effective stress on the isotropic consolidation line (point $I$, Fig. E6.9). You should note that $\{(p'_c)_o, (q_c)_o\}$ lies on the initial yield surface (point $A$, Fig. E6.9). Find $p'_c$ using Eq. (6.4), that is

$$(p'_c)^2_o + (p'_c)_o p'_c + \frac{(q_c)^2_o}{M^2} = 0$$

Therefore, $17^2 + 17p'_c + (10.6)^2/(1.03)^2 = 0$ and solving for $p'_c$ we get $p'_c = 23.2$ kPa.

The yield stresses (point $C$, Fig. E6.9) are found from Eq. (6.51); that is,

$$q_y = Mp'_o\sqrt{\frac{p'_c}{p'_o} - 1} = 1.03 \times 15\sqrt{\frac{23.2}{15} - 1} = 11.4 \text{ kPa}$$

$$p'_y = p'_o = 15 \text{ kPa}, \quad \Delta q_y = q_y - q_o = 11.4 - 7.6 = 3.8 \text{ kPa}$$

The excess pore water pressure at yield is

$$\Delta u_y = \Delta p_y = \frac{\Delta q_y}{1.42} = \frac{3.89}{1.42} = 2.7 \text{ kPa}$$

The vertical effective stress and vertical total stresses are

$$(\Delta \sigma'_z)_y = \Delta p' + \tfrac{2}{3}\Delta q = 0 + \tfrac{2}{3} \times 3.8 = 2.5 \text{ kPa}$$

$$(\Delta \sigma_z)_y = (\Delta \sigma'_z)_y + \Delta u_y = 2.5 + 2.7 = 5.2 \text{ kPa}$$

**Step 4:** Calculate the equivalent surface stress.

$$\Delta q_s = \frac{(\Delta\sigma_z)_y}{0.78} = \frac{5.2}{0.78} = 6.7 \text{ kPa}$$

The vertical surface stress from the dead load of the tank is 7 kPa, which is greater than 6.7 kPa. Therefore, under the dead load the soil will yield.

**Step 5:** Calculate the failure stresses.
Failure occurs at point $D$, Fig. E6.9.

$$\Delta q_f = M \exp\left(\frac{e_\Gamma - e_o}{\lambda}\right) = 1.03 \exp\left(\frac{2.08 - 1.49}{0.32}\right) = 6.5 \text{ kPa}$$

$$\Delta p_f' = \frac{\Delta q_f}{M} = \frac{6.5}{1.03} = 6.3 \text{ kPa}$$

$$\Delta u_f = \Delta p_f' + \frac{\Delta q_f}{1.42} = 6.3 + \frac{6.5}{1.42} = 10.9 \text{ kPa}$$

$$(\Delta\sigma_z')_f = \Delta p_f' + \tfrac{2}{3}\Delta q_f = 6.3 + \tfrac{2}{3} \times 6.5 = 10.6 \text{ kPa}$$

$$(\Delta\sigma_z)_f = (\Delta\sigma_z')_f + \Delta u_f = 10.6 + 10.9 = 21.5 \text{ kPa}$$

**Step 6:** Calculate the height of water to bring the soil to failure.

$$\text{Equivalent surface stress:} \quad \Delta q_s = \frac{(\Delta\sigma_z)_f}{0.78} = \frac{21.5}{0.78} = 27.6 \text{ kPa}$$

The vertical surface stress from the dead load of the tank is 7 kPa. Therefore, the equivalent vertical surface stress from water is $27.6 - 7 = 20.6$ kPa.

$$\text{Height of water:} \quad h_w = \frac{20.6}{\gamma_w} = \frac{20.6}{9.8} = 2.1 \text{ m}$$

Therefore, you cannot fill the tank to capacity. You will have to fill the tank with water to a height less than 2.1 m, allow the soil to consolidate, and then increase the height of water gradually.

**Step 7:** Determine the failure stresses after consolidation.
The soil is consolidated along a stress path of slope $1.42:1$ up to point $E$, Fig. E6.9. Loading from $E$ under undrained conditions (TSP has a slope of $1.42:1$) will cause yielding immediately ($E$ lies on the yield surface) and failure will occur at $F$ (Fig. E6.9).

$$p_E' = p_o' + \Delta p = 15 + 22.4 = 37.4 \text{ kPa}$$

$$e_E = e_o - \lambda \ln \frac{p_E'}{p_o'} = 1.49 - 0.32 \ln \frac{37.4}{15} = 1.20$$

$$\Delta q_F = M \exp\left(\frac{e_\Gamma - e_E}{\lambda}\right) = 1.03 \exp\left(\frac{2.08 - 1.20}{0.32}\right) = 16.1 \text{ kPa}$$

$$\Delta p_F' = \frac{\Delta q_F}{M} = \frac{16.1}{1.03} = 15.6 \text{ kPa}$$

$$\Delta u_F = \Delta p_F' + \frac{\Delta q_F}{1.42} = 15.6 + \frac{16.1}{1.42} = 26.9 \text{ kPa}$$

$$(\Delta\sigma_z')_F = \Delta p_F' + \tfrac{2}{3}\Delta q_F = 15.6 + \tfrac{2}{3} \times 16.1 = 26.4 \text{ kPa}$$

$$(\Delta\sigma_z)_F = (\Delta\sigma_z')_F + \Delta u_F = 26.4 + 26.9 = 53.4 \text{ kPa}$$

**Step 8:**   Calculate the equivalent surface stress and load.

Equivalent surface stress:   $\Delta q_s = \dfrac{(\Delta\sigma_z)_F}{0.78} = \dfrac{53.4}{0.78} = 68.5$ kPa

Surface load applied during consolidation:   $350 + H\gamma_w A = 350 + 5 \times 9.8 \times 50.27$
$$= 2813.2 \text{ kN}$$

Possible additional surface load:   $68.5 \times A = 68.5 \times 50.27 = 3443.6$ kN

Total surface load:   $2813.2 + 3443.6 = 6256.8$ kN

**Step 9:**   Find the additional height to bring the soil to failure after consolidation. Let $\Delta h$ be the additional height.

$$(5 + \Delta h)\gamma_{\text{oil}}A + 350 + \text{additional load per meter} \times \Delta h = 6256.8$$
$$\therefore (5 + \Delta h) \times 8.5 \times 50.27 + 350 + 40\Delta h = 6256.8$$

and $\Delta h = 8.1$ m.

**Step 10:**   Calculate the mean effective stress to cause 75 mm settlement.

$$\rho = \frac{\Delta e}{1 + e_o} H = \frac{H}{1 + e_E} \lambda \ln \frac{p'}{p'_E}$$

where $H$ is the thickness of the very soft clay layer. Therefore,

$$75 = \frac{6000}{1 + 1.20} \times 0.32 \ln \frac{p'}{37.4}$$

$\therefore p' = 40.8$ kPa

$\Delta p' = p' - p'_E = 40.8 - 37.4 = 3.4$ kPa,   $\Delta q = 1.42 \times \Delta p' = 1.42 \times 3.4 = 4.8$ kPa

$\Delta u = \Delta p' + \dfrac{\Delta q}{1.42} = 3.4 + \dfrac{4.8}{1.42} = 6.8$ kPa

$\Delta\sigma'_z = \Delta p' + \frac{2}{3}\Delta q = 3.4 + \frac{2}{3} \times 4.8 = 6.6$ kPa,

$\Delta\sigma_z = \Delta\sigma_z + \Delta u = 6.6 + 6.8 = 13.4$ kPa

**Step 11:**   Calculate the height of oil for 75 mm settlement.

Equivalent surface stress:   $\Delta q_s = \dfrac{\Delta\sigma_z}{0.78} = \dfrac{13.4}{0.78} = 17.2$ kPa

Additional height of tank:   $\Delta h = \dfrac{17.2}{8.5} = 2.0$ m

Since the tank was preloaded with water and water is heavier than the oil, it is possible to get a further increase in height by $(9.8/8.5 - 1)5 = 0.76$ m. To be conservative, because the analysis only gives an estimate, you should recommend an additional height of 2.0 m.   ∎

## EXAMPLE 6.10

You requested a laboratory to carry out soil tests on samples of soils extracted at different depths from a borehole. The laboratory results are shown in Table E6.10a. The tests at depth 5.2 m were repeated and the differences in results were about 10%. The average results are reported for this depth. Are any of the results suspect? If so, which are?

**TABLE E6.10a**

| Depth | $w$ (%) | $w_{PL}$ (%) | $w_{LL}$ (%) | $s_u$ (kPa) | $\lambda$ |
|-------|---------|--------------|--------------|-------------|-----------|
| 2.1   | 22      | 12           | 32           | 102         | 0.14      |
| 3     | 24      | 15           | 31           | 10          | 0.12      |
| 4.2   | 29      | 15           | 29           | 10          | 0.09      |
| 5.2   | 24      | 17           | 35           | 35          | 0.1       |
| 6.4   | 17      | 13           | 22           | 47          | 0.07      |
| 8.1   | 23      | 12           | 27           | 85          | 0.1       |

***Strategy*** It appears that the results at depth 5.2 m are accurate. Use the equations in Section 6.10 to predict $\lambda$ and $s_u$ and then compare the predicted with the laboratory test results.

## Solution 6.10

**Step 1:** Prepare a table and calculate $\lambda$ and $s_u$.
Use Eq. (6.65) to predict $\lambda$ and Eq. (6.67) to predict $s_u$. See Table E6.10b.

**Step 2:** Compare laboratory test results with predicted results.
The $s_u$ value at 2.1 m is suspect because all the other values seem reasonable. The predicted value of $s_u$ at depth 4.2 m is low in comparison with the laboratory test results. However, the water content at this depth is the highest reported but the plasticity index is about average. If the water content were about 24% (the average of the water content just above and below 4.2 m), the predicted $s_u$ is 10.4 kPa compared with 10 kPa from laboratory tests. The water content at 4.2 m is therefore suspect.

The $s_u$ value at 6.4 m, water content, and liquid limit appear suspicious. Even if the water content were taken as the average for

**TABLE E6.10b**

| Depth (m) | Laboratory results | | | | | Calculated results | | |
|-----------|---------|--------------|--------------|-------------|-----------|--------|--------|-----------|-------------|
|           | $w$ (%) | $w_{PL}$ (%) | $w_{LL}$ (%) | $s_u$ (kPa) | $\lambda$ | $I_p$  | $I_L$  | $\lambda$ | $s_u$ (kPa) |
| 2.1       | 22      | 12           | 32           | 102         | 0.14      | 20     | 0.50   | 0.12      | 20.1        |
| 3         | 24      | 15           | 31           | 10          | 0.12      | 16     | 0.56   | 0.096     | 15.0        |
| 4.2       | 29      | 15           | 29           | 10          | 0.09      | 14     | 1.00   | 0.084     | 2.0         |
| 5.2       | 24      | 17           | 35           | 35          | 0.1       | 18     | 0.39   | 0.108     | 33.4        |
| 6.4       | 17      | 13           | 22           | 47          | 0.07      | 9      | 0.44   | 0.054     | 25.9        |
| 8.1       | 23      | 12           | 27           | 85          | 0.1       | 15     | 0.73   | 0.09      | 6.9         |
| Average   | 23.2    | 14.0         | 29.3         |             |           |        |        |           |             |
| STD[a]    | 3.5     | 1.8          | 4.1          |             |           |        |        |           |             |

[a]STD is standard deviation.

the borehole, the $s_u$ values predicted ($\cong 1$ kPa) would be much lower than the laboratory results. You should repeat the tests for the sample taken at 6.4 m. The $s_u$ value at 8.1 m is suspect because all the other values seem reasonable at these depths. ∎

# EXERCISES

Assume $G_s = 2.7$, where necessary.

## Theory

**6.1** Prove that

$$R_o = \frac{1 + 2K_o^{nc}}{1 + 2K_o^{oc}} \text{OCR}$$

**6.2** Prove that

$$K_o^{nc} = \frac{6 - 2M_c}{6 + M_c}$$

**6.3** Show that the effective stress path in one-dimensional consolidation is

$$\frac{q}{p'} = \frac{3M_c}{6 - M_c}$$

**6.4** Show, for an isotropically heavily overconsolidated clay, that $s_u = 0.5Mp_o'(0.5R_o)^{(\lambda-\kappa)/\lambda}$.

**6.5** Show that $e_\Gamma = e_c - (\lambda - \kappa)\ln 2$, where $e_\Gamma$ is the void ratio on the critical state line when $p' = 1$ kPa and $e_c$ is the void ratio on the normal consolidation line corresponding to $p' = 1$ kPa.

**6.6** The water content of a soil is 55% and $\lambda = 0.15$. The soil is to be isotropically consolidated. Plot the expected volume changes against mean effective stress if the load increment ratios are (a) $\Delta p/p = 1$ and (b) $\Delta p/p = 2$.

**6.7** Plot the variation of Skempton's pore water pressure coefficient at failure, $A_f$, with overconsolidation ratio using the CSM for two clays: one with $\phi_{cs}' = 21°$ and the other with $\phi_{cs}' = 32°$. Assume $K = 0.2\lambda$ and the peak shear stress is the failure shear stress.

**6.8** A fill of height 5m with $\gamma_{sat} = 18$ kN/m³ is constructed to preconsolidate a site consisting of a soft normally consolidated soil. Test at a depth of 2 m in the soil gave the following results: $w = 45\%$, $\phi_{cs}' = 23.5°$, $\lambda = 0.25$, and $\kappa = 0.05$. Groundwater is at the ground surface.

**(a)** Show that the current stress state of the soil prior to loading lies on the yield surface given by

$$F = (p')^2 - p'p_c' + \frac{q^2}{M^2} = 0$$

**(b)** The fill is rapidly placed in lifts of 1 m. The excess pore water pressure is allowed to dissipate before the next lift is placed. Show how the soil will behave in $(q, p')$ space and in $(e, p')$ space.

## Problem Solving

6.9 The following data were obtained from a consolidation test on a clay soil. Determine $\lambda$ and $\kappa$.

| $p'$ (kPa) | 25 | 50 | 200 | 400 | 800 | 1600 | 800 | 400 | 200 |
|---|---|---|---|---|---|---|---|---|---|
| $e$ | 1.65 | 1.64 | 1.62 | 1.57 | 1.51 | 1.44 | 1.45 | 1.46 | 1.47 |

6.10 The water content of a sample of saturated soil at a mean effective stress of 10 kPa is 85%. The sample was then isotropically consolidated using a mean effective stress of 150 kPa. At the end of the consolidation the water content was 50%. The sample was then isotropically unloaded to a mean effective stress of 100 kPa and the water content increased by 1%. (a) Draw the normal consolidation line and the unloading/reloading line and (b) draw the initial yield surface and the critical state line in $(q, p')$, $(e, p')$, and $(e, \ln p')$ spaces if $\phi'_{cs} = 25°$.

6.11 Determine the failure stresses under (a) a CU test and (b) a CD test for the conditions described in Exercise 6.10.

6.12 A CU triaxial test was conducted on a normally consolidated sample of a saturated clay. The water content of the clay was 50% and its undrained shear strength was 22 kPa. Estimate the undrained shear strength of a sample of this clay if $R_o = 15$, $w = 30\%$, and the initial stresses were the same as the sample that was tested. The parameters for the normally consolidated clay are $\lambda = 0.28$, $\kappa = 0.06$, and $\phi'_{cs} = 25.3°$.

6.13 Two samples of a soft clay are to be tested in a conventional triaxial apparatus. Both samples were isotropically consolidated under a cell pressure of 250 kPa and then allowed to swell back to a mean effective stress of 175 kPa. Sample A is to be tested under drained conditions while sample B is to be tested under undrained conditions. Estimate the stress–strain, volumetric strain (sample A), and excess pore water pressure (sample B) responses for the two samples. The soil parameters are $\lambda = 0.15$, $\kappa = 0.04$, $\phi'_{cs} = 26.7°$, $e_o = 1.08$, and $v' = 0.3$.

6.14 Determine and plot the stress–strain ($q$ versus $\varepsilon_1$) and volume change ($\varepsilon_p$ versus $\varepsilon_1$) responses for an overconsolidated soil under a CD test. The soil parameters are $\lambda = 0.17$, $\kappa = 0.04$, $\phi'_{cs} = 25°$, $v' = 0.3$, $e_o = 0.92$, $p'_c = 280$ kPa, and OCR = 8.

6.15 Repeat Exercise 6.14 for an undrained triaxial compression (CU) test and compare the results with the undrained triaxial extension test.

6.16 A sample of a clay is isotropically consolidated to a mean effective stress of 300 kPa and is isotropically unloaded to a mean effective stress of 250 kPa. An undrained triaxial extension test is to be carried out by keeping the axial stress constant and increasing the radial stress. Predict and plot the stress–strain ($q$ versus $\varepsilon_1$) and the excess pore water pressure ($\Delta u$ versus $\varepsilon_1$) responses up to failure. The soil parameters are $\lambda = 0.23$, $\kappa = 0.07$, $\phi'_{cs} = 24°$, $v' = 0.3$, and $e_o = 1.32$.

## Practical

6.17 A tank of diameter 5 m is to be located on a deep deposit of normally consolidated homogeneous clay, 25 m thick. The vertical stress imposed by the tank at the surface is 75 kPa. Calculate the excess pore water pressure at depths of 2, 5, 10, and 20 m if the vertical stress were to be applied instantaneously. The soil parameters are $\lambda = 0.26$, $\kappa = 0.06$, and $\phi'_{cs} = 24°$. The average water content is 42% and groundwater level is at 1 m below the ground surface.

## CHAPTER 7

# BEARING CAPACITY OF SOILS AND SETTLEMENT OF SHALLOW FOUNDATIONS

| ABET | ES | ED |
|------|----|----|
|      | 75 | 25 |

## 7.0 INTRODUCTION

In this chapter, we will consider bearing capacity of soils and settlement of shallow foundations. Loads from a structure are transferred to the soil through a foundation. A foundation itself is a structure, often constructed from concrete, steel, or wood. A geotechnical engineer must ensure that a foundation satisfies the following two stability conditions:

**1.** The foundation must not collapse or become unstable under any conceivable loading.
**2.** Settlement of the structure must be within tolerable limits.

Both requirements must be satisfied. Often, it is settlement that governs the design of shallow foundations.

We will also consider, in this chapter, the limit equilibrium method of analysis. This is the analysis that Coulomb performed in analyzing the lateral force on the fortresses from soil placed behind them (Chapter 1). The limit equilibrium method is used to find solutions for a variety of problems including bearing capacity of foundations, stability of retaining walls, and slopes. When you complete this chapter, you should be able to:

- Calculate the safe bearing capacity of soils
- Estimate the settlement of shallow foundations
- Estimate the size of shallow foundations to satisfy bearing capacity and settlement criteria

You will use the following concepts learned from previous chapters and from your courses in mechanics.

- Statics
- Stresses and strains—Chapter 3
- Shear strength—Chapter 5

*Sample Practical Situation*   The loads from a building are to be transferred to the soil by shallow foundations. You are required to recommend the sizes of shallow foundations so that there is a margin of safety against soil failure and

**FIGURE 7.1**  Construction of a shallow foundation. (Photo courtesy of John Cernica.)

settlements of the building are within tolerable limits. The construction of a shallow foundation is shown in Fig. 7.1.

## 7.1  DEFINITIONS OF KEY TERMS

*Foundation* is a structure that transmits loads to the underlying soils.

*Footing* is a foundation consisting of a small slab for transmitting the structural load to the underlying soil. Footings can be individual slabs supporting single columns (Fig. 7.2a) or combined to support two or more columns (Fig. 7.2b), or be a long strip of concrete slab (Fig. 7.2c, width $B$ to length $L$ ratio is small, i.e., it approaches zero) supporting a load bearing wall, or a mat (Fig. 7.2d).

*Embedment depth ($D_f$)* is the depth below the ground surface where the base of the foundation rests.

*Shallow foundation* is one in which the ratio of the embedment depth to the minimum plan dimension, which is usually the width, is $D_f/B \leq 2.5$.

*Ultimate bearing capacity* is the maximum pressure that the soil can support.

*Ultimate net bearing capacity ($q_{ult}$)* is the maximum pressure that the soil can support above its current overburden pressure.

*Allowable bearing capacity* or *safe bearing capacity ($q_a$)* is the working pressure that would ensure a margin of safety against collapse of the structure from shear failure. The allowable bearing capacity is usually a fraction of the ultimate net bearing capacity.

*Factor of safety* or *safety factor (FS)* is the ratio of the ultimate net bearing capacity to the allowable net bearing capacity or to the applied maximum net vertical stress. In geotechnical engineering, a factor of safety between 1.5 and 5 is used to calculate the allowable bearing capacity.

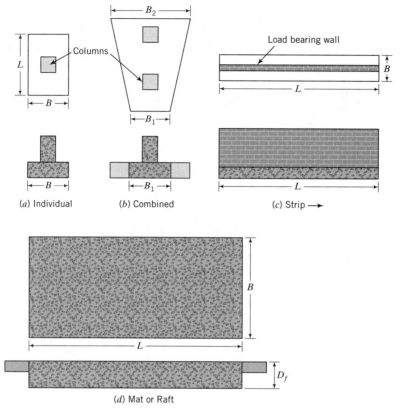

**FIGURE 7.2**  Types of shallow footings.

*Ultimate limit state* defines a limiting stress or force that should not be exceeded by any conceivable or anticipated loading during the design life of a foundation or any geotechnical system.

*Serviceability limit state* defines a limiting deformation or settlement of a foundation, which, if exceeded, will impair the function of the structure that it supports.

## 7.2   QUESTIONS TO GUIDE YOUR READING

1. What are the ultimate net bearing capacity and the allowable bearing capacity of shallow footings?
2. What are the differences between the various methods for calculating a soil's ultimate net bearing capacity?
3. How do I determine the allowable bearing capacity for shallow footings?
4. What are the assumptions made in bearing capacity analyses?
5. What soil parameters are needed to calculate its bearing capacity?
6. What effects do groundwater and eccentric loads have on bearing capacity?
7. How do I determine the size of a footing to satisfy ultimate and serviceability limit states?

## 7.3 BASIC CONCEPTS

### 7.3.1 Collapse and Failure Loads

In developing the basic concepts we will use a generic friction angle, $\phi'$, and later discuss whether to use $\phi'_p$ or $\phi'_{cs}$. Failure in the context of bearing capacity means the ultimate net bearing capacity. For dilating soils, failure corresponds to the peak shear stress, while for nondilating soils failure corresponds to the critical state shear stress. To distinguish these two states, we will refer to the failure load in dilating soils as the collapse load and reserve the term failure load for nondilating soils. Thus collapse load is the load at peak shear stress while failure load is the load at critical state. Collapse means a sudden decrease in the bearing capacity of a soil.

Let us consider two separate soil layers. One is a deep deposit of a dense sand; the other is a deep deposit of the same sand but in a loose state. We will now take two similar square blocks of concrete of width $B$ and place one each on the surfaces of the two soil layers as shown in Fig. 7.3. We are going to add the same magnitude of vertical load at the center of each block and measure their settlements using displacement gauges.

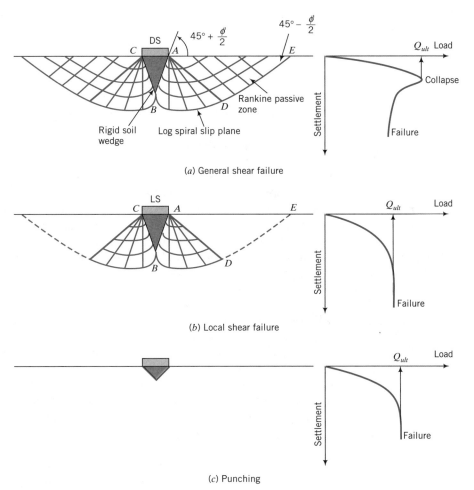

FIGURE 7.3  Failure mechanisms.

Let us represent the concrete block on the dense sand layer as DS and that on the loose sand as LS. As we load the blocks, which block will settle more? Intuitively, and from your experience playing with sand, you may have guessed correctly that the loose sand would.

Here is a list of expected responses as we load the blocks.

1. Block LS will settle more than block DS for a given load before soil failure.

2. Block DS would collapse while block LS would not as illustrated in the graphs of settlement versus vertical load on the right side of Fig. 7.3.

3. Based on the Mohr–Coulomb failure criterion (Chapter 5), the soil will fail along slip planes at $\theta_1 = 45° + \phi'/2$ to the horizontal plane and $\theta_2 = 45° - \phi'/2$ to the vertical plane.

The pressure from the load on each block that brings the soil to collapse or failure is the soil's ultimate bearing capacity. Would you, in practice, design a foundation so that the applied load is equal to the ultimate load, $Q_{ult}$? Congratulations, if your answer is no. Because of spurious soil variations, uncertainty in soil test values and structural and environmental loads, and difficulties in quality control during construction, geotechnical engineers usually divide the ultimate bearing capacity by a factor of safety and call the resulting load the allowable bearing capacity.

## 7.3.2 Failure Surface

Prandtl (1920) showed theoretically that a wedge of material is trapped below a rigid plate when it is subjected to concentric loads. Terzaghi (1943) applied Prandtl's theory to a strip footing (length is much longer than its width) with the assumption that the soil is a semi-infinite, homogeneous, isotropic, weightless rigid-plastic material. Based on Prandtl's theory, failure of the footing occurs by a wedge of soil below the footing pushing its way downward into the soil.

Why is the development of a wedge of soil under a shallow foundation a reasonable assumption for soils? You will recall from Chapter 5 that, according to the Mohr–Coulomb criterion, slip planes form when soils are sheared to failure. No slip plane, however, can pass through the rigid footing, so none can develop in the soil just below the footing. The slip planes form two zones around the rigid wedge—each zone symmetrical about a vertical plane (parallel to the length of the footing) through the center of the footing. One zone, $ABD$ (Fig. 7.3), is a fan with radial slip planes stopping on a logarithm spiral slip plane. The other zone, $ADE$, consists of slip planes oriented at angles of $45° + \phi'/2$ and $45° - \phi'/2$ to the horizontal and vertical planes, respectively, as we found in Chapter 5. Zone $ADE$ is called the Rankine passive zone. In Chapter 10, we will discuss Rankine passive zone and also Rankine active zone in connection with retaining walls. Surfaces $AB$ and $AD$ are frictional sliding surfaces and their actions are similar to rough walls being pushed into the soil. The pressure exerted is called passive lateral earth pressure. If we can find the value of this pressure, we can determine the ultimate bearing capacity from considering the equilibrium of wedge $ABC$.

For the dense sand, the slip planes are expected to reach the ground surface. But, for the loose sand, the slip planes, if they are developed, are expected to lie within the soil layer below the base of the footing and extend laterally. The collapse in the dense sand is termed general shear failure, as the shear planes are fully developed. The failure in the loose sand is termed local shear failure, as the shear planes are not fully developed. Another type of failure is possible. For very loose soil, the failure surfaces may be confined to the surfaces of the rigid wedge. This type of failure is termed punching shear.

> *The essential points are:*
>
> 1. *Dense soils fail suddenly along well-defined slip planes resulting in a general shear failure.*
> 2. *Loose soils do not fail suddenly and the slip planes are not well defined, resulting in a local shear failure.*
> 3. *Very loose soils can fail by punching shear.*
> 4. *More settlement is expected in loose soils than in dense soils.*
> 5. *The expected failure surface for general shear failure consists of a rigid wedge of soil trapped beneath the footing bordering radial shear zones under Rankine passive zones.*

*What's next . . .* The bearing capacity equations that we will discuss in this chapter were derived by making alterations to the failure surface found by Prandtl. Before we consider these bearing capacity equations, we will find the collapse load for a strip footing resting on a clay soil using a popular analytical technique called the limit equilibrium method.

## 7.4   COLLAPSE LOAD FROM LIMIT EQUILIBRIUM

The bearing capacity equations that are in general use in engineering practice were derived using an analytical method called the limit equilibrium method. We will illustrate how to use this method by finding the collapse load ($P_u$) for a strip footing. The essential steps in the limit equilibrium method are (1) selection of a plausible failure mechanism or failure surface, (2) determination of the forces acting on the failure surface, and (3) use of the equilibrium equations that you learned in statics to determine the collapse or failure load.

Let us consider a strip footing of width $B$, resting on the surface of a homogeneous, saturated clay whose undrained strength is $s_u$ (Fig. 7.4). For simplic-

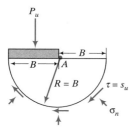

**FIGURE 7.4**  Circular failure mechanism.

ity, we will neglect the weight of the soil. Step 1 of the limit equilibrium method requires that we either know or speculate on the failure mechanism. Since we do not know what the failure mechanism is because we have not done any testing, we will speculate that the footing will fail by rotating about the inner edge $A$ (Fig. 7.4), so that the failure surface is a semicircle with radius $B$.

Step 2 is to determine the forces on the failure surface. Along the circumference of the semicircle, there would be shear stresses ($\tau$) and normal stresses ($\sigma'_n$). We do not know whether these stresses are uniformly distributed over the circumference, but we will assume that this is so; otherwise, we have to perform experiments or guess plausible distributions. Since failure occurred, the maximum shear strength of the soil is mobilized and therefore the shear stresses are equal to the shear strength of the soil. Now we are ready to move to Step 3. The moment due to the normal force acting on the semicircle about $A$ is zero since its line of action passes through $A$. The moment equilibrium equation is then

$$P_u \times \frac{B}{2} - s_u \, \pi B \times B = 0 \tag{7.1}$$

and the collapse load is

$$\boxed{P_u = 6.28 B s_u} \tag{7.2}$$

We are unsure that the collapse load we calculated is the correct one since we guessed a failure mechanism. We can repeat the three steps above by choosing a different failure mechanism. For example, we may suppose that the point of rotation is not $A$ but some point $O$ above the footing such that the radius of the failure surface is $R$ (Fig. 7.5).

Taking moments about $O$, we get

$$P_u(R \cos \theta - B/2) - s_u[(\pi - 2\theta)R]R = 0 \tag{7.3}$$

By rearranging Eq. (7.3), we get

$$P_u = \frac{s_u[(\pi - 2\theta)R]R}{(R \cos \theta - B/2)} = \frac{s_u(\pi - 2\theta)R}{(\cos \theta - B/2R)} \tag{7.4}$$

The collapse load now depends on two variables, $R$ and $\theta$, and as such there is a family of failure mechanisms. We must then find the least load that will produce collapse. We do this by searching for extrema (minima and maxima in curves) by taking partial derivatives of Eq. (7.4) with respect to $R$ and $\theta$. Thus,

$$\frac{\partial P_u}{\partial R} = \frac{4 s_u R(\pi - 2\theta)(R \cos \theta - B)}{(2R \cos \theta - B)^2} = 0 \tag{7.5}$$

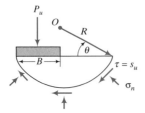

**FIGURE 7.5** Circular arc failure mechanism.

and

$$\frac{\partial P_u}{\partial \theta} = \frac{4s_u R^2 (B - 2R \cos \theta + \pi R \sin \theta - 2R\theta \sin \theta)}{(2R \cos \theta - B)^2} = 0 \qquad (7.6)$$

The solutions of Eqs. (7.5) and (7.6) are $\theta = 23.2°$ and $R = B \sec \theta$; that is, point $O$ is directly above $A$. Substituting these values in Eq. (7.4), we obtain the collapse load as

$$P_u = 5.52 B s_u \qquad (7.7)$$

This is a better solution because the collapse load is smaller than Eq. (7.2) but we need to investigate other possible mechanisms, which may yield yet a smaller value of $P_u$. The exact solution to our problem, using more complex analysis than the limit equilibrium method, gives

$$\boxed{P_u = 5.14 B s_u} \qquad (7.8)$$

which is about 9% lower than our second mechanism.

*What's next* . . .We have just learned the rudiments of the limit equilibrium method. It is an iterative method in which you speculate on possible failure mechanisms and then use statics to find the collapse load. The bearing capacity equations that we will discuss next were derived using the limit equilibrium method. We will not derive the bearing capacity equations because no new concept will be learned. You can refer to the published literature mentioned later if you want to pursue the derivations.

## 7.5  BEARING CAPACITY EQUATIONS

There is a plethora of bearing capacity equations available. We will discuss only three such equations, all of which were obtained by limit equilibrium analyses using the failure mechanism proposed by Prandtl or modifications to it. These equations have found general use in geotechnical practice.

We will consider long-term bearing capacity (drained condition) and short-term bearing capacity (undrained condition). An effective stress analysis (ESA) is used for the long-term bearing capacity while a total stress analysis (TSA) is used for the short-term bearing capacity. The bearing capacity equations to be presented were modified by the author from their original form to separate long-term and short-term bearing capacity.

### 7.5.1 Terzaghi's Bearing Capacity Equations

Terzaghi (1943) derived bearing capacity equations for a footing at a depth $D_f$ below the ground level of a homogeneous soil. For most shallow footings, the depth, $D_f$, called the embedment depth, accounts for frost action, freezing, thawing, and so on. Building codes provide guidance as to the minimum depth of embedment for footings. Mat foundations can be embedded at a depth $D_f$ such that the pressure of the soil removed is equal to all or part of the applied stress.

For such a case, the mat foundation is called a compensated mat or a raft foundation. Terzaghi assumed the following:

1. The embedment depth is not greater than the width of the footing ($D_f <$ $B$).
2. General shear failure occurs.
3. The angle $\theta$ in the wedge (Fig. 7.6) is $\phi'$.
4. The shear strength of the soil above the footing base is negligible.
5. The soil above the footing base can be replaced by a surcharge stress ($= \gamma D_f$).
6. The base of the footing is rough.

Later, it was found (Vesic, 1973) that $\theta = 45° + \phi'/2$. Through limit equilibrium analyses, using $\theta = 45° + \phi'/2$, and modifications to account for the shapes of footings, the ultimate net bearing capacity equations are

$$\boxed{\text{TSA:} \quad q_{ult} = 5.14 s_u s_c} \tag{7.9}$$

$$\boxed{\text{ESA:} \quad q_{ult} = \gamma D_f (N_q - 1) s_q + 0.5 \gamma B N_\gamma s_\gamma} \tag{7.10}$$

where $q_{ult}$ is the ultimate net bearing capacity; $N_q$ and $N_\gamma$ are bearing capacity factors that are functions of the friction angle, $\phi'$; $s_u$ is the undrained shear strength, and $s_c$, $s_q$, and $s_\gamma$ are shape factors for the foundation. The bearing capacity factors are calculated from the following equations:

$$\boxed{N_q = e^{\pi \tan \phi'} \tan^2(45° + \phi'/2)} \tag{7.11}$$

$$\boxed{N_\gamma = 2(N_q + 1) \tan \phi'} \tag{7.12}$$

The shape factors are

$$\boxed{s_c = 1 + 0.2 \frac{B}{L}, \quad s_q = 1 + \frac{B}{L} \tan \phi', \quad s_\gamma = 1 - 0.4 \frac{B}{L}} \tag{7.13}$$

The ultimate net bearing capacity is the ultimate pressure that the soil can support above its current overburden pressure.

For circular footings, the width $B$ in Eq. (7.10) is replaced by the diameter $D$. For square and circular footings, $B/L = 1$; for strip footings, $B/L = 0$. We will call Eqs. (7.9) and (7.10) the Terzaghi bearing capacity equations. These equations are not the original equations proposed by Terzaghi; others have modified

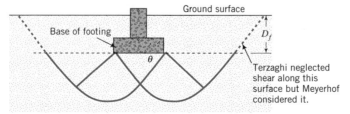

**FIGURE 7.6** Failure surface assumed by Terzaghi.

it, especially Vesic (1973). The factor $s_\gamma$ was proposed by deBeer (1967, 1970).

### 7.5.2 Skempton's Bearing Capacity Equation

Skempton (1951) proposed a bearing capacity equation, based on a TSA, for rectangular and square footings resting on a clay. Skempton's equation is

$$q_{ult} = 5s_u\left(1 + 0.2\,\frac{D_f}{B}\right)\left(1 + 0.2\,\frac{B}{L}\right); \quad \frac{D_f}{B} \le 2.5 \tag{7.14}$$

Skempton's equation was obtained by comparing field measurements with Terzaghi bearing capacity equation and making modifications to it.

### 7.5.3 Meyerhof's Bearing Capacity Equation

Meyerhof (1951, 1953, 1963) followed a similar approach to Terzaghi but included the shearing resistance of the soil above the footing base. He assumed that the failure surface (Fig. 7.6) extends to the ground surface. Hansen (1961, 1970) proposed a general equation that allows us to determine the bearing capacity of footings of any shape and size, and any inclination of loading or bearing surface. We will present only Meyerhof's (1963) equations.

**Vertical load**

$$\text{TSA:} \quad q_{ult} = 5.14 s_u s_c\, d_c \tag{7.15}$$

$$\text{ESA:} \quad q_{ult} = \gamma D_f(N_q - 1)s_q\, d_q + 0.5\gamma B N_\gamma s_\gamma\, d_\gamma \tag{7.16}$$

**Inclined load**

$$\text{TSA:} \quad q_{ult} = 5.14 s_u\, d_c i_c \tag{7.17}$$

$$\text{ESA:} \quad q_{ult} = \gamma D_f(N_q - 1)\, d_q i_q + 0.5\gamma B N_\gamma\, d_\gamma i_\gamma \tag{7.18}$$

Meyerhof's bearing capacity factors are as follows:

$$N_q = e^{\pi \tan \phi'} \tan^2(45° + \phi'/2); \quad N_q \text{ is the same as Terzaghi's}$$

$$N_\gamma = (N_q - 1) \tan(1.4\phi') \tag{7.19}$$

The shape ($s$), depth ($d$), and load inclination ($i$) factors are

$$s_c = 1 + 0.2\,\frac{B}{L}, \quad s_q = s_\gamma = 1 + 0.1K_p\,\frac{B}{L} \tag{7.20}$$

$$d_c = 1 + 0.2\,\frac{D_f}{B}, \quad d_q = d_\gamma = 1 + 0.1\sqrt{K_p}\,\frac{D_f}{B} \tag{7.21}$$

where

$$K_p = \tan^2\left(45° + \frac{\phi'}{2}\right) = \frac{1 + \sin \phi'}{1 - \sin \phi'} \tag{7.22}$$

For loads inclined at an angle $\theta$ to the vertical in the direction of the footing width, the inclination factors are

$$i_c = i_q = \left(1 - \frac{\theta}{90°}\right)^2, \quad i_\gamma = \left(1 - \frac{\theta}{\phi'}\right)^2 \tag{7.23}$$

For loads inclined at an angle $\theta$ to the vertical in the direction of the footing length for a surface footing ($D_f = 0$), the inclination factors (Meyerhof and Koumoto, 1987) are

$$i_c = \cos\theta\left[1 - \left(1 - \frac{\alpha_a}{\pi + 2}\right)\sin\theta\right], \quad i_q = i_\gamma = \cos\theta\left(1 - \frac{\sin\theta}{\sin\phi'}\right) \tag{7.24}$$

where $\alpha_a$, the adhesion factor, is usually $\frac{2}{3}$ to $\frac{1}{2}$ for short-term loading.

*What's next* . . .We have considered three bearing capacity equations. The question that arises is: Which equation should be used or which one is the best? There is no simple answer. It depends on a geotechnical engineer's familiarity with a given method, soil conditions, and the importance of the structure. In the next section, some general guidelines are suggested to help you make a choice among the equations and the appropriate friction angle to use.

## 7.6   CHOICE OF BEARING CAPACITY EQUATIONS AND FRICTION ANGLES

The major differences among the various bearing capacity equations are the values of $N_\gamma$ and the geometric factors (shape, depth and load inclination factors). For values of $\phi' < 35°$, the differences between $N_\gamma$ among the bearing capacity equations are not practically significant. However, for $\phi' > 35°$, the differences are substantial. Fortunately, most soils have $\phi' < 35°$ so that for many practical problems there would not be large differences between the predictions of the various methods. The crucial choice is the value of $\phi'$. Should it be $\phi_p'$ or $\phi_{cs}'$?

The bearing capacity equations have been derived based on the existence of slip planes. Slip planes, however, have been only observed in dense sands and heavily overconsolidated clays. Most foundations are constructed on compacted soils (exceptions include lightly loaded foundations, for example a foundation for a single story house, on normally consolidated clays or medium sands). It is expected that for these compacted soils slip planes would develop and the appropriate value of $\phi'$ to calculate the collapse load should be $\phi_p'$. However, the attainment of $\phi_p'$ depends on the ability of the soil to dilate, which can be suppressed by large normal effective stresses. Since neither the loads nor the stresses induced by the loads on the soil mass are certain, the use of $\phi_p'$ is then uncertain.

Slip planes, if they develop, do not occur suddenly. They are likely to develop progressively at different levels of stresses. If you reexamine the slip plane for our shallow footing on dense sand (Fig. 7.3), you would realize that the normal effective stress at $A$ is much larger than at $E$. Indeed, at $E$, the normal effective stress is zero. The implication is that while the peak shear stress is being mobilized near $E$, it has already been surpassed at $A$; that is, the critical state shear stress is attained at $A$. One is then tempted to use $\phi'_{cs}$, which is the only fundamental soil parameter. If $\phi'_{cs}$ is used rather than $\phi'_p$ (assuming it is mobilized) for a footing on a compacted soil, you will get a lower ultimate bearing capacity and consequently a more costly foundation.

The predictions from the bearing capacity equations have not been validated by experiments. For example, Zadroga (1994) compared $N_\gamma$ from a number of bearing capacity equations including those of Meyerhof, Terzaghi and Hansen with results from model footings on coarse-grained soils and showed that $N_\gamma$ from the equations are much lower than the experimental results. Some bearing capacity equations under-predict the ultimate bearing capacity by more than twice. The bearing capacity equations were derived based on a simplified soil behavior and they should not be expected to match reality accurately. Also, one has to be cautious in accepting model test results because the actual value of $\phi'$ mobilized in the tests is unknown and it may be substantially different from the $\phi'$, which is normally determined from direct shear on triaxial tests, used in calculating the ultimate bearing capacity.

For fine-grained soils, the ultimate bearing capacity under short-term loading is calculated in addition to long-term loading. Skempton (1951) and Peck and Byrant (1953) used the bearing capacity failure of the Transcona Grain Elevator (Chapter 1) to establish the accuracy of the bearing capacity equations for short-term loading on a fine-grained soil. The soil parameter required for short-term loading is $s_u$. However, $s_u$ is dependent on the initial void ratio or the initial confining pressure, which is uncertain in the field. The implication is that to get an accurate prediction for short-term loading, $s_u$ must be obtained at or very close to the initial void ratio or the initial confining pressure of the soil in the field.

You should by now appreciate the dilemma on deciding on which bearing capacity equation to use and the appropriate friction angle. The least uncertain soil parameter is $\phi'_{cs}$. You should use this value in the bearing capacity equations unless your experience indicates otherwise. For preliminary calculations and for simple foundation systems, the Terzaghi equations are adequate; otherwise you should use the Meyerhof equations.

**What's next . . .** So far, we have studied how to calculate the ultimate net bearing capacity. If a footing is designed based on the calculated ultimate net bearing capacity value, it will be in imminent danger of failing. And, of course, we do not want that to happen. We then have to obtain a design bearing capacity value called the allowable bearing capacity, $q_a$. The allowable bearing capacity allows a margin of safety against collapse or failure. We will examine how to calculate the allowable bearing capacity next.

# 7.7 ALLOWABLE BEARING CAPACITY AND FACTOR OF SAFETY

### 7.7.1 Calculation of Allowable Bearing Capacity

The allowable bearing capacity is calculated by dividing the ultimate bearing factor capacity by a factor, called the factor of safety, FS. The factor of safety or

**TABLE 7.1   Allowable Bearing Capacity (UBC, 1991)**

| Soil type | $q_a$ (kPa) |
|---|---|
| Sandy gravel/gravel (GW and GP) | 96 |
| Sand, silty sand, clayey sand, silty gravel (SW, SP, SM, SC, GM, GC) | 72 |
| Clay, sandy clay, silty clay, and clayey silt (CL, ML, MH, and CH) | 48 |

safety factor is intended to compensate for assumptions made in developing the bearing capacity equations, soil variability, inaccurate soil data, and uncertainties of loads. The allowable bearing capacity is

$$q_a = \frac{q_{\text{ult}}}{\text{FS}} + \gamma D_f \qquad (7.25)$$

Alternatively, if the maximum applied foundation stress $(\sigma_a)_{\text{max}}$ is known and the dimension of the footing is also known then you can find a factor of safety by replacing $q_a$ by $(\sigma_a)_{\text{max}}$ in Eq. (7.25) and solve for FS, giving

$$\text{FS} = \frac{q_{\text{ult}}}{(\sigma_a)_{\text{max}} - \gamma D_f}; \quad \gamma D_f < (\sigma_a)_{\text{max}} \qquad (7.26)$$

The maximum applied foundation stress must include the weight of the footing, backfill soil weight, dead and live loads. The factor of safety is not applied to $\gamma D_f$ because this is the overburden pressure (the pressure of the soil removed to place the footing). The factor of safety using $\phi'_{cs}$ in the bearing capacity equations ranges from 1.5 to 2.5. Because of the greater uncertainty in $\phi'_p$ (if you decide to use it) compared to $\phi'_{cs}$, a higher range of factor of safety, FS = 3 to 5, is used. The lower values in the above ranges are used for homogeneous soil deposits.

### 7.7.2 Building Codes Bearing Capacity Values

Building codes usually provide recommended values of bearing capacity for local conditions. You can use these values for preliminary design but you should check these values using soil test data and the analytical methods of Section 7.7.1. Table 7.1 shows the allowable bearing capacity values for general soil types (UBC, 1991).

## 7.8   EFFECTS OF GROUNDWATER

For all the bearing capacity equations, you will have to make some adjustment for the groundwater condition. The term $\gamma D_f$ in the bearing capacity equations refers to the vertical stress of the soil above the base of the foundation. The last term $\gamma B$ refers to the vertical stress of a soil mass of thickness $B$, below the base of the footing. You need to check which one of three groundwater situations is applicable to your project.

**Situation 1:** Groundwater level at a depth $B$ below the base of the footing. If the groundwater level is at a depth $B$ below the base of the foundation, no modification of the bearing capacity equations is required.

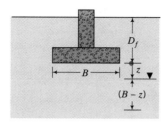

(a) Groundwater within a depth $B$ below base

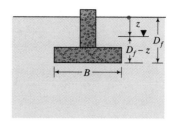

(b) Groundwater within embedment depth

**FIGURE 7.7   Groundwater effects.**

**Situation 2:** Groundwater level within a depth $B$ below the base of the footing. If the groundwater level is at a depth $z$ below the base, such that $z < B$ (Fig. 7.7a), then the term $\gamma B$ is $\gamma z + \gamma'(B - z)$ or $\gamma_{sat}z + \gamma'(B - z)$. The latter equation is used if the soil above the ground water level is also saturated. The term $\gamma D_f$ remains unchanged.

**Situation 3:** Groundwater level within the embedment depth. If the groundwater is at a depth $z$ within the embedment depth (Fig. 7.7b), then the term $\gamma D_f$ is $\gamma z + \gamma'(D_f - z)$ or $\gamma_{sat}z + \gamma'(D_f - z)$. The latter equation is used if the soil above the ground water level is also saturated. The term $\gamma B$ becomes $\gamma'B$.

## EXAMPLE 7.1

A footing 2 m square is located at a depth of 1.0 m below the ground surface in a deep deposit of compacted sand ($\phi'_p = 35°$, $\phi'_{cs} = 30°$ and $\gamma_{sat} = 18 \text{ kN/m}^3$). The groundwater level is 5 m below the ground surface but you should assume that the soil above the groundwater is saturated. Determine the allowable bearing capacity.

***Strategy***   It is a good policy to sketch a diagram illustrating the conditions given (see Fig. E7.1). The groundwater level is located at $(5 \text{ m} - 1 \text{ m}) = 4$ m from the footing base. That is, the groundwater level is more than $B = 2$ m below the base. We can neglect the effects of groundwater. You need to choose a method. We will use Terzaghi's method and use both $\phi'_p$ and $\phi'_{cs}$ for comparison.

## Solution 7.1

**Step 1:** Calculate the bearing capacity factors and geometric factors for Terzaghi's method.

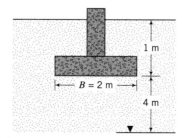

**FIGURE E7.1**

Use a table for ease of calculation and checking.

| Parameter | $\phi' = \phi'_p = 35°$ | $\phi' = \phi'_{cs} = 30°$ |
|---|---|---|
| $N_q = e^{\pi \tan \phi'} \tan^2\left(45° + \dfrac{\phi'}{2}\right)$ | 33.3 | 18.4 |
| $N_q - 1$ | 32.3 | 17.4 |
| $N_\gamma = 2(N_q + 1)\tan\phi'$ | 48.0 | 22.4 |
| $s_q = 1 + \dfrac{B}{L}\tan\phi'$ | 1.70 | 1.58 |
| $s_\gamma = 1 - 0.4\dfrac{B}{L}$ | 0.6 | 0.6 |

**Step 2:** Calculate the ultimate net bearing capacity.

$$q_{\text{ult}} = \gamma D_f(N_q - 1)s_q + 0.5\gamma BN_\gamma s_\gamma$$

$$q_a = \frac{q_{\text{ult}}}{\text{FS}} + \gamma D_f$$

$\phi' = \phi'_p = 35°$: $q_{\text{ult}} = 18 \times 1 \times 32.3 \times 1.7 + 0.5 \times 18 \times 2 \times 48 \times 0.6$
$$= 1507 \text{ kPa}$$

For FS $= 3$

$$q_a = \frac{1507}{3} + 18 \times 1 = 520 \text{ kPa}$$

$\phi' = \phi'_{cs} = 30°$: $q_{\text{ult}} = 18 \times 1 \times 17.4 \times 1.58 + 0.5 \times 18 \times 2 \times 22.4 \times 0.6$
$$= 737 \text{ kPa}$$

Using FS $= 1.5$

$$q_a = \frac{737}{1.5} + 18 \times 1 = 509 \text{ kPa}$$

For this problem, the allowable bearing capacity of the soil is about the same using $\phi'_p$ with FS $= 3$ and using $\phi'_{cs}$ with FS $= 1.5$. ∎

## EXAMPLE 7.2

Compare the ultimate net bearing capacity for Example 7.1 using $\phi'_p = 35°$ when the groundwater is located (a) at 5 m below the base, (b) at the ground surface, (c) at the base of the footing, and (d) at 1 m below the base. Use Meyerhof's method.

***Strategy*** The trick here is to use the appropriate value of the unit weight in the bearing capacity equation.

## Solution 7.2

**Step 1:** Calculate bearing capacity numbers and, shape and depth factors.

$N_q = e^{\pi \tan \phi'} \tan^2(45° + \phi'/2) = e^{\pi \tan 35°} \tan^2(45° + 35°/2) = 33.3$
$N_q - 1 = 33.3 - 1 = 32.3$
$N_\gamma = (N_q - 1)\tan(1.4\phi') = 32.3 \tan(1.4 \times 35°) = 37.2$

$$K_p = \tan^2(45° + \phi'/2) = \tan^2(45° + 35°/2) = 3.7$$

$$s_q = s_\gamma = 1 + 0.1K_p \frac{B}{L} = 1 + 0.1 \times 3.7 \times \frac{2}{2} = 1.37$$

$$d_q = d_\gamma = 1 + 0.1\sqrt{K_p} \frac{D_f}{B} = 1 + 0.1\sqrt{3.7}\,\frac{1}{2} = 1.09$$

**Step 2:** Substitute values from Step 1 into Meyerhof's equation.

(a) *Groundwater level at 5 m below the surface.* The groundwater level is 4 m below the base, which is greater than the width of the footing. Therefore, groundwater has no effect and $\gamma = \gamma_{sat}$.

$$
\begin{aligned}
q_{ult} &= \gamma D_f(N_q - 1)s_q\, d_q + 0.5\gamma B N_\gamma s_\gamma\, d_\gamma \\
&= 18 \times 1 \times 32.3 \times 1.37 \times 1.09 + 0.5 \times 18 \times 2 \times 37.2 \times 1.37 \times 1.09 \\
&= 1868 \text{ kPa}
\end{aligned}
$$

(b) *Groundwater level at the ground surface.* In this case, the groundwater level will affect the bearing capacity. You should use

$$\gamma' = \gamma_{sat} - \gamma_w = 18 - 9.8 = 8.2 \text{ kN/m}^3$$

$$
\begin{aligned}
q_{ult} &= \gamma' D_f(N_q - 1)s_q\, d_q + 0.5\gamma' B N_\gamma s_\gamma\, d_\gamma \\
&= 8.2 \times 1 \times 32.3 \times 1.37 \times 1.09 + 0.5 \times 8.2 \times 2 \times 37.2 \times 1.37 \times 1.09 \\
&= 851 \text{ kPa}
\end{aligned}
$$

Alternatively, since the change in the unit weight is the same for both terms of the bearing capacity equation, we can simply find $q_{ult}$ by taking the ratio $\gamma'/\gamma$, that is,

$$q_{ult} = 1868 \times \frac{8.2}{18} = 851 \text{ kPa}$$

(c) *Groundwater level at the base.* In this case, the groundwater level will affect the last term in the bearing capacity, where you should use

$$\gamma' = \gamma_{sat} - \gamma_w = 18 - 9.8 = 8.2 \text{ kN/m}^3$$

Thus,

$$
\begin{aligned}
q_{ult} &= \gamma D_f(N_q - 1)s_q\, d_q + 0.5\gamma' B N_\gamma s_\gamma\, d_\gamma \\
&= 18 \times 1 \times 32.3 \times 1.37 \times 1.09 + 0.5 \times 8.2 \times 2 \times 37.2 \times 1.37 \times 1.09 \\
&= 1323.7 \text{ kPa}
\end{aligned}
$$

(d) *Groundwater level at 1 m below the base.* In this case, the groundwater level is within a depth $B$ below the base and will affect the last term in the bearing capacity, where you should use

$$\gamma' B = \gamma_{sat} z + \gamma'(B - z) = 18 \times 1 + 8.2 \times (2 - 1) = 26.2 \text{ kN/m}^3$$

Thus,

$$
\begin{aligned}
q_{ult} &= \gamma D_f(N_q - 1)s_q\, d_q + 0.5(\gamma' B)N_\gamma s_\gamma\, d_\gamma \\
&= 18 \times 1 \times 32.3 \times 1.37 \times 1.09 + 0.5 \times (26.2) \times 37.2 \times 1.37 \times 1.09 \\
&= 1596.9 \text{ kPa}
\end{aligned}
$$

**Step 3:**   Compare results.
We will compare the results by dividing (normalizing) each ultimate net bearing capacity by the ultimate net bearing capacity of case (a).

| Groundwater level at | $\dfrac{q_{ult}}{(q_{ult})_a} \times 100$ |
|---|---|
| (b) Ground surface | $\dfrac{851}{1868} \times 100 \approx 46\%$ |
| (c) Base | $\dfrac{1323.7}{1868} \times 100 = 71\%$ |
| (d) 1 m below base | $\dfrac{1596.9}{1868} \times 100 = 85\%$ |

The groundwater level rising to the surface will reduce the bearing capacity by more than one-half.   ∎

## EXAMPLE 7.3

A footing 1.8 m × 2.5 m is located at a depth of 1.5 m below the ground surface in a deep deposit of an overconsolidated clay. The groundwater level is 2 m below the ground surface. The undrained shear strength is 120 kPa and $\gamma_{sat}$ = 20 kN/m³. Determine the allowable bearing capacity, assuming a factor of safety of 3.

**Strategy**   The only strategy here is to select a method. Skempton's equation is a very good choice for the short-term bearing capacity of clays. Since the groundwater level is below the width of the footing from the base, you do not need to consider the effect of groundwater.

## Solution 7.3

**Step 1:**   Select a bearing capacity equation and substitute the necessary values. Use Skempton's equation.

$$q_a = 5\,\frac{s_u}{FS}\left(1 + 0.2\,\frac{D_f}{B}\right)\left(1 + 0.2\,\frac{B}{L}\right) + \gamma D_f = 5 \times \frac{120}{3}$$
$$\times \left(1 + 0.2\,\frac{1.5}{1.8}\right)\left(1 + 0.2\,\frac{1.8}{2.5}\right) + 20 \times 1.5 = 297 \text{ kPa}$$

∎

## EXAMPLE 7.4

Determine the size of a rectangular footing to support a load of 1800 kN. The soil properties are $\phi'_p = 38°$, $\phi'_{cs} = 32°$, and $\gamma_{sat} = 18$ kN/m³. The footing is to be located at 1 m below the ground surface. Groundwater level is 6 m below the ground surface.

**Strategy**   You have to select a method. A good choice is Meyerhof's method. Neither the footing width nor the length is given. Both of these are required to find $q_a$. You can fix a length to width ratio and then assume a width ($B$). Solve for $q_a$ and if it is not satisfactory $[q_a \geqslant (\sigma_a)_{max}]$ then reiterate using a different $B$ value. We will use $\phi' = \phi'_{cs}$ and a factor of safety of 1.5.

## Solution 7.4

**Step 1:** Calculate bearing capacity numbers and shape and depth factors. Assume, $L/B = 1.5$; that is, $B/L = 0.67$, and $B = 1.5$ m.

Using $\phi' = \phi'_{cs} = 32°$,

$$N_q = e^{\pi \tan 32°} \tan^2(45° + 32°/2) = 23.2$$

$$N_q - 1 = 23.2 - 1 = 22.2$$

$$N_\gamma = (N_q - 1) \tan(1.4\phi') = 22.2 \tan(1.4 \times 32°) = 22$$

$$K_p = \tan^2(45° + \phi'/2) = \tan^2(45° + 32°/2) = 3.25$$

$$s_q = s_\gamma = 1 + 0.1 K_p \frac{B}{L} = 1 + 0.1 \times 3.25 \times 0.67 = 1.22$$

$$d_q = d_\gamma = 1 + 0.1\sqrt{K_p} \frac{D_f}{B} = 1 + 0.1\sqrt{3.25}\,\frac{1}{1.5} = 1.12$$

**Step 2:** Substitute the above values into Meyerhof's equation. The groundwater level is located more than $B$ below the base. Therefore, groundwater will not affect the bearing capacity.

$$q_{ult} = \gamma D_f (N_q - 1)s_q\, d_q + 0.5\gamma B N_\gamma s_\gamma\, d_\gamma$$
$$= 18 \times 1 \times 22.2 \times 1.22 \times 1.12 + 0.5 \times 18$$
$$\times 1.5 \times 22 \times 1.22 \times 1.12 = 952 \text{ kPa}$$

$$q_a = \frac{q_{ult}}{FS} + \gamma D_f = \frac{952}{1.5} + 18 \times 1 = 653 \text{ kPa}$$

$$(\sigma_a)_{max} = \frac{\text{Applied load}}{\text{Area}} = \frac{1800}{1.5 \times (1.5 \times 1.5)} = 533 \text{ kPa} < q_a \,(= 653 \text{ kPa})$$

The difference between $(\sigma_a)_{max}$ and $q_a$ is almost 18%. In practice this is an acceptable difference and the selected footing size would normally be used. However, we will try another footing size to illustrate the iterative nature of design.

**Step 3:** Try another width.
We need to try a smaller $B$. Try $B = 1.4$ m. The changes in the depth factors for this case are small and can be neglected or you can recalculate them as

$$d_q = d_\gamma = 1 + 0.1\sqrt{K_p}\frac{D_f}{B} = 1 + 0.1\sqrt{3.25}\,\frac{1}{1.4} = 1.13$$

The new allowable bearing capacity is

$$q_a = \frac{\gamma D_f (N_q - 1)s_q\, d_q + 0.5\gamma B N_\gamma s_\gamma\, d_q}{FS} + \gamma D_f$$

$$= \frac{18 \times 1 \times 22.2 \times 1.22 \times 1.13 + 0.5 \times 18 \times 1.4 \times 22 \times 1.22 \times 1.13}{1.5}$$

$$+ 18 \times 1 = 640 \text{ kPa}$$

$$(\sigma_a)_{max} = \frac{\text{Applied load}}{\text{Area}} = \frac{1800}{1.4 \times (1.5 \times 1.4)} = 612 \text{ kPa} < q_a.$$

The footing size is then 1.4 m × 2.1 m.  ∎

## EXAMPLE 7.5

Using the footing geometry of Example 7.1, determine $q_a$ with the load inclined at 20° to the vertical along the footing width, and $\phi' = \phi'_{cs}$. Assume the groundwater level is at 5m below ground surface and FS = 1.5.

*Strategy*   You need to use Meyerhof's equation for inclined loads. The bearing capacity numbers and shape factors remained unchanged. You need to calculate the depth factors and inclination factors.

### Solution 7.5

**Step 1:**   Calculate the inclination factors and depth factors.

$$i_q = (1 - 20°/90°)^2 = 0.6, \quad i_\gamma = (1 - 20°/35°)^2 = 0.18$$
$$K_p = \tan^2(45° + 30°/2) = 3$$
$$d_q = d_\gamma = 1 + 0.1\sqrt{K_p}\,\frac{D_f}{B} = 1 + 0.1\sqrt{3}\,\tfrac{1}{2} = 1.09$$

**Step 2:**   Calculate the ultimate net bearing capacity.

$$q_{ult} = \gamma D_f(N_q - 1)\,d_q i_q + 0.5\gamma B N_\gamma\,d_\gamma i_\gamma = 18 \times 1 \times 17.4 \times 1.09 \times 0.6$$
$$+ 0.5 \times 18 \times 2 \times 22.4 \times 1.09 \times 0.18 = 284 \text{ kPa}$$

$$q_a = \frac{q_{ult}}{FS} + \gamma D_f = \frac{284}{1.5} + 18 \times 1 = 207 \text{ kPa} \qquad ■$$

*What's next . . .*The bearing capacity equations have been derived for a footing resting on a layer of homogeneous soil and for loads applied at the center of the footing. In reality, soils are deposited in layers and loads are not always at the center of the footing. We will examine next how to modify the above equations to account for off-centered loads and then discuss layered soils.

## 7.9   ECCENTRIC LOADS

Theoretical solutions for eccentric (off-centered) loads are very complicated. Meyerhof (1963) proposed an approximate method to account for loads that are located off-centered. He proposed that, for a rectangular footing of width $B$ and length $L$, the base area should be modified with the following dimensions:

$$B' = B - 2e_B \quad \text{and} \quad L' = L - 2e_L \qquad (7.27)$$

where $B'$ and $L'$ are the modified width and length, and $e_B$ and $e_L$ are the eccentricities in the directions of the width and length, respectively (Fig. 7.8). From your course in mechanics, you should recall that

$$e_B = \frac{M_y}{P} \quad \text{and} \quad e_L = \frac{M_x}{P} \qquad (7.28)$$

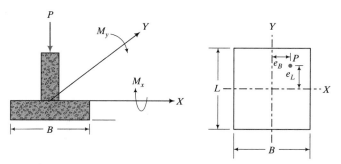

**FIGURE 7.8** Footing subjected to a vertical load and moments.

where $P$ is the vertical load and $M_y$ and $M_x$ are the moments about the $Y$ and $X$ axes, respectively, as shown in Fig. 7.8. The stresses due to a vertical load at an eccentricity $e$ (moment = $Pe$) are (from your knowledge of mechanics)

$$\sigma = \frac{P}{A} \pm \frac{My}{I} = \frac{P}{A} \pm \frac{Pey}{I} = \frac{P}{A} \pm \frac{Pe}{Z} \tag{7.29}$$

where $I$ is the second moment of the area, $y$ is the distance from the neutral axis to the outer edge, $A$ is the cross-sectional area, and $Z$ is the section modulus. For a rectangular section,

$$Z = \frac{I}{y} = \frac{B^3 L/12}{B/2} = \frac{B^2 L}{6} \quad \text{or} \quad \frac{BL^2}{6}$$

depending on whether you are considering a moment about the $Y$ axis or the $X$ axis.

The maximum and minimum vertical stresses along the $X$ axis are

$$\sigma_{\max} = \frac{P}{A} + \frac{Pe}{Z} = \frac{P}{BL}\left(1 + \frac{6e_B}{B}\right); \quad \sigma_{\min} = \frac{P}{A} - \frac{Pe}{Z} = \frac{P}{BL}\left(1 - \frac{6e_B}{B}\right) \tag{7.30}$$

and along the $Y$ axis are

$$\sigma_{\max} = \frac{P}{A} + \frac{Pe}{Z} = \frac{P}{BL}\left(1 + \frac{6e_L}{L}\right); \quad \sigma_{\min} = \frac{P}{A} - \frac{Pe}{Z} = \frac{P}{BL}\left(1 - \frac{6e_L}{L}\right) \tag{7.31}$$

Let us examine $\sigma_{\min}$. If $e_B = B/6$ or $e_L = L/6$, then $\sigma_{\min} = 0$. If, however, $e_B > B/6$ or $e_L > L/6$, then $\sigma_{\min} < 0$, and tension develops. Since the tensile strength of soil is approximately zero, part of the footing will not transmit loads to the soil. You should try to avoid this situation by designing the footing such that $e_B < B/6$ and $e_L < L/6$. The bearing capacity equations are modified for off-centered loads by replacing $B$ with $B'$. The ultimate load is

$$P_{\text{ult}} = q_{\text{ult}} B' L' \tag{7.32}$$

## EXAMPLE 7.6

Redo Example 7.2 using Meyerhof's method. The footing is subjected to a vertical load of 500 kN and a moment about the $Y$ axis of 125 kN · m. The groundwater level is 5 m below the ground surface. Calculate the factor of safety.

*Strategy*   Since we are only given the moment about the $Y$ axis, we only need to find the eccentricity, $e_B$. The bearing capacity factors are the same as those in Example 7.2.

### Solution 7.6

**Step 1:**   Draw a sketch of the problem and calculate $e_B$.
See Fig. E7.6 for a sketch.

$$e_B = \frac{125}{500} = 0.25 \text{ m}$$

**Step 2:**   Check if tension develops.

$$\frac{B}{6} = \frac{2}{6} = 0.33 \text{ m} > 0.25 \text{ m}$$

Therefore, tension will not occur.

**Step 3:**   Calculate the maximum vertical stress.

$$\sigma_{max} = \frac{P}{BL}\left(1 + \frac{6e_B}{B}\right) = \frac{500}{2 \times 2}\left(1 + \frac{6 \times 0.25}{2}\right) = 219 \text{ kPa}$$

**Step 4:**   Calculate reduced footing size.

$$B' = 2 - 2(0.25) = 1.5 \text{ m}$$

**Step 5:**   Calculate the shape and depth factors.

$$s_q = s_\gamma = 1 + 0.1K_p\frac{B'}{L} = 1 + 0.1 \times 3.7 \times \frac{1.5}{2} = 1.28$$

$$d_q = d_\gamma = 1 + 0.1\sqrt{K_p}\frac{D_f}{B'} = 1 + 0.1\sqrt{3.7}\,\frac{1}{1.5} = 1.13$$

**Step 6:**   Substitute the appropriate values into Meyerhof's equation.

$$q_{ult} = \gamma D_f(N_q - 1)s_q\,d_q + 0.5\gamma B'N_\gamma s_\gamma\,d_\gamma$$
$$= 18 \times 1 \times 32.3 \times 1.28 \times 1.13 + 0.5 \times 18 \times 1.5 \times 37.2 \times 1.28 \times 1.13$$
$$= 1567 \text{ kPa}$$

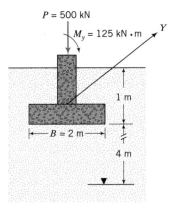

$P = 500$ kN

$M_y = 125$ kN $\cdot$ m

$Y$

1 m

$B = 2$ m

4 m

**FIGURE E7.6**

**Step 7:** Calculate the factor of safety.

$$\text{FS} = \frac{q_{\text{ult}}}{(\sigma_a)_{\text{max}} - \gamma D_f} = \frac{1567}{219 - 18 \times 1} = 7.8 \qquad \blacksquare$$

## 7.10 BEARING CAPACITY OF LAYERED SOILS

No simple, satisfactory, theoretical method is currently available to determine the bearing capacity of layered soils. A geotechnical engineer can apply a set of practical guidelines or use numerical tools such as the finite element method. The basic problem lies in determining and defining the soil properties for layered soils. We will resort to some practical guidelines for three common cases—a soft clay over a stiff clay, a stiff clay over a soft clay, and thinly stratified soils.

**Soft clay over stiff clay:** In general, shallow foundations on soft clays should be avoided except lightly loaded structures such as houses and one-story buildings. You should investigate removing the soft clay and replacing it with compacted fills. An inexperienced geotechnical engineer should calculate the bearing capacity using the methods described previously before making a decision to replace the soft clay.

**Stiff clay over soft clay:** The bearing capacity for this case is the smaller value of (1) treating the stiff clay as if the soft clay layer does not exist and (2) assuming that the footing punches through the stiff clay and is supported on the soft clay. The bearing capacity is the sum of the shear required to punch through a vertical plane in the stiff clay and the bearing capacity of the soft clay layer. Only a fraction, about $\frac{2}{3}$ to $\frac{1}{2}$, of the undrained shear strength should be used in computing the shear resistance on the vertical plane in the stiff clay layer.

Another method is to place an imaginary footing on the soft clay layer with dimensions $(B + t_{\text{sc}}) \times (L + t_{\text{sc}})$, where $B$ and $L$ are the real width and length of the footing and $t_{\text{sc}}$ is the thickness of the stiff clay layer, and then calculate the ultimate net bearing capacity using the bearing capacity equations.

**Thinly stratified soils:** In this type of deposit, deep foundations should be used. If deep foundations are uneconomical, then the bearing capacity can be calculated by using the shear strength parameters for the weakest layer. Alternatively, harmonic mean values for $s_u$ and $\phi'$ can be calculated (see Chapter 4) and then these values can be used to calculate the bearing capacity.

### EXAMPLE 7.7

The soil profile at a site is shown in Fig. E7.7. A square footing 2.5 m wide is located at 1.5 m below ground level in the stiff clay. Determine the safety factor for short-term loading for an applied load of 1000 kN.

*Strategy* Skempton's equation is an appropriate choice for clays, especially stiff clays. Check the factor of safety of the stiff clay assuming that the soft clay layer does not exist. We can then use an artificial footing on top of the soft clay and calculate the factor of safety.

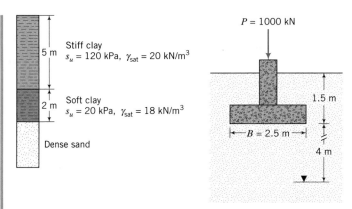

**FIGURE E7.7**

## Solution 7.7

**Step 1:** Calculate the factor of safety for the stiff clay.

$$q_{ult} = 5s_u\left(1 + 0.2\frac{D_f}{B}\right)\left(1 + 0.2\frac{B}{L}\right)$$

$$= 5 \times 120 \times \left(1 + 0.2\frac{1.5}{2.5}\right)\left(1 + 0.2\frac{2.5}{2.5}\right) = 806 \text{ kPa}$$

$$(\sigma_a)_{max} = \frac{P}{A} = \frac{1000}{2.5 \times 2.5} = 160 \text{ kPa}$$

$$FS = \frac{q_{ult}}{(\sigma_a)_{max} - \gamma D_f} = \frac{806}{160 - 20 \times 1.5} = 6.2$$

**Step 2:** Check the safety factor for the soft clay.

Artificial footing size:   $B + t_{sc} = 2.5 + 5 = 7.5$ m;   $L + t_{sc} = 2.5 + 5 = 7.5$ m

$$q_{ult} = 5s_u\left(1 + 0.2\frac{D_f}{B}\right)\left(1 + 0.2\frac{B}{L}\right) = 5 \times 20 \times \left(1 + 0.2\frac{5}{7.5}\right)\left(1 + 0.2\frac{7.5}{7.5}\right)$$

$$= 136 \text{ kPa}$$

$$(\Delta\sigma_a)_{max} = \frac{P}{A} = \frac{1000}{7.5 \times 7.5} = 17.8 \text{ kPa}$$

$$FS = \frac{136}{17.8} = 7.6$$

∎

*What's next* . . . The size of many shallow foundations is governed by settlement rather than bearing capacity considerations. That is, serviceability limit state governs the design rather than ultimate limit state. Next, we will consider how to determine settlement for shallow foundations.

## 7.11   SETTLEMENT

It is practically impossible to prevent settlement of shallow foundations. At least, elastic settlement will occur. Your task as a geotechnical engineer is to prevent

**TABLE 7.2    Serviceability Limit States**

| Serviceability limit state | Examples |
| --- | --- |
| Architectural damage (damage to appearance) | Tilting of structures (chimneys, retaining walls) and cracking in walls |
| Loss of serviceability | Cracked floors, misalignment of machinery, dislocation of pipe joints, jammed doors and windows |
| Structural damage (collapse) | Severe differential settlement of footings causing buckling of columns and overstressing beams |

the foundation system from reaching a serviceability limit state. A description of some serviceability limit states is given in Table 7.2.

Foundation settlement can be divided in three basic types: rigid block or uniform settlement (Fig. 7.9a), tilt or distortion (Fig. 7.9b) and non-uniform settlement (Fig. 7.9c). Most damage from uniform settlement is limited to surrounding drainage systems, attached buildings, and utilities. Distortion and non-uniform settlements are caused by differential movement and may cause serious problems especially in tall buildings. Distortion produces bending and is the cause of most cracking in structures. Distortion is quantified by $\delta/\ell$, where $\delta$ is the maximum differential settlement, and $\ell$ is the length over which the settlement occurs. This distortion is an angular measurement (radians). If we were able to accurately model soil behavior and be certain of the loading conditions and soil variations, then we would be able to calculate distortions that would cause damage to structures. Unfortunately, methods of calculating settlements have been found to be unreliable and we have to resort to experiments using models and collecting and analyzing data from field observations.

Skempton and MacDonald (1956) surveyed 98 buildings and found 40 to be damaged. They then correlated the occurrence or nonoccurrence of damage with the maximum distortion and recommended the limits shown in Table 7.3.

The data of Skempton and MacDonald (1956) do not apply to buildings that settled from the presence of deep seated clay layers (e.g., a clay layer overlaid by a thick deposit of sand). All buildings surveyed by Skempton and MacDonald (1956) had steel or reinforced concrete frames with load bearing walls.

It is desirable to get zero distortion but this is practically impossible because (1) the properties of building materials and the loading conditions are not known accurately, and (2) the variability of the soils at a site and the effects of construction methods are uncertain. Even if we do know items (1) and (2), the calculations

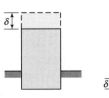

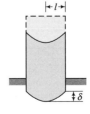

(a) Uniform settlement    (b) Tilt or distortion    (c) Nonuniform settlement

**FIGURE 7.9    Types of settlement.**

**TABLE 7.3    Distortion Limits**

| Type of damage | Distortion, $\dfrac{\delta}{\ell}$ |
| --- | --- |
| Architectural | $\dfrac{1}{300}$ |
| Structural | $\dfrac{1}{150}$ |

**TABLE 7.4   *R* Values and Maximum Allowable Settlement**[a]

|  |  | Foundation type | |
|---|---|---|---|
| Soil type | Values | Isolated footing | Rafts or mats |
| Clay | $R$ | 22,500 | 30,000 |
|  | $\rho_{max}$ (mm) | 75 | 100 |
| Sand | $R$ | 15,000 | 18,000 |
|  | $\rho_{max}$ (mm) | 50 | 60 |

[a]The original $R$ and $\rho_{max}$ values were for use in English units. The author converted these values for SI and adjusted them for ease of use in practice.

would be very complex. One practical solution is to correlate distortion to the maximum settlement. Skempton and MacDonald (1956) proposed that the maximum tolerable settlement could be calculated from

$$\rho_{max} = R\,\frac{\delta}{\ell} \tag{7.33}$$

where $R$ is a correlation factor that depends on the foundation and soil types. They also suggested limiting values of $\rho_{max}$. The values for $R$ and $\rho_{max}$ suggested by Skempton and MacDonald (1956) are listed in Table 7.4.

You should use the limits in Table 7.4 with caution. For important structures, a thorough site investigation should be conducted to uncover soil inhomogeneities that may result in intolerable differential settlement.

> *The **essential points** are:*
> 1. *Distortion caused by differential settlement is crucial in design because it is responsible for cracking and damage to structures.*
> 2. *The distortion limits were observed for old structures and may not be applicable to modern structures.*
> 3. *To estimate the total maximum allowable settlement, multiply the appropriate **R** values in Table 7.4 by the distortion value.*

*What's next . . .* In the next section, we are going to discuss methods to calculate settlement of foundations.

## 7.12   SETTLEMENT CALCULATIONS

The settlement of shallow foundations is divided into three segments—immediate or elastic settlement, primary consolidation settlement, and secondary consolidation settlement (creep). We have already considered elastic settlement (Chapter 3) and consolidation settlement (Chapter 4). However, we have to make some modifications to the methods described in those chapters for calculating settlement of shallow foundations. These modifications are made to the method of calculating elastic and primary consolidation settlements.

### 7.12.1 Immediate Settlement

We can use the theory of elasticity to determine the immediate or elastic settlement of shallow foundations. In the case of a uniform rectangular flexible load, we can use Eq. (3.84). However, the elastic equations do not account for the shape of the footing (not just $L/B$ ratio) and the depth of embedment, which significantly influence settlement. An embedded foundation has the following effects in comparison with a surface footing:

1. Soil stiffness generally increases with depth, so the footing loads will be transmitted to a stiffer soil than the surface soil. This can result in smaller settlements.

2. Normal stresses from the soil above the footing level have been shown (Eden, 1974; Gazetas and Stoke, 1991) to reduce the settlement by providing increased confinement on the deforming halfspace. This is called the trench effect or embedment effect.

3. Part of the load on the footing may also be transmitted through the side walls depending on the amount of shear resistance mobilized at the soil–wall interface. The accommodation of part of the load by side resistance reduces the vertical settlement. This has been called the side wall-soil contact effect.

Gazetas et al. (1985) considered an arbitrarily shaped rigid footing embedded in a deep homogeneous soil (Fig. 7.10) and proposed the following equation for the elastic settlement:

$$\rho_e = \frac{P}{E_u L} (1 - v_u^2)\mu_s\mu_{\mathrm{emb}}\mu_{\mathrm{wall}} \tag{7.34}$$

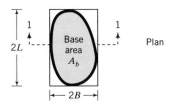

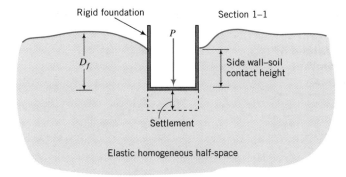

**FIGURE 7.10** Geometry to calculate elastic settlement of shallow footings. (On behalf of the Institution of Civil Engineers.)

where $P$ is total vertical load, $E_u$ is the undrained elastic modulus of the soil, $L$ is one-half the length of a circumscribed rectangle, $v_u$ is Poisson's ratio for the undrained condition, and $\mu_s$, $\mu_{emb}$, and $\mu_{wall}$ are shape, embedment (trench), and side wall factors given as

$$\mu_s = 0.45\left(\frac{A_b}{4L^2}\right)^{-0.38} \tag{7.35}$$

$$\mu_{emb} = 1 - 0.04\,\frac{D_f}{B}\left[1 + \frac{4}{3}\left(\frac{A_b}{4L^2}\right)\right] \tag{7.36}$$

$$\mu_{wall} = 1 - 0.16\left(\frac{A_w}{A_b}\right)^{0.54} \tag{7.37}$$

$A_b$ is the actual area of the base of the foundation and $A_w$ is the actual area of the wall in contact with the embedded portion of the footing. The length and width of the circumscribed rectangle are $2L$ and $2B$, respectively. The dimensionless shape parameter, $A_b/4L^2$, has the values for common footing geometry shown in Table 7.5.

The equations proposed by Gazetas et al. (1985) apply to a foundation of arbitrary shape on a deep homogeneous soil. There is no clear definition of what signifies "deep." The author suggests that the equations of Gazetas et al. (1985) can be used when the thickness of the soil layer is such that 90% of the applied stresses are distributed within it. For a rectangular area of actual width $B_r$, the thickness of the soil layer should be at least $3B_r$.

Equation (3.84) can be modified to account for embedment as

$$\rho_e = \frac{q_s B_r(1 - v_u^2)}{E_u}\,I_s\mu'_{emb} \tag{7.38}$$

where

$$\mu'_{emb} = 1 - 0.08\,\frac{D_f}{B_r}\left(1 + \frac{4}{3}\frac{B_r}{L_r}\right) \tag{7.39}$$

where $B_r$ and $L_r$ are the actual width and length, respectively.

The accuracy of any elastic equation for soils depends particularly on the accuracy of the elastic modulus. It is common laboratory practice to determine a secant $E_u$ from undrained triaxial tests or unconfined compression tests at a

**TABLE 7.5   Values of $A_b/4L^2$ for Common Footing Shapes**

| Footing shape | $\dfrac{A_b}{4L^2}$ |
| --- | --- |
| Square | 1 |
| Rectangle | $B/L$ |
| Circle | 0.785 |
| Strip | 0 |

deviatoric stress equal to one-half the maximum shear strength. However, for immediate settlement it is better to determine $E_u$ over the range of deviatoric stress pertaining to the problem. In addition, the elastic modulus is strongly dependent on depth while Eqs. (7.34) and (7.38) are cast in terms of a single value of $E_u$. One possible solution is to divide the soil into sublayers and use a weighted harmonic mean value of $E_u$ (Chapter 4).

The full wall resistance will only be mobilized if sufficient settlement occurs. It is difficult to ascertain the quality of the soil–wall adhesion. Consequently, you should be cautious in relying on the reduction of settlement resulting from the wall factor. If wall friction and embedment are neglected, then $\mu_{wall} = 1$ and $\mu_{emb} = 1$.

Equations (7.34) and (7.38) strictly apply to fine-grained soils under short-term loading. For long-term loading in fine-grained soils and for coarse-grained soils, you should use $E'$ and $\nu'$ instead of $E_u$ and $\nu_u$.

## EXAMPLE 7.8

Determine the immediate settlement of a rectangular footing 4 m $\times$ 6 m embedded in a deep deposit of homogeneous clay as shown in Fig. E7.8.

*Strategy*  You have sufficient information to directly apply Eq. (7.34) or Eq. (7.38). The side wall effect should not be considered, that is, $\mu_{wall} = 1$, since there really is no wall.

## Solution 7.8

**Step 1:**  Determine geometric parameters.

$$A_b = 4 \times 6 = 24 \text{ m}^2, \quad L = \tfrac{6}{2} = 3 \text{ m}, \quad B = \tfrac{4}{2} = 2 \text{ m}$$

$$\frac{A_b}{4L^2} = 0.67 \quad \left(\text{Proof:} \frac{A_b}{4L^2} = \frac{2B \times 2L}{4 \times L^2} = \frac{B}{L} = \frac{4}{6} = 0.67\right)$$

**Step 2:**  Calculate the shape and embedment factors.

$$\mu_s = 0.45\left(\frac{A_b}{4L^2}\right)^{-0.38} = 0.45(0.67)^{-0.38} = 0.52$$

$$\mu_{emb} = 1 - 0.04\frac{D_f}{B}\left[1 + \frac{4}{3}\left(\frac{A_b}{4L^2}\right)\right] = 1 - 0.04 \times \frac{3}{4}\left[1 + \frac{4}{3}(0.67)\right] = 0.94$$

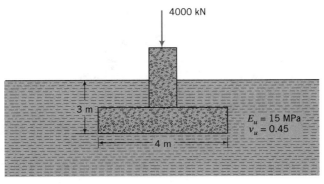

**FIGURE E7.8**

**Step 3:**    Calculate the immediate settlement.

$$\rho_e = \frac{P}{E_u L}\,(1 - v_u^2)\mu_s\mu_{emb}\mu_{wall} = \frac{4000}{15,000 \times 6}\,(1 - 0.45^2)$$
$$\times\ 0.52 \times 0.94 \times 1 = 0.018 \text{ m} = 18 \text{ mm} \qquad\blacksquare$$

## EXAMPLE 7.9

Determine the immediate settlement of the foundation shown in Fig. E7.9. The undrained elastic modulus varies with depth as shown in the figure and $v_u = 0.45$.

**_Strategy_**    You have to determine the length ($2L$) and width ($2B$) of a circumscribed rectangle. The undrained elastic modulus varies with depth, so you need to consider the average value of $E_u$ for each of the layers and then find the harmonic mean. You also need to find the shape parameter $A_b/4L^2$.

## Solution 7.9

**Step 1:**    Determine the length and width of the circumscribed rectangle.

$$2L = 8 + 4 = 12 \text{ m}; \quad L = 6 \text{ m}$$
$$2B = 3 + 3 + 4 = 10 \text{ m}; \quad B = 5 \text{ m}$$

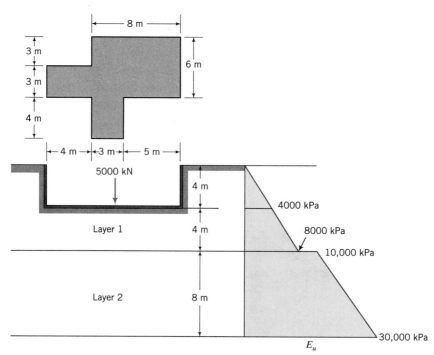

**FIGURE E7.9**

**Step 2:** Determine $E_u$.

**Layer 1**

$E_u$ at base level $= \dfrac{4}{8} \times 8000 = 4000$ kPa; $\quad E_u$ at bottom of layer $= 8000$ kPa

$$(E_u)_{\text{avg}} = \frac{4000 + 8000}{2} = 6000 \text{ kPa}$$

**Layer 2**

$E_u$ at top of layer $= 10{,}000$ kPa; $\quad E_u$ at bottom of layer $= 30{,}000$ kPa

$$(E_u)_{\text{avg}} = \frac{10{,}000 + 30{,}000}{2} = 20{,}000 \text{ kPa}$$

**Step 3:** Find the weighted harmonic mean $E_u$.

$$E_u = \frac{2(6000) + 1(20{,}000)}{3} = 10{,}667 \text{ kPa}$$

**Step 4:** Find the shape parameter $A_b/4L^2$.

$$A_b = (3 \times 4) + (3 \times 10) + (6 \times 5) = 72 \text{ m}^2; \quad \frac{A_b}{4L^2} = \frac{72}{4 \times 6^2} = 0.5$$

**Step 5:** Find the shape, embedment, and wall factors.

$$\mu_s = 0.4(0.5)^{-0.38} = 0.52; \quad \mu_{\text{emb}} = 1 - 0.4 \frac{4}{5} \left(1 + \frac{4}{3} \times 0.5\right) = 0.47$$

$$A_w = \text{Perimeter} \times \text{depth} = (3 + 4 + 5 + 6 + 8 + 3 + 4 + 3 + 4 + 4) \times 4$$
$$= 176 \text{ m}^2$$

$$\frac{A_w}{A_b} = \frac{176}{72} = 2.44; \quad \mu_{\text{wall}} = 1 - 0.16(2.44)^{0.54} = 0.74$$

**Step 6:** Calculate the immediate settlement.

$$\rho_e = \frac{P}{E_u L} (1 - v_u^2) \mu_s \mu_{\text{emb}} \mu_{\text{wall}}$$

$$= \frac{5000}{10667 \times 6} (1 - 0.45^2) \times 0.52 \times 0.47 \times 0.74 = 0.011 \text{ m} = 11 \text{ mm}$$

∎

## 7.12.2 Primary Consolidation Settlement

The method described in Chapter 4 can be used to calculate the primary consolidation settlement of clays below the footing. However, these equations were obtained for one-dimensional consolidation where the lateral strain is zero. In practice, lateral strains are significant except for very thin layers of clays or for situations when the ratio of the layer thickness to the lateral dimension of the loaded area is small (approaches zero). We also assumed that the initial excess pore water pressure is equal to the change in applied stress at the instant the load is applied. Theoretically, this is possible if the lateral stresses are equal to the vertical stresses. If the lateral strains are zero, then under undrained condition (at the instant the load is applied), the vertical settlement is zero.

Skempton and Bjerrum (1957) proposed a method to modify the one-dimensional consolidation equation to account for lateral stresses but not lateral strains. They proposed the following equation:

$$(\rho_{pc})_{SB} = \int_0^{H_o} m_v \, \Delta u \, dz \tag{7.40}$$

where $\Delta u$ is the excess pore water pressure and $H_o$ is the thickness of the soil layer. Skempton and Bjerrum (1957) suggested that the error in neglecting the lateral strains could lead to an error of up to 20% in the estimation of the consolidation settlement. Skempton's equation (Eq. 5.44) for the excess pore water pressure in a saturated soil under axisymmetric loading can be algebraically manipulated to yield

$$\Delta u = \Delta \sigma_1 \left( A + \frac{\Delta \sigma_3}{\Delta \sigma_1} (1 - A) \right) \tag{7.41}$$

By substituting Eq. (7.41) into Eq. (7.40) we get

$$(\rho_{pc})_{SB} = \int_0^{H_o} m_v \, \Delta \sigma_1 \left( A + \frac{\Delta \sigma_3}{\Delta \sigma_1} (1 - A) \right) dz = \sum (m_v \, \Delta \sigma_z \, H_o) \mu_{SB} = \sum \rho_{pc} \mu_{SB} \tag{7.42}$$

where $\rho_{pc}$ is the one-dimensional primary consolidation settlement (Chapter 4), $\mu_{SB} = A + \alpha_{SB}(1 - A)$ is a settlement coefficient to account for the effects of the lateral stresses, and $\alpha_{SB} = (\int \Delta \sigma_3 \, dz)/(\int \Delta \sigma_1 \, dz)$. The values of $\mu_{SB}$ typically vary from 0.6 to 1.0 for soft clays and from 0.3 to 0.8 for overconsolidated clays. Values of $\mu_{SB}$ for circular and strip footings are shown in Fig. 7.11. For square or rectangular footings use an equivalent circular footing of diameter $D =$

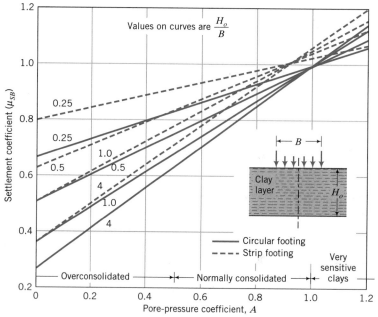

**FIGURE 7.11**   Values of $\mu_{SB}$ for circular and strip footings. (Redrawn from Scott, 1963.)

$\sqrt{A/\pi}$, where $A$ is area of the rectangle or the square. Equation (7.42) must be used appropriately. It is obtained from triaxial conditions and only applies to situations where axial symmetry occurs, such as under the center of a circular footing.

## EXAMPLE 7.10

Determine the primary consolidation settlement under the square footing shown in Fig. E7.10 using the Skempton–Bjerrum method.

*Strategy*  First you have to calculate the one-dimensional primary consolidation settlement and then determine $\mu_{SB}$ from the chart. Since the clay layer is finite, you should calculate the vertical stress increase using the coefficients in Appendix B.

## Solution 7.10

**Step 1:**  Calculate applied stress increase at the center of each layer below the base of the footing.

$$\frac{L}{B} = 1, \frac{H_o}{B} = \frac{H_1}{B} = 2, \frac{z}{B} = 0.5$$

Interpolation from Table B1.2 gives $I_{zp} = 0.72$

$$\Delta\sigma_z = I_{zp}q_s = 0.72 \times 100 = 72 \text{ kPa}$$

**Step 2:**  Determine $\mu_{SB}$.

Area of base $= 2 \times 2 = 4 \text{ m}^2$

Equivalent diameter of circular footing:  $D = \sqrt{\dfrac{4}{\pi}} = 1.13 \text{ m}$

$$\frac{H_o}{B} = \frac{4}{1.13} = 3.55, \quad A = 0.2; \quad \text{from Fig. 7.11, } \mu_{SB} = 0.41$$

**Step 3:**  Calculate the primary consolidation settlement.

$$(\rho_{pc})_{SB} = m_v \,\Delta\sigma_z \,H_o\mu_{SB} = 0.0001 \times 72 \times 4 \times 0.41 = 0.012 \text{ m} = 12 \text{ mm}$$

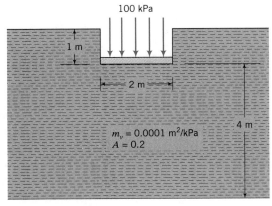

**FIGURE E7.10**

*What's next* . . .Often, the recovery of soils, especially coarse-grained soils, for laboratory testing is difficult and one has to use results from field tests to determine the bearing capacity and settlement of shallow foundations. Some of the field methods used for coarse-grained soils are presented in the next section.

# 7.13 DETERMINATION OF BEARING CAPACITY AND SETTLEMENT OF COARSE-GRAINED SOILS FROM FIELD TESTS

### 7.13.1 Bearing Capacity and Settlement from the Standard Penetration Test (SPT)

It is difficult to obtain undisturbed samples of coarse-grained soils for testing in the laboratory. Consequently, the allowable bearing capacity and settlement of footings on coarse-grained soils are often based on empirical methods using test data from field tests. One popular method utilizes results from the standard penetration test (SPT). It is customary to correct the $N$ values for various effects such as overburden pressure and energy transfer. Various correction factors have been suggested by a number of investigators. Only two such suggestions for correcting $N$ values for overburden pressure are included in this text. These are

$$c_N = \left(\frac{95.8}{\sigma'_{zo}}\right)^{1/2}; \quad c_N \le 2 \quad \text{(Liao and Whitman, 1985)} \qquad (7.43)$$

$$c_N = 0.77 \log_{10}\left(\frac{1916}{\sigma'_{zo}}\right); \quad c_N \le 2, \quad \sigma'_{zo} > 24 \text{ kPa} \quad \text{(Peck et al., 1974)} \qquad (7.44)$$

where $c_N$ is a correction factor for overburden pressures, and $\sigma'_{zo}$ is the effective overburden pressure in kPa. A further correction factor is imposed on $N$ values if the groundwater level is within a depth $B$ below the base of the footing. The groundwater correction factor is

$$c_W = \frac{1}{2} + \frac{z}{2(D_f + B)} \qquad (7.45)$$

where $z$ is the depth to the groundwater table, $D_f$ is the footing depth, and $B$ is the footing width. If the depth of the groundwater level is beyond $B$ from the footing base, $c_W = 1$.

The corrected $N$ value is

$$N_{cor} = c_N c_W N \qquad (7.46)$$

The allowable bearing capacity for a shallow footing $(q_a)$ is

$$q_a = 0.41 N_{cor} \rho_a \text{ kPa} \qquad (7.47)$$

where $\rho_a$ is the allowable settlement in mm. In practice, each value of $N$ in a soil layer up to a depth $B$ below the footing base is corrected and an average value of $N_{cor}$ is used in Eq. (7.47).

Meyerhof (1965) proposed that no correction should be applied to $N$ values for the effects of groundwater as these are already incorporated in the measurement. Furthermore, he suggested that $q_a$ calculated from Eq. (7.47) using $N_{cor} = c_N N$ be increased by 50%.

Burland and Burbridge (1985) did a statistical analysis of settlement records from 200 footings located in sand and in gravel. They proposed the following equation for the settlement of a footing in a normally consolidated sand at the end of construction:

$$\rho = f_s f_1 \sigma_a B^{0.7} I_c \tag{7.48}$$

where $\rho$ is the settlement (mm),

$$f_s = \text{Shape factor} = \left( \frac{1.25 L/B}{L/B + 0.25} \right)^2$$

$f_1 = (H_o/z_1)(z - H_o/z_1)$ is a correction factor if the thickness ($H_o$) of the sand stratum below the footing base is less than an influence depth $z_1$, $z$ is depth (m) from the ground surface, $\sigma_a$ is the vertical stress applied by the footing or allowable bearing capacity (kPa), $B$ and $L$ are the width and length of the footing (m), respectively,

$$I_c = \text{Compressibility index} = \frac{1.71}{N^{1.4}} \tag{7.49}$$

and $N$ is the uncorrected $N$ value. The influence depth is the depth below the footing that will influence the settlement and bearing capacity. If $N$ increases with depth or $N$ is approximately constant, the influence depth is taken as $z_1 = B^{0.763}$. If $N$ tends to decrease with depth, the influence depth is $z_1 = 2B$.

If the sand is overconsolidated,

$$\rho = f_1 f_s (\sigma_a - \tfrac{2}{3}\sigma'_{zc}) B^{0.7} I_c, \quad \text{if } \sigma_a > \sigma'_{zc} \tag{7.50}$$

$$\rho = q B^{0.7} \frac{I_c}{3}, \quad \text{if } \sigma_a < \sigma'_{zc} \tag{7.51}$$

The procedure for the Burland–Burbridge method is as follows:

1. Determine the influence depth $z_1$.
2. Find the average $N$ value within the depth $z_1$ below the footing.
3. Calculate $I_c$ from Eq. (7.49).
4. Determine $\rho$ from the appropriate equation [Eq. (7.49) or (7.50) or (7.51)] or, if $\rho$ is specified, you can determine $\sigma_a$.

### EXAMPLE 7.11

The SPT results at various depths in a soil are shown in Table E7.11a.

**TABLE E7.11a**

| Depth (m) | 0.6 | 0.9 | 1.2 | 1.5 | 2.1 | 2.7 | 3 | 3.3 | 4.2 |
|---|---|---|---|---|---|---|---|---|---|
| N (blows/ft) | 25 | 28 | 33 | 29 | 28 | 29 | 31 | 35 | 41 |

Determine the allowable bearing capacity for a footing located at 0.6 m below the surface. The tolerable settlement is 25 mm. The groundwater level is deep and its effects can be neglected.

*Strategy* The question that arises is what value of $N$ to use. If the size of the footing were known, then you could estimate the thickness of the soil below the footing that will be stressed significantly (>10% of applied stress) and take an average value of $N$ within that layer. But this is not the case here. We can make a judgment as to what depth would be effective. Here, one can say 3 m or 5 m or another reasonable value. The unit weight is not given so we have to estimate this based on the description and the $N$ values.

### Solution 7.11

**Step 1:** Determine $N_{cor}$.

Assume an average bulk unit weight of 16 kN/m$^3$; calculate $\sigma'_{zo}$ and the correction factor $c_N$ using either Eq. (7.43) or (7.44). Use a spreadsheet to do the calculation as shown in Table E7.11b.

**TABLE E7.11b**

| Depth (m) | Vertical effective stress (kPa) | $c_N$ Eq. (7.43) | | N | $N_{cor}$ |
|---|---|---|---|---|---|
| 0.6 | 9.6 | 3.2 | Use 2 | 25 | 50 |
| 0.9 | 14.4 | 2.6 | Use 2 | 28 | 56 |
| 1.2 | 19.2 | 2.2 | Use 2 | 33 | 66 |
| 1.5 | 24 | 2.0 | | 29 | 58 |
| 2.1 | 33.6 | 1.7 | | 28 | 47 |
| 2.7 | 43.2 | 1.5 | | 29 | 43 |
| 3 | 48 | 1.4 | | 31 | 44 |
| 3.3 | 52.8 | 1.3 | | 35 | 47 |
| 4.2 | 67.2 | 1.2 | | 41 | 49 |

**Step 2:** Calculate $q_a$.

$$q_a = 0.41 N_{cor}\rho_a = 0.41 \times 51 \times 25 = 523 \text{ kPa} \qquad \blacksquare$$

### EXAMPLE 7.12

Redo Example 7.11 using the Burland–Burbridge method for a footing 3 m × 4 m.

***Strategy*** You have to determine whether the sand is normally consolidated or overconsolidated. No direct evidence is provided to allow you to make a decision as to the consolidation state of the sand. One way around this problem is to use Table 5.3 to make an estimate of the consolidation state.

### Solution 7.12

**Step 1:** Determine the consolidation state and find $z_1$.

Within a depth equal to $B$ (3 m), the average $N$ value is 29. From Table 5.3, the sand can be classified as medium ($N$ in the range 10–30). A reasonable estimate of the consolidation state is normally consolidated.

$$z_1 = B^{0.763} = 3^{0.763} = 2.3 \text{ m}$$

**Step 2:** Find an average $N$ for a depth 2.3 m below the base.
Average $N$ value over a depth of 2.3 m below the base is 29. (*Note:* 2.3 m below the base is equivalent to a depth of 2.9 m, so use the $N$ values up to 3 m.)

**Step 3:** Calculate $I_c$.

$$I_c = \frac{1.71}{N^{1.4}} = \frac{1.71}{29^{1.4}} = 0.015$$

**Step 4:** Calculate $q_a$.

$$\frac{L}{B} = \frac{4}{3} = 1.33; \quad f_s = \left(\frac{1.25L/B}{L/B + 0.25}\right)^2 = \left(\frac{1.25 \times 1.33}{1.33 + 0.25}\right)^2 = 1.11$$

$$f_1 = 1 \quad \text{(thickness of sand stratum greater than 2.3 m)}$$

$$\sigma_a = \frac{\rho}{f_s f_1 B^{0.7} I_c} = \frac{25}{1.11 \times 1 \times 3^{0.7} \times 0.015} = 696 \text{ kPa} \qquad \blacksquare$$

## 7.13.2 Settlement from Cone Penetration Test

Schmertmann (1970) and Schmertmann et al. (1978) proposed a methodology to determine settlement from the quasi-static cone test data for sands. They assumed that the sand is a linearly elastic material and only stress changes within depths of $2B$ for axisymmetric conditions and $4B$ for plane strain conditions influence the settlement. Settlement is calculated by integrating the vertical strains; that is,

$$\rho = \int \varepsilon_z \, dz$$

The equation proposed for settlement (mm) by Schmertmann et al. (1978) is

$$\rho = \frac{c_D c_t}{\beta} q_{\text{net}} \sum_{i=1}^{n} \frac{(I_{co})_i}{(q_c)_i} \Delta z_i \tag{7.52}$$

where

$$c_D = \text{Depth factor} = 1 - 0.5 \frac{\sigma_z'}{q_{\text{net}}} \geq 0.5$$

$$c_t = \text{Creep factor} = 1.0 + A \log_{10} \left|\frac{t}{0.1}\right|,$$

$\beta$ is cone factor [$\beta = 2.5$ for square footing (axisymmetric condition), $\beta = 3.5$ for strip footing (plane strain condition)], $q_{\text{net}}$ is the net footing pressure in kPa (applied stress minus soil pressure above the base of footing), $\sigma_z'$ is the vertical effective stress in kPa, $t$ is time in years ($t \geq 0.1$), $A$ is an empirical factor taken as 0.2, $\Delta z_i$ is the thickness of the $i$th layer, $(I_{co})_i$ is the influence factor of the $i$th layer given as:

**Axisymmetric**

$$I_{co} = 0.1 + 0.8 \frac{z}{B} \quad \text{for } \frac{z}{B} \leq \frac{1}{2} \tag{7.53}$$

$$I_{co} = \frac{2}{3} - \frac{z}{3B} \quad \text{for } 2 \geq \frac{z}{B} > \frac{1}{2} \tag{7.54}$$

**Plane strain**

$$I_{co} = 0.2 + 0.3 \frac{z}{B} \quad \text{for } \frac{z}{B} \leq 1 \qquad (7.55)$$

$$I_{co} = \frac{2}{3} - \frac{z}{6B} \quad \text{for } 4 \geq \frac{z}{B} > 1 \qquad (7.56)$$

$(q_c)_i$ is the cone tip resistance for the *i*th layer, and $n$ is the number of sublayers. The units of $B$ is meter.

The procedure to determine the settlement from cone data is as follows:

1. Divide the soil below the footing into a number of sublayers. For square footings, the total depth of the sublayers is $2B$ and a reasonable number of sublayers is four. For strip footing, the total depth is $4B$ and a reasonable number of sublayers is eight.
2. Determine the average value of $(q_c)_i$ for each sublayer from the field data of $q_c$ versus depth.
3. Find $I_{co}$ at the center of each sublayer.
4. Estimate $\rho$ using Eq. (7.52)

## EXAMPLE 7.13

A representative set of cone data at a site is shown in Fig. E7.13. A square footing 3 m wide imposing a net applied stress of 200 kPa is to be located 1 m below ground level at this site. Determine the settlement of the footing one year after construction. The bulk unit weight of the sand is 17 kN/m³. Groundwater level is 8 m below the ground surface.

*Strategy*  For a square footing, the influence depth is $2B$. You need to divide this depth into soil layers and then use Eq. (7.52).

## Solution 7.13

**Step 1:**  Divide the influence depth into layers.

$$\text{Influence depth} = 2B = 6 \text{ m}$$

$$\text{Thickness of each of 4 sublayers:} \quad \Delta z = \frac{6}{4} = 1.5 \text{ m}$$

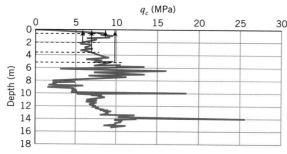

**FIGURE E7.13**

**Step 2:** For each sublayer, find the average value of $q_c$, $I_{co}$, and $\rho$. Use a spreadsheet program.

| Layer | $z^a$ | $\dfrac{z}{B}$ | $I_{co}{}^a$ | $q_c$ | $\dfrac{I_{co}}{q_c}\Delta z$ |
|---|---|---|---|---|---|
| 1 | 0.75 | 0.25 | 0.3 | 5.8 | 0.08 |
| 2 | 2.25 | 0.75 | 0.42 | 7 | 0.09 |
| 3 | 3.75 | 1.25 | 0.25 | 8.5 | 0.04 |
| 4 | 5.25 | 1.75 | 0.09 | 9.7 | 0.01 |
| | | | | | $\Sigma 0.22$ |

$^a$At center of layer.

$$q_{net} = 200 \text{ kPa}$$

$$\sigma'_z = 1 \times 17 = 17 \text{ kPa}; \quad c_D = 1 - 0.5 \times \frac{17}{200} = 0.96;$$

$$c_t = 1 + 0.2 \log_{10}\left|\frac{1}{0.1}\right| = 1.2$$

**Step 3:** Calculate the settlement.

$$\rho = \frac{c_D c_t}{\beta} q_{net} \sum_{i=1}^{n} \frac{(I_{co})_i}{(q_c)_i} \Delta z = \frac{0.96 \times 1.2}{2.5} \times 200 \times 0.22 = 20.3 \approx 20 \text{ mm} \quad \blacksquare$$

### 7.13.3 Settlement from Plate Load Tests

Tests on full sized footings are desirable but expensive. The alternative is to carry out plate load tests (Fig. 7.12) to simulate the load–settlement behavior of a real footing. The plates are made from steel with sizes varying from 150 to 760 mm. Two common plate sizes are used in practice. One is a square plate of width 300 mm and the other is a circular plate of diameter 300 mm. The test is carried out in a pit of depth at least 1.5 m. Loads are applied in increments of 10% to 20% of the estimated allowable bearing capacity. Each load increment is held until settlement ceases. The final settlement at the end of each loading increment is recorded. Loading is continued until the soil fails or settlements are in excess of 10% of the plate diameter. The maximum load should be at least 1.5 times the estimated allowable bearing capacity.

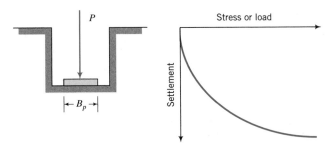

**FIGURE 7.12**  Plate load test.

If the sand were to behave like an elastic material, then the settlement can be calculated from

$$\rho_p = q_{ap} B_p \frac{1 - (\nu')^2}{E'} I_p \tag{7.57}$$

where $\rho_p$ is the plate settlement, $\sigma_{ap}$ is the applied stress, $B_p$ is the width or diameter of the plate, $\nu'$ is Poisson's ratio, $E'$ is the elastic modulus, and $I_p$ is an influence factor (0.82 for a rigid plate). The settlement of the real footing ($\rho$) of width $B$ is related to the plate settlement by

$$\rho = \rho_p \left( \frac{2}{1 + B_p/B} \right)^2 \tag{7.58}$$

In the limit $B_p/B \rightarrow \infty$, $\rho/\rho_p \rightarrow 4$.

Equation (7.57) is only valid if the strains are small (infinitesimal). There are several problems associated with the plate load test.

1. The test is reliable only if the sand layer is thick and homogeneous.

2. The depth of sand that is stressed below the plate is significantly lower than the real footing. A weak soil layer below the plate may not affect the test results because it may be at a depth at which the stresses imposed on the weak layer by the plate loads may be insignificant. However, this weak layer can have significant effect on the bearing capacity and settlement of the real footing.

3. Local conditions such as a pocket of weak soil near the surface of the plate can affect the test results but these may have no significant effect on the real footing.

4. The correlation between plate load test results and the real footing is problematic. Settlement in a sand depends on the size of the plate. Settlement increases with increases in the plate size. Bjerrum and Eggestad (1963) found that there is significant scatter in the relationship between plate size and settlement for a given applied stress. Bjerrum and Eggestad (1963) also reported that field evidence indicates that the limit of $\rho/\rho_p$ ranges between 3 and 5 rather than a fixed value of 4.

5. Performance of the test is difficult. On excavating sand to make a pit, the soil below the plate invariably becomes looser and this has considerable influence on the test results. Good contact must be achieved between the plate and the sand surface but this is often difficult. If the plate were just above the groundwater, your results would be affected by negative pore water pressure.

*What's next* . . .We have considered only vertical settlement under vertical and inclined loads. Horizontal displacements and rotations are important for structures subjected to significant horizontal loads. Next we will consider horizontal displacements and rotations.

## 7.14  HORIZONTAL ELASTIC DISPLACEMENT AND ROTATION

Structures such as radar towers and communication transmission towers are subjected to significant horizontal loads from wind, which can lead to intolerable lateral displacement and rotation of their foundations. Gazetas and Hatzikonstantinou (1988) proposed equations based on an isotopic linearly elastic soil to determine the elastic horizontal displacement and rotation of an arbitrarily loaded foundation. The equations were obtained by curve fitting theoretical elastic solutions for $\dfrac{D_f}{B} \leq 2$. A summary of these equations are presented in Table 7.6 You must be cautious (see Section 7.12.1) in using the equations in Table 7.6.

**TABLE 7.6   Equations for Estimating the Horizontal Displacement and Rotation of Arbitrarily Shaped Foundations (Source: Gazetas and Hatzikonstantinou, 1988).**

| Loading | Surface foundation | | Foundation placed in open trench |
|---|---|---|---|
| | Lateral direction $y$     Longitudinal direction $x$ <br> $L > B$, $I_y > I_x$ | | **Both directions** |
| Horizontal | $\rho_e = \dfrac{H_i}{EL}(2 - \nu)(1 + \nu)\mu_s$ | | $(\rho_e)_{emb} \approx \rho_e \mu_{emb}$ |
| | $\mu_{sy} \approx 0.5 - 0.28\left[\dfrac{A_b}{4L^2}\right]^{0.45}$ | $\mu_{sx} \approx \left[1 + 1.12\left(\dfrac{1 - B/L}{1 - \nu}\right)\right]\mu_{sy}$ | $\mu_{emb} \approx 1 - 0.14\left(\dfrac{D_f}{B}\right)^{0.35}$ |
| Moment | $\theta_{ry} = \dfrac{M}{E}\dfrac{2(1 - \nu^2)}{I_y^{0.75}}(\mu_s)_{ry}$ | $\theta_{rx} = \dfrac{M}{E}\dfrac{2(1 - \nu^2)}{I_x^{0.75}}\left(\dfrac{B}{L}\right)^{0.25}(\mu_s)_{rx}$ | $\theta_{emb} \approx \theta_r$ |
| | $(\mu_s)_{ry} \approx 0.43 - 0.10\dfrac{B}{L}$ | $(\mu_s)_{rx} \approx 0.33\left(\dfrac{B}{L}\right)^{0.15}$ | |

| | Embedded foundation with sidewall-soil contact surface of total areas $A_w$ | |
|---|---|---|
| | | |
| Loading | Lateral direction $y$ | Longitudinal direction $x$ |
| Horizontal | $\rho_e = (\rho_e)_{emb}\mu_{wall}$ <br> $(\mu_{wall})_y \approx 1 - 0.35\left[\dfrac{h}{B}\dfrac{A_w}{L^2}\right]^{0.2} \approx (\mu_{wall})_x$ | |
| Moment | $\theta = \theta_{emb}(\mu_{wall})_r$ | |
| | $(\mu_{wall})_{ry} \approx \left\{1 + 0.92\left(\dfrac{d}{L}\right)^{0.6}\left[1.5 + \left(\dfrac{d}{L}\right)^{1.9}\left(\dfrac{D_f}{d}\right)^{0.6}\right]\right\}^{-1}$ | $(\mu_{wall})_{rx} \approx \left\{1 + 1.26\dfrac{d}{B}\left[1 + \dfrac{d}{B}\left(\dfrac{D_f}{d}\right)^{0.2}\left(\dfrac{B}{L}\right)^{0.5}\right]\right\}^{-1}$ |

NOTES: Substitute the appropriate values for $H_i$ and $M$ in the above equations. For example, if you are considering the $X$ direction, $H_i = H_x$ and $M = M_x$. Also use the appropriate values for $\mu_s$ and $\theta_r$. For example, in the $X$ direction, $\mu_s = \mu_{sx}$ and $\theta_r = \theta_{rx}$. For short term loading use the undrained values of $E$ and $\nu$ while for long term loading use the effective values. The terms $I_x$ and $I_y$ are the second moment of areas about the $X$ and $Y$ axes respectively and $\theta$ is the rotation caused by the moment.

## 7.15   SUMMARY

In this chapter, we described ways in which you can calculate the bearing capacity and settlement of shallow foundations. The critical criterion for design of most shallow foundations is the serviceability limit state. Because of sampling difficulties, the bearing capacity and settlement of coarse-grained soils are often determined from field tests. We showed how to use the results of the SPT, the cone test, and the plate test to estimate the bearing capacity and settlement of shallow foundations on coarse-grained soils.

*Practical Examples*

### EXAMPLE 7.14

Determine the size of a square footing to carry a vertical load of 500 kN. The soil at the site is shown in Fig. E7.14. The tolerable total settlement is 20 mm. The groundwater level is 3 m below the ground surface and the depth of the footing is 1.5 m. However, the groundwater level is expected to seasonally rise to the surface. You may assume that the clay layer is thin so that one-dimensional consolidation takes place.

**Strategy**   The presence of the soft clay layer gives a clue that settlement may govern the design. In this case, we should determine the width required to satisfy settlement and then check the bearing capacity. One can assume a width, calculate the settlement, and reiterate until the settlement criterion is met. Since this is a multilayer soil profile, it is more accurate to calculate the increase in vertical stress for multilayer soils. However, for simplicity, we will use Boussinesq's method for a uniform soil.

### Solution 7.14

**Step 1:**   Assume a width and a shape.
Try $B = 3$m, and assume a square footing.

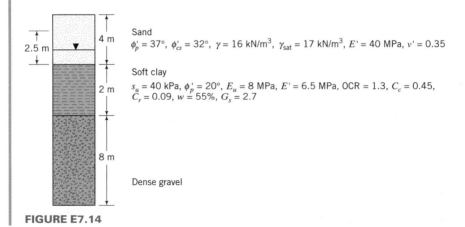

Sand
$\phi'_p = 37°$, $\phi'_{cs} = 32°$, $\gamma = 16$ kN/m³, $\gamma_{sat} = 17$ kN/m³, $E' = 40$ MPa, $v' = 0.35$

Soft clay
$s_u = 40$ kPa, $\phi'_p = 20°$, $E_u = 8$ MPa, $E' = 6.5$ MPa, OCR = 1.3, $C_c = 0.45$, $C_r = 0.09$, $w = 55\%$, $G_s = 2.7$

Dense gravel

**FIGURE E7.14**

**Step 2:** Calculate the elastic settlement.

**Sand**

Neglect side wall effects; that is, $\mu_{wall} = 1$

$$\mu'_{emb} = 1 - 0.08\frac{D_f}{B}\left(1 + \frac{4}{3}\frac{B}{L}\right) = 1 - 0.08\frac{1.5}{3}\left(1 + \frac{4}{3} \times 1\right) = 0.91$$

$$I_s = 0.62\ln(L/B) + 1.12 = 1.12$$

$$\rho_e = \frac{q_sB[1 - (v')^2]}{E'}I_s\mu'_{emb} = \frac{P[1 - (v')^2]}{E'L}I_s\mu'_{emb}$$

$$= \frac{500(1 - 0.35^2)}{40 \times 10^3 \times 3} \times 1.12 \times 0.91 = 3.7 \times 10^{-3} = 3.7 \text{ mm}$$

**Clay**

Find the equivalent footing size at the top of the clay layer. Let $z_1$ be the depth from the base of the footing to the top of the clay layer.

Equivalent width and length of footing at top of clay:   $B + z_1 = 3 + 2.5 = 5.5$ m

$$\mu_{emb} = 1 - 0.08\frac{2.5}{5.5}\left(1 + \frac{4}{3} \times 1\right) = 0.92$$

For immediate settlement in clays, use undrained conditions with $v = v_u = 0.5$.

$$\rho_e = \frac{P(1 - v_u^2)}{E_uL}I_s\mu'_{emb} = \frac{500(1 - 0.5)}{8000 \times 5.5} \times 1.12 \times 0.92 = 8.8 \times 10^{-3} = 8.8 \text{ mm}$$

**Step 3:** Calculate the consolidation settlement of the clay.

$$e = wG_s = 0.55 \times 2.7 = 1.49$$

$$\gamma_{sat} = \frac{G_s + e}{1 + e}\gamma_w = \frac{2.7 + 1.49}{1 + 1.49}9.8 = 16.5 \text{ kN/m}^3$$

Calculate the current vertical effective stress (overburden pressure) at the center of the clay layer.

$$\sigma'_{zo} = 3 \times 16 + 1(17 - 9.8) + 1(16.5 - 9.8) = 61.9 \text{ kPa}$$

Calculate the stress increase at the center of the clay layer ($z = 3.5$ m).

$$m = n = \frac{\left(\frac{3}{2}\right)}{3.5} = 0.43; \quad \text{from stress chart (Fig. 3.25), } I_z = 0.068$$

$$\Delta\sigma_z = 4 \times \frac{500}{3^2} \times 0.068 = 15.1 \text{ kPa}$$

$$\sigma'_{zo} + \Delta\sigma_z = 61.9 + 15.1 = 77 \text{ kPa}$$

$$\sigma'_{zc} = \text{OCR} \times \sigma'_{zo} = 1.3 \times 61.9 = 80.5 \text{ kPa} > \sigma'_{zo} + \Delta\sigma_z$$

$$\rho_{pc} = \frac{H_o}{1 + e_o}C_r\log\frac{\sigma'_{zo} + \Delta\sigma_z}{\sigma'_{zo}}$$

$$= \frac{2}{1 + 1.49} \times 0.09\log\frac{77}{61.9} = 6.9 \times 10^{-3} \text{ m} = 6.9 \text{ mm}$$

**Step 4:**    Find the total settlement.

$$\text{Total settlement:} \quad \rho = (\rho_e)_{\text{sand}} + (\rho_c)_{\text{clay}} + \rho_{\text{pc}}$$
$$= 3.7 + 8.8 + 6.9 = 19.4 \text{ mm} < 20 \text{ mm}$$

**Step 5:**    Check the bearing capacity.
The groundwater table is less than $B = 3$ m below the footing base so groundwater effects must be taken into account.
Calculate the bearing capacity using Meyerhof's method.

**ESA (sand)**

$$K_p = \frac{1 + \sin 32°}{1 - \sin 32°} = 3.25; \, s_q = s_\gamma = 1 + 0.1 \times 3.25 = 1.33;$$

$$d_q = d_\gamma = 1 + 0.1\sqrt{3.25} \times \frac{1.5}{3} = 1.09$$

$$N_q = e^{\pi \tan 32°} \tan^2\left(45 + \frac{32}{2}\right) = 23.2; \, N_q - 1 = 23.2 - 1 = 22.2$$

$$N_\gamma = (N_q - 1)\tan(1.4 \times 32) = 22.0$$

Calculate the bearing capacity for the worst case scenario—groundwater level at surface.

$$q_{\text{ult}} = \gamma D_f(N_q - 1)s_q \, d_q + 0.5(\gamma B)N_\gamma s_\gamma \, d_\gamma; \text{ use } \gamma = \gamma' = 16 - 9.8 = 6.2 \text{ kN/m}^3$$
$$q_{\text{ult}} = 7.2 \times 1.5 \times 22.2 \times 1.33 \times 1.09 + 0.5$$
$$\times 7.2 \times 3 \times 22 \times 1.33 \times 1.09 = 692 \text{ kPa}$$

$$(\sigma_a)_{\text{max}} = \frac{\text{Applied load}}{\text{Area}} = \frac{500}{3^2} = 55.6 \text{ kPa}$$

$$\text{FS} = \frac{q_{\text{ult}}}{(\sigma_a)_{\text{max}} - \gamma D_f} = \frac{692}{55.6 - 7.2 \times 1.5} = 15.4 > 1.5 \text{ (okay)}$$

Check the bearing capacity of the clay.

$$\text{Equivalent footing width:} \quad B + z_1 = 3 + 2.5 = 5.5 \text{ m}$$

$$(\Delta\sigma)_{\text{max}} = \frac{500}{5.5^2} = 16.5 \text{ kPa}$$

**TSA (clay)**

Use Skempton's equation (7.14).

$$q_{\text{ult}} = 5s_u\left(1 + 0.2 \frac{D_f}{B}\right)\left(1 + 0.2 \frac{B}{L}\right)$$

$$= 5 \times 40 \times \left(1 + 0.2 \frac{4}{3}\right)\left(1 + 0.2 \frac{3}{3}\right) = 304 \text{ kPa}$$

$$\text{FS} = \frac{304}{16.5} = 18.4 > 3 \text{ (okay)}$$

**ESA (clay)**

Use Meyerhof's method.

$$K_p = \frac{1 + \sin 20°}{1 - \sin 20°} = 2;$$

$$s_q = s_\gamma = 1 + 0.1 \times 2 = 1.2; \; d_q = d_\gamma = 1 + 0.1\sqrt{2} \times \frac{4}{3} = 1.19$$

$$N_q = e^{\pi \tan 20°} \tan^2\left(45 + \frac{20}{2}\right) = 6.4; \quad N_q - 1 = 6.4 - 1 = 5.4$$

$$N_\gamma = (N_q - 1) \tan(1.4 \times 20) = 2.9$$

$$\gamma' = \gamma_{sat} - \gamma_w = 16.5 - 9.8 = 6.7 \; \text{KN/m}^3$$

$$q_{ult} = 6.7 \times 4 \times 5.4 \times 1.2 \times 1.19 + 0.5$$
$$\times 6.7 \times 2.9 \times 1.2 \times 1.19 = 221 \; \text{kPa}$$

$$FS = \frac{221}{16.5} = 13.4 > 1.5 \; \text{(okay)}$$

Settlement governs the design. ∎

## EXAMPLE 7.15

Figure E7.15 shows two isolated footings at the two ends of a building. Your local code requires a maximum distortion of $\delta/\ell = 350$ and a maximum total settlement of 50 mm. Determine the most economical size for each footing to satisfy ultimate and serviceability limit states. Groundwater level is 5 m below the ground surface. Creep effects in the soil are negligible.

**Strategy**  The clay layer is very thick and only a certain thickness below each footing would be stressed. From Chapter 3, you know that the vertical stress increase below a depth of 3B for a square footing is less than 10%. So you can use a thickness of 3B as the effective thickness of the clay. The Skempton–Bjerrum method for calculating consolidation settlement is suitable for this problem.

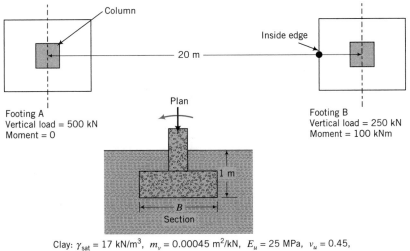

Clay: $\gamma_{sat} = 17$ kN/m³,  $m_v = 0.00045$ m²/kN,  $E_u = 25$ MPa,  $v_u = 0.45$,
$\phi'_{cs} = 24°$,  $s_u = 80$ kPa

**FIGURE E7.15**

## Solution 7.15

**Footing A**

**Step 1:** Assume a width.

Assume a square footing of width $B = 3$ m.

**Step 2:** Calculate the elastic settlement.

From Eqs. (7.35) to (7.37): $\dfrac{A}{4L^2} = 1$, $\mu_{\text{wall}} = 1$, $\mu_s = 0.45\left(\dfrac{A_b}{4L^2}\right)^{-0.38}$

$$= 0.45 \times 1 = 0.45$$

$$\mu_{\text{emb}} = 1 - 0.04\,\dfrac{D_f}{B}\left[1 + \dfrac{4}{3}\left(\dfrac{A_b}{4L^2}\right)\right]$$

$$= 1 - 0.04\,\dfrac{1}{1.5}\left[1 + \dfrac{4}{3}(1)\right] = 0.94$$

From Eq. (7.34): $\rho_e = \dfrac{P}{E_u L}(1 - v_u^2)\mu_s\mu_{\text{emb}}\mu_{\text{wall}} = \dfrac{500}{25{,}000 \times 1.5}(1 - 0.45^2)$

$$\times\ 0.45 \times 0.94 \times 1 = 4.5 \times 10^{-3} = 4.5 \text{ mm}$$

**Step 3:** Compute the consolidation settlement.

Determine the stress increase over a depth $3B = 9$ m (effective depth) below the footing. Divide the effective depth into three layers of 3 m each and find the stress increase at the center of each layer under the center of the footing. Stress overlap is unlikely as the footings are greater than $B$ apart.

$$q_s = \dfrac{500}{3^2} = 55.6 \text{ kPa}$$

| Layer | Depth to center of layer, $z$ (m) | $m = n = \dfrac{(3/2)}{z}$ | $I_z$ | $\Delta\sigma_z = 4q_s I_z$ (kPa) |
|-------|-----------------------------------|----------------------------|-------|-----------------------------------|
| 1 | 1.5 | 1 | 0.175 | 38.9 |
| 2 | 4.5 | 0.33 | 0.045 | 10.0 |
| 3 | 7.5 | 0.2 | 0.018 | 4.0 |
| | | | | $\Sigma 52.9$ |

Equivalent diameter of footing: $D = \sqrt{\dfrac{A}{\pi}} = \sqrt{\dfrac{9}{\pi}} = 1.69$ m

$$\dfrac{H_o}{B} = \dfrac{H_o}{D} = \dfrac{9}{1.69} = 5.3$$

$H_o/B = 5.3$ is outside the plotted limits in Fig. 7.11. We will use $H_o/B = 4$, which will result in an overestimation of the primary consolidation settlement. From Fig. 7.11, $\mu_{\text{SB}} = 0.58$.

$$\rho_{\text{pc}} = \Sigma m_v\,\Delta\sigma_z\,H_o\mu_{\text{SB}} = 0.00045 \times 52.9 \times 3 \times 0.58$$

$$= 41.4 \times 10^{-3} \text{ m} = 41.4 \text{ mm}$$

Total settlement $= 4.5 + 41.4 = 45.9$ mm $< 50$ mm

Total settlement of footing A is okay.

**Footing B**

**Step 4:**  Calculate the eccentricity.

$$e = \frac{M}{P} = \frac{100}{250} = 0.4 \text{ m}$$

**Step 5:**  Assume a width.
For no tension $e/B < 6$; that is, $B > 6e$.

$$B_{\min} = 6 \times 0.4 = 2.4 \text{ m}; \quad \text{try } B = 2.5 \text{ m}$$

**Step 6:**  Calculate the elastic settlement

$$\mu_{emb} = 1 - 0.04 \times \frac{1}{1.25}\left(1 + \frac{4}{3}(1)\right) = 0.93$$

$$\rho_e = \frac{500}{25{,}000 \times 1.25}(1 - 0.45^2)0.45 \times 0.93 \times 1 = 5.3 \times 10^{-3} \text{ m} = 5.3 \text{ mm}$$

**Step 7:**  Calculate the consolidation settlement.
Because of the eccentric loading, the vertical stress distribution under footing B would be nonuniform. The maximum vertical stress occurs at the inside edge of magnitude (Eq. 7.30) is

$$q_s = \frac{250}{2.5^2}\left(1 + \frac{6 \times 0.4}{2.5}\right) = 78.4 \text{ kPa}$$

We need then to find the stress increase under the center of the inside edge of footing B (shown by a circle in Fig. E7.15). A table is used for ease of computation and checking.

| Layer | Depth to center of layer, $z$ (m) | $m = \dfrac{B}{z} = \dfrac{2.5}{z}$ | $n = \dfrac{L}{z} = \dfrac{1.25}{z}$ | $I_z$ | $\Delta\sigma_z = 2q_sI_z$ |
|---|---|---|---|---|---|
| 1 | 3 | 0.83 | 0.42 | 0.10 | 15.7 |
| 2 | 4.5 | 0.56 | 0.28 | 0.055 | 8.6 |
| 3 | 7.5 | 0.33 | 0.17 | 0.025 | 3.9 |
| | | | | | $\Sigma 28.2$ |

Equivalent diameter of footing: $D = \sqrt{\dfrac{A}{\pi}} = \sqrt{\dfrac{2.5^2}{\pi}} = 1.41 \text{ m}$

$$\frac{H_o}{B} = \frac{H_o}{D} = \frac{9}{1.41} = 6.4$$

$H_o/B = 6.4$ is outside the plotted limits in Fig. 7.11. We will use $H_o/B = 4$, which will result in an overestimation of the primary consolidation settlement. From Fig. 7.11, $\mu_{SB} = 0.58$.

$$\rho_{pc} = \Sigma m_v \, \Delta\sigma_z \, H_o\mu_{SB} = 0.00045 \times 28.2 \times 3 \times 0.58$$
$$= 22.1 \times 10^{-3} \text{ m} = 22.1 \text{ mm}$$

**Step 8:**  Calculate the total settlement.

$$\rho = \rho_e + \rho_{pc} = 5.3 + 22.1 = 27.4 \text{ mm} < 50 \text{ mm}$$

Total settlement of footing B is okay.

**Step 9:** Calculate the distortion.

$$\delta = 45.9 - 27.4 = 18.5 \text{ mm}$$

$$\frac{\delta}{\ell} = \frac{18.5}{(20 - 1.25) \times 10^3} = \frac{1}{1014} < \frac{1}{350}; \quad \text{therefore, distortion is okay}$$

**Step 10:** Check the bearing capacity.
Use Meyerhof's method because of the eccentric loading of footing B.

**Footing A**

**Factors**

$$K_p = \tan^2(45° + 24°/2) = 2.37$$

Equation (7.20):   $s_c = 1 + 0.2\left(\dfrac{1}{1}\right) = 1.2$

$$s_q = s_\gamma = 1 + 0.1 \times 2.37\left(\frac{3}{3}\right) = 1.24$$

Equation (7.21):   $d_c = 1 + 0.2\left(\dfrac{1}{3}\right) = 1.07;$

$$d_q = d_\gamma = 1 + 0.1\sqrt{2.37}\left(\frac{1}{3}\right) = 1.05$$

**TSA**

Equation (7.15):   $q_{\text{ult}} = 5.14 s_u s_c\, d_c = 5.14 \times 80 \times 1.2 \times 1.07 = 528 \text{ kPa}$

$$\text{FS} = \frac{528}{55.6 - 17 \times 1} = 13.7 > 3 \text{ (okay)}$$

**ESA**

$$N_q = e^{\pi \tan 24°} \tan^2(45° + 24°/2) = 9.6$$
$$N_\gamma = (N_q - 1)\tan(1.4\phi') = (9.6 - 1)\tan(1.4 \times 24°) = 5.7$$
$$q_{\text{ult}} = \gamma D_f(N_q - 1)s_q\, d_q + 0.5\gamma B N_\gamma s_\gamma\, d_\gamma$$
$$q_{\text{ult}} = (17 \times 1 \times 8.6 \times 1.24 \times 1.05)$$
$$+ (0.5 \times 17 \times 3 \times 5.7 \times 1.24 \times 1.05) = 380 \text{ kPa}$$

$$\text{FS} = \frac{380}{55.6 - 17 \times 1} = 9.8 > 1.5 \text{ (okay)}$$

**Footing B**

$$B' = B - 2e = 2.5 - 2 \times 0.4 = 1.7 \text{ m}$$

The only factors that would change are the depth factors.

$$d_c = 1 + 0.2\left(\frac{1}{1.7}\right) = 1.12; \quad d_q = d_\gamma = 1 + 0.1\sqrt{2.37}\left(\frac{1}{1.7}\right) = 1.09$$

**TSA**

$$q_{\text{ult}} = 5.14 \times 80 \times 1.2 \times 1.12 = 553 \text{ kPa}$$

Equation (7.26):   $\text{FS} = \dfrac{553}{78.4 - 17 \times 1} = 9 > 3 \text{ (okay)}$

**ESA**

$$q_{ult} = \gamma D_f(N_{q-1})s_q \, d_q + 0.5\gamma B' N_\gamma s_\gamma \, d_\gamma$$
$$= (17 \times 1 \times 8.6 \times 1.24 \times 1.09)$$
$$+ (0.5 \times 17 \times 1.7 \times 5.7 \times 1.24 \times 1.09) = 310 \text{ kPa}$$

$$\text{FS} = \frac{310}{78.4 - 17 \times 1} = 5 > 1.5 \text{ (okay)}$$

**Step 11:** Recommend footing sizes.

Footing A: 3 m × 3 m; $\rho \approx 46$ mm
Footing B: 2.5 m × 2.5 m; $\rho \approx 27$ mm ∎

# EXERCISES

## Theory

7.1 Show that the ultimate load for a strip footing under long-term conditions using the two triangle failure surfaces shown in Fig. P7.1 is $P_u = \frac{1}{2}\gamma B^2 N_\gamma$, where

$$N_\gamma = \frac{2(1 + \tan^2 \phi')}{[2 - (1 + \tan^2 \phi')](1 - \tan \phi')}$$

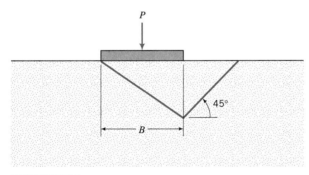

**FIGURE P7.1**

7.2 A strip footing, 5 m wide, is founded just below the surface of a deep deposit of clay. The undrained shear strength of the clay increases linearly from 3 kPa at the footing base to 10 kPa at a depth of 5 m. Estimate the vertical ultimate load assuming that the load is applied at an eccentricity of 0.75 m from the center of the footing's width. [*Hint:* Try a circular failure surface, determine the equation for the distribution of shear strength with depth, and integrate the shear strength over the circumference to find the shear force.]

7.3 The centroid of a square foundation of side 5 m is located 10 m away from the edge of a vertical cut of depth 4 m. The soil is a stiff clay whose undrained strength is 20 kPa and whose unit weight of 16 kN/m³. Calculate the vertical ultimate load using the limit equilibrium method.

## Problem Solving

7.4 Calculate the ultimate net bearing capacity of (a) a strip footing 1 m wide, (b) a square footing 3 m × 3 m, and (c) a circular footing 3 m in diameter using Terzaghi's and Meyerhof's methods. All footings are located on the ground surface and the groundwater level is at the ground surface. The soil is coarse-grained with $\gamma_{sat} = 17$ kN/m³, $\phi_{cs} = 30°$. If a factor of safety of 1.5 is desired, which footing is most effective?

**7.5** A strip footing, founded on dense sand ($\phi_p' = 35°$ and $\gamma_{sat} = 17$ kN/m³), is to be designed to support a vertical load of 60 kN per meter length. Determine a suitable width for this footing for FS = 3. The footing is located 1 m below the ground surface. The groundwater level is 10 m below the ground surface.

**7.6** A square footing, 3 m wide, is located 1.5 m below the surface of a stiff clay. Determine the allowable bearing capacity if $s_u = 100$ kPa, and $\gamma_{sat} = 20$ kN/m³. If the footing were located on the surface, what would be the allowable bearing capacity? Use FS = 3.

**7.7** A column carrying a load of 750 kN is to be founded on a rectangular footing at a depth of 2 m below the ground surface in a deep clay stratum. What will be the size of the footing for FS = 1.5 for an ESA and FS = 3 for a TSA? The soil properties are $\phi_{cs} = 28°$, $\gamma_{sat} = 18.5$ kN/m³, and $s_u = 55$ kPa. The groundwater level is at the base of the footing but it is expected to rise to the ground surface during rainy seasons.

**7.8** Repeat Exercise 7.7 with a moment of 250 kN.m about the $Y$ axis in addition to the vertical load.

**7.9** A square footing is required to carry a load of 500 kN inclined at 15° to the vertical plane. The building code requires an embedment depth of 1.2 m. Groundwater level is at 1 m below the ground surface. Calculate the size of the footing for $\phi_p' = 37°$, $\gamma_{sat} = 18.5$ kN/m³ and FS = 3.

**7.10** The footing for a bridge pier is to be founded in sand, as shown in Fig. P7.10. The clay layer is normally consolidated with $C_c = 0.25$. Determine the factor of safety against bearing capacity failure and the total settlement (elastic compression and primary consolidation) of the pier.

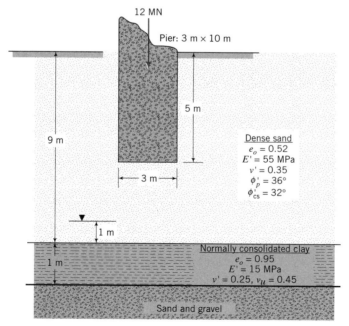

12 MN

Pier: 3 m × 10 m

5 m

9 m

Dense sand
$e_o = 0.52$
$E' = 55$ MPa
$v' = 0.35$
$\phi_p' = 36°$
$\phi_{cs}' = 32°$

3 m

1 m

1 m

Normally consolidated clay
$e_o = 0.95$
$E' = 15$ MPa
$v' = 0.25$, $v_u = 0.45$

Sand and gravel

**FIGURE P7.10**

**7.11** A multilevel building is supported on a footing 58 m wide × 75 m long × 3 m thick resting on a very stiff deposit of saturated clay. The footing is located at 3 m below ground level. The average stress at the base of the footing is 350 kPa. Groundwater level is at 12 m below the surface. Field and laboratory tests gave the following results:

| Depth (m) | 0.5 | 6 | 25 |
|-----------|-----|-----|-----|
| $s_u$ (kPa) | 58 | 122 | 156 |

$e_o = 0.57$, $C_c = 0.16$, $C_r = 0.035$, OCR = 10, $\phi'_p = 28°$, $\phi'_{cs} = 24°$, $E_u = 100$ MPa, $\nu_u = 0.45$, $E' = 90$ MPa, and $\nu' = 0.3$. Determine the total settlement and the safety factor against bearing capacity failure.

## Practical

**7.12** A circular foundation of diameter 8 m supports a tank. The base of the foundation is at 1 m from the ground surface. The maximum applied stress is 110 kPa. The tank foundation was designed for short-term loading conditions ($s_u = 60$ kPa and $\gamma_{sat} = 19$ kN/m³). The groundwater level when the tank was initially designed was at 4 m below the ground surface. It was assumed that the groundwater level was stable. Fourteen months after the tank was constructed and during a week of intense rainfall, the tank foundation failed. It was speculated that failure occurred by bearing capacity failure. Establish whether this is so or not. The friction angles are $\phi'_p = 27°$ and $\phi'_{cs} = 21°$.

## CHAPTER 8

# PILE FOUNDATIONS

## 8.0 INTRODUCTION

In the previous chapter, we studied the analysis of shallow foundations. In some cases, shallow foundations are inadequate to support the structural loads and deep foundations are required. When the term deep foundations is used, it invariably means pile foundations. A pile is a slender, structural member installed in the ground to transfer the structural loads to soils at some significant depth below the base of the structure. Structural loads include axial loads, lateral loads, and moments.

Pile foundations are used when:

- The soil near the surface does not have sufficient bearing capacity to support the structural loads
- The estimated settlement of the soil exceeds tolerable limits (i.e., settlement greater than the serviceability limit state)
- Differential settlement due to soil variability or nonuniform structural loads is excessive
- The structural loads consist of lateral loads and/or uplift forces
- Excavations to construct a shallow foundation on a firm soil layer are difficult or expensive

In this chapter, you will study the bearing capacity (load capacity) and settlement of single and group piles under axial loads. When you complete this chapter, you should be able to:

- Determine the allowable axial load capacity of single piles and pile groups
- Determine the settlement of single piles and pile groups

You would need to recall the following:

- Effective stresses—Chapter 3
- Consolidation—Chapter 4
- Statics

*Sample Practical Situation*   A structure is to be constructed on a deposit of soft soil. Shallow foundations were ruled out because the estimated settlement exceeded the tolerable settlement. The structure is to be supported on piles. You are required to determine the type of piles to be used, the configuration of the

**FIGURE 8.1**   Concrete piles used for the foundation of a skyscraper. (© Fritz Henle/Photo Researchers.)

piles (single piles or pile groups) and the length of piles required to safely support the structure without exceeding ultimate and serviceability limit states. An example of a pile foundation under construction is shown in Fig. 8.1.

## 8.1   DEFINITIONS OF KEY TERMS

*Pile* —a slender, structural member consisting of steel or concrete or timber.

*Skin friction stress* or *shaft friction stress* or *adhesive stress* $(f_s)$ is the frictional or adhesive stress on the shaft of a pile.

*End bearing stress* or *point resistance stress* $(f_b)$ is the stress at the base or tip of a pile.

*Ultimate load capacity* $(Q_{ult})$ is the maximum load that a pile can sustain before soil failure occurs.

*Ultimate group load capacity* $[(Q_{ult})_g]$ is the maximum load that a group of piles can sustain before soil failure occurs.

*Skin friction* or *shaft friction* or *side shear* $(Q_f)$ is the frictional force generated on the shaft of a pile.

*End bearing* or *point resistance* $(Q_b)$ is the resistance generated at the base or tip of a pile.

*End bearing* or *point bearing pile* is one that transfers almost all the structural load to the soil at the bottom end of the pile.

*Friction pile* is one that transfers almost all the structural load to the soil by skin friction along a substantial length of the pile.

*Floating pile* is a friction pile in which the end bearing resistance is neglected.

## 8.2  QUESTIONS TO GUIDE YOUR READING

1. What are the differences between the different types of piles?
2. How is a pile installed in the ground?
3. How do I estimate the allowable load capacity of a single pile and pile groups?
4. How do I estimate the settlement of a single pile and pile groups?
5. Is the analysis of the load capacity of a pile exact or is it an approximation?

## 8.3  TYPES OF PILES AND INSTALLATION

Piles are made from concrete or steel or timber (Fig. 8.2). The selection of the type of pile required for a project depends on what type is readily available, the

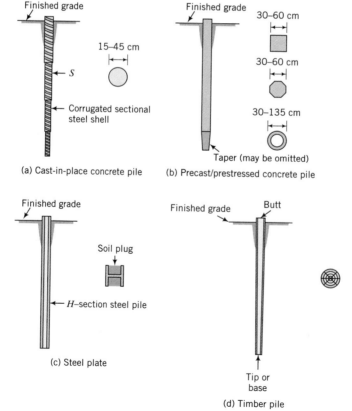

**FIGURE 8.2**  Pile types.

magnitude of the loading, the soil type, and the environment in which the pile will be installed, for example, a corrosive environment or a marine environment.

### 8.3.1 Concrete Piles

There are several types of concrete piles that are commonly used. These include cast-in-place concrete piles, precast concrete piles, and drilled shafts. Cast-in-place concrete piles are formed by driving a cylindrical steel shell into the ground to the desired depth and then filling the cavity of the shell with fluid concrete. The steel shell is for construction convenience and does not contribute to the load transfer capacity of the pile. Its purpose is to open a hole in the ground and keep it open to facilitate the construction of the concrete pile. Plain concrete is used when the structural load is only compressive. If moments and lateral loads are to be transferred, then a steel reinforcement cage is used in the upper part of the pile. Vigilant quality control and good construction practice are necessary to ensure the integrity of cast-in-place piles.

Precast concrete piles usually have square or circular or octagonal cross sections and are fabricated in a construction yard from reinforced or prestressed concrete. They are preferred when the pile length is known in advance. The disadvantages of precast piles are problems in transporting long piles, cutting, and lengthening. A very popular type of precast concrete pile is the Raymond cylindrical prestressed pile. This pile comes in sections, and lengths up to 70 m can be obtained by stacking the sections. Typical design loads are greater than 2 MN.

### 8.3.2 Steel Piles

Steel piles come in various shapes and sizes and include cylindrical, tapered, and H-piles. Steel H-piles are rolled steel sections. Steel pipe piles are seamless pipes that can be welded to yield lengths up to 70 m. They are usually driven with open ends into the soil. A conical tip is used where the piles have to penetrate boulders and rocks. To increase the load capacity of steel pipe piles, the soil plug (Fig. 8.2c) is excavated and replaced by concrete. These piles are called concrete-filled steel piles.

### 8.3.3 Timber Piles

Timber piles have been used since ancient times. The lengths of timber piles depend on the types of trees used to harvest the piles, but common lengths are about 12 m. Longer lengths can be obtained by splicing several piles. Timber piles are susceptible to termites, marine organisms, and rot within zones exposed to seasonal changes.

A comparative summary of the different pile types is given in Table 8.1.

### 8.3.4 Pile Installation

Piles can either be driven into the ground (driven piles) or be installed in a predrilled hole (bored piles or drilled shafts). A variety of driving equipment is used in pile installations. The key components are the leads and the hammer. The leads are used to align the hammer to strike the pile squarely (Fig. 8.3).

**TABLE 8.1** **Comparisons of Different Piles**

| Pile type | Section (m) | Common lengths (m) | Average load (kN) | Allowable stress (MPa) | Advantages | Disadvantages |
|---|---|---|---|---|---|---|
| Cast-in-place concrete | 0.15–0.45 (diameter) | ≤35 | 600 | 4.5–8.5 | Can sustain hard driving, resistant to marine organisms, easily inspected, length can be changed easily, easy to handle and ship | Concrete can arch during placement, can be damaged if adjacent piles are driven before concrete sets |
| Precast concrete | 0.15–0.25 (width) | ≤35 | 750 | 4.5–7 | Economical for specified length, higher capacity than timber | Cutting and lengthening of piles can be expensive, handling is a problem, shipping long piles is expensive, may crack during driving |
| Raymond cylinder | 0.6–2.3 (diameter) | <60 | 2000 | 40–70 | Can ship in sections, high capacity, long length | Cost (expensive) |
| Steel pipe | 0.2–1 (diameter) | <35 | 900 | 59–83 | High axial and lateral capacity, can take hard driving, easy inspection and handling, length can be changed easily, resistant to deterioration | Needs treatment for corrosive environment |
| Concrete-filled pipe | 0.2–1 (diameter) | <35 | 900 | Concrete: 4.5–8.5 Steel: 62–83 | Similar to steel pile | Similar to steel pile |
| Steel H-pile | Webs: 1–3 Flange: 0.2–0.35 | <60 | 900 | 59–83 | "Nondisplacement," can take hard driving, easy handling, high axial and lateral capacity, length can be changed easily | Needs treatment for corrosive environment |
| Timber | 0.125–0.45 (diameter) | 12–35 | 250 | 5.5–8.5 | Low cost, availability | Low capacity, can deteriorate if not protected, cannot take hard driving, cannot be inspected after driving |

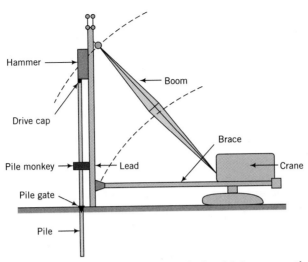

**FIGURE 8.3**   Key components of pile-driving operation.

Hammers can be simple drop hammers of weights between 2.5 and 15 kN or modern steam/pneumatic hammers. Two popular types of steam/pneumatic hammers are shown in Fig. 8.4. The single-acting hammer utilizes steam or air to lift the ram and its accessories (cushion, drive caps, etc.). The double-acting hammer is used to increase the number of blows/minute and utilizes steam/air to lift the ram and force it down.

> **The** essential points *are:*
>
> *1.* **The selection of a pile type depends on the structural loads, availability, and the environment at the site.**
>
> *2.* **Piles can be installed using simple drop hammers but, most often, they are installed using steam or pneumatic hammers.**

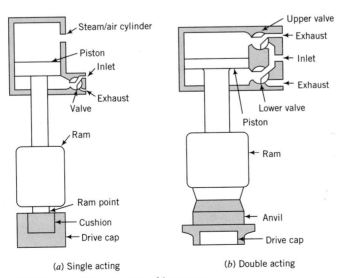

(a) Single acting     (b) Double acting

**FIGURE 8.4**   Two types of hammer.

*What's next . . .*We have described the various types of piles that are in common use. In the next section, methods of estimating the load capacity are presented. The load capacity of piles has been studied extensively. But no single satisfactory method of determining the load capacity has evolved. Pile load capacity is mostly based on empirical equations, experience, and judgment.

# 8.4   LOAD CAPACITY OF SINGLE PILES

Analysis of the bearing capacity, called load capacity, of piles can be classified into three categories depending on the degree of sophistication required and the importance of the structure. These three categories are summarized in Table 8.2. In this book, we will deal only with category 1 and category 2.

All analyses of pile load capacity are approximations because it is difficult, if not impossible, to account for the variability of soil types and the differences in the quality of construction practice. Even if the highest quality construction practice is followed, the installation of piles invariably changes the soil characteristics in ways that we cannot, at least simply, take into account.

The ultimate load capacity, $Q_{ult}$, of a pile consists of two parts. One part is due to friction (Fig. 8.5), called skin friction or shaft friction or side shear, $Q_f$, and the other is due to end bearing at the base or tip of the pile, $Q_b$. If the skin friction is greater than about 80% of the end bearing load capacity, the pile is

**TABLE 8.2   Categories of Analysis/Design Procedures**

| Category | Subdivision | Characteristics | Method of parameter determination |
|---|---|---|---|
| 1 | | Empirical—not based on soil mechanics principles | Simple in situ or laboratory tests with correlations |
| 2 | 2A | Based on simplified theory or charts; uses soil mechanics principles; amenable to hand calculation. Theory is linear elastic (deformation) or rigid plastic (stability). | Routine, relevant in situ tests—may require some correlations |
| | 2B | As for 2A, but theory is nonlinear (deformation) or elastoplastic (stability). | |
| 3 | 3A | Based on theory using site-specific analysis, uses soil mechanics principles. Theory is linear elastic (deformation) or rigid plastic (stability). | Careful laboratory and/or in situ tests that follow the appropriate stress paths |
| | 3B | As for 3A, but nonlinearity is allowed for in a relatively simple manner. | |
| | 3C | As for 3A, but nonlinearity is allowed for by way of proper constitutive models of soil behavior. | |

SOURCE: On behalf of the Institution of Civil Engineers.

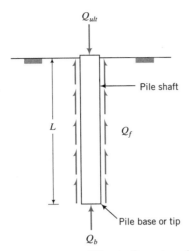

**FIGURE 8.5** Pile shaft and end bearing resistance.

deemed a friction pile and, if the reverse, an end bearing pile. If the end bearing is neglected, the pile is called a floating pile.

From statics,

$$Q_{ult} = Q_f + Q_b \qquad (8.1)$$

The conventional allowable load capacity is expressed as

$$Q_a = \frac{Q_{ult}}{FS} \qquad (8.2)$$

where FS is a gross factor of safety usually greater than 2.

The manner in which skin friction is transferred to the adjacent soil depends on the soil type. In fine-grained soils, the load transfer is nonlinear and decreases with depth, as illustrated in Fig. 8.6. As a result, elastic compression of the pile is not uniform; more compression occurs on the top part than on the bottom part

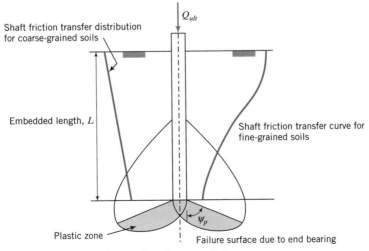

**FIGURE 8.6** Load transfer characteristics.

of the pile. For coarse-grained soils, the load transfer is approximately linear with depth (higher loads at the top and lower loads at the bottom).

In order to mobilize skin friction and end bearing, some movement of the pile is necessary. Field tests have revealed that to mobilize the full skin friction a vertical displacement of 5 to 10 mm is required. The actual vertical displacement depends on the strength of the soil and is independent of the pile length and pile diameter. The full end bearing resistance is mobilized in driven piles when the vertical displacement is about 10% of the pile tip diameter. For bored piles or drilled shafts, a vertical displacement of about 30% of the pile tip diameter is required. The full end bearing resistance is mobilized when slip or failure zones similar to shallow foundations are formed (Fig. 8.6). The end bearing resistance can then be calculated by analogy with shallow foundations. The important bearing capacity factor is $N_q$.

The full skin friction and full end bearing are not mobilized at the same displacement. The skin friction is mobilized at about one-tenth of the displacement required to mobilize the end bearing resistance. This is important in deciding on the factor of safety to be applied to the ultimate load. Depending on the tolerable settlement, different factors of safety can be applied to skin friction and to end bearing.

Generally, piles driven into loose, coarse-grained soils tend to densify the adjacent soil. When piles are driven into dense, coarse-grained soils, the soil adjacent to the pile becomes loose. Pile driving usually remolds fine-grained soils near the pile shaft. The implication of pile installation is that the intact shear strength of the soil is changed and you must account for this change in your estimations of the load capacity.

Except for loose, coarse-grained soils, the soil mass adjacent to a pile reaches the critical state or close to it. You should then use critical state shear strength parameters in determining the long-term load capacity of piles. Loose, coarse-grained soils adjacent to driven piles are densified and $\phi' = \phi'_p$. The actual value of $\phi'_p$ depends on the magnitude of the normal effective stress. Since the magnitude of the normal effective stress is uncertain, you should be cautious on using $\phi'_p$ in estimating the long-term load capacity. It would be prudent to use $\phi' = \phi'_{cs}$ in all cases except for some overconsolidated clays with a predominance of flat particles, where it is advisable to use $\phi' = \phi'_r$. The formulas for load capacity are developed or stated for a generic $\phi'$. You must, however, use the appropriate value of $\phi'$ in problem solving.

*The essential points are:*

1. *Pile load capacity depends on the soil type, method of installation, and construction practice.*
2. *The critical state friction angle should be used in estimating the long-term load capacity of piles, except you should use the residual friction angle for overconsolidated clays with a predominance of flat particles.*

**What's next . . .** A variety of methods are available to determine $Q_f$ and $Q_b$. We will deal with four methods:

- Pile load test
- Statics—$\alpha$- and $\beta$-methods

- Pile-driving formulas
- Wave analysis

We begin with the pile load test.

## 8.5 PILE LOAD TEST

The purposes of a pile load test are:

- To determine the axial load capacity of a single pile
- To determine the settlement of a single pile at working loads
- To verify estimated axial load capacity
- To obtain information on load transfer in skin friction and end bearing

In a typical pile load test, the test pile is driven to the desired depth, loads are applied incrementally, and the settlement of the pile is recorded. The axial loads can be applied by stacking sand bags on a loading frame attached to the pile or, more popularly, by jacking against a reaction beam and reaction piles (Fig. 8.7). If load transfer information is required, the pile must be instrumented to ascertain the internal load of the pile shaft.

The interpretation of the load capacity depends on the method of loading. Two loading methods are popular. In one method, called the constant rate of penetration (CRP) test, the load is applied at a constant rate of penetration of 0.75 mm/min in fine-grained soils and 1.5 mm/min in coarse-grained soils. In the other method, called the quick maintained load (QML), increments of load, about 15% of the design load, are applied at intervals of about 2.5 min. At the

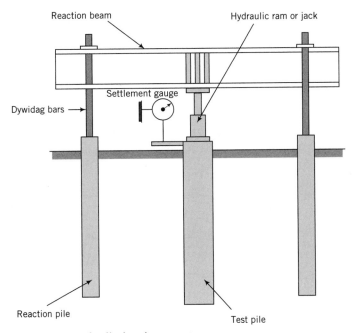

**FIGURE 8.7**  A pile load test setup.

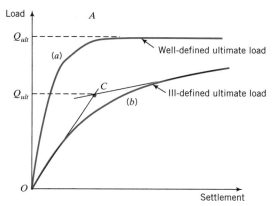

**FIGURE 8.8**    Load–settlement curves.

end of each load increment, the load and settlement are recorded. Schematic variations of pile load test plots are shown in Fig. 8.8.

The ultimate load is not always well defined. Load–settlement curve (a) in Fig. 8.8 shows a well-defined ultimate load while curve (b) does not. To obtain the ultimate load from curve (b) various empirical procedures have been suggested. One simple method is to find the intersection of the tangents of the two parts of the curve. The value at the ordinate of the intersection ($C$ in Fig. 8.8) is $Q_{ult}$. The allowable load capacity is found by dividing the ultimate load by a factor of safety, usually 2. An alternative criterion is to determine the allowable pile load capacity for a desired serviceability limit state, for example, a settlement of 10% of the pile diameter. The settlement at the allowable (working) load capacity is readily determined from the load–settlement plot (Fig. 8.8).

> *The* **essential points** *are:*
>
> 1. *A pile load test provides the load capacity and settlement of a pile at the working load at a particular location in a job site.*
>
> 2. *Various criteria and techniques are used to determine the allowable load capacity from pile load tests.*

## EXAMPLE 8.1

The results of a load test on a 0.45 m diameter pile are shown in the table below. Determine (a) the ultimate pile load capacity, (b) the allowable load for a factor of safety of 2, and (c) the allowable load capacity at 10% pile displacement.

| Displacement (mm) | 0.0 | 1.3 | 2.5 | 5.1 | 7.6 | 10.2 | 12.7 | 15.2 | 17.8 | 20.3 | 22.9 | 25.4 | 27.9 | 30.5 | 33.0 |
|---|---|---|---|---|---|---|---|---|---|---|---|---|---|---|---|
| Load (kN) | 0 | 200 | 350 | 670 | 870 | 1070 | 1250 | 1400 | 1500 | 1600 | 1700 | 1750 | 1780 | 1810 | 1830 |

| Displacement (mm) | 35.6 | 38.1 | 40.6 | 43.2 | 45.7 | 47.0 |
|---|---|---|---|---|---|---|
| Load (kN) | 1860 | 1870 | 1890 | 1890 | 1900 | 1905 |

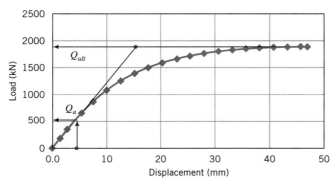

**FIGURE E8.1**

***Strategy*** Plot a graph of displacement versus load and then follow the procedure in Section 8.4.

## Solution 8.1

**Step 1:** Plot a displacement–load graph.
See Fig. E8.1.

**Step 2:** Determine the ultimate pile load capacity.
The failure load is ill defined. Locate the intersection of the tangents at the beginning and the end of the curve. The ordinate of this intersection is the ultimate pile load capacity.

$$Q_{ult} = 1780 \text{ kN}$$

**Step 3:** Determine the allowable pile load capacity.

$$FS = 2$$

$$Q_a = \frac{Q_{ult}}{FS} = \frac{1780}{2} = 890 \text{ kN}$$

**Step 4:** Determine the pile load capacity at 10% pile diameter.

$$\text{Displacement} = 450 \times 0.01 = 4.5 \text{ mm}$$
$$\text{From Fig. E8.1:} \quad Q_a = 510 \text{ kN} \qquad \blacksquare$$

***What's next . . .***Pile load tests are expensive and are not often conducted in the preliminary stages of a design. To get an estimate of the pile load capacity, at least in the preliminary design stages, recourse is made to statics and to correlations using soil test results. In the next section, we will examine how statics is applied to obtain an estimate of pile load capacity and some correlations between field soil test results and measured pile load capacity.

## 8.6 METHODS USING STATICS

### 8.6.1 α-Method

The α-method is based on a total stress analysis (TSA) and is normally used to estimate the short-term load capacity of piles embedded in fine-grained soils. In the $\alpha$-method, a coefficient, $\alpha_u$, is used to relate the undrained shear strength, $s_u$, to the adhesive stress ($f_s$) along the pile shaft. The skin friction, $Q_f$, over the embedded length of the pile is the product of the adhesive stress ($f_s = \alpha_u s_u$) and the surface area of the shaft (perimeter $\times$ embedded length). Thus,

$$Q_f = \sum_{i=1}^{j} (\alpha_u)_i (s_u)_i \times (\text{Perimeter})_i \times (\text{Length})_i \qquad (8.3)$$

where $j$ is the number of soil layers within the embedded length of the pile. For a cylindrical pile of uniform cross section and diameter, $D$, penetrating a homogeneous soil, $Q_f$ is given by

$$Q_f = \alpha_u s_u \pi D L \qquad (8.4)$$

where $L$ is the embedded length of the pile.

The value of $\alpha_u$ to use in determining the load capacity of piles is a subject of much debate and testing. Most tests to determine $\alpha_u$ are laboratory tests on

**TABLE 8.3  Skin Friction Factors $\alpha_u$ and $\beta$ for Driven Piles from Laboratory Tests**

| Soil type | Equation | Skin friction factors | Reference |
|---|---|---|---|
| Clay | $f_s = \alpha_u s_u$ | $\alpha_u = 1.0$ ($s_u \le 25$ kPa)<br>$\alpha_u = 0.5$ ($s_u \ge 70$ kPa)<br>$\alpha_u = 1 - \left(\dfrac{s_u - 25}{90}\right)$ for 25 kPa $< s_u <$ 70 kPa | API (1984) |
|  |  | $\alpha_u = 1.0$ ($s_u \le 35$ kPa)<br>$\alpha_u = 0.5$ ($s_u \ge 80$ kPa)<br>$\alpha_u = 1 - \left(\dfrac{s_u - 35}{90}\right)$ for 35 kPa $< s_u <$ 80 kPa | Semple and Rigden (1984) |
|  |  | $\alpha_u = \left(\dfrac{s_u}{\sigma'_z}\right)_{nc}^{0.5} \left(\dfrac{s_u}{\sigma'_z}\right)^{-0.5}$ for $(s_u/\sigma'_{zo}) \le 1$<br>$\alpha_u = \left(\dfrac{s_u}{\sigma'_z}\right)_{nc}^{0.5} \left(\dfrac{s_u}{\sigma'_z}\right)^{-0.25}$ for $(s_u/\sigma'_{zo}) \ge 1$ | Fleming et al. (1985) |
|  | $f_s = \beta \sigma'_z$ | $\beta = (1 - \sin \phi')(\tan \phi')(OCR)^{0.5}$ | Burland (1973) |
| Sand | $f_s = \beta \sigma'_z$ | $\beta = 0.15$–$0.35$ (compression)<br>$\quad\ 0.10$–$0.24$ (tension) | McClelland (1974) |
|  |  | $\beta = 0.44$ for $\phi' = 28°$<br>$\quad\ 0.75$ for $\phi' = 35°$<br>$\quad\ 1.2$ for $\phi' = 37°$ | Meyerhof (1976) |
|  |  | $\beta = (K/K_o) K_o \tan(\phi' \phi'_i/\phi')$<br>$\phi'_i/\phi'$ depends on installation method (range 0.5–1.0)<br>$K/K_o$ depends on installation method (range 0.5–2.0)<br>$K_o$ = coefficient of earth pressure at rest and is a function of OCR | Stas and Kulhawy (1984) |
| Uncemented calcareous sand | $f_s = \beta \sigma'_z$ | $\beta = 0.05$–$0.1$ | Poulos (1988) |

SOURCE: On behalf of the Institution of Civil Engineers.

model piles installed in a uniform deposit of soil. The major problems with these laboratory tests are:

1. It is difficult to scale up the laboratory model test results to real piles.
2. The soils in the field are mostly nonuniform compared with carefully prepared uniform soils in the laboratory.
3. Pile installation in the field strongly influences $\alpha_u$, which cannot be accurately duplicated in the laboratory.

Full-scale field tests on real piles are preferred, but such tests are expensive and the results may apply only to the site where the tests are performed. The results from cone penetrometers and the SPT have been linked to $\alpha_u$ but these are found from statistical correlations with rather low coefficient of correlation. Some popular ranges of values of $\alpha_u$ used in design are summarized in Tables 8.3 and 8.4. You should use $\alpha_u$ from at least two sources to estimate $Q_f$. As you gain field experience and correlate your estimated $Q_f$ with field tests on full-scale piles, you should be able to judge which source of $\alpha_u$ best represents your field condition.

The end bearing capacity is found by analogy with shallow foundations and is expressed as

$$Q_b = f_b A_b = N_c (s_u)_b A_b \tag{8.5}$$

where $f_b$ is the base resistance; $N_c$ is a bearing capacity coefficient, usually 9; $(s_u)_b$ is the undrained shear strength of the soil at the base of the pile; and $A_b$ is

**TABLE 8.4   Skin Friction Factors $\alpha_u$ and $\beta$ for Bored Piles or Drilled Shafts**

| Soil type | Equation | Skin friction factors | Reference |
|---|---|---|---|
| Clay | $f_s = \alpha_u s_u$ | $\alpha_u = 0.45$ (London clay) | Skempton (1959) |
| | | $\alpha_u = 0.7$ times value for driven displacement pile | Fleming et al. (1985) |
| | | $\alpha_u = 0$  for $z \leq 1.5$ m and $z > L - D$ | Reese and O'Neill (1988) |
| | | $\alpha_u = 0.55$ for all other points | |
| | $f_s = (K \tan \phi'_i) \sigma'_z$ | $K$ is lesser of $K_o$ or $0.5(1 + K_o)$ | Fleming et al. (1985) |
| | | $K/K_o = \frac{2}{3}$ to 1; $K_o$ is a function of OCR; $\phi'_i$ depends on interface materials | Stas and Kulhawy (1984) |
| Sand | $f_s = \beta \sigma'_z$ | $\beta = 0.1$  for $\phi' = 33°$ $0.2$  for $\phi' = 35°$ $0.35$  for $\phi' = 37°$ | Meyerhof (1976) |
| | | $\beta = F \tan(\phi' - 5°)$ where $F = 0.7$  (compression) $= 0.5$  (tension) | Kraft and Lyons (1974) |
| Uncemented calcareous sand | $f_s = \beta \sigma'_z$ $60$ kPa $\leq f_s \leq 100$ kPa | $\beta = 0.5 - 0.8$ | Poulos (1988) |

SOURCE: On behalf of the Institution of Civil Engineers.

**TABLE 8.5   End Bearing or Base Resistance Factors, $N_c$ and $N_q$**

| Soil type | Equation | End bearing factors | Reference |
|---|---|---|---|
| Clay | $f_b = N_c(s_u)_b$ | $N_c = 9$   for $L/D \geq 3$ | Skempton (1959) |
| | | $N_c = 6[1 + 0.2(L/D_b)]$;   $D_b$ is the diameter at the base of the pile, $N_c \leq 9$, and $f_b \leq 3.8$ MPa | Reese and O'Neill (1988) |
| | $f_b = N_q(\sigma_z')_b$ | $N_q = (\tan \phi' + \sqrt{1 + \tan^2 \phi'})^2 \exp(2\psi_p \tan \phi')$ | Janbu (1976) |
| | | $\psi_p = \pi/3$ to $0.58\pi$; lower values in the range are for soft, fine-grained soils, higher values are for dense, coarse-grained soils and overconsolidated fine-grained soils | |
| Sand | $f_b = N_q(\sigma_z')_b \leq f_{blim}$ | $N_q = 40$ | API (1984) |
| | | $N_q$ plotted against $\phi'$ (applicable to dense sand)—see Fig. 8.9 | Berezantzev et al. (1961) |
| | $f_{blim} = 10 - 15$ MPa | $N_q$ related to $\phi'$, relative density, and mean effective stress | Fleming et al. (1985) |
| | | $N_q$ from cavity expansion theory, as a function of $\phi'$ and volume compressibility [see Eq. (8.12)] | Vesic (1972) |
| Uncemented calcareous sand | $f_b = N_q\sigma_z' \leq f_{blim}$ | $N_q = 20$ | Datta et al. (1980) |
| | | Typical range of $N_q = 8$–$20$ | Poulos (1988) |
| | $f_{blim} = 3 - 5$ MPa | $N_q$ determined for reduced value of $\phi'$ (e.g., 18°) | Dutt and Ingram (1984) |

SOURCE: On behalf of the Institution of Civil Engineers.

the cross-sectional area of the base of the pile. Ranges of values for $N_c$ are shown in Table 8.5.

### 8.6.2 β-Method

The β-method is based on an effective stress analysis and is used to determine the short-term and long-term pile load capacities. The friction along the pile shaft is found using Coulomb's friction law, where the frictional stress is given by $f_s = \mu\sigma_x' = \sigma_x' \tan \phi_i'$, and where $\mu$ is the coefficient of friction, $\sigma_x'$ is the lateral effective stress, and $\phi_i'$ is the interfacial effective friction angle. The skin friction is expressed as

$$Q_f = \sum_{i=1}^{j} (\sigma_x')_i \tan \phi_i' \times (\text{Perimeter})_i \times (\text{Length})_i \qquad (8.6)$$

The lateral effective stress is proportional to the vertical effective stress ($\sigma_z'$) by a coefficient $K$ (Chapters 3 and 4). Therefore, we can write Eq. (8.6) as

$$Q_f = \sum_{i=1}^{j} K_i(\sigma_z)_i \tan \phi_i' \times (\text{Perimeter})_i \times (\text{Length})_i \qquad (8.7)$$

We can replace the two coefficients $K$ and $\tan \phi'_i$ by a single factor $\beta$ to yield

$$Q_f = \sum_{i=1}^{j} \beta_i (\sigma'_z)_i \times (\text{Perimeter})_i \times (\text{Length})_i \qquad (8.8)$$

Recall from Chapters 3 and 4 that for normally consolidated soils

$$K = K_o^{\text{nc}} = 1 - \sin \phi'_{\text{cs}}$$

and for overconsolidated soils

$$K = K_o^{\text{oc}} = (1 - \sin \phi'_{\text{cs}})(\text{OCR})^{0.5}$$

A general expression for $\beta$ is

$$\beta = K \tan \phi'_i = K_o^{\text{oc}} \tan \phi'_i = (1 - \sin \phi'_{\text{cs}})(\text{OCR})^{0.5} \tan \phi'_i \qquad (8.9)$$

The value of $\beta$ is also a subject of many debates, especially for coarse-grained soils. A short list of recommended values is shown in Tables 8.3 and 8.4. Because of the remolding or loosening of the soil adjacent to piles during installation, the value of $\phi'_i$ to use in determining $\beta$ should not be greater than $\phi'_{\text{cs}}$ or $\phi'_r$. The latter friction value should be used for overconsolidated clays with a predominance of flat platy particles. The vertical effective stress, $\sigma'_z$, is normally calculated at the center of each soil layer. You should estimate $Q_f$ using at least two sources of $\beta$.

The end bearing capacity is calculated by analogy with the bearing capacity of shallow footings and is determined from

$$Q_b = f_b A_b = N_q (\sigma'_z)_b A_b \qquad (8.10)$$

where $f_b = N_q (\sigma'_z)_b$ is the base resistance, $N_q$ is a bearing capacity coefficient that is a function of $\phi'$, $(\sigma'_z)_b$ is the vertical effective stress at the base, and $A_b$ is the cross-sectional area of the base.

Janbu (1976) proposed an equation for $N_q$ as a function of $\phi'$ and the angle $\psi_p$ (called the angle of pastification, as shown in Fig. 8.6). Janbu's equation is

$$N_q = (\tan \phi' + \sqrt{1 + \tan^2 \phi'})^2 \exp(2\psi_p \tan \phi') \qquad (8.11)$$

The value of $\psi_p$ varies from $\psi_p \leq \pi/3$ for soft, fine-grained soils to $\psi_p \leq 0.58\pi$ for dense, coarse-grained soils and overconsolidated fine-grained soils. For soft, compressible soils, the pile tip may easily penetrate the soil without causing significantly large plastic zones. For these types of soils, $\psi_p$ should not exceed $\pi/3$. In dense, coarse-grained soils, the plastic zones could be substantial but it is recommended (Janbu, 1976) that for these soils $\psi_p$ should not exceed $\pi/2$. Berezantzev et al. (1961) proposed a relationship between $N_q$ and $\phi'$ as shown in

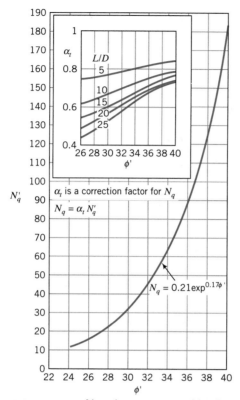

**FIGURE 8.9**   $N_q$ values proposed by Berezantzev et al. (1961).

Fig. 8.9; $N_q$ is corrected by a factor $\alpha_t$, which depends on the length to diameter ratio (see inset diagram, Fig. 8.9).

Vesic (1975) proposed another equation for $N_q$ from theoretical considerations of the expansion of a cylindrical cavity. Vesic's equation for $N_q$ is

$$N_q = \frac{3}{3 - \sin \phi'} \left\{ \exp\left[ \left( \frac{\pi}{2} - \phi' \right) \tan \phi' \right] \tan^2\left( \frac{\pi}{4} + \frac{\phi'}{2} \right) I_{rr}^{\frac{4}{3} \sin \phi'}{1 + \sin \phi} \right\} \tag{8.12}$$

where $I_{rr}$ is a rigidity index expressed as

$$I_{rr} = \frac{I_r}{1 + \varepsilon_p I_r} \tag{8.13}$$

and

$$I_r = \frac{G}{(\sigma_z')_b \tan \phi'} \tag{8.14}$$

where $G$ is the shear modulus, $\varepsilon_p$ is the volumetric strain, and $(\sigma_z')_b$ is the vertical effective stress at the base. The original equation for $N_q$ proposed by Vesic (1975) contained $s_u$ in the denominator of Eq. (8.14). The author removed $s_u$ from the original equation since we are considering an effective stress analysis. A list giving values for $N_q$ proposed by various investigators is shown in Table 8.5.

Coarse-grained soils are difficult to sample. A number of correlations be-

**TABLE 8.6    Correlations[a] Between Skin Frictional Stress, $f_s$, and SPT Values**

| Pile type | Soil type | A | B | Remarks | Reference |
|---|---|---|---|---|---|
| Driven displacement | Coarse-grained | 0 | 2.0 | $f_s$ = average value over shaft | Meyerhof (1956) |
| | | | | $N$ = average SPT along shaft Halve $f_s$ for small displacement pile | Shioi and Fukui (1982) |
| | Coarse-grained and fine-grained | 10 | 3.3 | Pile type not specified $50 \geq N \geq 3$ $f_s \leq 170$ kPa | Decourt (1982) |
| | Fine-grained | 0 | 10 | | Shioi and Fukui (1982) |
| Cast in place | Coarse-grained | 30 | 2.0 | $f_s \leq 200$ kPa | Yamashita et al. (1987) |
| | | 0 | 5.0 | | Shioi and Fukui (1982) |
| | Fine-grained | 0 | 5.0 | $f_s \leq 150$ kPa | Yamashita et al. (1982) |
| | | 0 | 10.0 | | Shioi and Fukui (1982) |
| Bored | Coarse-grained | 0 | 1.0 | | Findlay (1984) Shioi and Fukui (1982) |
| | | 0 | 3.3 | | Wright and Reese (1979) |
| | Fine-grained | 0 | 5.0 | | Shioi and Fukui (1982) |
| | Fine-grained | 10 | 3.3 | Piles cast under pentonite $50 \leq N \leq 3$ $f_s \leq 170$ kPa | Decourt (1982) |
| | Chalk | −125 | 12.5 | $30 > N > 15$ $f_s \leq 250$ kPa | Fletcher and Mizon (1984) |

[a]Correlation equation is $f_s = A + BN$ (kPa); $N$ values in blows/ft or blows/0.31 m. SOURCE: On behalf of the Institution of Civil Engineers.

tween field soil tests and pile load capacities have been proposed for coarse-grained soils. Some of these correlations for the SPT are shown in Tables 8.6 and 8.7. Other correlations exist for some penetrometers and pressure meters.

> *The essential points are:*
>
> 1. *The α-method is based on a TSA and is used to estimate the short-term pile load capacity in fine-grained soils.*
> 2. *The β-method is based on an ESA and is used to estimate the short-term and long-term pile load capacities in all soil types.*
> 3. *The actual values of $\alpha_u$, β, and $N_q$ are uncertain.*

**TABLE 8.7** **Correlations[a] Between End Bearing Resistance, $f_b$, and SPT Values**

| Pile type | Soil type | C | Remarks | Reference |
|---|---|---|---|---|
| Driven displacement | Sand | 0.45 | $N$ = average SPT value in local failure zone | Martin et al. (1987) |
| | Sand | 0.40 | | Decourt (1982) |
| | Sand | $0.04\dfrac{L_s}{D}$ | $L_s$ = length of pile in sand, $D$ = diameter or width. $N$ is in vicinity of base; $C \leq 0.4$ | Meyerhof (1976) |
| | Silt, sandy silts | 0.35 | | Martin et al. (1987) |
| | Glacial coarse to fine silt deposits | 0.25 | | Thorburn and MacVicar (1971) |
| | Residual sandy silts | 0.25 | | Decourt (1982) |
| | Residual clayey silts | 0.20 | | Decourt (1982) |
| | Clay | 0.20 | | Martin et al. (1987) |
| | Clay | 0.12 | | Decourt (1982) |
| | All soils | 0.30 | For $L/D \geq 5$ If $L/D < 5$, $C = 0.1 + 0.04\ L/D$ (closed-end piles) or $C = 0.06\ L/D$ (open-end piles) | Shioi and Fukui (1982) |
| Cast in place | Coarse-grained soils | | $f_b$ = 3.0 MPa | Shioi and Fukui (1982) |
| | | 0.15 | $f_b \leq 7.5$ MPa | Yamashita et al. (1987) |
| | Fine-grained soils | | $f_b = 0.09\ (1 + 0.16L_t)$ where $L_t$ = tip depth (m) | Yamashita et al. (1987) |
| Bored | Sand | 0.1 | | Shioi and Fukui (1982) |
| | Clay | 0.15 | | Shioi and Fukui (1982) |
| | Chalk | 0.25 0.20 | $N < 30$ $N > 40$ | Hobbs (1977) |

[a]Correlation equation is $f_b = CN$ (MPa) SOURCE: On behalf of the Institution of Civil Engineers.

## EXAMPLE 8.2

A cylindrical pile of diameter 400 mm is driven to a depth of 10 m into a clay with $s_u$ = 40 kPa, $\phi'_{cs}$ = 28°, OCR = 2 and $\gamma_{sat}$ = 18 kN/m³. Groundwater level is at the surface. Estimate the allowable load capacity for a factor of safety of 3. Is the pile a friction pile?

**Strategy** The solution is a straightforward application of the pile load capacity equations.

## Solution 8.2

**Step 1:** Select $\alpha_u$ and $\beta$.

$$\text{Table 8.3:} \quad \text{(API)} \ \alpha_u = 1 - \left(\frac{s_u - 25}{90}\right) = 1 - \left(\frac{40 - 25}{90}\right) = 0.83$$

$$\beta = (1 - \sin \phi'_{cs})(\text{OCR})^{0.5} \tan \phi'$$

For $\text{OCR} = 2$ and $\phi' = \phi'_{cs}$, we get

$$\beta = (1 - \sin 28°)(2)^{0.5} \tan 28° = 0.4$$

**Step 2:** Calculate $Q_a$ using a TSA.

$$L = 10 \text{ m}, \quad \text{Perimeter} = \pi D = \pi \times 0.4 = 1.26 \text{ m},$$

$$A_b = \frac{\pi D^2}{4} = \frac{\pi \times 0.4^2}{4} = 0.126 \text{ m}^2$$

$$Q_f = \alpha_u s_u (\pi D) L = 0.83 \times 40 \times (1.26) \times 10 = 418 \text{ kN}$$

$$Q_b = 9 s_u A_b = 9 \times 40 \times 0.126 = 45 \text{ kN}$$

$$Q_{ult} = Q_f + Q_b = 418 + 45 = 463 \text{ kN}$$

$$Q_a = \frac{Q_{ult}}{\text{FS}} = \frac{463}{3} = 154 \text{ kN}$$

**Step 3:** Calculate $Q_a$ using an ESA.

$$Q_f = \beta \sigma'_z (\pi D L)$$

Use an average value of $\sigma'_z$ over the embedded length.

$$\sigma'_z = \gamma' \frac{L}{2} = (18 - 9.8) \times \frac{10}{2} = 41 \text{ kPa}$$

$$Q_f = 0.4 \times 41 \times 1.26 \times 10 = 207 \text{ kN}$$

$$Q_b = N_q (\sigma'_z)_b A_b = N_q \gamma' L A_b = N_q \times (18 - 9.8)10 \times 0.126 = 10.3 N_q$$

Use Janbu's equation for $N_q$.

$$N_q = (\tan \phi'_{cs} + \sqrt{1 + \tan^2 \phi'_{cs}})^2 \exp(2\psi_p \tan \phi'_{cs})$$

For $\phi'_{cs} = 28°$ and assuming $\psi_p = \pi/3$ (soft clay), we find

$$N_q = \{\tan 28° + \sqrt{1 + \tan^2(28°)}\}^2 \exp\left(\frac{2\pi}{3} \tan 28°\right) = 8.4$$

$$Q_b = 10.3 \times 8.4 = 87 \text{ kN}$$

$$Q_{ult} = Q_f + Q_b = 207 + 87 = 294 \text{ kN}$$

$$Q_a = \frac{Q_{ult}}{3} = \frac{294}{3} = 98 \text{ kN}$$

$$\frac{Q_f}{Q_{ult}} = \frac{207}{294} = 0.7; \text{ pile not solely a friction pile.}$$

The load capacity from an ESA is less than that from a TSA. Therefore, use the allowable pile load capacity from the ESA, that is, $Q_a = 98$ kN. ∎

## EXAMPLE 8.3

A square concrete pile 0.3 m × 0.3 m is required to support a load of 150 kN with a factor of safety of 3. The soil stratification consists of 5 m of medium clay ($s_u = 40$ kPa, $\phi'_{cs} = 26°$, OCR = 2, $\gamma_{sat} = 18$ kN/m³) underlain by a deep deposit of overconsolidated clay ($s_u = 80$ kPa, $\phi'_{cs} = 24°$, OCR = 4, $\gamma_{sat} = 18.5$ kN/m³). Groundwater level is at 2 m below the ground surface. You may assume that the soil above the groundwater level is saturated. Estimate the length of pile required.

***Strategy*** You are given $Q_a$ and FS, so you can calculate $Q_{ult} = Q_a \times$ FS. Assume an embedment depth, $L_1$, into the stiff clay and then solve for $L_1$ since you know $Q_{ult}$. You should consider both short-term and long-term conditions and select the value of $L_1$ that is larger.

## Solution 8.3

**Step 1:** Determine the ultimate load.

$$Q_a = 150 \text{ kN}$$
$$Q_{ult} = Q_a \times \text{FS} = 150 \times 3 = 450 \text{ kN}$$

**Step 2:** Draw a sketch of the soil profile and pile, and calculate the required coefficients.
See Fig. E8.3.

Table 8.3:  $s_u = 40$ kPa,  $\alpha_u = 0.83$ [see Example 8.2]
$\qquad\qquad s_u = 80$ kPa,  $\alpha_u = 0.5$ (stiff clay)

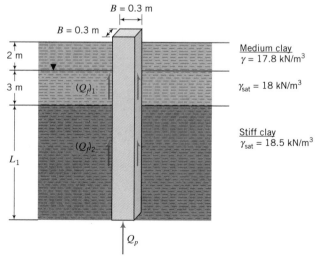

**FIGURE E8.3**

For $\phi' = \phi'_i = \phi'_{cs}$, we find the following:

$$\text{Medium clay:} \quad \beta = (1 - \sin \phi'_{cs})(OCR)^{0.5} \tan \phi'_{cs}$$
$$= (1 - \sin 26°)(2)^{0.5} \tan 26° = 0.39$$
$$\text{Stiff clay:} \quad \beta = (1 - \sin 24°)(4)^{0.5} \tan 24° = 0.53$$
$$\text{Perimeter} = 2(B + \overline{B}) = 2(0.3 + 0.3) = 1.2 \text{ m}$$
$$A_b = 0.3 \times 0.3 = 0.09 \text{ m}^2$$

**Step 3:** Calculate the required length using a TSA.

**Medium clay**

$$\text{Length in medium clay} = 5 \text{ m}$$
$$(Q_f)_{\text{medium clay}} = \alpha_u s_u \times \text{Perimeter} \times \text{Length}$$
$$= 0.83 \times 40 \times 200 \times 5 = 200 \text{ kN}$$

**Stiff clay**

$$(Q_f)_{\text{stiff clay}} = 0.5 \times 80 \times 1.2 \times L_1 = 48L_1 \text{ kN}$$
$$Q_p = 9s_u A_b = 9 \times 80 \times 0.09 = 65 \text{ kN}$$

**Ultimate load capacity**

$$Q_{ult} = (Q_f)_{\text{medium clay}} + (Q_f)_{\text{stiff clay}} + Q_p$$

That is,

$$450 = 200 + 48L_1 + 65$$

Solving, we get

$$L_1 = 3.9 \text{ m}$$

**Step 4:** Calculate the required length using an ESA.

Medium clay: Average value of $\sigma'_z = (2 \times 18) + 0.5(18 - 9.8) = 40.1$ kPa

Stiff clay: Average value of $\sigma'_z$ over length $L_1$:

$$\sigma'_z = (2 \times 18) + 3(18 - 9.8) + \frac{L_1}{2}(18.5 - 9.8) = 60.6 + 4.35L_1$$

$$Q_f = \beta \sigma'_z \times \text{Perimeter} \times \text{Length}$$

Medium clay: $(Q_f)_{\text{medium clay}} = 0.39 \times 40.1 \times 1.2 \times 5 = 94$ kN

Stiff clay: $(Q_f)_{\text{stiff clay}} = 0.53 \times (60.6 + 4.35L_1) \times 1.2 \times (5 + L_1)$
$$= 2.8(L_1)^2 + 52.4L_1 + 192.7$$

**End bearing resistance in stiff clay**

Using Janbu's equation with $\phi' = \phi'_{cs} = 24°$ and $\psi_p = \pi/2$ gives

$$N_q = [\tan 24° + \sqrt{1 + \tan^2(24°)}]^2 \exp\left(\frac{2\pi}{2} \tan 24°\right) = 9.6$$

$$(\sigma'_z)_b = 2 \times 18 + 3(18 - 9.8) + L_1(18.5 - 9.8) = 60.6 + 8.7L_1$$
$$Q_p = N_q(\sigma'_z)_b A_b = 9.6 \times (60.6 + 8.7L_1) \times 0.09 = 52.4 + 7.5L_1$$

**Ultimate load capacity**

$$Q_{ult} = (Q_f)_{\text{soft clay}} + (Q_f)_{\text{stiff clay}} + Q_p$$

That is,

$$450 = 94 + 2.8(L_1)^2 + 52.4L_1 + 192.7 + (52.4 + 7.5L_1)$$

Rearranging and simplifying, we get

$$450 = 2.8L_1^2 + 59.9L_1 + 339.1$$
$$2.8L_1^2 + 59.9L_1 - 110.9 = 0$$

Solution is $L_1 = 1.7$ m.

**Step 5:** Decide on the length required.

$$\text{TSA:}\quad L_1 = 3.9 \text{ m}$$
$$\text{ESA:}\quad L_1 = 1.7 \text{ m}$$

Use the larger of the two values,

$$L_1 = 3.9 \text{ m}$$
$$\text{Total length} = 5 + L_1 = 8.9 \text{ m};\quad \text{use } L = 9 \text{ m} \quad \blacksquare$$

## EXAMPLE 8.4

A pile 450 mm in diameter and 15 m long is driven into a clay. The undrained strength of the soil varies linearly with depth such that $s_u = 0.22\sigma_z'$. Determine the allowable pile load capacity using a TSA. The factor of safety required is 2 and $\gamma_{\text{sat}} = 17$ kN/m³. Groundwater is at the surface.

**Strategy** Consider an element of thickness $dz$ and perform integration to find the skin friction. You need to check if $s_u > 25$ kPa. If it is not, then $\alpha_u = 1$ or else you would have to use $\alpha_u$ as a linear variable of $s_u$.

## Solution 8.4

**Step 1:** Make a sketch of the problem.
See Fig. E8.4.

**Step 2:** Calculate the skin friction.

$$\gamma' = 17 - 9.8 = 7.2 \text{ kN/m}^3$$
$$(s_u)_b = 0.22\gamma'L = 0.22 \times 7.2 \times 15 = 23.8 \text{ kPa}$$
$$\text{Table 8.3:}\quad \alpha_u = 1 \quad \text{for } (s_u)_b < 25 \text{ kPa}$$

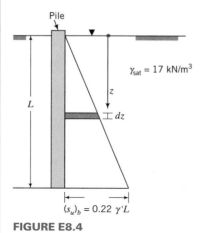

**FIGURE E8.4**

Consider an element of thickness $dz$. The surface area is $2\pi r\, dz$ and the skin friction is

$$Q_f = 2\pi r\alpha_u \int_0^L (0.22\gamma z)\, dz = 2.24\left[\frac{z^2}{2}\right]_0^{15} = 252 \text{ kN}$$

**Step 3:** Calculate the end bearing capacity.

$$A_b = \frac{\pi \times D^2}{4} = \frac{\pi \times 0.45^2}{4} = 0.159 \text{ m}^2$$

$$Q_b = 9(s_u)_b A_b = 9 \times 23.8 \times 0.159 = 34 \text{ kN}$$

**Step 4:** Calculate the allowable load capacity.

$$Q_{ult} = Q_f + Q_b = 252 + 34 = 286 \text{ kN}$$

$$Q_a = \frac{Q_{ult}}{\text{FS}} = \frac{286}{3} = 95 \text{ kN}$$ ∎

## EXAMPLE 8.5

A 450 mm diameter pile is driven into a sand profile to depth of 10 m. The SPT results are shown in the table below. Estimate the allowable load capacity for a factor of safety of 3.

| Depth (m) | 1 | 3 | 5 | 6 | 8 | 10 | 11 | 13 |
|---|---|---|---|---|---|---|---|---|
| $N$ (blows/ft) | 22 | 18 | 25 | 20 | 30 | 36 | 39 | 45 |

***Strategy*** The $N$ values are blows/ft not blows/m. Use one of the correlations shown in Tables 8.6 and 8.7.

## Solution 8.5

**Step 1:** Determine the average value of $N$ over the depth.

$$N = N_{av} = \frac{22 + 18 + 25 + 20 + 30 + 36}{6} = 25$$

**Step 2:** Determine the load capacity.

**Skin friction**

$$\text{Perimeter} = \pi D = \pi \times 0.45 = 1.41$$

Table 8.6: $\quad Q_f = (A + BN) \times \text{Perimeter} \times \text{Length}$

Using Meyerhof's method, $A = 0, .B = 2$.

$$Q_f = (0 + 2 \times 25) \times 1.41 \times 10 = 705 \text{ kN}$$

**End bearing**

$$A_b = \frac{\pi D^2}{4} = \frac{\pi \times 0.45^2}{4} = 0.159 \text{ m}^2$$

Using Table 8.7 and Meyerhof's method,

$$Q_b = 0.04N \frac{L}{D} A_b \leq 0.4NA_b, \text{ where } N \text{ is the SPT value at the base.}$$

$$Q_b = 0.04 \times 36 \times \frac{10}{0.45} \times 0.159 = 5.09 \text{ MN}$$

Check that $Q_b \leq 0.4NA_b$:   $0.4NA_b = 0.4 \times 36 \times 0.159 = 2.29 \text{ MN} < Q_b.$

$$\therefore Q_b = 2.29 \text{ MN} = 2290 \text{ kN}$$
$$Q_{ult} = Q_f + Q_b = 705 + 2290 = 2995 \text{ kN}$$
$$Q_a = \frac{Q_{ult}}{\text{FS}} = \frac{2995}{3} = 998.3 \text{ kN}$$ ∎

***What's next*** . . .Piles are rarely used singly. They are often clustered into groups. Next, we will discuss how to determine the load capacity of pile groups.

## 8.7   PILE GROUPS

In most practical situations, piles are used in groups. They are arranged in geometric patterns (squares, rectangles, circles, and octagons) at a spacing, $s$ (center to center distance), not less than $2D$ (where $D$ is the diameter or width of the pile). The piles are connected at their heads by a concrete pile cap, which may or may not be in contact with the ground (Fig. 8.10). If the pile cap is in contact with the ground, part of the load will be transferred directly to the soil.

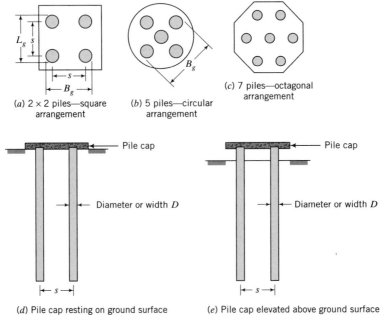

(a) 2 × 2 piles—square arrangement

(b) 5 piles—circular arrangement

(c) 7 piles—octagonal arrangement

(d) Pile cap resting on ground surface

(e) Pile cap elevated above ground surface

**FIGURE 8.10**   Pile groups.

The load capacity for a pile group is not necessarily the load capacity of a single pile multiplied by the number of piles. In fine-grained soils, the outer piles tend to carry more loads than the piles in the center of the group. In coarse-grained soils, the piles in the center take more loads than the outer piles.

The ratio of the load capacity of a pile group, $(Q_{ult})_g$, to the total load capacity of the piles acting as individual piles ($nQ_{ult}$) is called the efficiency factor, $\eta_e$; that is,

$$\eta_e = \frac{(Q_{ult})_g}{nQ_{ult}} \tag{8.15}$$

where $n$ is the number of piles in the group and $Q_{ult}$ is the ultimate load capacity of a single pile. The efficiency factor is usually less than 1. However, piles driven into a loose, coarse-grained soil tend to densify the soil around the piles and $\eta_e$ could exceed 1.

Two modes of soil failure are normally investigated to determine the load capacity of a pile group. One mode, called block failure (Fig. 8.11), may occur when the spacing of the piles is small enough to cause the pile group to fail as a unit. The group load capacity for block failure mode is

**ESA**

$$(Q_{ult})_{gb} = \sum_{i=1}^{j} \{\beta_i(\sigma_z')_i \times (\text{Perimeter})_{ig} \times L_i\} + N_q(\sigma_z')_b(A_b)_g \tag{8.16}$$

**TSA**

$$(Q_{ult})_{gb} = \sum_{i=1}^{j} \{(\alpha_u)_i(s_u)_i \times (\text{Perimeter})_{ig} \times L_i\} + N_c(s_u)_b(A_b)_g \tag{8.17}$$

where the subscript $gb$ denotes block mode of failure for the group.

The other failure mode is failure of individual piles called single pile failure mode or punching failure mode. The key assumption in the single pile failure mode is that each pile mobilizes its full load capacity. Thus, the group load capacity is

$$(Q_{ult})_g = nQ_{ult} \tag{8.18}$$

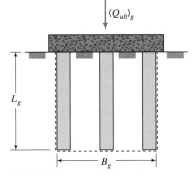
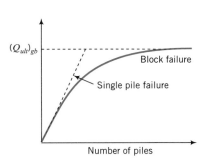

**FIGURE 8.11** Block failure mode.

The group efficiency in fine-grained soils is defined as

$$\boxed{\eta_e = (Q_{ult})_{gb}[(Q_{ult})_{gb}^2 + (nQ_{ult})^2]^{-1/2}} \tag{8.19}$$

The values of $\psi_p$ to use in determining $N_q$ in Janbu's equation depend on the $s/D$ ratio and the friction angle. Janbu (1976) showed that

$$\boxed{\frac{s}{D} = 1 + 2\sin\psi_p(\tan\phi' + \sqrt{1 + \tan^2\phi'})\exp(\psi_p\tan\phi')} \tag{8.20}$$

The value of $\psi_p$ is not significantly affected by $s/D \leq 2.5$.

> **The essential points are:**
> 1. **The ultimate load capacity of a pile group is not necessarily the ultimate load capacity of a single pile multiplied by the number of piles in the group.**
> 2. **A pile group can either fail by the group failing as a single unit, called block failure mode, or as individual piles, called single pile failure mode.**

### EXAMPLE 8.6

A pile group consisting of 9 piles, each 0.4 m in diameter, is arranged in a $3 \times 3$ matrix at a spacing of 1.2 m. The piles penetrate a clay soil ($s_u = 40$ kPa, $\phi'_{cs} = 30°, \gamma = 17$ kN/m³, $\gamma_{sat} = 18$ kN/m³, OCR = 2) of thickness 8 m and are embedded 2 m in a stiff clay ($s_u = 90$ kPa, $\phi'_{cs} = 28°, \gamma = 17.5$ kN/m³, $\gamma_{sat} = 18.5$ kN/m³, OCR = 5). Calculate the group allowable load capacity for a factor of safety of 3 and the efficiency factor. Groundwater level (GWL) is at 2 m below the surface but can rise to the surface due to seasonal changes.

**Strategy** You need to calculate the ultimate load capacity assuming (a) block failure mode and (b) single pile failure mode. Use a sketch to illustrate the problem.

### Solution 8.6

**Step 1:** Draw a sketch and calculate the geometric properties. See Fig. E8.6.

Single pile: $D = 0.4$ m; Perimeter $= \pi D = \pi \times 0.4 = 1.26$ m;

$$A_b = \frac{\pi D^2}{4} = \pi \times \frac{0.4^2}{4} = 0.126 \text{ m}^2$$

Group: Perimeter $= 4(2s + D) = 4[2(1.2) + 0.4] = 11.2$ m;

Base area $= (A_b)_g = (2s + D)^2 = 2.8^2 = 7.84$ m²

**Step 2:** Calculate the ultimate load capacity using TSA.

#### TSA—Block Failure Mode

##### Soft clay—skin friction

Table 8.3: $\alpha_u = 0.83$ for $s_u = 40$ kPa [see Example 8.2]

$(Q_f)_{\text{soft clay}} = \alpha_u s_u \times (\text{Perimeter})_g \times \text{Length}$

$\qquad = 0.83 \times 40 \times 11.2 \times 8 = 2975$ kN

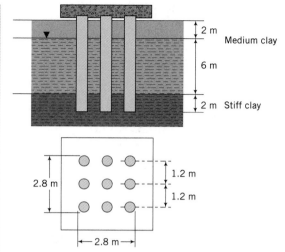

**FIGURE E8.6**

**Stiff clay—skin friction**

Table 8.3:   $\alpha_u = 0.5$   for $s_u = 90$ kPa

$$(Q_f)_{\text{stiff clay}} = 0.5 \times 90 \times 11.2 \times 2 = 1008 \text{ kN}$$

**Stiff clay—end bearing**

$$(Q_p)_{\text{stiff clay}} = 9s_u(A_b)_g = 9 \times 90 \times 7.84 = 6350 \text{ kN}$$

**Group load capacity**

$$(Q_{ult})_{gb} = (Q_f)_{\text{medium clay}} + (Q_f)_{\text{stiff clay}} + (Q_p)_{\text{stiff clay}}$$
$$(Q_{ult})_{gb} = 2975 + 1008 + 6350 = 10333 \text{ kN}$$

## TSA—Single Pile Failure Mode

**Medium clay—skin friction**

$$\begin{aligned}(Q_f)_{\text{medium clay}} &= \alpha_u s_u \times \text{Perimeter} \times \text{Length} \\ &= 0.83 \times 40 \times 1.26 \times 8 = 335 \text{ kN}\end{aligned}$$

**Stiff clay—skin friction**

$$(Q_f)_{\text{stiff clay}} = 0.5 \times 90 \times 1.26 \times 2 = 113 \text{ kN}$$

**Stiff clay—end bearing**

$$(Q_b)_{\text{stiff clay}} = 9s_u A_b = 9 \times 90 \times 0.126 = 102 \text{ kN}$$

**Single pile load capacity**

$$Q_{ult} = (Q_f)_{\text{medium clay}} + (Q_f)_{\text{stiff clay}} + (Q_p)_{\text{stiff clay}}$$
$$Q_{ult} = 335 + 113 + 102 = 550 \text{ kN}$$

**Group load capacity**

Number of piles:   $n = 9$

$$(Q_{ult})_g = nQ_{ult} = 9 \times 550 = 4950 \text{ kN}$$

**Step 3:** Calculate the ultimate load capacity using an ESA.

### ESA—Block Failure Mode

Assume GWL will rise to the surface and $\phi'_i = \phi'_{cs}$.

Table 8.3: $\beta = (1 - \sin \phi'_{cs})(OCR)^{0.5} \tan \phi'_{cs}$

Medium clay: $\gamma' = 18 - 9.8 = 8.2 \text{ kN/m}^3$

Stiff clay: $\gamma' = 18.5 - 9.8 = 8.7 \text{ kN/m}^3$

Medium clay: $\beta = (1 - \sin 30°)(2)^{0.5} \tan 30° = 0.41$

Stiff clay: $\beta = (1 - \sin 28°)(5)^{0.5} \tan 28° = 0.63$

#### Skin friction

$Q_f = \beta\sigma'_z \times (\text{Perimeter})_g \times L$; $\sigma'_z$ is calculated at mid-depth of the layer

Medium clay: $Q_f = 0.41 \times 8.2 \times \dfrac{8}{2} \times 11.2 \times 8 = 1205 \text{ kN}$

Stiff clay: $Q_f = 0.63 \times (8.2 \times 8 + 8.7 \times 1) \times 11.2 \times 2 = 1049 \text{ kN}$

#### End bearing—stiff clay

$$Q_b = N_q(\sigma'_z)_b(A_b)_g$$

Use Janbu's equation with $\phi' = \phi'_{cs}$ and $\psi_p = \pi/2$.

$$N_q = [\tan 28° + \sqrt{1 + \tan^2(28°)}]^2 \exp\left(2\,\frac{\pi}{2} \tan 28°\right) = 14.7$$

$(\sigma'_z)_b = 8 \times 8.2 + 2 \times 8.7 = 83 \text{ kPa}$

$Q_b = 14.7 \times 83 \times 7.84 = 9566 \text{ kN}$

#### Group load capacity

$$(Q_{ult})_{gb} = 1205 + 1049 + 9566 = 11820 \text{ kN}$$

### ESA—Single Pile Failure Mode

#### Skin friction

Use proportion since the only difference in the calculation is the perimeter.

Medium clay: $Q_f = 1205 \times \dfrac{136}{11.2} = 136 \text{ kN}$

Stiff clay: $Q_f = 1049 \times \dfrac{1.26}{11.2} = 118 \text{ kN}$

#### End bearing—stiff clay

$$Q_b = 9566 \times \frac{0.126}{7.84} = 154 \text{ kN}$$

#### Group load capacity

$$Q_{ult} = 136 + 118 + 154 = 408 \text{ kN}$$
$$(Q_{ult})_g = nQ_{ult} = 9 \times 408 = 3672 \text{ kN}$$

**Step 4:** Decide which failure mode and conditions govern.

| Analysis | Load capacity (kN) | |
|---|---|---|
| | **Block mode** | **Single pile mode** |
| TSA | 10333 | 4950 |
| ESA | 11820 | 3672 |

The lowest ultimate load capacity is 3672 kN for an ESA.

$$\therefore Q_a = \frac{(Q_{ult})_g}{FS} = \frac{3672}{3} = 1224 \text{ kN} \qquad \blacksquare$$

*What's next* . . .We have discussed methods to determine the ultimate load capacity, that is, the ultimate limit state, which is only one of two limit states required in analyses and design of geotechnical systems. The other limit state is serviceability limit state (settlement). Next, we will examine settlement of piles.

## 8.8 ELASTIC SETTLEMENT OF PILES

The elastic settlement of a single pile depends on the relative stiffness of the pile and the soil, the length to diameter ratio of the pile, the relative stiffness of the soil at the base and of the soil over the pile length, and the distribution of elastic modulus of the soil along the pile length. The relative stiffness of the pile to the soil is $K_{ps} = E_p/E_{so}$, where $E_p$ is the elastic modulus of an equivalent solid cross section of the pile and $E_{so}$ is the elastic modulus of the soil. The elastic modulus of the soil at the base or tip of the pile will be denoted by $E_{sb}$. Usually, the secant elastic modulus is used in design practice. For short-term loading in fine-grained soils, $(E_{so})_u$ is used.

Various analyses have been proposed to calculate the settlement of single piles and pile groups. Poulos (1989) provided an excellent discussion on the various numerical procedures to calculate settlement of piles. The settlement consists of three components—skin friction, end bearing, and elastic shortening.

Skin friction tends to deform the soil near the shaft, as illustrated in Fig. 8.12. The deformation mode near the shaft is analogous to simple shear strain (Chapter 3), and the shear strain, $\gamma_{zx}$, is

$$\boxed{\gamma_{zx} = \frac{\tau}{G} = \frac{f_s}{G}} \qquad (8.21)$$

where $G$ is the shear modulus, $\tau$ is shear stress and $f_s$ is the skin frictional stress. The shear strains can be integrated over the pile length to give the elastic settlement ($\rho_{es}$) resulting from skin friction; that is,

$$\rho_{es} = \frac{1}{G(z)} \int_0^L \tau(z) \, dz \qquad (8.22)$$

where $(z)$ means that the parameter in front of it varies with depth. To solve Eq. (8.22), we need to know how $G$ and $\tau$ vary with depth but this we do not know.

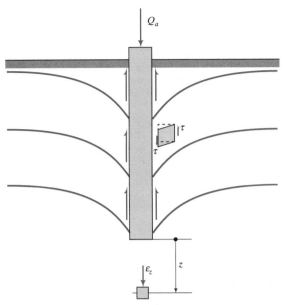

**FIGURE 8.12** Shear deformation from skin friction and compression from end bearing.

Therefore, we have to speculate on their variations and then solve Eq. (8.22). A further complication arises in that the boundary conditions for a pile problem are complex. Thus, we have to solve Eq. (8.22) using numerical procedures. For example, we can assume that

$$\tau(z) = \int F(z, i)\tau(i) \; di \tag{8.23}$$

where $F(z, i)$ is a stress function and $\tau(i)$ is the shear stress at an ordinate $i$. We can use finite element or boundary element to solve Eqs. (8.22) and (8.23). Stress functions for a point load within a half-space were developed by Mindlin (1936).

For a homogeneous soil ($E_{so}$ is constant with depth), a solution of Eq. (8.22) using Eq. (8.23) for the elastic settlement of a single pile is

$$\rho_{es} = \frac{Q_a}{E_{so}L} I \tag{8.24}$$

where $I$ is an influence factor and $Q_a$ is the design load transferred as skin friction. An approximate equation for $I$ is

$$I = 0.5 + \log\left(\frac{L}{D}\right) \tag{8.25}$$

Soft soils tend to have elastic moduli that vary linearly with depth; that is,

$$E_{so} = mz \tag{8.26}$$

where $m$ is the rate of increase of $E_{so}$ with depth (units: kPa/m). For soft soils, the elastic settlement is

$$(\rho_{es})_{so} = \frac{Q_a}{mL^2} I_{so} \tag{8.27}$$

where

$$I_{so} = 2.0 \log\left(\frac{L}{D}\right)$$ (8.28)

Poulos (1989) developed another solution for the elastic settlement. He showed that $\rho_{es}$ for a floating pile can be determined from

$$\rho_{es} = \frac{Q_a}{E_{so}D} I_\rho$$ (8.29)

where $I_\rho$ is an influence factor that depends on the $L/D$ ratio and $K_{ps}$ as shown in Fig. 8.13.

The soil mass under a pile is subjected to compression from end bearing. We can use elastic analyses (described in Chapter 3) to determine the elastic compression under the pile. For friction piles, the settlement due to end bearing is small in comparison to skin friction and is often neglected.

The elastic shortening of the pile ($\rho_p$) is found for column theory, which for a soil embedded in a homogeneous soil is

$$\rho_p = C \frac{Q_a L}{E_p A_p}$$ (8.30)

where $C$ is a reduction factor to account for the fact that the vertical strain reduces with the embedded length of the pile. For most soils, $C \approx 0.5$ except for soft clay soils for which $C \approx 0.7$. Elastic shortening is only significant for slender piles ($E_p/E_{so} < 500$). The total elastic settlement for a floating pile is $\rho_{et} = \rho_{es} + \rho_p$.

Field observations indicate that at design loads, the elastic settlement of a single driven pile is within the range 0.9% to 1.25% of the pile diameter. Hull (1987) found, from numerical analyses, that embedding a pile beyond a certain critical length does not reduce the settlement. The critical length normalized to the diameter (width) is

$$\frac{L_c}{D} = \sqrt{\frac{\pi E_p A_p}{E_{so}D^2}} = \sqrt{\frac{\pi K_{ps} A_p}{D^2}}$$ (8.31)

where $L_c$ is the critical length and $A_p$ is the area of the cross section of the pile.

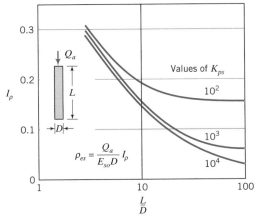

**FIGURE 8.13** Influence factor for vertical settlement of a single floating pile. (After Poulos, 1989.)

Piles in a group tend to interact with each other depending on the spacing between them. The smaller the spacing, the greater the interaction and the larger the settlement. Pile group settlement is influenced by spacing to diameter (or width) ratio $(s/D)$, the number of piles $(n)$ in the group, and the length to diameter ratio $(L/D)$. For convenience, pile group settlement is related to a single pile settlement through a group settlement factor $R_s$ defined as

$$R_s = \frac{\text{Settlement of group}}{\text{Settlement of single pile at same average load}} \tag{8.32}$$

An empirical relationship between $R_s$ and $n$ (Fleming et al., 1985) is

$$\boxed{R_s = n^\Phi} \tag{8.33}$$

where the exponent $\Phi$ is between 0.4 and 0.6.

## 8.9  CONSOLIDATION SETTLEMENT UNDER A PILE GROUP

Sometimes, a pile group may be embedded above a soft clay layer and transfer sufficient load to it (soft clay) to cause consolidation settlement. To estimate the consolidation settlement, the full design load is assumed to act at a depth of $\frac{2}{3}L$ and is then distributed in the ratio of 2:1 (vertical:horizontal). The increase in vertical stress at a depth $z$ in the soft clay layer shown in Fig. 8.14 is

$$\boxed{\Delta\sigma_z = \frac{Q_{ag}}{(B_g + z)(L_g + z)}} \tag{8.34}$$

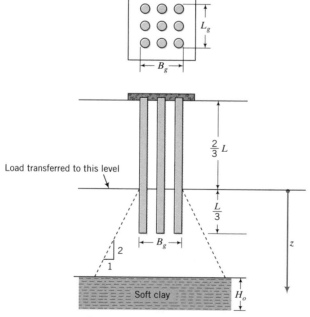

**FIGURE 8.14** Assumed distribution of load for calculating settlement of a pile group.

where $Q_{ag}$ is the design group load, $B_g$ is the width of the group, $L_g$ is the length of the group, and $z$ is the depth from the load transfer point on the pile group to the location in the clay layer at which the increase in vertical stress is desired. The primary consolidation settlement is then calculated using the procedure described in Section 4.4.

# 8.10  PROCEDURE TO ESTIMATE SETTLEMENT OF SINGLE AND GROUP PILES

The procedure to estimate the settlement of single piles and pile groups is:

1. Obtain the required parameters $Q_a$ or $Q_{ag}$, $E_{so}$, and $E_p$.
2. Calculate the elastic settlement, $\rho_{es}$, for a single pile using Eq. (8.24) or (8.27) or (8.29).
3. Calculate the elastic shortening ($\rho_p$) using Eq. (8.30).
4. Calculate the total elastic settlement, $\rho_{et} = \rho_{es} + \rho_p$.

For group piles, skip procedure 4 and continue as follows:

5. Calculate $R_s$ using an estimated value of $\Phi$.
6. Calculate the group settlement from $(\rho_e)_g = R_s\rho_{es} + \rho_p$.
7. Calculate the consolidation settlement, $\rho_{pc}$, if necessary.
8. Add the elastic and the consolidation settlement to obtain the total group settlement.

> *The* **essential points** *are:*
> 1. **Settlement of piles is determined using numerical analyses assuming the soil is an elastic material.**
> 2. **The equations given for pile settlement will only give an estimated settlement.**

## EXAMPLE 8.7

Determine the settlement of the pile in Example 8.2 using $E_p = 20 \times 10^6$ kPa, $E_{so} = 5000$ kPa, and a design shaft friction load $Q_a = 70$ kN.

**Strategy**  This is a straightforward application of the procedure in Section 8.10.

### Solution 8.7

**Step 1:**  Determine the influence factor.

$$\frac{L}{D} = \frac{10}{0.4} = 25$$

Equation (8.25):  $I = 0.5 + \log\left(\frac{L}{D}\right) = 0.5 + \log(25) = 1.9$

$$K_{ps} = \frac{E_p}{E_{so}} = \frac{20 \times 10^6}{5000} = 4000$$

From Fig. 8.13, $I_\rho = 0.09$ for $K_{ps} = 4000$ and $L/D = 25$.

**Step 2:**   Calculate the elastic settlement.

Equation (8.24):   $\rho_{es} = \dfrac{Q_a}{E_{so}L} I = \dfrac{70}{5000 \times 10} \times 1.9$

$= 2.7 \times 10^{-3}$ m $= 2.7$ mm

Equation (8.29):   $\rho_{es} = \dfrac{Q_a}{E_{so}D} I_\rho = \dfrac{70}{5000 \times 0.4} \times 0.09$

$= 3.2 \times 10^{-3}$ m $= 3.2$ mm

Difference between the two solutions is 0.5 mm. Use an average value of 3 mm.

**Step 3:**   Calculate the elastic shortening of the pile.

Equation (8.30):   $\rho_p = C \dfrac{Q_a L}{E_p A_p}$

$A_p = \dfrac{\pi \times 0.4^2}{4} = 0.126$ m$^2$;   $C = 0.7$

$\rho_p = 0.7 \dfrac{70 \times 10}{20 \times 10^6 \times 0.126}$

$= 277.8 \times 10^{-6}$ m $\cong 278 \times 10^{-3}$ mm

The elastic shortening of the pile is extremely small and can be neglected.

**Step 4:**   Determine the total elastic settlement.

$\rho_{et} = \rho_{es} + \rho_p = 3.0 + (\approx 0) = 3.0$ mm

The estimated total elastic settlement is approximately 3 mm.

$$\dfrac{\rho_t}{D} = \dfrac{3}{400} \times 100 = 0.75\%$$   ∎

## EXAMPLE 8.8

A $3 \times 3$ pile group with a pile spacing of 1 m and pile diameter of 0.4 m supports a load of 3 MN (Fig. E8.8). (a) Determine the factor of safety of the pile group. (b) Calculate the total settlement of the pile group.

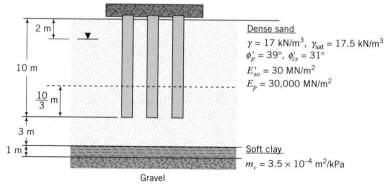

FIGURE E8.8

***Strategy*** Since the sand is dense, driving of the piles will likely loosen it around the piles. You should use $\phi'_{cs}$ in calculating the pile load capacity.

## Solution 8.8

**Step 1:** Determine the geometric parameters and $\beta$ and $N_q$.

$$D = 0.4 \text{ m}, \frac{L}{D} = \frac{10}{0.4} = 2s; \quad n = 9 \text{ piles}, \ s = 1 \text{ m};$$

$$K_{ps} = \frac{E_p}{E'_{so}} = \frac{30 \times 10^6}{30,000} = 1000$$

Single pile:   Perimeter $= \pi D = \pi \times 0.4 = 1.26 \text{ m};$

$$A_b = \frac{\pi D^2}{4} = \frac{\pi \times 0.4^2}{4} = 0.126 \text{ m}^2$$

Group piles:   $L_g = B_g = 2s + D = 2 \times 1 + 0.4 = 2.4;$

Perimeter for group $= 2.4 \times 4 = 9.6 \text{ m}; \quad (A_b)_g = 2.4^2 = 5.76 \text{ m}^2$

Table 8.3:   $\beta = 0.44 + \left( \dfrac{0.75 - 0.44}{7} \times 3 \right) = 0.57$

[by linear interpolation of Meyerhof (1976) values]

Figure 8.9:   $N'_q = 40; \quad \alpha_t = 0.55$ for $\dfrac{L}{D} = 25;$

$$N_q = \alpha_t N'_q = 0.55 \times 40 = 22$$

**Step 2:** Determine the single pile failure mode load capacity.

$\gamma' = 17.5 - 9.8 = 7.7 \text{ k/m}^3$

Center of sand layer within embedment length of the pile:

$\sigma'_z = 2 \times 17 + 3 \times 7.7 = 57.1 \text{ kPa}$

At base:   $(\sigma'_z)_b = 2 \times 17 + 8 \times 7.7 = 95.6 \text{ kPa}$

Skin friction:   $Q_f = \beta \sigma'_z \times$ Perimeter $\times$ Length

$= 0.57 \times 57.1 \times 1.26 \times 10 = 410.1 \text{ kN}$

End bearing:   $Q_b = N_q (\sigma'_z)_b A_b = 22 \times 95.6 \times 0.126 = 265 \text{ kN}$

Ultimate load capacity:   $Q_{ult} = Q_f + Q_b = 410.1 + 265 = 675.1 \text{ kN}$

Assume $\eta_e = 1$.

$$(Q_{ult})_g = n Q_{ult} = 9 \times 675.1 = 6076 \text{ kN}$$

**Step 3:** Determine the block failure mode load capacity by proportion.

Skin friction:   $(Q_f)_g = 410.1 \times \dfrac{9.6}{1.26} = 3124.6 \text{ kN}$

End bearing:   $(Q_b)_g = 265 \times \dfrac{5.76}{0.126} = 12114.3 \text{ kN}$

Ultimate load capacity: $(Q_{ult})_g$

$= (Q_f)_g + (Q_b)_g = 3124.6 + 12114.3 = 15238.9 \text{ kN}$

**Step 4:** Calculate the factor of safety.
The single pile failure mode governs.

$$\therefore \text{FS} = \frac{6076}{3000} \cong 2$$

**Step 5:**  Calculate the elastic settlement of the pile.
Assume the full design load is carried by skin friction and the load is equally shared by each pile.

$$Q_a = \frac{3000}{9} = 333.3 \text{ kN}$$

Equation (8.25):   For $\frac{L}{D} = 25$,   $I = 0.5 + \log(25) = 1.9$

Equation (8.24):   $\rho_{es} = \frac{Q_a}{E_{so}L} I = \frac{333.3}{30,000 \times 10} \times 1.9 = 2.1 \times 10^{-3} = 2.1 \text{ mm}$

Neglect elastic shortening of the pile since $K_{ps} > 500$.
Assume $\Phi = 0.5$.

$$R_s = n^\Phi = 9^{0.5} = 3$$
$$(\rho_{es})_g = 2.1 \times 3 = 6.3 \text{ mm}$$

**Step 6:**  Calculate the consolidation settlement.
The design load is transferred to $\frac{2}{3}L$ from the surface. The distance of the load transfer plane to the center of the clay is

$$\frac{10}{3} + 3 + \frac{1}{2} = 6.83 \text{ m}$$

$$\Delta\sigma_z = \frac{Q_{ag}}{(B_g + z)^2} = \frac{3000}{(2.4 + 6.83)^2} = 35.2 \text{ kPa}$$

$$\rho_{pc} = m_v H_o \Delta\sigma_z = 3.5 \times 10^{-4} \times 1 \times 35.2$$
$$= 123.2 \times 10^{-4} \text{ m} = 12.3 \text{ mm}$$

**Step 7:**  Compute the total settlement.

$$\rho_t = (\rho_{es})_g + \rho_{pc} = 6.3 + 12.3 = 18.6 \text{ mm} \qquad \blacksquare$$

# 8.11  PILES SUBJECTED TO NEGATIVE SKIN FRICTION

Piles located in settling soil layers (e.g., soft clays or fills) are subjected to negative skin friction called downdrag (Fig. 8.15). The settlement of the soil layer causes the friction forces to act in the same direction as the loading on the pile. Rather than providing resistance, the negative friction imposes additional loads on the pile. The net effect is that the pile load capacity is reduced and pile settlement increases. The allowable load capacity is given, with reference to Fig. 8.15, as

$$\boxed{Q_a = \frac{Q_b + Q_f}{\text{FS}} - Q_{\text{nf}}} \qquad (8.35)$$

For a soft, normally consolidated soil, the negative skin friction is usually calculated over one-half its thickness. Negative skin friction should be computed for long-term condition; that is, you should use an ESA.

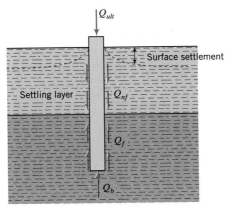

**FIGURE 8.15** Negative skin friction.

## EXAMPLE 8.9

Determine the allowable load capacity of the 0.4 m diameter pile shown in Fig. E8.9. The fill is recent and unconsolidated. To eliminate negative skin friction, a steel shell is proposed around the pile within the fill. The groundwater level is at 1 m below the fill in the soft clay but is expected to rise to the surface. A factor of safety of 3 is required.

*Strategy*   The trick here is to think about what would happen or is happening to the soft clay. Under the load of the fill, the soft clay will settle, dragging the pile down. Therefore, we have to consider negative skin friction imposed by the soft clay. You should use an ESA. Since the pile would likely loosen the dense sand adjacent to it, you should use $\phi'_{cs}$ rather than $\phi'_p$.

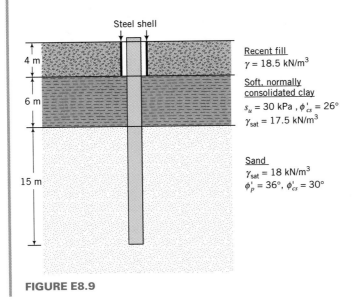

**FIGURE E8.9**

## Solution 8.9

**Step 1:**   Determine $\beta$, $N_q$, and other relevant parameters.

$$\frac{L}{D} = \frac{25}{0.4} = 62.5; \text{ Perimeter } = \pi D = 0.4\pi = 1.26 \text{ m};$$

$$A_b = \frac{\pi D^2}{4} = \frac{\pi \times 0.4^2}{4} = 0.126 \text{ m}^2$$

**Clay**

$$\beta = (1 - \sin \phi'_{cs}) \tan \phi'_{cs}(OCR)^{0.5} = (1 - \sin 26°) \tan(26°)(1)^{0.5} = 0.27$$

**Sand**

Table 8.3: Use Meyerhof's (1976) proposed values of $\beta$

For $\phi'_{cs} = 30°$:   $\beta \approx 0.52$   (linear interpolation)

From Fig. 8.9, $N'_q = 32$, for $\phi'_{cs} = 30°$. The inset curve for $\alpha_t$ stops at $L/D = 25$; therefore, you have to make an estimate for $\alpha_t$. It is unsafe to extrapolate results, especially experimental results. You should make a conservative estimate of $\alpha_t$. Using $\alpha_t = 0.4$, we get $N_q = \alpha_t N'_q = 0.4 \times 32 = 12.8$.

**Step 2:**   Calculate the negative skin friction for the clay layer.
Assume the groundwater will rise to the surface.

$$\gamma'_{fill} = 18.5 - 9.8 = 8.7 \text{ kN/m}^3$$
$$\gamma'_{clay} = 17.5 - 9.8 = 7.7 \text{ kN/m}^3$$

At center of the clay layer:   $\sigma'_z = 8.7 \times 4 + 7.7 \times 3 = 57.9 \text{ kPa}$

$Q_{nf} = \beta\sigma'_z \times \text{Perimeter} \times \text{Length} = 0.27 \times 57.9 \times 1.26 \times 6 = 118.2 \text{ kN}$

**Step 3:**   Calculate the skin friction and end bearing in sand.

$$\gamma'_{sand} = 18 - 9.8 = 8.2 \text{ kN/m}^3$$

At center of sand layer:   $\sigma'_z = 8.7 \times 4 + 7.7 \times 6 + 8.2 \times 7.5 = 142.5 \text{ kPa}$

$Q_f = 0.52 \times 142.5 \times 1.26 \times 15 = 1400.5 \text{ kN}$

$(\sigma'_z)_b = 8.7 \times 4 + 7.7 \times 6 + 8.2 \times 15 = 204 \text{ kPa}$

$Q_b = N_q(\sigma'_z)_b A_b = 12.8 \times 204 \times 0.126 = 329 \text{ kN}$

**Step 4:**   Calculate the allowable load capacity.

$$Q_a = \frac{(Q_f)_{sand} + (Q_b)_{sand}}{FS} - Q_{nf} = \frac{1400.5 + 329}{3} - 118.2 = 458.3 \text{ kN} \quad \blacksquare$$

# 8.12   PILE-DRIVING FORMULAS AND WAVE EQUATION

A number of empirical equations have been proposed to relate the energy delivered by a hammer during pile driving and the pile load capacity. One of the earliest equations is the ENR (*Engineering News Record*) equation, given as

$$Q_{ult} = \frac{W_R h}{s_1 + C_1} \tag{8.36}$$

**TABLE 8.8 Hammer Efficiency**

| Hammer type | $\eta_1$ |
|---|---|
| Drop hammer | 0.75–1.0 |
| Single-acting hammer | 0.75–0.85 |
| Double-acting hammer | 0.85 |
| Diesel hammer | 0.85–1.0 |

where $W_R$ is the weight of the ram, $h$ is the height of fall, $s_1$ is the penetration per blow, and $C_1$ is a constant. For drop hammers, $C_1 \cong 25$ mm; for steam hammers, $C_1 \cong 2.5$ mm. The units of $h$ and $s_1$ must be the same. The ENR equation was developed for timber piles using drop hammers for installation but is used to estimate or check the ultimate load capacity of all types of piles. Equation (8.36) does not account for energy losses due to the elastic compression of the pile. This equation can be modified using an efficiency factor, $\eta_1$; that is,

$$Q_{ult} = \frac{\eta_1 W_R h}{s_1 + C_1} \tag{8.37}$$

Suggested ranges of values for $\eta_1$ are shown in Table 8.8.

When a hammer strikes a pile the shock sets up a stress wave, which moves down the pile at the speed of sound. From elastic theory, the change in force on the pile over an infinitesimal length $\Delta z$, assuming the pile to be a rod, is

$$\frac{\partial F}{\partial z} = E_p A_p \frac{\partial}{\partial z} (\varepsilon_z) = E_p A_p \frac{\partial^2 u}{\partial z^2} \tag{8.38}$$

where $\varepsilon_z = \partial u / \partial z$ is the vertical strain, $E_p$ is Young's modulus of the pile, $A_p$ is the cross-sectional area of the pile, and $u$ is the pile displacement. From Newton's second law,

$$\frac{\partial F}{\partial z} = A_p \frac{\gamma}{g} \frac{\partial^2 u}{\partial t^2} \tag{8.39}$$

where $g$ is the acceleration due to gravity and $t$ is time. Setting Eq. (8.39) equal to Eq. (8.38), we get

$$E_p A_p \frac{\partial^2 u}{\partial z^2} = \frac{\gamma}{g} A_p \frac{\partial^2 u}{\partial t^2} \tag{8.40}$$

which, by simplification, leads to

$$\boxed{\frac{\partial^2 u}{\partial t^2} = V_c^2 \frac{\partial^2 u}{\partial z^2}} \tag{8.41}$$

where

$$\boxed{V_c = \sqrt{\frac{E_p g}{\gamma}}} \tag{8.42}$$

is the vertical wave propagation velocity in the pile.

The solution of Eq. (8.41) is found using appropriate boundary conditions and can be modified to account for soil resistance. Computer programs (e.g.,

WEAP) have been developed for routine wave analysis of pile-driving operations. These programs are beyond the scope of this book.

Driving records can provide useful information on the consistency of a soil at a site. For example, if the number of blows to drive a pile at a certain depth is $A$ blows and $B$ blows at another location at the same site, then the soil stratification is different. You should re-examine your design and the soil boring records and make the necessary adjustments, for example, increase or decrease the pile length.

> **The essential points *are*:**
>
> 1. *A number of empirical equations are available to estimate the pile load capacity from driving records.*
> 2. *Driving records can provide some useful information regarding the character of the soil at a site.*
> 3. *Careful judgment and significant experience are required to rely on pile load capacity from pile-driving operations.*

## 8.13 SUMMARY

Piles are used to support structural loads that cannot be supported on shallow foundations. The predominant types of pile material are steel, concrete, and timber. The selection of a particular type of pile depends on availability, environmental conditions, pile installation methods, and cost. Pile load capacity cannot be determined accurately because the method of installation invariably changes the soil properties near the pile. We do not know the extent of these changes. The equations for pile load capacities and settlement are, at best, estimates. Load capacities from pile load tests are preferred but these tests are expensive and may only be cost effective for large projects.

*Practical Example*

### EXAMPLE 8.10

A fish port facility is to be constructed near a waterfront area as shown in Figs. E8.10a,b. A soil investigation shows two predominant deposits. One is very soft, normally consolidated clay and the other is a stiff, overconsolidated clay. Soil data on the two deposits are shown in Fig. E8.10c. Determine the pile configuration (single or group piles), the pile length, and the expected settlement to support a design column load (working load) of 500 kN at $A$ (Fig. E8.10). Timber piles of average diameter 0.38 m and average length 18 m are readily available. The elastic modulus of the pile is $E_p = 20{,}000$ MPa. The settlement should not exceed 0.5% of the pile diameter. The pile shafts within the tidal zone will be treated to prevent rot. From experience on this site, it is difficult to drive piles beyond 8 m in the stiff, overconsolidated clay. Driving tends to damage the pile head. You should allow for a 0.25 m cut out from the pile head.

*Strategy* The very soft silty clay is likely to cause downdrag on the pile. You should determine the single pile capacity assuming the full available length of

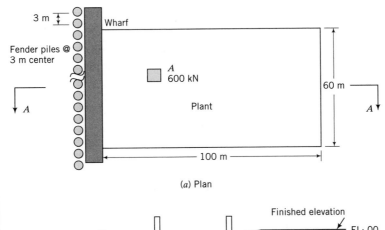

(a) Plan

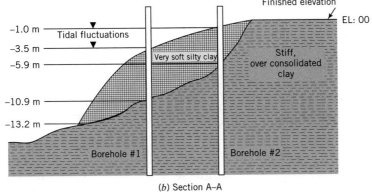

(b) Section A–A

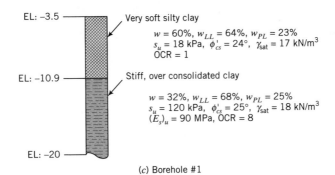

(c) Borehole #1

**FIGURE E8.10**

the pile will be used. If a single pile is not capable of carrying the load, then you should design a pile group. Use an ESA when you are considering downdrag.

## Solution 8.9

**Step 1:** Determine the geometric parameters, $\alpha$, $\beta$, and $N_q$.

$$D = 0.38 \text{ m}; \quad \text{Perimeter} = \pi D = \pi \times 0.38 = 1.19 \text{ m};$$

$$A_b = \frac{\pi D^2}{4} = \frac{\pi \times 0.38^2}{4} = 0.113 \text{ m}^2$$

From borehole #1, the length of pile in soft clay:  $L_1 = 7.4$ m

Length of pile in stiff clay:  $L_2 = 18 - 7.4 - 3.5 - 0.25 = 6.85$ m.

The value 0.25 m is the cut-off length from the pile head and 3.5 m is the distance from the finished elevation to the surface of the soft soil.

Soft clay, Table 8.3:  $\beta = (1 - \sin 24°) \tan 24°(1)^{1/2} = 0.26$

Stiff clay, Table 8.3:  $\beta = (1 - \sin 25°) \tan 25°(8)^{1/2} = 0.76$

Assume $\psi_p = \pi/2$. Then

$$N_q = [\tan 25° + \sqrt{1 + \tan^2(25°)}]^2 \exp\left(2\,\frac{\pi}{2}\,\tan 25°\right) = 10.7$$

**Step 2:** Determine the negative skin friction for a single pile.
A conservative estimate is to assume negative skin friction over the whole length of the pile in the very soft clay.

Soft clay: $\gamma' = 17 - 9.8 = 7.2$ kN/m$^3$

Center of top half of soft clay:  $\sigma'_z = 7.2 \times \dfrac{3.7}{2} = 13.3$ kPa

$Q_{nf} = \beta \sigma'_z \times$ Perimeter $\times L_1$
$Q_{nf} = 0.27 \times 13.3 \times 1.19 \times 7.4 = 31.6$kN

**Step 3:** Determine the load capacity in the stiff clay for a single pile.

Stiff clay:  $\gamma' = 18 - 9.8 = 8.2$ kN/m$^3$

Center of stiff clay:  $\sigma'_z = 7.2 \times 7.4 + 8.2 \times \dfrac{6.85}{2} = 81.4$ kPa

Base of stiff clay:  $(\sigma'_z)_b = 7.2 \times 7.4 + 8.2 \times 6.85 = 109.5$ kPa
$Q_f = \beta\,\sigma'_z \times$ Perimeter $\times L_2 = 0.76 \times 81.4 \times 1.19 \times 6.85 = 504.3$ kN
$Q_b = N_q(\sigma'_z)_b A_b = 10.7 \times 109.5 \times 0.113 = 132.4$ kN

**Step 4:** Determine the allowable load capacity of a single pile.

$$Q_a = \frac{Q_f + Q_b}{3} - Q_{nf} = \frac{504.3 + 132.4}{3} - 31.6 = 180.6 \text{ kN}$$

A single pile is inadequate for the load.

**Step 5:** Determine the block failure for the pile group.
Assume $\eta_e = 1$. Therefore,

$$\text{Number of piles required} = \frac{600}{180.6} = 3.3$$

Try 4 piles in a 2 × 2 matrix at a spacing of 1 m:

$\dfrac{s}{D} = \dfrac{1}{0.38} = 2.6$

$B_g = L_g = s + D = 1 + 0.38 = 1.38$ m;

Perimeter $= (1.38 + 1.38) \times 2 = 5.52$ m;  $A_b = 1.38^2 = 1.9$ m$^2$

Soft clay:  $Q_{nf} = 0.27 \times 13.3 \times 5.52 \times 7.4 = 146.7$ kN

Stiff clay:  $(Q_f)_g = 0.76 \times 81.4 \times 5.52 \times 6.85 = 2339.2$ kN

Stiff clay:  $(Q_b)_g = 10.7 \times 109.5 \times 1.9 = 2226.1$ kN

**Step 6:** Calculate the allowable load capacity for block failure mode.

$$(Q_a)_g = \frac{(Q_f)_g + (Q_b)_g}{3} - Q_{nf} = \frac{2339.2 + 2226.1}{3} - 146.7 = 1375 \text{ kN}$$

**Step 7:** Calculate the allowable load capacity for single pile failure mode.

$$n = 4 \text{ piles}$$
$$(Q_a)_g = nQ_a = 4 \times 180.6 = 722.4 \text{ kN}$$

Therefore, a $2 \times 2$ pile group is adequate. Single pile failure mode governs the design.

**Step 8:** Calculate the settlement.
Assume the full design load of $Q_a = 600$ kN will be carried by skin friction (floating pile) within the stiff clay.

$$\text{Load per pile } (Q_{ap}) = \frac{600}{4} = 150 \text{ kN}$$

$$\frac{L}{D} = \frac{L_2}{D} = \frac{6.85}{0.38} = 18$$

Using Eq. (8.25) to calculate the elastic settlement, we get

$$I = 0.5 + \log(18) \approx 1.8$$

$$\rho_{es} = \frac{Q_{ap}}{(E_{so})_u L_2} I = \frac{150}{90,000 \times 6.85} \times 1.8 = 0.43 \times 10^{-3} \text{ m} \approx 0.4 \text{ mm}$$

$$\rho_p = \frac{0.5 Q_{ap} L}{E_p A_p} = \frac{0.5 \times 150 \times 6.85}{20 \times 10^6 \times 0.113} = 227 \times 10^{-6} \text{ m} \approx 0.2 \text{ mm}$$

From Eq. (8.33):   $R_s = n^\Phi$

Assume $\Phi = 0.5$:   $R_s = 4^{0.5} = 2$

Group piles:   $(\rho_t)_g = R_s \rho_t + \rho_p = 2 \times 0.4 + 0.2 = 1.0 \text{ mm}$

$$\frac{(\rho_t)_g}{D} = \frac{1.0}{380} = 2.6 \times 10^{-3} \approx 0.2\%$$

■

# EXERCISES

## Problem Solving

8.1   Determine the length, $L$, required to support the load shown in Fig. P8.1. The diameter of the pile is 450 mm. A factor of safety of 3 is required.

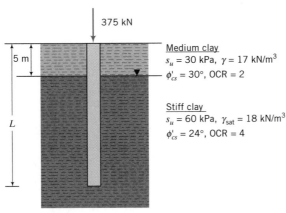

375 kN

5 m

$L$

Medium clay
$s_u = 30$ kPa, $\gamma = 17$ kN/m$^3$
$\phi'_{cs} = 30°$, OCR = 2

Stiff clay
$s_u = 60$ kPa, $\gamma_{sat} = 18$ kN/m$^3$
$\phi'_{cs} = 24°$, OCR = 4

**FIGURE P8.1**

8.2 Determine the allowable load for a pile, 0.4 m in diameter, driven 20 m into the soil profile shown in Fig. P8.2. Groundwater is at 2 m below the surface but you can assume it will rise to the surface. A factor of safety of 3 is required.

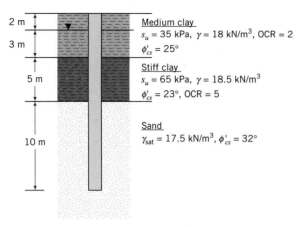

**FIGURE P8.2**

8.3 A square precast concrete pile of sides 0.4 m is to be driven 12 m into the soil strata shown in Fig. P8.3. Estimate the allowable load capacity for a factor of safety of 3. Owing to changes in design requirements, the pile must support 20% more load. Determine the additional embedment depth required.

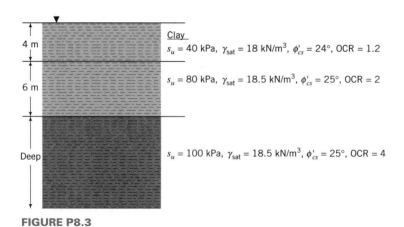

**FIGURE P8.3**

8.4 Estimate the allowable load capacity of a 0.5 m diameter pile embedded 17 m in the soil profile shown in Fig. P8.4. The factor of safety required is 3. The $N$ values are blows/ft (blows/0.31 m).

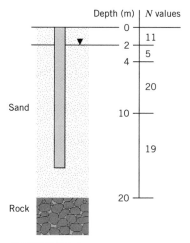

Depth (m) | *N* values

| | |
|---|---|
| 0 | |
| 2 | 11 |
| 4 | 5 |
| | 20 |
| 10 | |
| | 19 |
| 20 | |

Sand

Rock

**FIGURE P8.4**

8.5 The soil profile at a site for an offshore structure is shown in Fig. P8.5. Determine the allowable load for a driven pile with diameter 1.25 m and 50 m long. A factor of safety of 3 is required.

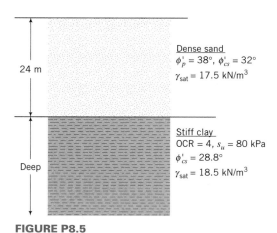

24 m

Dense sand
$\phi'_p = 38°$, $\phi'_{cs} = 32°$
$\gamma_{sat} = 17.5$ kN/m³

Deep

Stiff clay
OCR = 4, $s_u = 80$ kPa
$\phi'_{cs} = 28.8°$
$\gamma_{sat} = 18.5$ kN/m³

**FIGURE P8.5**

8.6 A soil profile consists of 3 m of a loose fill ($\phi'_{cs} = 15\frac{°}{2}$ $\gamma_{sat} = 18$ kN/m³) over a stratified overconsolidated clay. The shear strength of the overconsolidated clay increases linearly with depth. At the top of the clay, $s_u = 90$ kPa and at a depth of 20 m (17 m into the clay), $s_u = 180$ kPa. The critical state friction angle is $\phi'_{cs} = 28°$, OCR = 8 and $\gamma_{sat} = 18$ kN/m³. Determine the ultimate load capacity of a drilled shaft (bored pile) 1 m in diameter embedded 18 m into the clay. Groundwater is at 8 m below the ground surface. You may assume that the soil above the groundwater level is saturated.

8.7 Estimate the allowable load capacity of a 3 × 3 pile group. Each pile has a diameter of 0.4 m and is driven 15 m into a soft clay whose undrained shear strength varies linearly with depth according to $s_u = 0.25 \, \sigma'_z$. The critical state friction angle is 30° and $\gamma_{sat} = 17.5$ kN/m³. The spacing of the piles is 1.5 m. Groundwater level is at ground surface. A factor of safety of 3 is required.

## Practical

8.8   The soil profile and soil properties at a site are shown in the table below. A group of 12 concrete piles in a $3 \times 4$ matrix and of length 12 m is used to support a load. The pile diameter is 0.45 m and pile spacing is 1.5 m. Determine the allowable load capacity for a factor of safety of 3. Calculate the total settlement (elastic and consolidation) under the allowable load. Assume $E_p = 20 \times 10^6$ kPa.

| Depth (m) | Type of deposit | Soil test results |
|---|---|---|
| 0–3<br>3–groundwater level | Sand | $\gamma = 17$ kN/m³, $\phi'_{cs} = 28°$<br>$E'_{so} = 19$ MN/m² |
| 3–6 | Sand | $\gamma_{sat} = 17.5$ kN/m³, $\phi'_{cs} = 30°$<br>$E'_{so} = 18$ MN/m² |
| 6–15 | Clay | $\gamma_{sat} = 18.5$ kN/m³, $\phi'_{cs} = 27°$<br>$s_u = 30$ kPa<br>$C_c = 0.4$, $C_r = 0.06$,<br>OCR $= 1.5$<br>$E'_{so} = 30$ MN/m², $\nu' = 0.3$ |
| 15–17 | Soft clay | $\gamma_{sat} = 18$ kN/m³, $\phi'_{cs} = 24°$<br>$s_u = 20$ kPa<br>$C_c = 0.8$, OCR $= 1.0$<br>$E'_{so} = 10$ MN/m², $\nu' = 0.3$ |
| >17 | Rock | |

## CHAPTER 9

# TWO-DIMENSIONAL FLOW OF WATER THROUGH SOILS

## 9.0  INTRODUCTION

Many catastrophic failures in geotechnical engineering result from instability of soil masses due to groundwater flow. Lives are lost, infrastructures are damaged or destroyed, and major economic losses occur. In this chapter, you will study the basic principles of two-dimensional flow of water through soils. The topics covered here will help you to avoid pitfalls in the analyses and design of geotechnical systems where groundwater flow can lead to instability. The emphasis in this chapter is on gaining an understanding of the forces that provoke failures resulting from groundwater flow. You will learn methods to calculate flow, pore water pressure distribution, uplift forces, and seepage stresses for a few simple geotechnical systems.

When you complete this chapter, you should be able to:

- Calculate flow under and within earth structures
- Calculate seepage stresses, pore water pressure distribution, uplift forces, hydraulic gradients, and the critical hydraulic gradient
- Determine the stability of simple geotechnical systems subjected to two-dimensional flow of water

You will use the following principles learned from previous chapters and your courses in mechanics:

- Statics
- Hydraulic gradient, flow of water through soils (Chapter 2)
- Effective stresses and seepage (Chapter 3)

*Sample Practical Situation*  A deep excavation is required for the construction of a building. The soil is a silty sand with groundwater level just below ground level. The excavation cannot be made unless the sides are supported. You, a geotechnical engineer, are required to design the retaining structure for the excavation and to recommend a scheme to keep the inside of the excavation dry. Figure 9.1 shows the collapse of a sewer and a supported excavation by seepage forces. You should try to prevent such a collapse.

**FIGURE 9.1** Damage of a braced excavation by seepage forces. (Courtesy of George Tamaro, Mueser Rutledge Consulting Engineers.)

## 9.1 DEFINITIONS OF KEY TERMS

*Equipotential line* is a line representing constant head.

*Flow line* is the flow path of a particle of water.

*Flow net* is a graphical representation of a flow field.

*Seepage stress* is the stress (similar to frictional stress in pipes) imposed on a soil as water flows through it.

*Static liquefaction* is the behavior of a soil as a viscous fluid when seepage reduces the effective stress to zero.

## 9.2 QUESTIONS TO GUIDE YOUR READING

1. What is the governing equation for two-dimensional flow and what are the methods adopted for its solution for practical problems?
2. What are flow lines and equipotential lines?
3. What is a flow net?
4. How do I draw a flow net?
5. What are the practical uses of a flow net?
6. What is the critical hydraulic gradient?
7. How do I calculate the pore water pressure distribution near a retaining structure and under a dam?
8. What are uplift pressures and how can I calculate them?

9. What do the terms static liquefaction, heaving, quicksand, boiling, and piping mean?

10. What are the forces that lead to instability due to two-dimensional flow?

11. How does seepage affect the stability of an earth retaining structure?

# 9.3  TWO-DIMENSIONAL FLOW OF WATER THROUGH POROUS MEDIA

The flow of water through soils is described by Laplace's equation. Flow of water through soils is analogous to steady state heat flow and flow of current in homogeneous conductors. The popular form of Laplace's equation for two-dimensional flow of water through soils is

$$k_x \frac{\partial^2 H}{\partial x^2} + k_z \frac{\partial^2 H}{\partial z^2} = 0 \qquad (9.1)$$

where $H$ is the total head and $k_x$ and $k_z$ are the coefficients of permeability in the $X$ and $Z$ directions. Laplace's equation expresses the condition that the changes of hydraulic gradient in one direction are balanced by changes in the other directions.

The assumptions in Laplace's equation are:

- Darcy's law is valid.
- The soil is homogeneous and saturated.
- The soil and water are incompressible.
- No volume change occurs.

If the soil were an isotropic material then $k_x = k_z$ and Laplace's equation becomes

$$\frac{\partial^2 H}{\partial x^2} + \frac{\partial^2 H}{\partial z^2} = 0 \qquad (9.2)$$

The solution of any differential equation requires knowledge of the boundary conditions. The boundary conditions for most "real" structures are complex, so we cannot obtain an analytical solution or closed form solution for these structures. We have to resort to approximate solutions, which we can obtain using numerical methods such as finite difference, finite element, and boundary element. We can also use physical models to attempt to replicate the flow through the real structure.

In this chapter, we are going to consider two solution techniques for Laplace's equation. One is an approximate method called flow net sketching; the other is the finite difference technique, which you have encountered in Chapter 4. The flow net sketching technique is simple and flexible and conveys a picture of the flow regime. It is the method of choice among geotechnical engineers. But before we delve into these solution techniques, we will establish some key conditions that are needed to understand two-dimensional flow.

The solution of Eq. (9.1) depends only on the values of the total head within the flow field in the $XZ$ plane. Let us introduce a velocity potential ($\xi$), which describes the variation of total head in a soil mass as

$$\xi = kH \tag{9.3}$$

where $k$ is a generic coefficient of permeability. The velocities of flow in the $X$ and $Z$ directions are

$$v_x = k_x \frac{\partial H}{\partial x} = \frac{\partial \xi}{\partial x} \tag{9.4}$$

$$v_z = k_z \frac{\partial H}{\partial z} = \frac{\partial \xi}{\partial z} \tag{9.5}$$

The inference from Eqs. (9.4) and (9.5) is that the velocity of flow ($v$) is normal to lines of constant total head (also called constant piezometric head or equipotential lines) as illustrated in Fig. 9.2. The direction of $v$ is in the direction of decreasing total head. The head difference between two equipotential lines is called a potential drop or head loss.

If lines are drawn that are tangent to the velocity of flow at every point in the flow field in the $XZ$ plane, we will get a series of lines that are normal to the equipotential lines. These tangential lines are called streamlines or flow lines (Fig. 9.2). A flow line represents the flow path that a particle of water is expected to take in steady state flow. A family of streamlines is called a stream function, $\psi_s(x, z)$.

The components of velocity in the $X$ and $Z$ directions in terms of the stream function are

$$v_x = \frac{\partial \psi_s}{\partial z} \tag{9.6}$$

$$v_z = -\frac{\partial \psi_s}{\partial x} \tag{9.7}$$

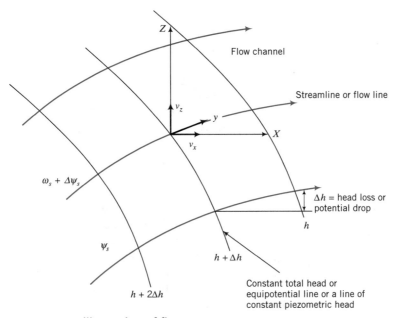

**FIGURE 9.2** Illustration of flow terms.

Since flow lines are normal to equipotential lines, there can be no flow across flow lines. The rate of flow between any two flow lines is constant. The area between two flow lines is called a flow channel (Fig. 9.2). Therefore, the rate of flow is constant in a flow channel.

> ***The*** **essential points** ***are:***
> 1. ***Streamlines or flow lines represent flow paths of particles of water.***
> 2. ***The area between two flow lines is called a flow channel.***
> 3. ***The rate of flow in a flow channel is constant.***
> 4. ***Flow cannot occur across flow lines.***
> 5. ***The velocity of flow is normal to the equipotential line.***
> 6. ***Flow lines and equipotential lines are orthogonal (perpendicular) to each other.***
> 7. ***The difference in head between two equipotential lines is called the potential drop or head loss.***

**What's next . . .** The flow conditions we established in the previous section allow us to use a graphical method to find solutions to two-dimensional flow problems. In the next section, we will describe flow net sketching and provide guidance in interpreting a flow net to determine flow through soils, the distribution of pore water pressures, and the hydraulic gradients.

##  9.4 FLOW NET SKETCHING

### 9.4.1 Criteria for Sketching Flow Nets

A flow net is a graphical representation of a flow field that satisfies Laplace's equation and comprises a family of flow lines and equipotential lines.

A flow net must meet the following criteria:

1. The boundary conditions must be satisfied.
2. Flow lines must intersect equipotential lines at right angles.
3. The area between flow lines and equipotential lines must be curvilinear squares. A curvilinear square has the property that an inscribed circle can be drawn to touch each side of the square and continuous bisection results, in the limit, in a point.
4. The quantity of flow through each flow channel is constant.
5. The head loss between each consecutive equipotential line is constant.
6. A flow line cannot intersect another flow line.
7. An equipotential line cannot intersect another equipotential line.

An infinite number of flow lines and equipotential lines can be drawn to satisfy Laplace's equation. However, only a few are required to obtain an accurate solution. The procedure for constructing a flow net is described next.

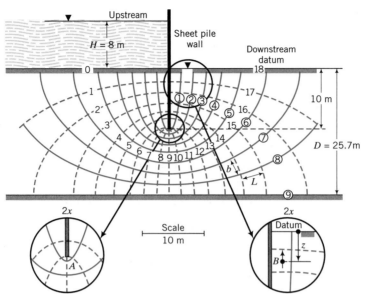

**FIGURE 9.3**   Flow net for a sheet pile.

### 9.4.2 Flow Net for Isotropic Soils

1. Draw the structure and soil mass to a suitable scale.

2. Identify impermeable and permeable boundaries. The soil–impermeable boundary interfaces are flow lines because water can flow along these interfaces. The soil–permeable boundary interfaces are equipotential lines because the total head is constant along these interfaces.

3. Sketch a series of flow lines (four or five) and then sketch an appropriate number of equipotential lines such that the area between a pair of flow lines and a pair of equipotential lines (cell) is approximately a curvilinear square. You would have to adjust the flow lines and equipotential lines to make curvilinear squares. You should check that the average width and the av-

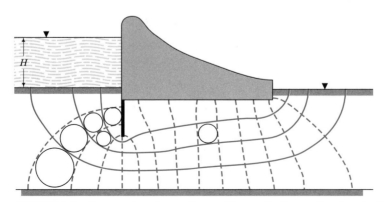

**FIGURE 9.4**   Flow net under a dam with a cutoff curtain (sheet pile) on the upstream end.

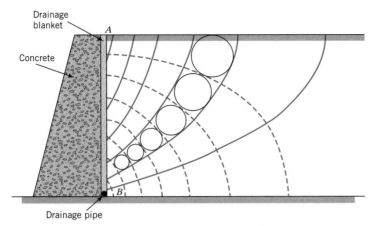

**FIGURE 9.5** Flow net in the backfill of a retaining wall with a vertical drainage blanket.

erage length of a cell are approximately equal. You should also sketch the entire flow net before making adjustments.

The flow net in confined areas between parallel boundaries usually consists of flow lines and equipotential lines that are elliptical in shape and symmetrical (Fig. 9.3). Try to avoid making sharp transitions between straight and curved sections of flow and equipotential lines. Transitions should be gradual and smooth. For some problems, portions of the flow net are enlarged and are not curvilinear squares, and they do not satisfy Laplace's equation. For example, the portion of the flow net below the bottom of the sheet pile in Fig. 9.3 does not consist of curvilinear squares. For an accurate flow net, you should check these portions to ensure that repeated bisection results in a point.

A few examples of flow nets are shown in Figs. 9.3 to 9.5. Figure 9.3 shows a flow net for a sheet pile wall; Fig. 9.4 shows a flow net beneath a dam and Fig. 9.5 shows a flow net in the backfill of a retaining wall. In the case of the retaining wall, the vertical drainage blanket of coarse-grained soil is used to transport excess pore water pressure from the backfill to prevent the imposition of a hydrostatic force on the wall. The interface boundary, $AB$, is neither an equipotential nor a flow line. The total head along the boundary $AB$ is equal to the elevation head.

### 9.4.3 Anisotropic Soil

Equation (9.1) is Laplace's two-dimensional equation for anisotropic soils (the permeabilities in the $X$ and $Z$ directions are different). Let us manipulate this equation to transform it into a form for which we can use a procedure similar to the one for isotropic soils to draw and interpret flow nets. Put $C = \sqrt{k_z/k_x}$ and $x_1 = Cx$. Then

$$\frac{\partial x_1}{\partial x} = C$$

$$\frac{\partial H}{\partial x} = \frac{\partial H}{\partial x_1}\frac{\partial x_1}{\partial x} = C\frac{\partial H}{\partial x_1}$$

and

$$\frac{\partial^2 H}{\partial x^2} = C^2 \frac{\partial^2 H}{\partial x_1^2}$$

Therefore, we can write Eq. (9.1) as

$$C^2 \frac{\partial^2 H}{\partial x_1^2} + C^2 \frac{\partial^2 H}{\partial z^2} = 0$$

which simplifies to

$$\frac{\partial^2 H}{\partial x_1^2} + \frac{\partial^2 H}{\partial z^2} = 0 \qquad (9.8)$$

Equation (9.8) indicates that for anisotropic soils we can use the procedure for flow net sketching described for isotropic soils by scaling the $x$ distance by $\sqrt{k_z/k_x}$. That is, you must draw the structure and flow domain by multiplying the horizontal distances by $\sqrt{k_z/k_x}$.

## 9.5 INTERPRETATION OF FLOW NET

### 9.5.1 Flow Rate

Let the total head loss across the flow domain be $\Delta H$, that is, the difference between upstream and downstream water level elevation. Then the head loss ($\Delta h$) between each consecutive pair of equipotential lines is

$$\Delta h = \frac{\Delta H}{N_d} \qquad (9.9)$$

where $N_d$ is the number of equipotential drops, that is, the number of equipotential lines minus one. In Fig. 9.3, $\Delta H = 8$ m and $N_d = 18$. From Darcy's law, the flow through each flow channel for an isotropic soil is

$$\Delta q = Aki = (b \times 1)k \frac{\Delta h}{L} = k \, \Delta h \frac{b}{L} = k \frac{\Delta H}{N_d} \frac{b}{L} \qquad (9.10)$$

where $b$ and $L$ are defined as shown in Fig. 9.3. By construction, $b/L = 1$, and therefore the total flow is

$$q = k \sum_{i=1}^{N_f} \left(\frac{\Delta H}{N_d}\right)_i = k \, \Delta H \frac{N_f}{N_d} \qquad (9.11)$$

where $N_f$ is the number of flow channels (number of flow lines minus one); in Fig. 9.3, $N_f = 9$. The ratio $N_f/N_d$ is called the shape factor. Finer discretization of the flow net by drawing more flow lines and equipotential lines does not significantly change the shape factor. Both $N_f$ and $N_d$ can be fractional. In the case of anisotropic soils, the quantity of flow is

$$q = \Delta H \frac{N_f}{N_d} \sqrt{k_x k_z} \qquad (9.12)$$

### 9.5.2 Hydraulic Gradient

You can find the hydraulic gradient over each square by dividing the head loss by the length, $L$, of the cell; that is,

$$i = \frac{\Delta h}{L} \tag{9.13}$$

You should notice from Fig. 9.3 that $L$ is not constant. Therefore, the hydraulic gradient is not constant. The maximum hydraulic gradient occurs where $L$ is a minimum; that is,

$$i_{\max} = \frac{\Delta h}{L_{\min}} \tag{9.14}$$

where $L_{\min}$ is the minimum length of the cells within the flow domain. Usually, $L_{\min}$ occurs at exit points or around corners (e.g., point $A$ in Fig. 9.3), and it is at these points that we can get the maximum hydraulic gradient.

### 9.5.3 Static Liquefaction, Heaving, Boiling, and Piping

Let us consider an element, $B$, of soil at a depth $z$ near the downstream end of the sheet pile wall structure shown in Fig. 9.3. Flow over this element is upward. Therefore, the vertical effective stress is

$$\sigma'_z = \gamma' z - i \gamma_w z \tag{9.15}$$

If the effective stress becomes zero, the soil loses its strength and behaves like a viscous fluid. The soil state at which the effective stress is zero is called static liquefaction. Various other names such as boiling, quicksand, piping, and heaving are used to describe specific events connected to the static liquefaction state. Boiling occurs when the upward seepage force exceeds the downward force of the soil. Piping refers to the subsurface "pipe-shaped" erosion that initiates near the toe of dams and similar structures. High localized hydraulic gradient statically liquefies the soil, which progresses to the water surface in the form of a pipe, and water then rushes beneath the structure through the pipe, leading to instability and failure. Quicksand is the existence of a mass of sand in a state of static liquefaction. Heaving occurs when seepage forces push the bottom of an excavation upward. A structure founded on a soil that statically liquefies will collapse. Liquefaction can also be produced by dynamic events such as earthquakes.

### 9.5.4 Critical Hydraulic Gradient

We can determine the hydraulic gradient that brings a soil mass (essentially, coarse-grained soils) to static liquefaction. Solving for $i$ in Eq. (9.15) when $\sigma'_z = 0$, we get

$$i = i_{\mathrm{cr}} = \frac{\gamma'}{\gamma_w} = \left( \frac{G_s - 1}{1 + e} \right) \frac{\gamma_w}{\gamma_w} = \frac{G_s - 1}{1 + e} \tag{9.16}$$

where $i_{\mathrm{cr}}$ is called the critical hydraulic gradient, $G_s$ is specific gravity, and $e$ is the void ratio. Since $G_s$ is constant, the critical hydraulic gradient is solely a function of the void ratio of the soil. In designing structures that are subjected to steady state seepage, it is absolutely essential to ensure that the critical hydraulic gradient cannot develop.

### 9.5.5 Pore Water Pressure Distribution

The pore water pressure at any point $j$ is calculated as follows:

1. Select a datum. Let us choose the downstream water level as the datum (Fig. 9.3).
2. Determine the total head at $j$: $H_j = \Delta H - (N_d)_j \Delta h$, where $(N_d)_j$ is the number of equipotential drops at point $j$; $(N_d)_j$ can be fractional. For example, at $B$, $H_B = \Delta H - 16.5 \Delta h$.
3. Subtract the elevation head at point $j$ from the total head $H_j$ to get the pressure head. For point $B$ (Fig. 9.3), the elevation head $h_z$ is $-z$ (point $B$ is below the datum). The pressure head is then

$$\boxed{(h_p)_j = \Delta H - (N_d)_j \Delta h - h_z} \tag{9.17}$$

   For point $B$, $(h_p)_B = \Delta H - 16.5 \Delta h - (-z) = \Delta H - 16.5 \Delta h + z$.
4. The pore water pressure is

$$\boxed{u_j = (h_p)_j \gamma_w} \tag{9.18}$$

### 9.5.6 Uplift Forces

Lateral and uplift forces due to groundwater flow can adversely affect the stability of structures such as dams and weirs. The uplift force per unit length (length is normal to the $XZ$ plane) is found by calculating the pore water pressure at discrete points along the base (in the $X$ direction, Fig. 9.4) and then finding the area under the pore water pressure distribution diagram, that is,

$$\boxed{P_w = \sum_{j=1}^{n} u_j \, \Delta x_j} \tag{9.19}$$

where $P_w$ is the uplift force per unit length, $u_j$ is the average pore water pressure over an interval $\Delta x_j$, and $n$ is the number of intervals. It is convenient to use Simpson's rule to calculate $P_w$:

$$\boxed{P_w = \frac{\Delta x}{3} \left( u_1 + u_n + 2 \sum_{\substack{i=3 \\ \text{odd}}}^{n} u_i + 4 \sum_{\substack{i=2 \\ \text{even}}}^{n} u_i \right)} \tag{9.20}$$

### EXAMPLE 9.1

An excavation is proposed for a site consisting of a homogeneous, isotropic layer of silty clay, 12.24 m thick, above a deep deposit of sand. The groundwater is 2 m below ground level (see Fig. E9.1). The void ratio of the silty clay is 0.62 and its specific gravity is 2.7. What is the limiting depth of the excavation to avoid heaving?

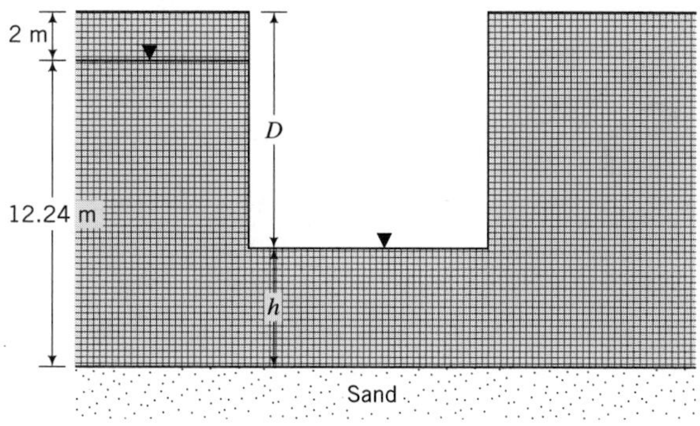

**FIGURE E9.1**

***Strategy*** Heaving will occur if $i > i_{cr}$. Find the critical hydraulic gradient from the void ratio given and then find the depth at which this hydraulic gradient is reached.

### Solution 9.1

**Step 1:** Calculate $i_{cr}$.

$$i_{cr} = \frac{G_s - 1}{1 + e} = \frac{2.7 - 1}{1 + 0.62} = 1.05$$

**Step 2:** Determine $D$.

Total head difference:   $\Delta H = H - h = 12.24 - h$

Average hydraulic gradient:   $i = \dfrac{\Delta H}{h}$

Putting $i = i_{cr}$, we get

$$1.05 = \frac{12.24 - h}{h}$$

Solving for $h$, we get $h = 5.97$ m.

$$D = (12.24 + 2) - h = 14.24 - 5.97 = 8.27 \text{ m} \qquad \blacksquare$$

### EXAMPLE 9.2

A dam, shown in Fig. E9.2a, retains 10 m of water. A sheet pile wall (cutoff curtain) on the upstream side, which is used to reduce seepage under the dam, penetrates 7 m into a 20.3 m thick silty sand stratum. Below the silty sand is a thick deposit of clay. The average coefficient of permeability of the silty sand is $2.0 \times 10^{-4}$ cm/s. Assume that the silty sand is homogeneous and isotropic.

**(a)** Draw the flow net under the dam.

**(b)** Calculate the flow, $q$.

**(c)** Calculate and draw the pore water pressure distribution at the base of the dam.

**(d)** Determine the uplift force.

**(e)** Determine and draw the pore water pressure distribution on the upstream and downstream faces of the sheet pile wall.

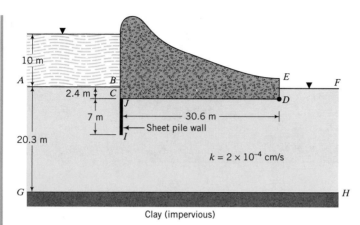

**FIGURE E9.2a**

(f) Determine the resultant lateral force on the sheet pile wall due to the pore water.

(g) Determine the maximum hydraulic gradient.

(h) Will piping occur if the void ratio of the silty sand is 0.8?

(i) What is the effect of reducing the depth of penetration of the sheet pile wall?

**Strategy**    Follow the procedures described in Section 9.4 to draw the flow net and calculate the required parameters.

## Solution 9.2

**Step 1:**    Draw the dam to scale.
See Fig. E9.2b.

**Step 2:**    Identify the impermeable and permeable boundaries.
With reference to Fig. E9.2a, $AB$ and $EF$ are permeable boundaries and are therefore equipotential lines. $BCIJDE$ and $GH$ are impermeable boundaries and are therefore flow lines.

**Step 3:**    Sketch the flow net.
Draw about three flow lines and then draw a suitable number of equipotential lines. Remember that flow lines are orthogonal to equipotential lines and the area between two consecutive flow lines and two consecutive equipotential lines is approximately a square. Use a circle template to assist you in estimating the square. Adjust/add/subtract flow lines and equipotential lines until you are satisfied that the flow net consists essentially of curvilinear squares. See sketch of flow net in Fig. E9.2b.

**Step 4:**    Calculate the flow.
Select the downstream end, $EF$, as the datum.

$\Delta H = 10$ m

$N_d = 14, N_f = 4$

Equation (9.11):    $q = k \, \Delta H \dfrac{N_f}{N_d} = 2 \times 10^{-4} \times (10 \times 10^2) \times \dfrac{4}{14} = 0.057$ cm³/s

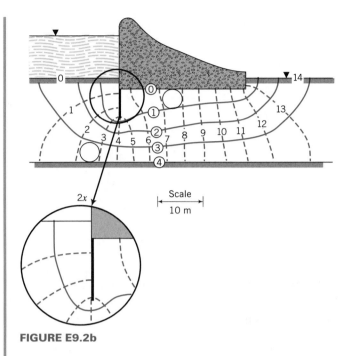

**FIGURE E9.2b**

**Step 5:** Determine the pore water pressure under the base of the dam. Divide the base into a convenient number of equal intervals. Let us use 10 intervals; that is

$$\Delta x = \frac{30.6}{10} = 3.06 \text{ m}$$

Determine the pore water pressure at each nodal point. Use a table for convenience or, better yet, use a spreadsheet.

$$\Delta h = \frac{\Delta H}{N_d} = \frac{10}{14} = 0.714 \text{ m}$$

The calculation in the table below was done using a spreadsheet program

Flow under a dam: $\Delta h = 0.714$ m

| Parameters | Under base of dam | | | | | | | | | | |
|---|---|---|---|---|---|---|---|---|---|---|---|
| $x$ (m) | 0 | 3.06 | 6.12 | 9.18 | 12.24 | 15.3 | 18.36 | 21.42 | 24.48 | 27.54 | 30.6 |
| $N_d$ (m) | 5.60 | 5.80 | 6.20 | 6.90 | 7.40 | 8.00 | 8.80 | 9.40 | 10.30 | 11.10 | 12.50 |
| $N_d\,\Delta h$ (m) | 4.00 | 4.14 | 4.43 | 4.93 | 5.28 | 5.71 | 6.28 | 6.71 | 7.35 | 7.93 | 8.93 |
| $h_z$ (m) | −2.40 | −2.40 | −2.40 | −2.40 | −2.40 | −2.40 | −2.40 | −2.40 | −2.40 | −2.40 | −2.40 |
| $h_p$ (m) $= \Delta H - N_d\,\Delta h - h_z$ | 8.40 | 8.26 | 7.97 | 7.47 | 7.12 | 6.69 | 6.12 | 5.69 | 5.05 | 4.47 | 3.48 |
| $u$ (kPa) $= h_p\gamma_w$ | 82.3 | 80.9 | 78.1 | 73.2 | 69.7 | 65.5 | 59.9 | 55.7 | 49.4 | 43.9 | 34.1 |

Plot the pore water pressure distribution. See Fig. E9.2c.

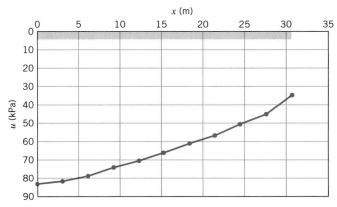

**FIGURE E9.2c**

**Step 6:** Calculate the uplift force and its location.
Using Simpson's rule [Eq. (9.20)], we find

$$P_w = \frac{3.06}{3}[82.3 + 34.1 + 2(78.1 + 69.7 + 59.9 + 49.4)$$
$$+ 4(80.9 + 73.2 + 65.5 + 55.7 + 43.9)]$$
$$= 1946.4 \text{ kN/m}$$

**Step 7:** Determine the pore water pressure distribution on the sheet pile wall. Divide the front face of the wall into six intervals of $7/6 = 1.17$ m and the back face into one interval. Six intervals were chosen because it is convenient for the scaling using the scale that was used to draw the flow net. The greater the intervals, the greater the accuracy. Only one interval is used for the back face of the wall because there are no equipotential lines that meet there. Use a spreadsheet to compute the pore water pressure distribution and the hydrostatic forces. The distributions of pore water pressure at the front and back of the wall are shown in Figs. E9.2d,e. Use Simpson's rule to calculate the hydrostatic force on the front face of the wall. The pore water pressure distribution at the back face is a trapezoid and the area is readily calculated.

| Parameters | Front of wall | | | | | | | Back of wall | |
|---|---|---|---|---|---|---|---|---|---|
| $z$ (m) | 0 | 1.17 | 2.33 | 3.50 | 4.67 | 5.83 | 7.00 | 7.00 | 0.00 |
| $N_d$ (m) | 0.70 | 1.00 | 1.30 | 1.60 | 1.90 | 2.40 | 3.00 | 5.00 | 5.60 |
| $N_d \Delta h$ (m) | 0.50 | 0.71 | 0.93 | 1.14 | 1.36 | 1.71 | 2.14 | 3.57 | 4.00 |
| $h_z$ (m) | −2.40 | −3.57 | −4.73 | −5.90 | −7.07 | −8.23 | −9.40 | −9.40 | −2.40 |
| $h_p$ (m) = $\Delta H - N_d \Delta h - h_z$ | 11.90 | 12.85 | 13.81 | 14.76 | 15.71 | 16.52 | 17.26 | 15.83 | 8.40 |
| $u$ (kPa) = $h_p \gamma_w$ | 116.6 | 126.0 | 135.3 | 144.6 | 154.0 | 161.9 | 169.1 | 155.1 | 82.3 |
| | Front | Back | Difference | | | | | | |
| $P_w$ (kN/m) | 1011.7 | 830.9 | 180.8 | | | | | | |

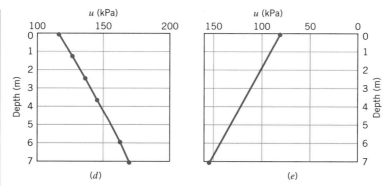

**FIGURE E9.2** Pore water pressure distribution (d) at front of wall and (e) at back of wall.

**Step 8:** Determine the maximum hydraulic gradient.
The smallest value of $L$ occurs at the exit. By measurement, $L_{min} = 2$ m.

$$i_{max} = \frac{\Delta h}{L_{min}} = \frac{0.714}{2} = 0.36$$

**Step 9:** Determine if piping would occur.

$$\text{Equation (9.16):} \quad i_{cr} = \frac{G_s - 1}{1 + e} = \frac{2.7 - 1}{1 + 0.8} = 0.89$$

Since $i_{max} < i_{cr}$, piping will not occur.

$$\text{Factor of safety against piping:} \quad \frac{0.89}{0.36} = 2.5$$

**Step 10:** State the effect of reducing the depth of penetration of the sheet pile wall.
If the depth is reduced, the value of $\Delta h$ increases and $i_{max}$ is likely to increase. ∎

## 9.6 FINITE DIFFERENCE SOLUTION FOR TWO-DIMENSIONAL FLOW

In Chapter 4, we used the finite difference technique to solve the governing one-dimensional partial differential equation to determine the spatial variation of excess pore water pressure. We will do the same to solve Laplace's equation to determine two-dimensional confined flow through soils. Let us consider a grid of a flow domain as shown in Fig. 9.6, where $(i, j)$ is a nodal point.
Using Taylor's theorem, we have

$$k_x \frac{\partial^2 H}{\partial x^2} + k_z \frac{\partial^2 H}{\partial z^2} = \frac{k_x}{\Delta x^2} (h_{i+1, j} + h_{i-1, j} - 2h_{i, j})$$

$$+ \frac{k_z}{\Delta z^2} (h_{i, j+1} + h_{i, j-1} - 2h_{i, j}) = 0 \quad (9.21)$$

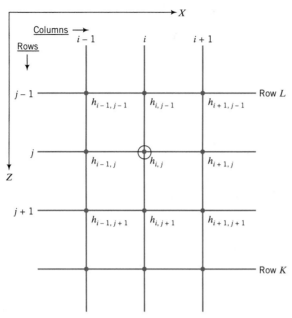

**FIGURE 9.6** A partial grid of the flow domain.

Let $\alpha = k_x/k_z$ and $\Delta x = \Delta z$ (i.e., we subdivide the flow domain into a square grid). Then solving for $h_{i,j}$ from Eq. (9.21) gives

$$h_{i,j} = \frac{1}{2(1 + \alpha)} \, (\alpha h_{i+1,j} + \alpha h_{i-1,j} + h_{i,j+1} + h_{i,j-1}) \qquad (9.22)$$

For isotropic conditions, $\alpha = 1$ ($k_x = k_z$) and Eq. (9.22) becomes

$$h_{i,j} = \tfrac{1}{4}(h_{i+1,j} + h_{i-1,j} + h_{i,j+1} + h_{i,j-1}) \qquad (9.23)$$

Since we are considering confined flow, one or more of the boundaries would be impermeable. Flow cannot cross impermeable boundaries and, therefore, for a horizontal impermeable surface,

$$\frac{\partial h}{\partial x} = 0 \qquad (9.24)$$

The finite difference form of Eq. (9.24) is

$$\frac{\partial h}{\partial x} = \frac{1}{2\Delta x} \, (h_{i,j+1} - h_{i,j-1}) = 0 \qquad (9.25)$$

Therefore, $h_{i,j+1} = h_{i,j-1}$ and, by substitution in Eq. (9.22), we get

$$h_{i,j} = \tfrac{1}{4}(h_{i+1,j} + h_{i-1,j} + 2h_{i,j-1}) \qquad (9.26)$$

Various types of geometry of impermeable boundaries are encountered in practice, three of which are shown in Fig. 9.7. For Figs. 9.7a,b the finite difference equation is

$$h_{i,j} = \tfrac{1}{2}(h_{i+1,j} + h_{i,j-1}) \qquad (9.27)$$

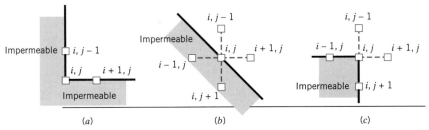

(a)　　　　　　　　　　　(b)　　　　　　　　　　　(c)

**FIGURE 9.7** Three types of boundary encountered in practice.

and, for Fig. 9.7c,

$$h_{i,j} = \tfrac{1}{3}(h_{i,j-1} + h_{i+1,j} + h_{i,j+1} + \tfrac{1}{2}h_{i-1,j} + \tfrac{1}{2}h_{i,j+1}) \tag{9.28}$$

The pore water pressure at any node $(u_{i,j})$ is

$$u_{i,j} = \gamma_w(h_{i,j} - z_{i,j}) \tag{9.29}$$

where $z_{i,j}$ is the elevation head.

Contours of potential heads can be drawn from the discrete values of $h_{i,j}$. The finite difference equations for flow lines are analogous to the potential lines; that is, $\psi_s$ replaces $h$ in the above equations and the boundary conditions for $\psi_s$ are specified rather than for $h$.

The horizontal velocity of flow at any node $(v_{i,j})$ is given by Darcy's law:

$$v_{i,j} = k_x i_{i,j}$$

where $i_{i,j}$ is the hydraulic gradient expressed as

$$i_{i,j} = \frac{\tfrac{1}{2}(h_{i+1,j} - h_{i-1,j})}{\Delta x} \tag{9.30}$$

Therefore,

$$v_{i,j} = \frac{k_x}{2\Delta x}(h_{i+1,j} - h_{i-1,j}) \tag{9.31}$$

The flow rate, $q$, is obtained by considering a vertical plane across the flow domain. Let $L$ be the top row and $K$ be the bottom row of a vertical plane defined by column $i$ (Fig. 9.6). Then the expression for $q$ is

$$q = \frac{k_x}{4}\left(h_{i+1,L} - h_{i-1,L} + 2\sum_{j=L+1}^{K-1}(h_{i+1,j} - h_{i-1,j}) + h_{i+1,K} - h_{i-1,K}\right) \tag{9.32}$$

The procedure to determine the distribution of potential head, flow, and pore water pressure using the finite difference method is as follows:

1. Divide the flow domain into a square grid. Remember from Chapter 4 that finer grids give more accurate solutions than coarse grids, but are more tedious to construct and require more computational time. If the problem is symmetrical, you only need to consider one-half of the flow domain. For example, the sheet pile wall shown in Fig. 9.8 is symmetrical about the wall

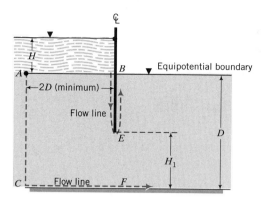

**FIGURE 9.8**  A sheet pile wall.

and only the left half may be considered. The total flow domain should have a width of at least four times the thickness of the soil layer. For example, if $D$ is the thickness of the soil layer (Fig. 9.8), then the minimum width of the left half of the flow domain is $2D$.

2. Identify boundary conditions, for example, impermeable boundaries (flow lines) and permeable boundaries (equipotential lines).

3. Determine the heads at the permeable or equipotential boundaries. For example, the heads along the equipotential boundary $AB$ (Fig. 9.8) is $\Delta H$. Therefore, all the nodes along this boundary will have a constant head of $\Delta H$. Because of symmetry, the head along nodes directly under the sheet pile wall ($EF$) is $\Delta H/2$.

4. Apply the known heads to corresponding nodes and assume reasonable initial values for the interior nodes. You can use linear interpolation for the potential heads of the interior nodes.

5. Apply Eq. (9.23), if the soil is isotropic, to each node except (a) at impermeable boundaries, where you should use Eq. (9.26), (b) at corners, where you should use Eqs. (9.27) and (9.28) for the corners shown in Figs. 9.8a–c, and (c) at nodes where the heads are known.

6. Repeat item 5 until the new value at a node differs from the old value by a small numerical tolerance, for example, 0.001 m.

7. Arbitrarily select a sequential set of nodes along a column of nodes and calculate the flow, $q$, using Eq. (9.32). It is best to calculate $q' = q$ for a unit permeability value to avoid too many decimal points in the calculations.

8. Repeat items 1 to 6, to find the flow distribution by replacing heads by flow $q'$. For example, the flow rate calculated in item 7 is applied to all nodes along $AC$ and $CF$ (Fig. 9.8). The flow rate at nodes along $BE$ is zero.

9. Calculate the pore water pressure distribution by using Eq. (9.29).

A spreadsheet program can be prepared to automatically carry out the above procedure. However, you should carry out "hand" calculations at selected nodes to verify that the spreadsheet values are correct.

## EXAMPLE 9.3

Determine the flow under the sheet pile wall (Fig. E9.3a) and the pore water pressure distribution using the finite difference method.

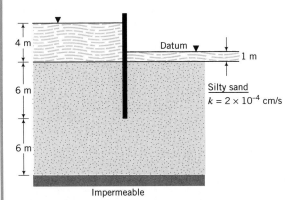

**FIGURE E9.3a**

*Strategy*   Use a spreadsheet and follow the procedures in Section 9.6.

## Solution 9.3

The finite difference solution is shown in Table E9.3 and in Fig. 9.3c.

**Step 1:**   Divide the flow domain into a grid.
See Fig. E9.3b.
Since the problem is symmetrical, perform calculations for only one-half of the domain. Use the left half of the flow domain of width $2D = 2 \times 12 = 24$ m. Use a grid 2 m $\times$ 2 m.

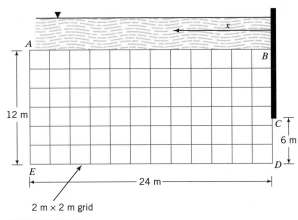

**FIGURE E9.3b**

## TABLE E9.3

| | A | B | C | D | E | F | G | H | I | J | K | L | M | N |
|---|---|---|---|---|---|---|---|---|---|---|---|---|---|---|
| 1 | Example 9.3 | | | | | | | | | | | | | |
| 2 | Flow under a sheet pile wall using finite difference method | | | | | | | | | | | | | |
| 3 | $H_1$ | | 4 m | | | | | | | | | | | |
| 4 | $H_2$ | | 1 m | | | | | | | | | | | |
| 5 | $\Delta H$ | | 3 m | | | | | | | | | | | |
| 6 | $H_3$ | | 6 m | | | | | | | | | | | |
| 7 | $D$ | | 12 m | | | | | | | | | | | |
| 8 | Cell size | | 2 m | | | | | | | | | | | |
| 9 | $k$ | | 0.0002 cm/s | | | | | | | | | | | |
| 10 | | | | Equipotentials | | | | | | | | | | |
| 11 | $z/x$ | 0 | 2 | 4 | 6 | 8 | 10 | 12 | 14 | 16 | 18 | 20 | 22 | 24 |
| 12 | 0 | 3.00 | 3.00 | 3.00 | 3.00 | 3.00 | 3.00 | 3.00 | 3.00 | 3.00 | 3.00 | 3.00 | 3.00 | 3.00 |
| 13 | 2 | 2.96 | 2.96 | 2.96 | 2.95 | 2.94 | 2.92 | 2.90 | 2.88 | 2.84 | 2.80 | 2.74 | 2.68 | 2.64 |
| 14 | 4 | 2.93 | 2.92 | 2.92 | 2.90 | 2.88 | 2.85 | 2.82 | 2.76 | 2.69 | 2.60 | 2.49 | 2.34 | 2.21 |
| 15 | 6 | 2.90 | 2.89 | 2.88 | 2.86 | 2.83 | 2.79 | 2.74 | 2.67 | 2.57 | 2.44 | 2.26 | 1.99 | 1.50 |
| 16 | 8 | 2.87 | 2.87 | 2.85 | 2.83 | 2.80 | 2.75 | 2.68 | 2.59 | 2.48 | 2.32 | 2.12 | 1.85 | 1.50 |
| 17 | 10 | 2.86 | 2.85 | 2.84 | 2.81 | 2.77 | 2.72 | 2.65 | 2.55 | 2.42 | 2.26 | 2.05 | 1.80 | 1.50 |
| 18 | 12 | 2.85 | 2.85 | 2.83 | 2.81 | 2.77 | 2.71 | 2.63 | 2.53 | 2.40 | 2.24 | 2.03 | 1.78 | 1.50 |
| 19 | | | | | | | | | | | | | | |
| 20 | | $q'=$ | 1.63 m³/s per unit value of $k$ | | | | | | | | | | | |
| 21 | | | Pore water pressure (kPa) | | | | | | | | | | | |
| 22 | $z/x$ | 0 | 2 | 4 | 6 | 8 | 10 | 12 | 14 | 16 | 18 | 20 | 22 | 24 |
| 23 | 0 | 39.2 | 39.2 | 39.2 | 39.2 | 39.2 | 39.2 | 39.2 | 39.2 | 39.2 | 39.2 | 39.2 | 39.2 | 39.2 |
| 24 | 2 | 58.4 | 58.4 | 58.4 | 58.3 | 58.2 | 58.1 | 57.9 | 57.6 | 57.3 | 56.8 | 56.3 | 55.7 | 55.3 |
| 25 | 4 | 77.7 | 77.7 | 77.6 | 77.4 | 77.2 | 77.0 | 76.6 | 76.1 | 75.4 | 74.5 | 73.4 | 71.9 | 70.6 |
| 26 | 6 | 97.0 | 96.9 | 96.8 | 96.6 | 96.4 | 96.0 | 95.4 | 94.7 | 93.8 | 92.5 | 90.7 | 88.1 | 83.3 |
| 27 | 8 | 116.3 | 116.3 | 116.2 | 115.9 | 115.6 | 115.1 | 114.5 | 113.6 | 112.5 | 111.0 | 109.0 | 106.3 | 102.9 |
| 28 | 10 | 135.8 | 135.8 | 135.6 | 135.4 | 135.0 | 134.4 | 133.7 | 132.8 | 131.5 | 129.9 | 127.9 | 125.4 | 122.5 |
| 29 | 12 | 155.4 | 155.3 | 155.2 | 154.9 | 154.5 | 153.9 | 153.2 | 152.2 | 151.0 | 149.3 | 147.3 | 144.8 | 142.1 |
| 30 | | | | | | | | | | | | | | |
| 31 | | Flow lines | | | | | | | | | | | | |
| 32 | | $q'$ (upstream) = 1.63 m³/s per value unit of $k$ | | | | | | | | | | | | |
| 33 | | $q'$ (downstream) = 0 | | | | | | | | | | | | |
| 34 | | | | | | | | | | | | | | |
| 35 | $z/x$ | 0 | 2 | 4 | 6 | 8 | 10 | 12 | 14 | 16 | 18 | 20 | 22 | 24 |
| 36 | 0 | 1.63 | 1.58 | 1.54 | 1.49 | 1.43 | 1.35 | 1.26 | 1.14 | 1.00 | 0.81 | 0.58 | 0.31 | 0.00 |
| 37 | 2 | 1.63 | 1.59 | 1.54 | 1.49 | 1.43 | 1.36 | 1.27 | 1.16 | 1.02 | 0.83 | 0.61 | 0.32 | 0.00 |
| 38 | 4 | 1.63 | 1.59 | 1.55 | 1.51 | 1.45 | 1.39 | 1.31 | 1.21 | 1.07 | 0.90 | 0.68 | 0.38 | 0.00 |
| 39 | 6 | 1.63 | 1.60 | 1.56 | 1.53 | 1.48 | 1.43 | 1.37 | 1.28 | 1.17 | 1.03 | 0.83 | 0.53 | 0.00 |
| 40 | 8 | 1.63 | 1.61 | 1.58 | 1.56 | 1.53 | 1.49 | 1.44 | 1.38 | 1.30 | 1.20 | 1.07 | 0.91 | 0.77 |
| 41 | 10 | 1.63 | 1.62 | 1.60 | 1.59 | 1.57 | 1.56 | 1.53 | 1.50 | 1.46 | 1.41 | 1.35 | 1.28 | 1.24 |
| 42 | 12 | 1.63 | 1.63 | 1.63 | 1.63 | 1.63 | 1.63 | 1.63 | 1.63 | 1.63 | 1.63 | 1.63 | 1.63 | 1.63 |

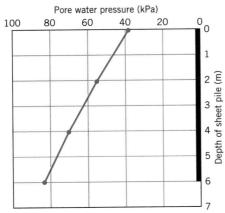

**FIGURE E9.3c** Pore water pressure distribution.

**Step 2:** Identify the boundary conditions.

> Permeable boundaries: $AB$ and $CD$ are equipotential lines.
> Impermeable boundaries: $BC$, $AE$, and $DE$, are flow lines.

**Step 3:** Determine the heads at equipotential boundaries.

> Along $AB$, the head difference is $4 - 1 = 3$ m.
> Along $CD$, the head difference is $3/2 = 1.5$ m.

**Step 4:** Insert the heads at the nodes.
Set up the initial parameters in column $B$, rows 3 to 9. Note that cell $B5$ is cell $B3$ − cell $B2$. In cells $B12$ to $N12$ (corresponding to the equipotential boundary $AB$), copy cell $B5$. In cells, $N15$ to $N18$, insert cell $B5/2$. Arbitrarily insert values in all other cells from $B13$ to $M18$, $N13$, and $N14$.

**Step 5:** Apply the appropriate equations.
On the impermeable boundaries—$B13$ to $B18$ (corresponding to $AE$), $C18$ to $C22$ (corresponding to $ED$), and $N13$ to $N14$ (corresponding to $BC$)—apply Eq. (9.26). You should note that some nodes (e.g., $B18$) are common. Apply Eq. (9.22) to all other cells except cells with known heads.

**Step 6:** Carry out the iterations.
In Excel, go to Tools → Options → Calculation. Select the following:

> **(i)** Automatic.
> **(ii)** Iteration, insert 100 in Maximum Iterations and 0.001 in Maximum change.
> **(iii)** Under Workbook Options, select Update remote reference, save external link values, accept labels in formulas. You can then click on Calculate Now (F9) or Calc.sheet.

**Step 7:**  Calculate $q$.
Use Eq. (9.32) to calculate $q'$ for a unit value of permeability. In the spreadsheet for this example, $q'$ is calculated in cell $C20$ as

```
{(B13 - B15) + N13 - N15 + 2) * (SUM(C13:M13)
- 2 * SUM(C15:M15)}/4
```

The actual value of $q$ is

$$q = kq' = 1.62 \times 2 \times 10^{-4} \times 10^{-2}$$
$$= 3.2 \times 10^{-6} \text{ m/s} \quad (10^{-2} \text{ is used to convert cm/s to m/s})$$

**Step 8:**  Calculate the flow for each cell (flow lines).
In cells $B36$ to $B42$ (corresponding to $AE$) and $C42$ to $N32$ (corresponding to $ED$), copy $q'$. The flow at the downstream end (cells $N36$ to $N39$) is zero. Apply Eq. (9.26) to cells $C36$ to $M41$ and $N40$ to $N41$. Apply Eq. (9.23) to all other cells except the cell with known values of $q'$. Carry out the reiterations.

**Step 9:**  Calculate the pore water pressures.
From the potential heads, you can calculate the pore water pressure using Eq. (9.29). A plot of the pore water pressure distribution is shown in Fig. E9.3c.  ∎

## 9.7  FLOW THROUGH EARTH DAMS

Flow through earth dams is an important design consideration. We need to ensure that the pore water pressure at the downstream end of the dam will not lead to instability and the exit hydraulic gradient does not lead to piping. The major exercise is to find the top flow line called the phreatic surface (Fig. 9.9). The pressure head on the phreatic surface is zero.

Casagrande (1937) showed that the phreatic surface can be approximated by a parabola with corrections at the points of entry and exit. The focus of the parabola is at the toe of the dam, point $F$ (Fig. 9.9). The assumed parabola representing the uncorrected phreatic surface is called the basic parabola. Recall from your geometry course that the basic property of a parabola is that every point on it is equidistant from its focus and a line called the directrix. To draw the basic parabola, we must know point $A$, the focus $F$, and $f$ (one-half the distance from the focus to the directrix). Casagrande (1937) recommended that point $C$ be located at a distance $0.3AB$, where $AB$ is the horizontal projection of

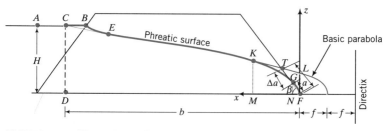

**FIGURE 9.9**  Phreatic surface within an earth dam.

the upstream slope at the water surface. From the basic property of a parabola, we get

$$2f = \sqrt{b^2 + H^2} - b \qquad (9.33)$$

The equation to construct the basic parabola is

$$\sqrt{x^2 + z^2} = x + 2f$$

Solving for $z$, we obtain

$$z^2 = 4f(f + x) \qquad (9.34)$$

or

$$\boxed{z = 2\sqrt{f(f + x)}} \qquad (9.35)$$

Since $H$ and $b$ are known from the geometry of the dam, the basic parabola can be constructed. We now have to make some corrections at the upstream entry point and the downstream exit point.

The upstream end is corrected by sketching a transition curve (BE) that blends smoothly with the basic parabola. The correction for the downstream end depends on the angle $\beta$ and the type of discharge face. Casagrande (1937) determined the length of the discharge face, $a$, for a homogeneous earth dam with no drainage blanket at the discharge face and $\beta \leq 30°$ as follows. He assumed that Dupuit's assumption, which states that the hydraulic gradient is equal to the slope, $dz/dx$, of the phreatic surface, is valid. If we consider two vertical sections— one is $KM$ of height $z$, and the other is $GN$ of height $a \sin \beta$—then the flow rate across $KM$ is

$$q_{KM} = Aki = (z \times 1)k\frac{dz}{dx} \qquad (9.36)$$

and across $GN$ is

$$q_{GN} = Aki = (a \sin \beta \times 1)k\frac{dz}{dx} = (a \sin \beta)k \tan \beta \qquad (9.37)$$

From the continuity condition at sections $KM$ and $GN$, $q_{KM} = q_{GN}$; that is,

$$zk\frac{dz}{dx} = (a \sin \beta)k \tan \beta \qquad (9.38)$$

which simplifies to

$$z\frac{dz}{dx} = a \sin \beta \tan \beta \qquad (9.39)$$

We now integrate Eq. (9.39) within the limits $x_1 = a \cos \beta$ and $x_2 = b$, $z_1 = a \sin \beta$ and $z_2 = H$.

$$\int_{a \sin \beta}^{H} z \, dz = a \sin \beta \tan \beta \int_{a \cos \beta}^{b} dx$$

$$\therefore H^2 - a^2 \sin^2 \beta = a \sin \beta \tan \beta(b - a \cos \beta)$$

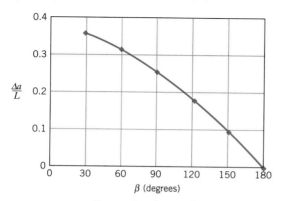

**FIGURE 9.10**  Correction factor for downstream face.

Simplification leads to

$$a = \frac{b}{\cos \beta} - \sqrt{\frac{b^2}{\cos^2 \beta} - \frac{H^2}{\sin^2 \beta}} = \frac{b}{\cos \beta} - \cos \beta \sqrt{b^2 - H^2 \cot^2 \beta} \qquad (9.40)$$

Casagrande (1937) produced a chart relating $\Delta a/L$ (see Fig. 9.9 for definitions of $\Delta a$ and $L$) with different values of $\beta$, as shown in Fig. 9.10.

The flow through the dam is obtained by substituting Eq. (9.40) into Eq. (9.37), giving

$$q = k \sin \beta \tan \beta \left( \frac{b}{\cos \beta} - \cos \beta \sqrt{b^2 - H^2 \cot^2 \beta} \right)$$
$$= k \sin^2 \beta \left( \frac{b}{\cos^2 \beta} - \sqrt{b^2 - H^2 \cot^2 \beta} \right) \qquad (9.41)$$

Because the exit hydraulic gradient is often large, drainage blankets are used at the downstream end of dams to avoid piping. Figure 9.11 shows a horizontal drainage blanket at the toe of an earth dam. Seepage is controlled by the gradation of the coarse-grained soils used for the drainage blanket. The phreatic surface for dams with drainage blankets is forced to intersect the drainage blanket

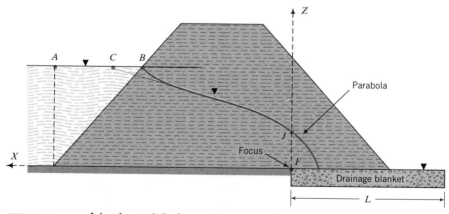

**FIGURE 9.11**  A horizontal drainage blanket at the toe of an earth dam.

and does not intersect the downstream face of the dam. Therefore, no correction to the basic parabola is required on the downstream end of the dam.

The flow through the dam is

$$q = Aki = Ak\frac{dz}{dx}$$

where $dz/dx$ is the slope of the basic parabola and the area $A = FJ \times 1$ (Fig. 9.11). From the geometry of the basic parabola, $FJ = 2f$ and the slope of the basic parabola at $J$ is, from Eq. (9.34),

$$\frac{dz}{dx} = \frac{2f}{z} = \frac{2f}{2f} = 1$$

Therefore, the flow through a dam with a horizontal drainage blanket is

$$q = 2f \times k \times 1 = 2fk \tag{9.42}$$

The procedure to draw a phreatic surface within an earth dam, with reference to Fig. 9.9, is as follows:

1. Draw the structure to scale.
2. Locate a point $A$ at the intersection of a vertical line from the bottom of the upstream face and the water surface, and a point $B$ where the waterline intersects the upstream face.
3. Locate point $C$, such that $BC = 0.3AB$.
4. Project a vertical line from $C$ to intersect the base of the dam at $D$.
5. Locate the focus of the basic parabola. The focus is located conveniently at the toe of the dam.
6. Calculate the focal distance, $f = (\sqrt{b^2 + H^2} - b)/2$, where $b$ is the distance $FD$ and $H$ is the height of water on the upstream face.
7. Construct the basic parabola from $z = 2\sqrt{f(f + x)}$.
8. Sketch in a transition section $BE$.
9. Calculate the length of the discharge face, $a$, using

$$a = \frac{b}{\cos \beta} - \cos \beta \sqrt{b^2 - H^2 \cot^2 \beta}; \quad \beta \le 30°.$$

For $\beta > 30°$, use Fig. 9.10 and (a) measure the distance $TF$, where $T$ is the intersection of the basic parabola with the downstream face; (b) for the known angle $\beta$, read the corresponding factor $\Delta a/L$ from the chart; and (c) find the distance $a = TF(1 - \Delta a/L)$.

10. Measure the distance $a$ from the toe of the dam along the downstream face to point $G$.
11. Sketch in a transition section, $GK$.
12. Calculate the flow using $q = ak \sin \beta \tan \beta$, where $k$ is the coefficient of permeability. If the downstream slope has a horizontal drainage blanket as shown in Fig. 9.11, the flow is calculated using $q = 2fk$.

## 9.8  SUMMARY

The flow of water through soils is a very important consideration in the analysis, design, and construction of many civil engineering systems. The governing equation for flow of water through soils is Laplace's equation. In this chapter, we examined two types of solution for Laplace's equation for two-dimensional flow. One is a graphical technique, called flow net sketching, which consists of a network of flow and equipotential lines. The network (flow net) consists of curvilinear squares in which flow lines and equipotential lines are orthogonal to each other. From the flow net, we can calculate the flow rate, the distribution of heads, pore water pressures, seepage forces, and the maximum hydraulic gradient. The other type of solution is based on the finite difference method and requires, in most cases, the use of a spreadsheet or a computer program. Instability of structures embedded in soils could occur if the maximum hydraulic gradient exceeds the critical hydraulic gradient.

*Practical Example*

### EXAMPLE 9.4

A bridge pier is to be constructed in a riverbed by constructing a cofferdam as shown in Fig. E9.4a. A cofferdam is a temporary enclosure consisting of long, slender elements of steel, concrete, or timber members to support the sides of the enclosure. After constructing the cofferdam, the water within it will be pumped out. Determine (a) the flow rate using $k = 1 \times 10^{-4}$ cm/s and (b) the factor of safety against piping. The void ratio of the sand is 0.59.

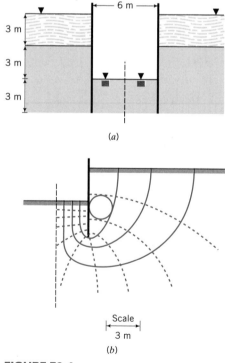

(a)

(b)

**FIGURE E9.4**

There was a long delay before construction began and a 100 mm layer of silty clay with $k = 1 \times 10^{-6}$ cm was deposited at the site. What effect would this silty clay layer have on the factor of safety against piping?

**Strategy** The key is to draw a flow net and determine whether the maximum hydraulic gradient is less than the critical hydraulic gradient. The presence of the silty clay would result in significant head loss within it, and consequently the factor of safety against piping is likely to increase. Since the cofferdam is symmetrical about a vertical plane, you only need to draw the flow net for one-half of the cofferdam.

## Solution 9.4

**Step 1:** Draw the cofferdam to scale and sketch the flow net.
See Fig. E9.4b.

**Step 2:** Determine the flow.

$$\Delta h = 6 \text{ m}; \quad N_f = 4, \quad N_d = 10$$

$$q = 2k \, \Delta H \frac{N_f}{N_d} = 2 \times 1 \times 10^{-4} \times 10^{-2} \times 6 \times \frac{4}{10} = 4.8 \times 10^{-6} \text{ m}^3/\text{s}$$

(*Note:* The factor 2 is needed because you have to consider both halves of the structure; the factor $10^{-2}$ is used to convert cm/s to m/s.)

**Step 3:** Determine the maximum hydraulic gradient.

$$L_{\min} \approx 0.3 \text{ m} \quad \text{(this is an average value of the flow length at the exit of the sheet pile)}$$

$$i_{\max} = \frac{\Delta h}{L_{\min}} = \frac{\Delta H}{N_d L_{\min}} = \frac{6}{10 \times 0.3} = 2$$

**Step 4:** Calculate the critical hydraulic gradient.

$$i_{cr} = \frac{G_s - 1}{1 + e} = \frac{2.7 - 1}{1 + 0.59} = 1.07$$

Since $i_{\max} > i_{cr}$, piping is likely to occur; the factor of safety is $1.07/2 \approx 0.5$.

**Step 5:** Determine the effects of the silty clay layer.
Consider the one-dimensional flow in the flow domain, as shown in Fig. E9.4c. The head loss through 9 m of sand (6 m outside and 3 m inside of excavation) is 6 m in the absence of the silt layer. Let us find the new head loss in the sand due to the presence of the silt layer.

From Darcy's law and the continuity condition: $\quad k_{\text{sand}} \, i_{\text{sand}} = k_{\text{silt}} \, i_{\text{silt}}$

$$\text{or} \quad k_{\text{sand}} \frac{\Delta h_{\text{sand}}}{L_{\text{sand}}} = k_{\text{silt}} \frac{\Delta h_{\text{silt}}}{L_{\text{silt}}}$$

Therefore,

$$\Delta h_{\text{sand}} = \frac{1 \times 10^{-6} \times 9\Delta h_{\text{silt}}}{1 \times 10^{-4} \times 0.1} = 0.9\Delta h_{\text{silt}}$$

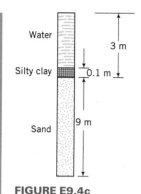

**FIGURE E9.4c**

But $\Delta h_{sand} + \Delta h_{silt} = 6$ m; that is, $0.9\Delta h_{silt} + \Delta h_{silt} = 6$ m and $\Delta h_{silt} = 3.16$ m. Therefore,

$$\Delta h_{sand} = 6 - 3.16 = 2.84 \text{ m}$$

Thus, the maximum hydraulic gradient would be reduced.  ■

# EXERCISES

## Theory

**9.1**   Derive a relationship between the critical hydraulic gradient, $i_{cr}$, and porosity, $n$.

## Problem Solving

**9.2**   A sheet pile wall supporting 6 m of water is shown in Fig. P9.2.

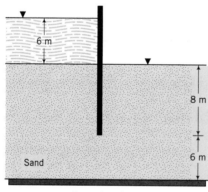

**FIGURE P9.2**

**(a)** Draw the flow net.
**(b)** Determine the flow rate if $k = 0.0019$ cm/s.
**(c)** Determine the pore water pressure distributions on the upstream and downstream faces of the wall.
**(d)** Would piping occur if $e = 0.55$?

**9.3**  Repeat Exercise 9.2 using the finite difference method.

**9.4**  The sheet pile wall supporting 6 m of water has a clay (almost impervious) blanket of 3 m on the downstream side, as shown in Fig. P9.4.

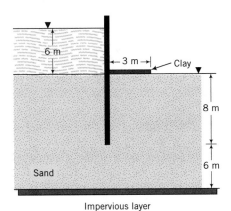

**FIGURE P9.4**

**(a)** Draw the flow net.
**(b)** Determine the flow rate if $k = 0.0019$ cm/s.
**(c)** Determine the pore water pressure distributions on the upstream and downstream faces of the wall.
**(d)** Would piping occur if $e = 0.55$?

**9.5**  A section of a dam constructed from a clay is shown in Fig. P9.5. The dam is supported on 10 m of sandy clay with $k_x = 0.00012$ cm/s and $k_z = 0.00002$ cm/s. Below the sandy clay is a thick layer of impervious clay.

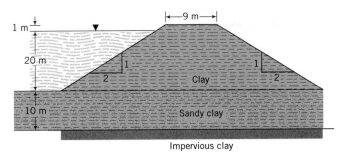

**FIGURE P9.5**

**(a)** Draw the flow net under the dam.
**(b)** Determine the pore water pressure distribution at the base of the dam.
**(c)** Calculate the resultant uplift force and its location from the upstream face of the dam.

**9.6**   Draw the phreatic surface for the earth dam shown in Fig. P9.5. Determine the flow rate within the dam, if $k = k_x = 1 \times 10^{-6}$ cm/s.

**9.7**   Draw the phreatic surfaces and determine the flow rates with and without the drainage blanket for the dam shown in Fig. P9.7. The coefficient of permeability of the clay is $k = k_x = 0.0000012$ cm/s.

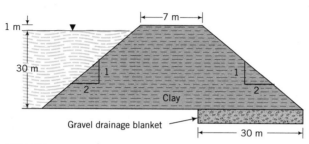

**FIGURE P9.7**

## Practical

**9.8**   Borings at a site for a road pavement show a water-bearing stratum of coarse-grained soil below an 8 m thick deposit of a mixture of sand, silt, and clay. The water-bearing stratum below the clay will create an artesian condition. The excess head is 1 m. One preliminary design proposal is to insert two trenches, as shown in Fig. P9.8, to reduce the excess head. Draw the flow net and determine the flow rate into each of the trenches, if $k = k_z = 1 \times 10^{-5}$ cm/s.

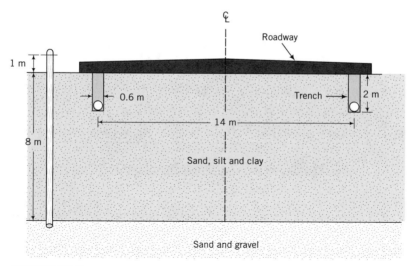

**FIGURE P9.8**

9.9  A retaining wall has a vertical drainage blanket (Fig. P9.9). After a heavy rainfall, a steady state seepage condition occurs. Draw the flow net and determine the pore water pressure distribution acting on a potential failure plane $AB$. The coefficient of permeability is $1.8 \times 10^{-4}$ cm/s.

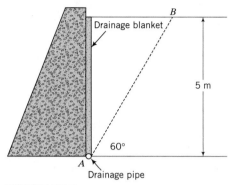

**FIGURE P9.9**

## CHAPTER 10

# STABILITY OF EARTH RETAINING STRUCTURES

| ABET | ED | ES |
|------|----|----|
|      | 75 | 25 |

## 10.0 INTRODUCTION

Earth retaining structures are ubiquitous in the man-made environment. These structures have the distinction of being the first to be analyzed using mechanics (remember we mentioned Coulomb's analysis of the lateral earth pressure on the fortresses protected by soil in Chapter 1). In this chapter, we will analyze some typical earth retaining structures to determine their stability. The emphases will be on gaining an understanding of the forces that provoke failures and methods of analysis of simple earth retaining structures.

You should recall that stability refers to a condition in which a geotechnical system will not fail or collapse under any conceivable loading (static and dynamic loads, fluid pressure, seepage forces). Stability is synonymous with ultimate limit state but serviceability limit state is also important. In many circumstances, the serviceability limit state is the deciding design limit state. The methods of analyses that we are going to discuss do not consider the serviceability limit state. The analyses involved to determine the serviceability limit state are beyond the scope of this book.

When you complete this chapter, you should be able to:

- Understand and determine lateral earth pressures
- Understand the forces that lead to instability of earth retaining structures
- Determine the stability of simple earth retaining structures

You will use the following principles learned from previous chapters and your courses in mechanics:

- Static equilibrium
- Effective stresses and seepage (Chapter 3)
- Mohr's circle (Chapter 3)
- Shear strength (Chapter 5)
- Two-dimensional flow of water through soils (Chapter 9)

*Sample Practical Situation* A retaining wall is required around a lake. You, a geotechnical engineer, are required to design the retaining wall. An example of a retaining structure in a waterfront area is shown in Fig. 10.1

**FIGURE 10.1** A flexible retaining wall under construction.

## 10.1   DEFINITIONS OF KEY TERMS

*Active earth pressure coefficient* $(K_a)$ is the ratio between the lateral and vertical principal effective stresses when an earth retaining structure moves away (by a small amount) from the retained soil.

*Passive earth pressure coefficient* $(K_p)$ is the ratio between the lateral and vertical principal effective stresses when an earth retaining structure is forced against a soil mass.

*Gravity retaining wall* is a massive concrete wall relying on its mass to resist the lateral forces from the retained soil mass.

*Flexible retaining wall* or a *sheet pile wall* is a long slender wall relying on passive resistance and anchors or props for its stability.

*Mechanical stabilized earth* is a gravity type retaining wall in which the soil is reinforced by thin reinforcing elements (steel, fabric, fibers, etc.).

## 10.2   QUESTIONS TO GUIDE YOUR READING

**1.** What is meant by the stability of earth retaining structures?

**2.** What are the factors that lead to instability?

**3.** What are the main assumptions in the theory of lateral earth pressures?

**4.** When shall I use either Rankine's theory or Coulomb's theory?

5. Does Coulomb's theory give an upper bound or a lower bound solution?

6. What is the effect of wall friction on the shape of slip planes?

7. What are the differences among gravity wall, a cantilever wall, a cantilever sheet pile wall, and an anchored sheet pile wall?

8. How do I analyze a retaining wall to check that it is stable?

9. How deep can I make a vertical cut without wall supports?

10. What are mechanically stabilized earth walls?

# 10.3  BASIC CONCEPTS ON LATERAL EARTH PRESSURES

We have discussed the lateral earth pressure at rest and the lateral increases in stresses on a semi-infinite, isotropic, homogeneous, elastic soil mass from surface loading in Chapter 3. We are now going to consider the lateral earth pressures on a vertical wall that retains a soil mass. We will deal with two theories: one proposed by Coulomb (1776) and the other by Rankine (1857). First, we will develop a basic understanding of lateral earth pressures using a generic $\phi'$ and make the following assumptions:

1. The earth retaining wall is vertical.

2. The interface between the wall and soil is frictionless.

3. The soil surface is horizontal and no shear stress acts on horizontal and vertical boundaries.

4. The wall is rigid and extends to an infinite depth in a dry, homogeneous, isotropic soil mass.

5. The soil is loose and initially in an at-rest state.

Consider the wall as shown in Fig. 10.2. If the wall remains rigid and no movement (not even an infinitesimal movement) occurs, then the vertical and

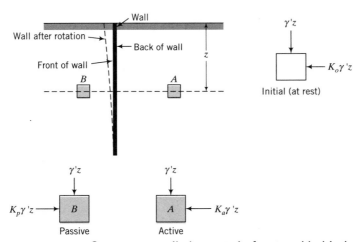

**FIGURE 10.2**  Stresses on soil elements in front and behind a retaining wall.

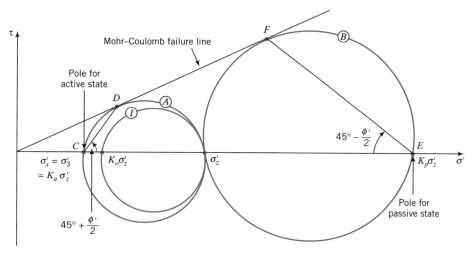

**FIGURE 10.3**  Mohr's circles at rest, active and passive states.

horizontal effective stresses at rest on elements A, at the back wall, and B, at the front wall (Fig. 10.2), are

$$\sigma'_z = \sigma'_1 = \gamma' z$$
$$\sigma'_x = \sigma'_3 = K_o \sigma'_1 = K_o \gamma' z$$

where $K_o$ is the lateral earth pressure at rest [Eq. (3.50)]. Mohr's circle for the at-rest state is shown by circle ① in Fig. 10.3.

Let us now assume a rotation about the bottom of the wall sufficient to produce slip planes in the soil mass behind and in front of the wall (Fig. 10.4). The rotation required, and consequently the lateral strains, to produce slip planes in front of the wall is much larger than that required for the back of the wall, as shown in Fig. 10.5. The soil mass at the back of the wall is assisting in producing failure while the soil mass at the front of the wall is resisting failure. In the latter, you have to rotate the wall against the soil to produce failure.

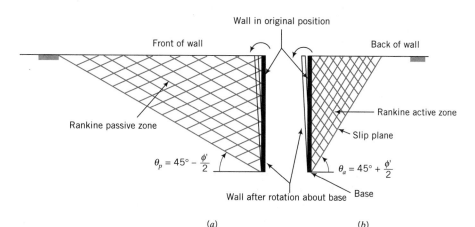

(a)                                        (b)

**FIGURE 10.4**  Slip planes within a soil mass near a retaining wall.

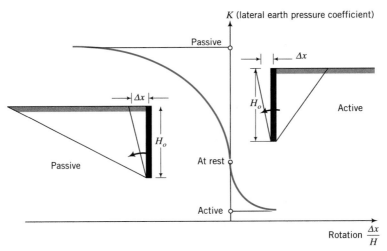

**FIGURE 10.5**   Rotation required to mobilize active and passive resistance.

What happens to the lateral effective stresses on elements A and B (Fig. 10.2) when the wall is rotated? The vertical stress will not change on either element but the lateral effective stress on element A will be reduced while that for element B will be increased. We can now plot two additional Mohr's circles: one to represent the stress state of element A (circle Ⓐ, Fig. 10.3) and the other to represent the stress state of element B (circle Ⓑ, Fig. 10.3). Both circles are drawn such that the decrease (element A) or increase (element B) in lateral effective stress is sufficient to bring the soil to the Mohr–Coulomb failure state. For element B to reach the failure state, the lateral effective stress must be greater than the vertical effective stress, as shown in Fig. 10.3.

The ratio of lateral principal effective stress to vertical principal effective stress is given by Eq. (5.22), which, for circle Ⓐ, is

$$\frac{(\sigma_3')_f}{(\sigma_1')_f} = \frac{(\sigma_x')_f}{(\sigma_z')_f} = \frac{1 - \sin \phi'}{1 + \sin \phi'} = \tan^2\left(45^\circ - \frac{\phi'}{2}\right) = K_a \qquad (10.1)$$

where $K_a$ is called the active lateral earth pressure coefficient. Similarly, for circle Ⓑ,

$$\frac{(\sigma_1')_f}{(\sigma_3')_f} = \frac{(\sigma_x')_f}{(\sigma_z')_f} = \frac{1 + \sin \phi'}{1 - \sin \phi'} = \tan^2\left(45^\circ + \frac{\phi'}{2}\right) = K_p \qquad (10.2)$$

where $K_p$ is the passive earth pressure coefficient. Therefore,

$$K_p = \frac{1}{K_a}$$

If, for example, $\phi' = 30^\circ$, then $K_a = \frac{1}{3}$ and $K_p = 3$.

The stress states of soil elements A and B are called the Rankine active state and the Rankine passive state, respectively (named after the original developer of this theory, Rankine, 1857). Each of these Rankine states is associated with a family of slip planes. For the Rankine active state, the slip planes are oriented at

$$\theta_a = 45° + \frac{\phi'}{2} \tag{10.3}$$

to the horizontal, as illustrated in Fig. 10.4b and proved in Chapter 5. For the Rankine passive state, the slip planes are oriented at

$$\theta_p = 45° - \frac{\phi'}{2} \tag{10.4}$$

to the horizontal, as illustrated in Fig. 10.4a.

The lateral earth pressure for the Rankine active state is

$$(\sigma'_x)_a = K_a \sigma'_z = K_a \gamma' z \tag{10.5}$$

and for the Rankine passive state it is

$$(\sigma'_x)_p = K_p \sigma'_z = K_p \gamma' z \tag{10.6}$$

Equations (10.5) and (10.6) indicate that, for a homogeneous soil layer, the lateral earth pressure varies linearly with depth, as shown in Figs. 10.6a,b.

The lateral earth force is the area of the lateral stress diagram, which, for the Rankine active state, is

$$P_a = \int_0^{H_o} K_a \gamma' z = \tfrac{1}{2} K_a \gamma' H_o^2 \tag{10.7}$$

and, for the Rankine passive state, is

$$P_p = \int_0^{H_o} K_p \gamma' z = \tfrac{1}{2} K_p \gamma' H_o^2 \tag{10.8}$$

These lateral forces, $P_a$ and $P_p$, are located at the centroid of the lateral earth pressure distribution diagram. In this case, the centroid is at $H_o/3$ from the base.

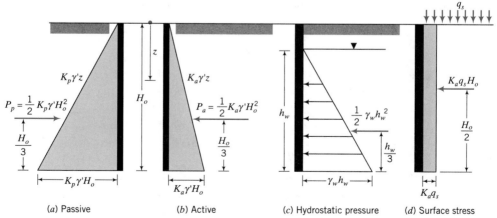

(a) Passive  (b) Active  (c) Hydrostatic pressure  (d) Surface stress

**FIGURE 10.6** Variation of active and passive lateral earth pressures, hydrostatic pressure, and a uniform surface stress with depth.

If groundwater is present, you need to add the hydrostatic pressure (pore water pressure) to the lateral earth pressure. For example, if the groundwater level is at distance $h_w$ from the base of the wall (Fig. 10.6c), the hydrostatic pressure is

$$u = \gamma_w h_w \tag{10.9}$$

and the hydrostatic force is

$$\boxed{P_w = \tfrac{1}{2}\gamma_w h_w^2} \tag{10.10}$$

Surface stresses also impose lateral earth pressures on retaining walls. A uniform surface stress, $q_s$, will transmit a uniform active lateral earth pressure of $K_a q_s$ (Fig. 10.6d) and a uniform passive lateral earth pressure of $K_p q_s$. The active and passive lateral earth pressures due to the soil, groundwater, and the uniform surface stresses are then

$$\boxed{(\sigma_x)_a = K_a \sigma_z' + K_a q_s + (u)_a} \tag{10.11}$$

and

$$\boxed{(\sigma_x)_p = K_p \sigma_z' + K_p q_s + (u)_p} \tag{10.12}$$

where $u$ is the hydrostatic pressure and the subscripts $a$ and $p$ denote active and passive states, respectively.

The lateral forces from a line load and a strip load are given in Chapter 3. For other types of surface loads you can consult Poulos and Davis (1974).

> *The* **essential points** *are:*
>
> *1. **The lateral earth pressures on retaining walls are related directly to the vertical effective stress through two coefficients. One is the active earth pressure coefficient,***
>
> $$\mathbf{K_a} = \frac{1 - \sin \phi'}{1 + \sin \phi'} = tan^2\left(45° - \frac{\phi'}{2}\right)$$
>
> *and the other is the passive earth pressure coefficient,*
>
> $$\mathbf{K_p} = \frac{1 + \sin \phi'}{1 - \sin \phi'} = tan^2\left(45° + \frac{\phi'}{2}\right) = \frac{1}{K_a}$$
>
> *2. **Substantially more movement is required to mobilize the full passive earth pressure than the full active earth pressure.***
>
> *3. **A family of slip planes occurs in the Rankine active and passive states. In the active state, the slip planes are oriented at $45° + \phi'/2$ to the horizontal, while for the passive case they are oriented at $45° - \phi'/2$ to the horizontal.***
>
> *4. **The lateral earth pressure coefficients, developed so far, are only valid for a smooth, vertical wall supporting a homogeneous soil mass with a horizontal surface.***
>
> *5. **The lateral earth pressure coefficients must be applied only to effective stresses.***

## EXAMPLE 10.1

Determine the active lateral earth pressure on the frictionless wall shown in Fig. E10.1a. Calculate the resultant force and its location from the base of the wall. Neglect seepage effects.

***Strategy*** The lateral earth pressure coefficients can only be applied to the effective stresses. You need to calculate the vertical effective stress, apply $K_a$, and then add the pore water pressure.

### Solution 10.1

**Step 1:** Calculate $K_a$.

$$K_a = \frac{1 - \sin \phi'}{1 + \sin \phi'} = \frac{1 - \sin(30°)}{1 + \sin(30°)}$$

$$= \frac{1}{3} \quad \text{or} \quad \tan^2\left(45 - \frac{\phi'}{2}\right) = \tan^2\left(45° - \frac{30°}{2}\right) = \frac{1}{3}$$

**Step 2:** Calculate the vertical effective stress.

$$\text{At the surface:} \quad \sigma'_z = 0, u = 0$$
$$\text{At the base:} \quad \sigma'_z = \gamma' H_o = (20 - 9.8) \times 5 = 51 \text{ kPa},$$
$$u = \gamma_w H_o = 9.8 \times 5 = 49 \text{ kPa}$$

**Step 3:** Calculate the lateral effective stress.

$$(\sigma'_x)_a = K_a \sigma'_z = \tfrac{1}{3} \times 51 = 17 \text{ kPa}$$

**Step 4:** Sketch the lateral earth pressure distributions.
See Figs. E10.1b,c.

**Step 5:** Calculate the lateral force.

$$P_a = P_s + P_w$$

where $P_s$ is the lateral force due to the soil solids and $P_w$ is the lateral force due to the pore water.

$$P_a = \tfrac{1}{2}(\sigma'_x)_a H_o + \tfrac{1}{2}u H_o = (\tfrac{1}{2} \times 17 \times 5) + (\tfrac{1}{2} \times 49 \times 5) = 165 \text{ kN}$$

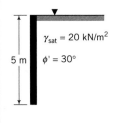

$\gamma_{sat} = 20$ kN/m$^2$

5 m    $\phi' = 30°$

*(a)* Wall

17 kPa

*(b)* Lateral pressure from soil

49 kPa

*(c)* Hydrostatic pressure

**FIGURE E10.1**

**Step 6:** Determine the location of the resultant.

Since both the lateral earth pressure and the pore water pressure distributions are triangular over the whole depth, the resultant is at the centroid of the triangle, that is, $\bar{z} = H_o/3 = 5/3 = 1.67$ m from the base of the wall. ∎

## EXAMPLE 10.2

For the frictionless wall shown Fig. E10.2a, determine the following:

**(a)** The active lateral earth pressure distribution with depth.

**(b)** The passive lateral earth pressure distribution with depth.

**(c)** The magnitudes and locations of the active and passive forces.

**(d)** The resultant force and its location.

**(e)** The ratio of passive moment to active moment.

***Strategy*** There are two layers. It is best to treat each layer separately. Neither $K_a$ nor $K_p$ should be applied to the pore water pressure. You do not need to calculate $K_p$ for the top soil layer. Since the water level on both sides of the wall is the same, the resultant hydrostatic force is zero. However, you are asked to determine the forces on each side of the wall; therefore, you have to consider the hydrostatic force. A table is helpful to solve this type of problem.

## Solution 10.2

**Step 1:** Calculate the active lateral earth pressure coefficients.

Top layer (0–2 m): $K_a = \tan^2\left(45° - \dfrac{\phi'}{2}\right) = \tan^2\left(45° - \dfrac{25°}{2}\right) = 0.41$

Bottom layer (2–6 m): $K_a = \tan^2\left(45° - \dfrac{\phi'}{2}\right) = \tan^2\left(45° - \dfrac{30°}{2}\right) = \dfrac{1}{3}$; $K_p = \dfrac{1}{K_a} = 3$

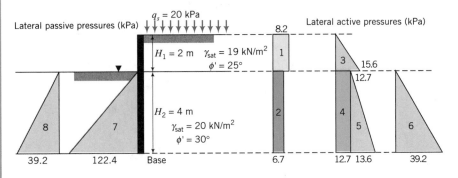

| (f) Pore water | (e) Soil | (a) Wall | | (b) Surface stresses | (c) Soil | (d) Pore water |

**FIGURE E10.2**

**Step 2:** Calculate the active and passive lateral earth pressures. Use a table as shown below to do the calculations or use a spreadsheet.

| Active | Depth[a] (m) | $u$ (kPa) | $\sigma_z$ (kPa) | $\sigma_z' = \sigma_z - u$ | $(\sigma_x')_a = K_a\sigma_z'$ (kPa) |
|---|---|---|---|---|---|
| Surcharge | 0 | 0 | 20 | 20 | $0.41 \times 20 = 8.2$ |
| | 2–6 | 0 | 20 | 20 | $\frac{1}{3} \times 20 = 6.7$ |
| Soil | 0 | 0 | 0 | 0 | 0 |
| | $2^-$ | 0 | $\gamma_1 H_1 = 19 \times 2 = 38$ | 38 | $0.41 \times 38 = 15.6$ |
| | $2^+$ | 0 | $\gamma_1 H_1 = 19 \times 2 = 38$ | 38 | $\frac{1}{3} \times 38 = 12.7$ |
| | 6 | $\gamma_w H_2 = 9.8 \times 4$ $= 39.2$ | $\gamma_1 H_1 + \gamma_2 H_2$ $= 19 \times 2 + 20 \times 4 = 118$ | 78.8 | $\frac{1}{3} \times 78.8 = 26.3$ |

| Passive | Depth (m) | $u$ (kPa) | $\sigma_z$ (kPa) | $\sigma_z' = \sigma_z - u$ | $(\sigma_x')_p = K_p\sigma_z'$ |
|---|---|---|---|---|---|
| Soil | 0 | 0 | 0 | 0 | 0 |
| | 4 | $\gamma_w H_2 = 9.8 \times 4$ $= 39.2$ | $\gamma_2 H_2 = 20 \times 4 = 80$ | 40.8 | $3 \times 40.8 = 122.4$ |

[a]The $-$ and $+$ superscripts indicate that you are calculating the stress just above $(-)$ and just below $(+)$ 2 m.

See Figs. E10.2b–e for the pressure distributions.

**Step 3:** Calculate the hydrostatic force.

$$P_w = \tfrac{1}{2}\gamma_w H_2^2 = \tfrac{1}{2} \times 9.8 \times 4^2 = 78.4 \text{ kN}$$

**Step 4:** Calculate the resultant lateral forces and their locations. See the table below for calculations. Active moments are assumed to be negative.

| Active Area | Depth (m) | Force (kN) | Moment arm from base (m) | Moment (kN · m) |
|---|---|---|---|---|
| 1 | 0–2 | $8.2 \times 2 = 16.4$ | $4 + 1 = 5 = 4.42$ | −82.0 |
| 2 | 2–6 | $6.7 \times 4 = 26.8$ | $\frac{4}{2} = 2$ | −53.6 |
| 3 | 0–2 | $\frac{1}{2} \times 15.6 \times 2 = 15.6$ | $\frac{2}{3} + 4 = 4.67$ | −72.9 |
| 4 | 2–6 | $12.7 \times 4 = 50.8$ | $\frac{4}{2} = 2$ | −101.6 |
| 5 | 2–6 | $\frac{1}{2} \times 13.6 \times 4 = 27.2$ | $\frac{4}{3}$ | −36.3 |
| 6 (water) | 2–6 | 78.4 | $\frac{4}{3}$ | −104.5 |
| | | Σ Active lateral forces = 215.2 | Σ Active moments = | −450.9 |

| Passive Area | Depth (m) | Force (kN) | Moment arm from base (m) | Moment (kN · m) |
|---|---|---|---|---|
| 7 | 2–6 | $\frac{1}{2} \times 122.4 \times 4 = 244.8$ | $\frac{4}{3}$ | 326.4 |
| 8 (water) | 2–6 | 78.4 | $\frac{4}{3}$ | 104.5 |
| | | Σ Passive forces = 323.2 | Σ Passive moments = | 430.9 |

Location of resultant active lateral earth force:

$$\bar{z}_a = \frac{\Sigma \text{ Moments}}{\Sigma \text{ Active lateral forces}} = \frac{450.9}{215.2} = 2.09 \text{ m}$$

Location of passive lateral force: $\bar{z}_p = \frac{4}{3} = 1.33$ m

**Step 5:** Calculate the resultant lateral force.

$$R_x = P_p - P_a = 323.2 - 215.2 = 108 \text{ kN/m}$$

**Step 6:** Calculate the ratio of moments.

$$\text{Ratio of moments} = \frac{\Sigma \text{ Passive moments}}{\Sigma \text{ Active moments}} = \frac{430.9}{450.9} = 0.96$$

Since the active moment is greater than the passive moment, the wall will rotate.                                                     ∎

*What's next* . . .The pioneer of earth pressure theory is Coulomb (1776). We are going to introduce his ideas in the next section.

## 10.4  COULOMB'S EARTH PRESSURE THEORY

Coulomb (1776) proposed that a condition of limit equilibrium exists through which a soil mass behind a vertical retaining wall will slip along a plane inclined an angle $\theta$ to the horizontal. He then determined the slip plane by searching for the plane on which the maximum thrust acts. We begin consideration of Coulomb's theory by reminding you of the basic tenets of limit equilibrium. The essential steps in the limit equilibrium method are (1) selection of a plausible failure mechanism, (2) determination of the forces acting on the failure surface, and (3) use of equilibrium equations to determine the maximum thrust.

Let us consider a vertical, frictionless wall of height $H_o$, supporting a soil mass with a horizontal surface (Fig. 10.7a). We are going to assume a dry, homogeneous soil mass and postulate that slip occurs on a plane inclined an angle

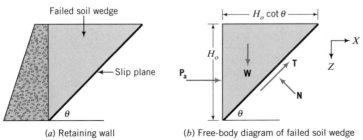

(a) Retaining wall          (b) Free-body diagram of failed soil wedge

**FIGURE 10.7**  Coulomb failure wedge.

$\theta$ to the horizontal. Since the soil is dry, $\gamma' = \gamma$. We can draw the free-body diagram as shown in Fig. 10.7b and solve for $P_a$ using statics as follows:

$$\Sigma F_x = P_a + T \cos \theta - N \sin \theta = 0$$
$$\Sigma F_z = W - T \sin \theta - N \cos \theta = 0$$

The weight of the sliding mass of soil is

$$W = \tfrac{1}{2}\gamma H_o^2 \cot \theta$$

At limit equilibrium,

$$T = N \tan \phi'$$

Solving for $P_a$ we get

$$P_a = \tfrac{1}{2}\gamma H_o^2 \cot \theta \tan(\theta - \phi') \tag{10.13}$$

To find the maximum thrust and the inclination of the slip plane, we use calculus to differentiate Eq. (10.13) with respect to $\theta$:

$$\frac{\partial P_a}{\partial \theta} = \tfrac{1}{2}\gamma H_o^2[\cot \theta \sec^2(\theta - \phi') - \csc^2 \theta \tan(\theta - \phi')] = 0$$

which leads to

$$\theta = \theta_{cr} = 45° + \frac{\phi'}{2} \tag{10.14}$$

Substituting this value of $\theta$ into Eq. (10.13) we get

$$\boxed{P_a = \frac{1}{2}\gamma H_o^2 \tan^2\left(45° - \frac{\phi'}{2}\right) = \frac{1}{2}K_a\gamma H_o^2} \tag{10.15}$$

This is the same result obtained earlier from considering Mohr's circle.

 The solution from a limit equilibrium method is analogous to an upper bound solution because it gives a solution that is usually greater than the true solution. The reason for this is that a more efficient failure mechanism may be possible than the one we postulated. For example, rather than a planar slip surface we could have postulated a circular slip surface or some other geometric form, and we could have obtained a maximum horizontal force lower than for the planar slip surface.

 For the Rankine active and passive states, we considered the stress states and obtained the distribution of lateral stresses on the wall. At no point in the soil mass did the stress state exceed the failure stress state and static equilibrium is satisfied. The solution for the lateral forces obtained using the Rankine active and passive states is analogous to a lower bound solution—the solution obtained is usually lower than the true solution because a more efficient distribution of stress could exist. If the lower bound solution and the upper bound solution are in agreement, we have a true solution, as is the case here.

 Poncelet (1840), using Coulomb's limit equilibrium approach, obtained expressions for $K_a$ and $K_p$ for cases where wall friction ($\delta$) is present, the wall face

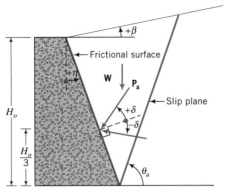

**FIGURE 10.8** Retaining wall with sloping back, wall friction, and sloping soil surface for use with Coulomb's method for active condition.

is inclined at an angle $\eta$ to the vertical, and the backfill is sloping at an angle $\beta$. With reference to Fig. 10.8, $K_{aC}$ and $K_{pC}$ are

$$K_{aC} = \frac{\cos^2(\phi' - \eta)}{\cos^2 \eta \, \cos(\eta + \delta)\left[1 + \left\{\dfrac{\sin(\phi' + \delta) \sin(\phi' - \beta)}{\cos(\eta + \delta) \cos(\eta - \beta)}\right\}^{1/2}\right]^2} \tag{10.16}$$

$$K_{pC} = \frac{\cos^2(\phi' + \eta)}{\cos^2 \eta \, \cos(\eta - \delta)\left[1 - \left\{\dfrac{\sin(\phi' + \delta) \sin(\phi' + \beta)}{\cos(\eta - \delta) \cos(\eta - \beta)}\right\}^{1/2}\right]^2} \tag{10.17}$$

The subscript C denotes Coulomb. You should note that $K_{pC} \neq 1/K_{aC}$. Recall that the lateral earth pressure coefficients are applied to the effective not total vertical stress. In the absence of groundwater, the total and effective vertical stresses are equal.

The inclination of the slip plane to the horizontal is

$$\tan \theta = \left[\frac{(\sin \phi' \cos \delta)^{1/2}}{\cos \phi'\{\sin(\phi' + \delta)\}^{1/2}}\right] \pm \tan \phi' \tag{10.18}$$

where the positive sign refers to the active state ($\theta_a$) and the negative sign refers to the passive state ($\theta_p$).

Wall friction causes the slip planes in both the active and passive states to be curved. The curvature in the active case is small in comparison to the passive case. The implication of the curved slip surface is that the values of $K_{aC}$ and $K_{pC}$ from Eqs. (10.16) and (10.17) are not accurate. In particular, the passive earth pressures are overestimated using Eq. (10.17). For the active state, the error is small and can be neglected. The error for the passive state is small if $\delta < \phi'/3$. In practice, $\delta$ is generally greater than $\phi'/3$.

Several investigators have attempted to find $K_a$ and $K_p$ using nonplanar slip surfaces. For example, Caquot and Kerisel (1948) used logarithm spiral slip surfaces while Packshaw (1969) used circular failure surfaces. The Caquot and Ker-

**TABLE 10.1  Correction Factors to be Applied to $K_{pC}$ to Approximate a Logarithm Spiral Slip Surface for a Backfill with a Horizontal Surface and Sloping Wall Face**

| $\phi'$ | $\delta/\phi'$ | | | | | | | |
|---|---|---|---|---|---|---|---|---|
| | −0.7 | −0.6 | −0.5 | −0.4 | −0.3 | −0.2 | −0.1 | 0.0 |
| 15 | 0.96 | 0.93 | 0.91 | 0.88 | 0.85 | 0.83 | 0.80 | 0.78 |
| 20 | 0.94 | 0.90 | 0.86 | 0.82 | 0.79 | 0.75 | 0.72 | 0.68 |
| 25 | 0.91 | 0.86 | 0.81 | 0.76 | 0.71 | 0.67 | 0.62 | 0.57 |
| 30 | 0.88 | 0.81 | 0.75 | 0.69 | 0.63 | 0.57 | 0.52 | 0.47 |
| 35 | 0.84 | 0.75 | 0.67 | 0.60 | 0.54 | 0.48 | 0.42 | 0.36 |
| 40 | 0.78 | 0.68 | 0.59 | 0.51 | 0.44 | 0.38 | 0.32 | 0.26 |

isel (1948) values of $K_p$ are generally used in practice. Table 10.1 lists factors that can be applied to $K_{pC}$ to correct it for a logarithm spiral slip surface.

The lateral forces are inclined at $\delta$ to the normal on the sloping wall face. The sign conventions for $\delta$ and $\beta$ are shown in Fig. 10.8. You must use the appropriate sign in determining $K_{aC}$ and $K_{pC}$. The direction of the frictional force on the wall depends on whether the wall moves relative to the soil or the soil moves relative to the wall. In general, the active wedge moves downward relative to the wall and the passive wedge moves upward relative to the wall. The frictional forces developed are shown in Fig. 10.9. The sense of the inclination of active lateral force is then positive while the sense of the inclination of passive lateral force is negative. Load bearing walls with large vertical loads tend to move downward relative to the soil. In such a situation the frictional force on both sides of the wall will be downward and the lateral earth pressures are increased on both sides of the wall. In practice, these increases in pressure are ignored and the sense of the active and lateral forces are taken as shown in Fig. 10.9.

The horizontal components of the lateral forces are $P_{ax} = P_a \cos \delta$ and $P_{px} = P_p \cos \delta$, and the vertical components are $P_{az} = P_a \sin \delta$ and $P_{pz} = P_p \sin \delta$. The point of application of these forces is $H_o/3$ from the base of the wall (Fig. 10.8). Typical values of $\delta$ for the interfaces of coarse-grained soils and concrete or steel walls range from $\frac{1}{2}\phi'$ to $\phi'$.

During compaction of the backfill soil, additional lateral stresses are imposed by compaction equipment. For walls with sloping backfill, heavy compac-

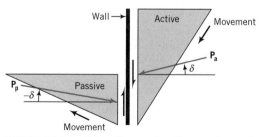

**FIGURE 10.9**  Directions of active and passive forces when wall friction is present.

tion equipment is generally used. The wall pressures from using heavy compaction equipment can be in excess of the active earth pressures. You should account for the lateral stresses in designing retaining walls. You may refer to Ingold (1979), who used elastic theory to estimate the lateral stresses imposed by construction equipment. Some practicing geotechnical engineers prefer to account for the additional compaction stresses by assuming that the resultant active lateral earth force acts at $0.4H_o$ or $0.5H_o$ rather than $\frac{1}{3}H_o$ from the base of the wall. Alternatively, you can multiply the active earth pressures by a factor ($\cong 1.20$) to account for compaction stresses.

> *The* **essential points** *are:*
>
> 1. *Coulomb's analysis of the lateral forces on a retaining structure is based on limit equilibrium.*
> 2. *Wall friction causes the slip planes to curve, which leads to an overestimation of the passive earth pressure using Coulomb's analysis.*
> 3. *For calculation of the lateral earth pressure coefficients you can use Eqs. (10.16) and (10.17), and correct $K_{pC}$ using the factors listed in Table 10.1.*
> 4. *The active and passive forces are inclined at an angle $\delta$ from the normal to the wall face.*

**What's next . . .**Coulomb's analysis is based on a postulated failure mechanism and the stresses within the soil mass are not considered. Rankine (1857) proposed an analysis based on the stress state of the soil. You have already encountered his approach in Section 10.3. Next, we introduce Rankine's solution for a wall with a sloping backfill and a sloping wall face.

## 10.5   RANKINE'S LATERAL EARTH PRESSURE FOR A SLOPING BACKFILL AND A SLOPING WALL FACE

Rankine (1857) established the principle of stress states or stress field in solving stability problems in soil mechanics. We have used Rankine's method in developing the lateral earth pressures for a vertical, frictionless wall supporting a dry, homogeneous soil with a horizontal surface. Rankine (1857) derived expressions for $K_a$ and $K_p$ for a soil mass with a sloping surface that was later extended to include a sloping wall face by Chu (1991). You can refer to Rankine's paper and Chu's paper for the mathematical details.

With reference to Fig. 10.10, the lateral earth pressure coefficients according to Rankine's analysis are

$$K_{aR} = \frac{\cos(\beta - \eta)\sqrt{1 + \sin^2 \phi' - 2 \sin \phi' \cos \theta_a}}{\cos^2 \eta(\cos \beta + \sqrt{\sin^2 \phi' - \sin^2 \beta})} \qquad (10.19)$$

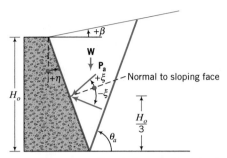

**FIGURE 10.10** Retaining wall with sloping soil surface, frictionless soil–wall interface, and sloping back for use with Rankine's method.

and

$$K_{pR} = \frac{\cos(\beta - \eta)\sqrt{1 + \sin^2 \phi' + 2 \sin \phi' \cos \theta_p}}{\cos^2 \eta(\cos \beta - \sqrt{\sin^2 \phi' - \sin^2 \beta})} \tag{10.20}$$

where the subscript R denotes Rankine, and

$$\theta_a = \frac{\pi}{4} + \frac{\phi'}{2} - \frac{\beta}{2} - \frac{1}{2} \sin^{-1}\left(\frac{\sin \beta}{\sin \phi'}\right) \tag{10.21}$$

$$\theta_p = \frac{\pi}{4} - \frac{\phi'}{2} + \frac{\beta}{2} + \frac{1}{2} \sin^{-1}\left(\frac{\sin \beta}{\sin \phi'}\right) \tag{10.22}$$

are the inclinations of the slip planes to the horizontal. The sign conventions for $\eta$ and $\beta$ are shown in Fig. 10.10; anticlockwise rotation is positive.

The active and passive lateral earth forces are

$$P_a = \tfrac{1}{2}K_{aR}\gamma' H_o^2 \quad \text{and} \quad P_p = \tfrac{1}{2}K_{pR}\gamma' H_o^2$$

These forces are inclined at

$$\xi_a = \tan^{-1}\left(\frac{\sin \phi' \sin \theta_a}{1 - \sin \phi' \cos \theta_a}\right) \tag{10.23}$$

$$\xi_p = \tan^{-1}\left(\frac{\sin \phi' \sin \theta_p}{1 + \sin \phi' \cos \theta_p}\right) \tag{10.24}$$

to the normal of the wall face. The angles $\xi_a$ and $\xi_p$ (reminder: the subscripts $a$ and $p$ denote active and passive) are not interface friction values.

In the case of a wall with a vertical face, $\eta = 0$, Eqs. (10.19) and (10.20) reduce to

$$K_{aR} = \frac{1}{K_{pR}} = \left(\frac{\cos \beta - \sqrt{\cos^2 \beta - \cos^2 \phi'}}{\cos \beta + \sqrt{\cos^2 \beta - \cos^2 \phi'}}\right) \tag{10.25}$$

and the active and passive lateral earth forces act in a direction parallel to the soil surface, that is, they are inclined at an angle $\beta$ to the horizontal.

> *The* **essential points** *are:*
>
> 1. *Rankine used stress states of a soil mass to determine the lateral earth pressures on a frictionless wall.*
> 2. *The active and passive lateral earth forces are inclined at $\xi_a$ and $\xi_p$, respectively, from the normal to the wall face. If the wall face is vertical, the active and passive lateral earth forces are parallel to the soil surface.*

**What's next** . . .We considered the lateral earth pressures for a dry soil mass, which is analogous to an effective stress analysis. Next, we will consider total stress analysis.

## 10.6  LATERAL EARTH PRESSURES FOR A TOTAL STRESS ANALYSIS

Figure 10.11 shows a smooth, vertical wall supporting a homogeneous soil mass under undrained conditions. Using the limit equilibrium method, we will assume, for the active state, that a slip plane is formed at an angle $\theta$ to the horizontal. The forces on the soil wedge are shown in Fig. 10.11.

Using static equilibrium, we obtain the sum of the forces along the slip plane:

$$P_a \cos \theta + T - W \sin \theta = 0 \tag{10.26}$$

But $T = s_u L = s_u(H_o/\sin \theta)$ and $W = \frac{1}{2}\gamma H_o^2 \cot \theta$. Equation (10.26) then yields

$$P_a = \frac{1}{2}\gamma H_o^2 - \frac{s_u H_o}{\sin \theta \cos \theta} = \frac{1}{2}\gamma H_o^2 - \frac{2 s_u H_o}{\sin 2\theta}$$

To find the maximum active lateral earth force, we differentiate $P_a$ with respect to $\theta$ and set the result equal to zero, giving

$$\frac{\partial P_a}{\partial \theta} = 4 s_u H_o \cot 2\theta \csc 2\theta = 0$$

The solution is $\theta = \theta_a = 45°$.

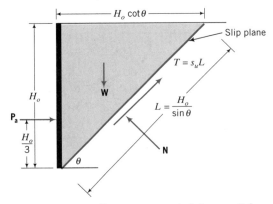

**FIGURE 10.11**   Forces on a retaining wall for a total stress analysis.

By substituting $\theta = 45°$ into the above equation for $P_a$, we get the maximum active lateral earth force as

$$P_a = \tfrac{1}{2}\gamma H_o^2 - 2s_u H_o \tag{10.27}$$

If we assume a uniform distribution of stresses on the slip plane, then the active lateral stress is

$$(\sigma_x)_a = \gamma z - 2s_u \tag{10.28}$$

Let us examine Eq. (10.28). If $(\sigma_x)_a = 0$, for example, when you make an excavation, then solving for $z$ in Eq. (10.28) gives

$$z = z_{cr} = \frac{2s_u}{\gamma} \tag{10.29}$$

Depth $z_{cr}$ is the depth at which tension cracks would extend into the soil (Fig. 10.12). If the tension crack is filled with water, the critical depth can extend to

$$z'_{cr} = \frac{2s_u}{\gamma'} \tag{10.30}$$

In addition, the soil in the vicinity of the crack is softened and a hydrostatic pressure, $\gamma_w z'_{cr}$, is imposed on the wall. Often, the critical depth of water-filled tension cracks in overconsolidated clays is greater than the wall height. For example, if $s_u = 80$ kPa, $\gamma = \gamma_{sat} = 18$ kN/m³, then $z'_{cr} = 19.5$ m. A wall height equivalent to the depth of the tension crack of 19.5 m is substantial. As a comparison, the average height of the Great Wall of China is about 12 m. When tension cracks occur they modify the slip plane, as shown in Fig. 10.12; no shearing resistance is available over the length of the slip plane above the depth of the tension cracks.

For an unsupported excavation, the active lateral force is also zero. From Eq. (10.27), we get

$$\tfrac{1}{2}\gamma H_o^2 - 2s_u H_o = 0$$

and, solving for $H_o$, we obtain

$$H_o = H_{cr} = \frac{4s_u}{\gamma} \tag{10.31}$$

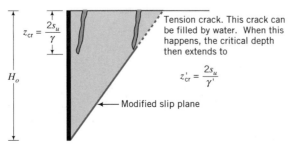

FIGURE 10.12   Tension cracks behind a retaining wall.

If the excavation is filled with water then

$$H'_{cr} = \frac{4s_u}{\gamma'}$$  (10.32)

We have two possible unsupported depths as given by Eqs. (10.29) and (10.31). The correct solution lies somewhere between these critical depths. In design practice, a value of

$$H_{cr} = \frac{3.8s_u}{\gamma}$$  (10.33)

is used for unsupported excavation in fine-grained soils. If the excavation is filled with water,

$$H'_{cr} = \frac{3.8s_u}{\gamma'}$$  (10.34)

The passive lateral earth force for a total stress analysis, following a procedure similar to that for the active state above, can be written as

$$P_p = \tfrac{1}{2}\gamma H_o^2 + 2s_u H_o$$  (10.35)

and the passive lateral pressure is

$$(\sigma_x)_p = \gamma z + 2s_u$$  (10.36)

We can write Eqs. (10.28) and (10.36) using apparent active and passive lateral earth pressures for the undrained condition as

$$(\sigma_x)_a = \sigma_z - K_{au}s_u$$  (10.37)

$$(\sigma_x)_p = \sigma_z + K_{pu}s_u$$  (10.38)

where $K_{au}$ and $K_{pu}$ are the undrained active and passive lateral earth pressure coefficients. In our case, for a smooth wall supporting a soil mass with a horizontal surface, $K_{au} = K_{pu} = 2$.

Walls that are embedded in fine-grained soils may be subjected to an adhesive stress $(s_w)$ at the wall face. The adhesive stress is analogous to a wall–soil interface friction for an effective stress analysis. The undrained lateral earth pressure coefficients are modified to account for adhesive stress as

$$K_{au} = K_{pu} = 2\sqrt{1 + \frac{s_w}{s_u}}$$  (10.39)

> **The essential points are:**
> 1. *Lateral earth pressures for a total stress analysis are found using apparent lateral earth pressure coefficients* **K**$_{au}$ *and* **K**$_{pu}$*. These coefficients are applied to the undrained shear strength. For smooth, vertical walls,* **K**$_{au}$ = **K**$_{pu}$ = 2.

2. *Tension cracks of theoretical depth $2s_u/\gamma$, or $2s_u/\gamma'$ if water fills the tension cracks, are usually formed in fine-grained soils and they modify the slip plane. If water fills the cracks, it softens the soil and a hydrostatic stress is imposed on the wall over the depth of the tension crack. You must pay particular attention to the possibility of the formation of tension cracks, and especially so if these cracks can be filled with water, because they can initiate failure of a retaining structure.*

3. *The theoretical maximum depth of an unsupported vertical cut in fine-grained soils is $\mathbf{H}_{cr} = 4s_u/\gamma$ or, if the cut is filled with water, $\mathbf{H}'_{cr} = 4s_u/\gamma'$.*

*What's next . . .* You were introduced to Coulomb's and Rankine's analyses of the active and passive lateral earth pressures. Concerns were raised regarding the accuracy of, in particular, $K_{pC}$ because wall friction causes the failure surface to diverge from a plane surface. Several investigators have proposed values of $K_p$, assuming curved failure surfaces. The question that arises is: "What values of $K_a$, $K_p$, $\phi'$, and $s_u$ should be used in the analyses of earth retaining structures?" Next, we will attempt to address this question. You are forewarned that the answer will not be definite. A geotechnical engineer usually has his/her preferences based on his/her experience with a particular method.

## 10.7 APPLICATION OF LATERAL EARTH PRESSURES TO RETAINING WALLS

Field and laboratory tests have not confirmed the Coulomb and Rankine theories. In particular, field and laboratory test results showed that both theories overestimate the passive lateral earth pressures. Values of $K_p$ obtained by Caquot and Kerisel (1948) lower the passive lateral earth pressures but they are still higher than experimental results. Other theories, for example, plasticity theory (Rosenfarb and Chen, 1972), have been proposed, but these theories are beyond the scope of this book and they too do not significantly change the Coulomb and Rankine passive lateral earth pressures for practical ranges of friction angle and wall geometry.

Rankine's theory was developed based on a semi-infinite "loose granular" soil mass for which the soil movement is uniform. Retaining walls do not support a semi-infinite mass but a soil mass of fixed depth. Strains, in general, are not uniform in the soil mass unless the wall rotates about its base to induce a state of plastic equilibrium.

The strains required to achieve the passive state are much larger than for the active case (Fig. 10.5). For sands, a decrease in lateral earth pressure of 40% of the at-rest lateral earth pressure can be sufficient to reach an active state, but an increase of several hundred percent in lateral earth pressure over the at-rest lateral earth pressure is required to bring the soil to a passive state. Because of the large strains that occur to achieve the passive state, it is customary to apply a factor of safety to the passive lateral earth pressure.

We have assumed a generic friction angle for the soil mass. Backfills are usually coarse-grained soils compacted to greater than 95% Proctor dry unit weight. If samples of the backfill were to be tested in shear tests in the laboratory at the desired degree of compaction, the samples may show peak shear stresses resulting in $\phi'_p$. If you use $\phi'_p$ to estimate the passive lateral earth pressure using either the Coulomb or Rankine method, you are likely to overestimate it because the shear strains required to develop the passive lateral earth pressure are much greater than those required to mobilize $\phi'_p$. The use of $\phi'_p$ in the Rankine or Coulomb equations is one reason for the disagreement between the predicted passive lateral earth pressures and experimental results.

Large shear strains ($\gamma > 10\%$) are required to mobilize $\phi'_{cs}$. For a backfill consisting of loose, coarse-grained soils, the displacement of the wall required to mobilize $\phi'_{cs}$ is intolerable in practice. You should then use conservative values of $\phi'$ in design. The maximum $\phi'_{design}$ should be $\phi'_{cs}$. In practice, factors of safety are applied to allow for uncertainties of loads and soil properties.

Wall friction causes the active lateral earth pressures to decrease and the passive lateral earth pressures to increase. For active lateral earth pressures, the Coulomb equation is sufficiently accurate for practice, but for passive lateral earth pressures you should use the Caquot and Kerisel (1948) values or similar values (e.g., Packshaw, 1969).

The total stress analysis should only be used in temporary works. But even for these works, you should be cautious in relying on the undrained shear strength. A total stress analysis should be used in conjunction with an effective stress analysis for retaining structures supporting fine-grained soils.

When a wet clay soil is excavated, negative excess pore water pressures develop and give the soil a greater undrained strength than prior to excavation. You should not rely on this gain in strength because the excess pore water pressures will dissipate with time. On inserting a wall, the soil at the wall–soil interface is remolded and you should use conservative values of wall adhesion. The maximum values of wall adhesion (Padfield and Mair, 1984) should be the lesser of

$$\boxed{\begin{array}{ll} \text{Active state:} & s_w = 0.5s_u \quad \text{or} \quad s_w \leq 50 \text{ kPa} \\ \text{Passive state:} & s_w = 0.5s_u \quad \text{or} \quad s_w \leq 25 \text{ kPa} \end{array}}$$

$$(10.40)$$
$$(10.41)$$

The interface friction between the wall face and the soil depends on the type of backfill used and construction methods. If the surface texture of the wall is rougher than $D_{50}$ of the backfill, the strength characteristics of backfill would control the interface friction. In such a case, the interface friction angle can be taken as equivalent to $\phi'_{cs}$. If the wall surface is smooth compared with $D_{50}$ of the backfill, the interface friction value can be assumed, in the absence of field measurements, to be between $\frac{2}{3}\phi'_{cs}$ and $\frac{1}{2}\phi'_{cs}$.

A layer of coarse-grained soil is often used in construction to rest the base of gravity retaining walls (see Section 10.8) founded on clays. The interface friction angle for sliding would then be the lesser of the interface friction between the layer of coarse-grained soil and the wall base, and the interface friction between the layer of coarse-grained soil and the clay.

***What's next . . .***In the next three sections, we will analyze retaining walls to determine their stability. We will consider an ESA (effective stress analysis) and a TSA (total stress analysis). We begin by considering the possible failure modes.

## 10.8   TYPES OF RETAINING WALLS AND MODES OF FAILURE

There are two general classes of retaining walls. One class is rigid and consists of concrete walls relying on gravity for stability (Fig. 10.13). The other class is flexible and consists of long slender members of either steel or concrete or wood and relies on passive soil resistance and anchors for stability (Fig. 10.14).

There are four modes of failures for rigid retaining walls—translational failure, rotation and bearing capacity failure, deep-seated failure, and structural failure (Fig. 10.15). Flexible walls, also called sheet pile walls, failed either by deep-seated failure, rotation at the base, rotation about the anchor or prop, failure of the anchor, bending of the wall, or seepage-induced failure (Fig. 10.16).

Seepage-induced failure is avoided in rigid retaining walls by providing adequate drainage systems, two of which are depicted in Fig. 10.17. The design of drainage systems is beyond the scope of this book. Flow nets discussed in Chapter 9 are used in designing drainage systems.

Flexible retaining walls are often used in waterfront structures and as temporary supports for excavations. Seepage forces are generally present and must be considered in evaluating the stability of these walls.

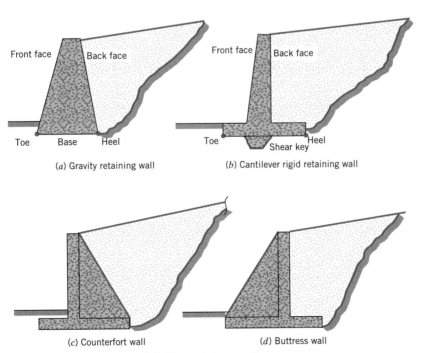

(a) Gravity retaining wall

(b) Cantilever rigid retaining wall

(c) Counterfort wall

(d) Buttress wall

**FIGURE 10.13**   Types of rigid retaining walls.

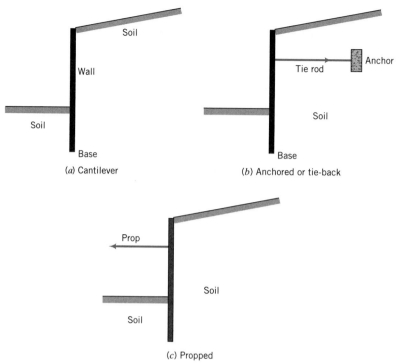

**FIGURE 10.14**    Types of flexible retaining walls.

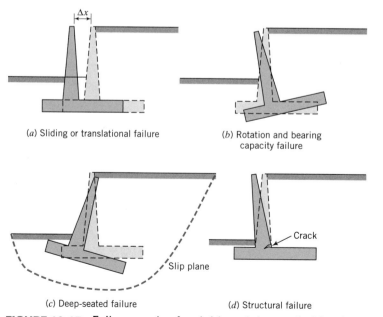

**FIGURE 10.15**    Failure modes for rigid retaining walls (the dotted lines show the original position of the wall).

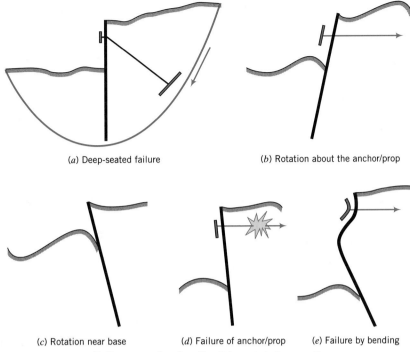

(*a*) Deep-seated failure

(*b*) Rotation about the anchor/prop

(*c*) Rotation near base   (*d*) Failure of anchor/prop   (*e*) Failure by bending

**FIGURE 10.16**   Failure modes for flexible retaining walls.

## 10.9   STABILITY OF RIGID RETAINING WALLS

Gravity retaining walls (Fig. 10.13a) are massive concrete walls. Their stability depends mainly on the self-weight of the walls. Cantilever walls (Fig. 10.13b) utilize the backfill to help mobilize stability and are generally more economical than gravity retaining walls. A rigid retaining wall must have an adequate factor

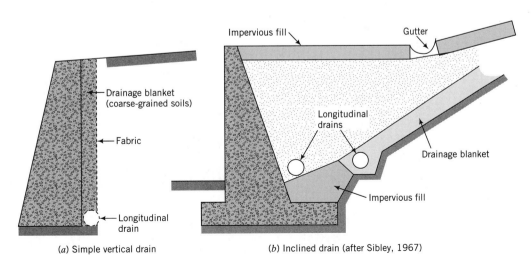

(*a*) Simple vertical drain

(*b*) Inclined drain (after Sibley, 1967)

**FIGURE 10.17**   Two types of drainage system behind rigid retaining walls.

of safety to prevent excessive translation, rotation, bearing capacity failure, deep-seated failure, and seepage-induced instability.

## 10.9.1 Translation

A rigid retaining wall must have adequate resistance against translation. That is, the sliding resistance of the base of the wall must be greater than the resultant lateral force pushing against the wall. The factor of safety against translation, $(FS)_T$, is

$$(FS)_T = \frac{T}{P_{ax}}; \quad (FS)_T \geq 1.5 \tag{10.42}$$

where $T$ is the sliding resistance at the base and $P_{ax}$ is the lateral force pushing against the wall. The sliding resistance is $T = R_z \tan \phi_b'$ for an ESA, and $T = s_w B$ for a TSA (if the base rests directly on fine-grained soils). $R_z$ is the resultant vertical force, $\phi_b'$ is the interfacial friction angle between the base of the wall and the soil

$$\phi_b' \approx \tfrac{1}{2}\phi_{cs}' \text{ to } \tfrac{2}{3}\phi_{cs}'$$

and $B$ is the projected horizontal width of the base. Typical sets of forces acting on gravity and cantilever rigid retaining walls are shown in Fig. 10.18.

Using statics, we obtain, for an ESA,

$$(FS)_T = \frac{[(W_w + W_s + P_{az}) \cos \theta_b - P_{ax} \sin \theta_b] \tan \phi_b'}{P_{ax} \cos \theta_b + (W_w + W_s + P_{az}) \sin \theta_b} \tag{10.43}$$

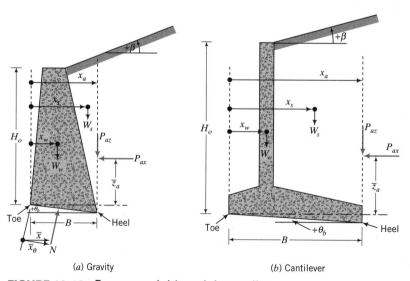

(a) Gravity          (b) Cantilever

**FIGURE 10.18**  Forces on rigid retaining walls.

where $W_w$ is the weight of the wall, $W_s$ is the weight of the soil wedge, $P_{az}$ and $P_{ax}$ are the vertical and horizontal components of the active lateral force, and $\theta_b$ is the inclination of the base to the horizontal ($\theta_b$ is positive if the inclination is counterclockwise, as shown in Fig. 10.18). If $\theta_b = 0$ (base is horizontal), then

$$(FS)_T = \frac{(W_w + W_s + P_{az}) \tan \phi_b'}{P_{ax}} \tag{10.44}$$

For a TSA,

$$(FS)_T = \frac{s_w B / \cos \theta_b}{P_{ax} \cos \theta_b + (W_w + W_s + P_{az}) \sin \theta_b} \tag{10.45}$$

If $\theta_b = 0$, then

$$(FS)_T = \frac{s_w B}{P_{ax}} \tag{10.46}$$

The embedment of rigid retaining walls is generally small and the passive lateral force is not taken into account. If base resistance is inadequate, the width $B$ of the wall can be increased. For cantilever walls, a shear key (Fig. 10.13b) can be constructed to provide additional base resistance against sliding.

### 10.9.2 Rotation

A rigid retaining wall must have adequate resistance against rotation. The rotation of the wall about its toe is satisfied if the resultant vertical force lies within the middle third of the base. Taking moments about the toe of the base, the resultant vertical force at the base is located at

$$\bar{x}_\theta = \frac{W_w x_w + W_s x_s + P_{az} x_a - P_{ax} \bar{z}_a}{(W_w + W_s + P_{az}) \cos \theta_b - P_{ax} \sin \theta_b} \tag{10.47}$$

where $\bar{z}_a$ is the location of the active lateral earth force from the toe. The wall is safe against rotation if $B/3 \leq \bar{x} \leq 2B/3$; that is, $e = |(B/2 - \bar{x})| \leq B/6$, where $e$ is the eccentricity of the resultant vertical load and $\bar{x} = \bar{x}_\theta \cos \theta_b$.

If $\theta_b = 0$, then

$$\bar{x} = \frac{W_w x_w + W_s x_s + P_{az} x_a - P_{ax} \bar{z}_a}{W_w + W_s + P_{az}} \tag{10.48}$$

### 10.9.3 Bearing Capacity

A rigid retaining wall must have a sufficient margin of safety against soil bearing capacity failure. The maximum pressure imposed on the soil at the base of the wall must not exceed the allowable soil bearing capacity; that is,

$$\sigma_{max} \leq q_a \tag{10.49}$$

where $\sigma_{max}$ is the maximum vertical stress imposed and $q_a$ is the allowable soil bearing capacity.

### 10.9.4 Deep-Seated Failure

A rigid retaining wall must not fail by deep-seated failure, whereby a slip surface encompasses the wall and the soil adjacent to it. In Chapter 11, we will discuss deep-seated failure.

### 10.9.5 Seepage

A rigid retaining wall must have adequate protection from groundwater seepage. The pore water pressures and the maximum hydraulic gradient developed under seepage must not cause any of the four stability criteria stated above to be violated and static liquefaction must not occur, that is, $i_{max} < i_{cr}$. Usually,

$$i_{max} \leq i_{cr}/(FS)_s$$

where $(FS)_s$ is a factor of safety for seepage and is conventionally greater than 3. To avoid seepage-related failures, adequate drainage should be installed in the backfill to dissipate excess pore water pressures quickly. Coarse-grained soils are preferable for the backfill because of their superior drainage characteristics compared with fine-grained soils.

### 10.9.6 Procedures to Analyze Gravity Retaining Walls

*The **essential steps** in determining the stability of rigid retaining walls are as follows:*

1. *Calculate the active lateral earth force and its components. If the wall is smooth, use Rankine's equations because they are simpler than Coulomb's equations to calculate the active lateral earth pressure coefficient.*
2. *Determine the weight of the wall and soil above the base.*
3. *Use Eq. (10.43) or Eq. (10.44) to find $(FS)_T$.*
4. *Use Eq. (10.47) or Eq. (10.48) to determine the location of the resultant vertical force, $R_z$, from the toe of the wall.*
5. *Check that the eccentricity is less than $B/6$. If it is, then the wall is unlikely to fail by rotation.*
6. *Determine the maximum soil pressure from $\sigma_{max} = (R_z/A)(1 + 6e/B)$.*
7. *Calculate the ultimate bearing capacity, $q_{ult}$, using one of the methods described in Chapter 7. Since, in most cases, $R_z$ would be eccentric, Meyerhof's bearing capacity equation is preferable.*
8. *Calculate the factor of safety against bearing capacity failure: $(FS)_B = q_{ult}/\sigma_{max}$.*

### EXAMPLE 10.3

A gravity retaining wall, shown in Fig. E10.3a, is required to retain 5 m of soil. The backfill is a coarse-grained soil with $\gamma_{sat} = 18$ kN/m³, $\phi'_{cs} = 30°$. The existing soil (below the base) has the following properties: $\gamma_{sat} = 20$ kN/m³, $\phi'_{cs} = 28°$.

The wall is embedded 1 m into the existing soil and a drainage system is provided as shown. The groundwater level is 4.5 m below the base of the wall. Determine the stability of the wall for the following conditions:

**(a)** Wall friction is zero.

**(b)** Wall friction is 20°.

**(c)** The drainage system becomes clogged during several days of a rainstorm and the groundwater rises to the surface. Neglect seepage forces.

The unit weight of concrete is $\gamma_c = 24$ kN/m³.

***Strategy*** For zero wall friction, you can use Rankine's method. But for wall friction, you should use Coulomb's method. The passive resistance is normally neglected in rigid retaining walls. Since only active lateral forces are considered, $K_a$ from the Rankine and Coulomb methods should be accurate enough. Since groundwater is below the base, $\gamma' = \gamma_{sat}$ over the wall depth.

## Solution 10.3

**Step 1:** Determine $K_a$.

Rankine:  $\delta = 0$,  $K_{aR} = \tan^2\left(45° - \dfrac{\phi'_{cs}}{2}\right) = \tan^2\left(45° - \dfrac{30°}{2}\right) = \dfrac{1}{3}$

Couloumb:  $\delta = 20°$,  $\phi' = \phi'_{cs}$, $\beta = 0$, $\eta = 0$;  and from Eq. (10.16)

$$K_{aC} = \frac{\cos^2(30° - 0°)}{\cos^2 0° \cos(0° + 20°)\left[1 + \left\{\dfrac{\sin(30° + 20°)\sin(30° - 0°)}{\cos(0° + 20°)\cos(0° - 0°)}\right\}^{1/2}\right]^2} = 0.3$$

**Step 2:** Determine the lateral forces.

Rankine:  $P_{aR} = \frac{1}{2}K_{aR}\gamma_{sat}H_o^2 = \frac{1}{2} \times \frac{1}{3} \times 18 \times 5^2 = 75$ kN

$P_{aR}$ acts horizontally because the ground surface is horizontal.

Coulomb:  $P_{aC} = \frac{1}{2}K_{aC}\gamma_{sat}H_o^2 = \frac{1}{2} \times 0.3 \times 18 \times 5^2 = 67.5$ kN

$P_{aC}$ acts at an angle $\delta = 20°$ to the horizontal (see Fig. E10.3b).

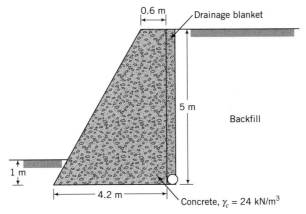

**FIGURE E10.3a**

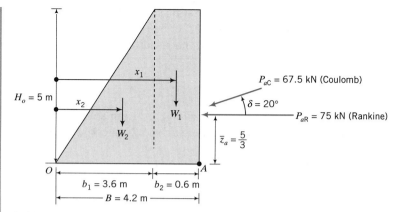

**FIGURE E10.3b**

Horizontal component of $P_{aC}$:   $(P_{ax})_C = P_{aC} \cos \delta = 67.5 \cos 20° = 63.4$ kN

Vertical component of $P_{aC}$:   $(P_{az})_C = P_{aC} \sin \delta = 67.5 \sin 20° = 23.1$ kN

(*Note:* Wall friction reduces $P_a$.)

**Step 3:**  Determine wall stability.

Consider a unit length of wall.

$$W_1 = b_2 H_o \gamma_c = 5 \times 0.6 \times 24 = 72 \text{ kN}$$
$$W_2 = \tfrac{1}{2} b_1 H_o \gamma_c = \tfrac{1}{2} \times 3.6 \times 5 \times 24 = 216 \text{ kN}$$
$$W = W_1 + W_2 = 72 + 216 = 288 \text{ kN}$$

*or*

$$W = \tfrac{1}{2}(B + b_2) H_o \gamma_c = \tfrac{1}{2}(4.2 + 0.6) \times 5 \times 24 = 288 \text{ kN}$$

Calculate the location of the resultant from $O$ (Fig. E10.3b).

Rankine:   $M_O = W_1 x_1 + W_2 x_2 - P_{aR} \bar{z}_a = 72(3.6 + 0.3) + 216 \times (\tfrac{2}{3} \times 3.6)$
$$- 75 \times \tfrac{5}{3} = 674.2 \text{ kN.m}$$

$R_z = W = 288$ kN

$\bar{x} = \dfrac{M_O}{R_z} = \dfrac{674.2}{288} = 2.34$ m

Coulomb:   $M_O = W_1 x_1 + W_2 x_2 + (P_{az})_C \times B - (P_{ax})_C \times \bar{z}_a$
$$= 72(3.6 + 0.3) + 216 \times (\tfrac{2}{3} \times 3.6) + 23.1 \times 4.2 - 63.4 \times \tfrac{5}{3} = 790.6 \text{ kN}$$

$R_z = W + (P_{az})_C = 288 + 23.1 = 311.1$ kN

$\bar{x} = \dfrac{M_O}{R_z} = \dfrac{790.6}{311.1} = 2.54$ m

Base resistance:   $T = R_z \tan \phi_b'$, where $R_z$ is the resultant vertical force

Assume $\phi_b' = \tfrac{2}{3} \phi_{cs}' = \tfrac{2}{3} \times 28 = 18.7°$.

Rankine:   $T = 288 \times \tan 18.7° = 97.5$ kN

Coulomb:   $T = (288 + 23.1) \times \tan 18.7° = 105.3$ kN

Rankine:   $(FS)_T = \dfrac{T}{P_{aR}} = \dfrac{97.5}{75} = 1.3 < 1.5$; therefore unsatisfactory

Coulomb:   $(FS)_T = \dfrac{T}{(P_{ax})_C} = \dfrac{105.3}{63.4} = 1.7 > 1.5;$ therefore satisfactory

With wall function, the factor of safety against translation is satisfactory.

### Determine Rotational Stability

$$\text{Rankine: } e = \left| \frac{B}{2} - \bar{x} \right| = \left| \frac{4.2}{2} - 2.34 \right| = 0.24 \text{ m}$$

$$\text{Coulomb: } e = \left| \frac{B}{2} - \bar{x} \right| = \left| \frac{4.2}{2} - 2.54 \right| = 0.44 \text{ m}$$

$$\frac{B}{6} = \frac{4.2}{6} = 0.7 > e$$

The resultant vertical forces for both Rankine and Coulomb method lie within the middle one-third of the base and, therefore, overturning is unlikely to occur

**Determining Factor of Safety Against Bearing Capacity Failure**   Since the resultant vertical force is located within the middle one-third, tension will not develop in the soil.

$$\sigma_{\max} = \frac{R_z}{A} \left( 1 + \frac{6e}{B} \right)$$

$$\text{Rankine: } (\sigma_{\max})_R = \frac{288}{4.2 \times 1} \left( 1 + \frac{6 \times 0.24}{4.2} \right) = 92.1 \text{ kPa}$$

$$\text{Coulomb: } (\sigma_{\max})_C = \frac{311.1}{4.2 \times 1} \left( 1 + \frac{6 \times 0.44}{4.2} \right) = 120.6 \text{ kPa}$$

The maximum stress occurs at $A$ (Fig. E10.3b) for both the Rankine and Coulomb methods. The base of the wall can be taken as a strip, surface foundation, that is, $B/L \to 0$, and $D_f = 0$. The groundwater level is below $B = 4.2$ m from the base, so groundwater would have no effect on the bearing capacity. Use Meyerhof's method [Eq. (7.16)] to determine the bearing capacity because the vertical load is eccentric.

$$q_{\text{ult}} \approx \tfrac{1}{2}\gamma B' N_\gamma, \quad \text{where } B' = B - 2e$$

For $\phi'_{cs} = 28°$, $N_\gamma = 11$.

Rankine:   $q_{\text{ult}} = \tfrac{1}{2} \times 20 \times (4.2 - 2 \times 0.24) \times 11 = 409.2$ kPa

$(FS)_B = \dfrac{q_{\text{ult}}}{(\sigma_{\max})_R} = \dfrac{409.2}{92.1} = 4.4 > 3;$   therefore satisfactory

Coulomb:   $q_{\text{ult}} = \tfrac{1}{2} \times 20 \times (4.2 - 2 \times 0.44) \times 11 = 365.2$ kPa

$(FS)_B = \dfrac{q_{\text{ult}}}{(\sigma_{\max})_C} = \dfrac{365.2}{120.6} \cong 3.0;$ therefore, bearing capacity is satisfactory

**Step 4:**   Determine the effects of water from the rainstorm.

**Using Rankine's method (zero wall friction)**

$$P_{aR} = \tfrac{1}{2}K_a\gamma' H_o^2 + \tfrac{1}{2}\gamma_w H_o^2 = \tfrac{1}{2} \times \tfrac{1}{3} \times (18 - 9.8) \times 5^2$$
$$+ \tfrac{1}{2} \times 9.8 \times 5^2 = 34.2 + 122.5 = 156.7 \text{ kN}$$

**Location of resultant from $O$**

$$M_O = W_1 x_1 + W_2 x_2 - P_{aR}\bar{z}_a = 72(3.6 + 0.3) + 216 \times (\tfrac{2}{3} \times 3.6)$$
$$- 156.7 \times \tfrac{5}{3} = 538 \text{ kN.m}$$

$$\bar{x} = \frac{538}{288} = 1.87 \text{ m}$$

**Translation**

$$(FS)_T = \frac{97.5}{156.7} = 0.62 < 1$$

The wall will fail by translation.

**Rotation**

$$e = \left| \frac{B}{2} - \bar{x} \right| = \left| \frac{4.2}{2} - 1.87 \right| = 0.23 < \frac{B}{6} \left( = \frac{4.2}{6} = 0.7 \right)$$

The wall is unlikely to fail by rotation.

**Bearing capacity**

$$q_{ult} = \tfrac{1}{2}\gamma' B' N_\gamma = \tfrac{1}{2} \times (18 - 9.8) \times (4.2 - 2 \times 0.23) \times 11 = 168.7 \text{ kPa}$$
$$\sigma_{max} = \frac{288}{4.2 \times 1} \left( 1 + \frac{6 \times 0.23}{4.2} \right) = 91.1 \text{ Pa}$$

The maximum stress now occurs at $O$ rather than at $A$.

$$(FS)_B = \frac{q_{ult}}{\sigma_{max}} = \frac{168.7}{99.1} = 1.9 < 3$$

The wall will not fail by bearing capacity failure but the factor of safety is inadequate. ∎

## EXAMPLE 10.4

Determine the stability of the cantilever, gravity retaining wall shown in Fig. E10.4a. The existing soil is a clay and the backfill is a coarse-grained soil. The base of the wall will rest on a 50 mm thick, compacted layer of the backfill. The interface friction between the base and the compacted layer of backfill is 25°.

***Strategy***   You should use Coulomb's method to determine the lateral earth pressure because of the presence of wall friction. The height of the wall for calculating the lateral earth pressure is the vertical height from the base of the wall to the soil surface. You should neglect the passive resistance of the 1.0 m of soil behind the wall.

## Solution 10.4

**Step 1:**   Determine the active lateral force and its location.
See Fig. E10.4b. We are given $\eta = 0$, $\beta = 8°$, $\delta = 15°$, and $\phi'_{cs} = 25°$. Therefore, from Eq. (10.16),

$$K_{aC} = \frac{\cos^2(25° - 0)}{\cos^2 0° \cos(0° + 15°) \left[ 1 + \left\{ \dfrac{\sin(25° + 15°) \sin(25° - 8°)}{\cos(0° + 15°) \cos(0° - 8°)} \right\}^{1/2} \right]^2} = 0.41$$

$$H_o = 0.9 + 6.1 + 3.0 \tan 8° = 7.42 \text{ m}$$

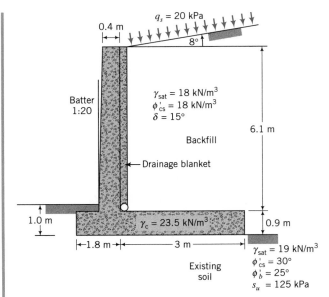

**FIGURE E10.4a**

**Soil mass**

All forces are per meter length of wall.

Lateral force from soil mass:   $P_{aC} = \frac{1}{2}K_{aC}\gamma_{sat}H_o^2 = \frac{1}{2} \times 0.41 \times 18$
$$\times 7.42^2 = 203.2 \text{ kN}$$

Horizontal component:   $F_{ax} = P_{aC} \cos \delta = 203.2 \cos 15° = 196.3 \text{ kN}$

Vertical component:   $F_{az} = P_{aC} \sin \delta = 203.2 \sin 15° = 52.6 \text{ kN}$

**Surcharge**

$$F_x = K_a q_s H_o \cos \delta = 0.41 \times 20 \times 7.42 \times \cos 15° = 58.8 \text{ kN}$$
$$F_z = K_a q_s H_o \sin \delta = 0.41 \times 20 \times 7.42 \times \sin 15° = 15.7 \text{ kN}$$

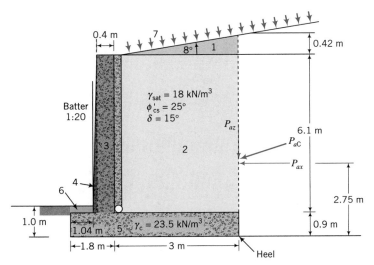

**FIGURE E10.4b**

**Resultant force components**

$$P_{ax} = F_{ax} + F_x = 196.3 + 58.8 = 255.1 \text{ kN}$$
$$P_{az} = F_{az} + F_z = 52.6 + 15.7 = 68.3 \text{ kN}$$

**Step 2:** Determine the resultant vertical force per unit length and its location. A table is useful to keep the calculation tidy and easy to check.

| Part | Force (kN/m) | Moment arm from toe (m) | Moment + kN · m |
|---|---|---|---|
| 1 | $0.5 \times 0.42 \times 3 \times 18 = 11.3$ | 3.80 | 42.9 |
| 2 | $3 \times 6.1 \times 18 = 329.4$ | 3.30 | 1087.0 |
| 3 | $0.4 \times 6.1 \times 23.5 = 57.3$ | 1.60 | 91.7 |
| 4 | $0.5 \times 0.36 \times 6.1 \times 23.5 = 25.8$ | 1.28 | 33.0 |
| 5 | $0.9 \times 4.8 \times 23.5 = 101.5$ | 2.40 | 243.6 |
| 6 | $0.1 \times 1.04 \times 23.5 = 2.4$ | 0.52 | 12.0 |
| 7 | $3 \times 20 = 60.00$ | 3.3 | 198.00 |
|  | 587.7 |  | 1708.2(+) |
| $P_{az}$ | 68.3 | 4.8 | 327.9(+) |
|  | $R_z = 656$ |  | $\Sigma$ Moments = 2036.1(+) |
| $P_{ax}$ | 255.1 | 2.75 | 701.5(−) |
|  |  |  | $\Sigma M_O = 1334.6(+)$ |

The location of the resultant horizontal component of force from the toe is

$$\bar{z} = \frac{F_{ax} \dfrac{H_o}{3} + F_x \dfrac{H_o}{2}}{F_{ax} + F_x} = \frac{196.3 \times \dfrac{7.42}{3} + 58.8 \times \dfrac{7.42}{2}}{196.3 + 58.8} = 2.75 \text{ m}$$

The location of the resultant vertical component of force from the toe is

$$\bar{x} = \frac{\Sigma M_O}{R_z} = \frac{1334.6}{656} = 2.03 \text{ m}$$

**Step 3:** Determine the eccentricity.

$$e = \frac{B}{2} - \bar{x} = \frac{4.8}{2} - 2.03 = 0.37 \text{ m}$$

**Step 4:** Determine the stability.

**Rotation**

$$\frac{B}{6} = \frac{4.8}{6} = 0.8 \text{ m} > e \ (= 0.37 \text{ m}); \quad \text{therefore rotation is satisfactory}$$

**Translation**

$$T = R_z \tan \phi'_b = 656 \times \tan 25° = 305.9 \text{ kN/m}$$

$$(\text{FS})_T = \frac{T}{P_{ax}} = \frac{305.9}{255.1} = 1.2 < 1.5; \quad \text{therefore, translation is not satisfactory}$$

In design, you can consider placing a key at the base to increase the factor of safety against translation.

**Bearing capacity**

$$\sigma_{max} = \frac{656}{4.8 \times 1}\left(1 + \frac{6 \times 0.37}{4.8}\right) = 199.9 \text{ kPa}$$

Use Meyerhof's equation.

For $\phi'_{cs} = 30°$, $N_\gamma = 15.7$;   $B' = B - 2e = 4.8 - 2 \times 0.37 = 4.06$ m

Groundwater level is within $B$ from the base; therefore, use $\gamma'$ in the bearing capacity equation.

$$q_{ult} = \tfrac{1}{2}\gamma'B'N_\gamma = \tfrac{1}{2} \times (19 - 9.8) \times 4.06 \times 15.7 = 293.2 \text{ kPa}$$

$$(FS)_B = \frac{q_{ult}}{\sigma_{max}} = \frac{293.2}{199.9} \approx 1.5 < 3; \quad \text{therefore, bearing capacity is not satisfactory}$$

∎

**What's next . . .** In the next section, we will study how to determine the stability of flexible retaining walls.

# 10.10 STABILITY OF FLEXIBLE RETAINING WALLS

Sheet pile walls are flexible and are constructed using steel or thin concrete slabs or wood. Two types of sheet pile walls are common. One is a cantilever wall commonly used to support soils to a height of less than 3 m (Fig. 10.14). The other is an anchored or propped sheet pile wall (Fig. 10.14) commonly used to support deep excavations and as waterfront retaining structures. Cantilever sheet pile walls rely on the passive soil resistance for their stability, while anchored sheet pile walls rely on a combination of anchors and passive soil resistance for their stability. The stability of sheet pile walls must satisfy all the criteria for rigid retaining walls described in Section 10.9 except bearing capacity. However, because sheet pile walls are used in situations where seepage may occur, it is necessary to pay particular attention to seepage-related instabilities.

## 10.10.1 Analysis of Sheet Pile Walls in Uniform Soils

In analyzing sheet pile walls, we are attempting to determine the depth of embedment, $d$, for stability. The analysis is not exact and various simplifications are made. The key static equilibrium condition is moment equilibrium. Once we determine $d$, the next step is to determine the size of the wall. This is done by calculating the maximum bending moment and then determining the section modulus by dividing the maximum bending moment by the allowable bending stress of the material constituting the sheet pile, for example, steel, concrete, or wood.

An effective stress analysis is generally used to analyze sheet pile walls and as such we must evaluate the pore water pressure distribution and seepage pressures. We can use flow net sketching or numerical methods to determine the pore water pressure distribution and seepage pressures. However, approximate meth-

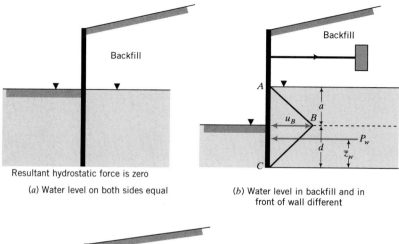

Resultant hydrostatic force is zero

(a) Water level on both sides equal

(b) Water level in backfill and in front of wall different

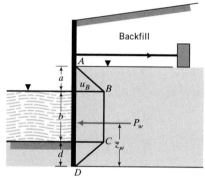

(c) Wall supports water in front of it and water level in backfill greater than water level in front of wall

**FIGURE 10.19** Approximate resultant pore water pressure distributions behind flexible retaining walls.

ods are often used in practice. If the groundwater level on both sides of a sheet pile wall is the same, then the resultant pore water pressures and seepage pressures are zero (Fig. 10.19a). You can then neglect the effects of groundwater in determining the stability of sheet pile walls. However, you must use effective stresses in your calculations of the lateral earth forces.

The approximate distribution of pore water pressures in front of and behind sheet pile walls for conditions in which the water tables are different is obtained by assuming a steady state seepage condition and uniform distribution of the total head. Approximate resultant pore water pressure distributions for some common conditions (Padfield and Mair, 1984) are shown in Fig. 10.19.

The maximum pore water pressures ($u_B$), maximum pore water forces ($P_w$) and their locations ($\bar{z}_w$), and the seepage force per unit volume ($j_s$) are as follows:

**Case (a)—Fig. 10.19a**

Resultant pore water pressure is zero and the seepage force is zero.

**Case (b)—Fig. 10.19b**

$$u_B = \frac{2ad}{a + 2d}\, \gamma_w$$

(10.50)

$$P_w = \frac{ad(a + d)}{a + 2d}\,\gamma_w \tag{10.51}$$

$$\overline{z}_w = \frac{a + 2d}{3} \tag{10.52}$$

$$j_s = \frac{a}{a + 2d}\,\gamma_w \tag{10.53}$$

**Case (c)—Fig. 10.19c**

$$u_B = u_C = \frac{a(b + 2d)}{a + b + 2d}\,\gamma_w \tag{10.54}$$

$$P_w = \frac{1}{2}\left[\frac{a(b + 2d)(a + 2b + d)}{a + b + 2d}\right]\gamma_w \tag{10.55}$$

$$\overline{z}_w = \frac{a^2 + 3a(b + d) + 3b(b + 2d) + 2d^2}{3(a + 2b + d)} \tag{10.56}$$

$$j_s = \frac{a}{a + b + 2d}\,\gamma_w \tag{10.57}$$

Recall that $j_s$ is the seepage pressure per unit volume and the resultant effective stress is increased when seepage is downward (behind the wall) and is decreased when seepage is upward (in front of the wall) as discussed in Chapter 3.

## 10.10.2 Analysis of Sheet Pile Walls in Mixed Soils

Sheet pile walls may penetrate different soil types. For example, Fig. 10.20 shows a sheet pile wall that supports a coarse-grained soil but is embedded in a fine-grained soil. In this case, you should consider a mixed analysis. For short-term condition, an effective stress analysis can be used for the coarse-grained soil but

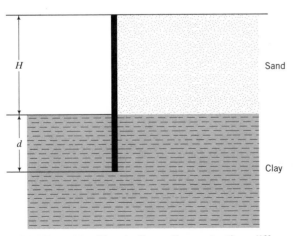

**FIGURE 10.20** Sheet pile wall penetrating different soils.

a total stress analysis should be used for the fine-grained soil. For long-term conditions, an effective stress analysis should be carried out for both soil types.

### 10.10.3 Consideration of Tension Cracks in Fine-Grained Soils

If a sheet pile wall supports fine-grained soils, you should consider the formation of tension cracks. The theoretical depth of a tension crack is

$$z_{cr} = \frac{2s_u}{\gamma} \quad \text{or} \quad \frac{2s_u}{\gamma'}$$

The latter is applicable when the tension cracks are filled with water. The depth of a tension crack is sometimes greater than the wall height. In this situation, you can assume a minimum active lateral effective pressure of $5z$ (kPa) as suggested by Padfield and Mair (1984), where $z$ is the depth measured from the top of the wall. When water fills the tension cracks, you should apply the full hydrostatic pressure to the wall over a depth equivalent to the depth of the tension crack or the wall height, whichever is smaller. In fine-grained soils, loss of moisture at the wall–soil interface can cause the soil to shrink, creating a gap. This gap can be filled with water. In this case, you should apply the full hydrostatic pressure over the depth of the wall.

### 10.10.4 Methods of Analyses

Several methods have been proposed to determine the stability of sheet pile walls. These methods differ in the way the lateral stresses are distributed on the wall and the way the factor of safety is applied in solving for the embedment depth. We will discuss three methods in this book. In the first method, called the factored moment method (FMM), you would determine an embedment depth to satisfy moment equilibrium by applying a factor of safety $(FS)_p$ on the passive resistance. The factor of safety, $(FS)_p$, is usually between 1.5 and 2.0.

In the second method, called the factored strength method (FSM), reduction factors are applied to the shear strength parameters. These reduction factors are called mobilization factors because they are intended to limit the shear strength parameters to values that are expected to be mobilized by the design loads. A mobilization factor, $F_\phi$, is applied to the friction angle, $\phi'_{cs}$, and a mobilization factor, $F_u$, is applied to $s_u$. The application of these mobilization factors results in a higher active pressure and a lower passive pressure than the unfactored soil strength parameters. The design parameters are

$$\phi'_{design} = \frac{\phi'_{cs}}{F_\phi}$$

and

$$(s_u)_{design} = \frac{s_u}{F_u}$$

where

$$F_\phi = 1.2 \text{ to } 1.5 \quad \text{and} \quad F_u = 1.5 \text{ to } 2$$

The results from the FSM are sensitive to $F_\phi$ and $F_u$.

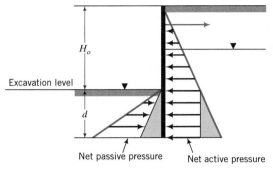

**FIGURE 10.21** Net pressures for the NPPM.

The third method, called the net passive pressure method (NPPM), utilizes a net available passive resistance (Burland et al., 1981). A vertical line is drawn from the active pressure at the excavation level to the base of the wall (Fig. 10.21). The shaded region of pressure on the active side (Fig. 10.21) is subtracted from the passive pressure to give the net passive pressure shown by the area hatched by arrows. The factor of safety for the NPPM is

$$(FS)_r = \frac{\Sigma \text{ Moments of net available passive resistance}}{\Sigma \text{ Moments of lateral forces causing rotation}};$$

$(FS)_r = 1.5$ to 2 with 2 most often used

The flexibility of sheet pile walls leads to lateral pressure distributions that do not correspond to the Rankine active and passive states. Expected lateral pressure distributions on a "stiff" flexible wall and a less "stiff" flexible wall are shown in Fig. 10.22. The net effect of the distribution of the lateral pressures as illustrated in Fig. 10.22 is a reduction in maximum bending moment from that calculated using Rankine active and passive states.

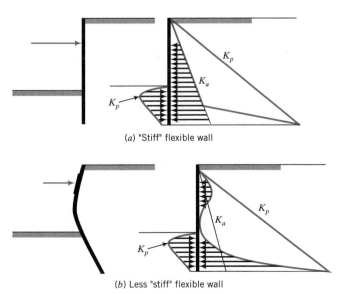

**FIGURE 10.22** Lateral pressure distributions expected on a "stiffer" flexible wall and less "stiff" flexible wall. (After Padfield and Mair, 1984.)

Rowe (1957) developed a method, based on laboratory tests, to reduce the calculated maximum bending moment to account for the effects of wall flexibility on the bending moment. Rowe's moment reduction is applicable when a factor of safety has been applied on the passive resistance as in the FMM. There is some debate on the applicability of Rowe's method. Some engineers prefer to calculate the maximum bending moment at limit equilibrium ($(FS)_p = F_\phi = (FS)_r = F_u = 1$) and use it as the design moment. This is the preferred method in this book.

To account for soil–wall interface friction, you need to use $K_{aC}$ and $K_{pC}$. However, the active and passive coefficients derived by Caquot and Kerisel (1948) are regarded as more accurate than those of Coulomb. For a flexible retaining wall, only the horizontal components of the lateral forces are important. In Appendix C, the horizontal components of the active and passive coefficients of Caquot and Kerisel (1948) as tabulated by Kerisel and Absi (1990) are plotted for some typical backfill slopes and soil–wall interface friction angles. We will use the values of the lateral earth pressure coefficients in Appendix C in some of the example problems in this chapter.

### 10.10.5 Analysis of Cantilever Sheet Pile Walls

Cantilever sheet pile walls are analyzed by assuming that rotation occurs at some point, $O$, just above the base of the wall (Fig. 10.23). The consequence of assuming rotation above the base is that, below the point of rotation, the lateral pressure is passive behind the wall and active in front of the wall (Fig. 10.23b). To simplify the analysis, a force $R$ (Fig. 10.23c) is used at the point of rotation to approximate the net passive resistance below it (the point of rotation). By taking moments about $O$, the unknown force $R$ is eliminated and we then obtain one equation with one unknown, that is, the unknown depth, $d_o$. To account for this simplification, the depth $d_o$ is increased by 20% to 30% to give the design embedment depth, $d$.

The general procedure for determining $d$ for stability and to determine the wall size is as follows:

1. Arbitrarily select a point $O$ at a distance $d_o$ from the excavation level.

2. Calculate the active and passive earth pressures using the FMM or FSM or NPPM.

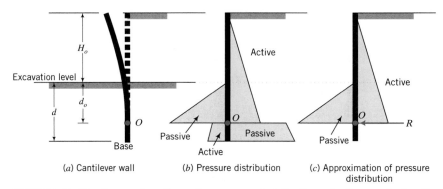

**FIGURE 10.23** Approximation of pressure distributions in the analysis of cantilever flexible retaining walls (Padfield and Mair, 1984).

3. Calculate the net pore water pressure ($u$) distribution and the seepage force per unit volume ($j_s$). The effective unit weight is increased by $j_s$ in the active zone and is decreased by $j_s$ in the passive zone.

4. Determine the unknown depth $d_o$ by summing moments about $O$.

5. Calculate $d$ by increasing $d_o$ by 20% or 30% to account for simplifications made in the analysis. The depth of penetration $d$ is therefore $1.2d_o$ or $1.3d_o$.

6. Calculate $R$ by summing forces horizontally over the depth $(H_o + d)$.

7. Calculate the net passive resistance, $(P_p)_{net}$, over the distance, $d - d_o$, below $O$.

8. Check that $R$ is less than $(P_p)_{net}$. If not, extend the depth of embedment and recalculate $R$.

9. Calculate the maximum bending moment ($M_{max}$) over the depth $(H_o + d_o)$ using unfactored passive resistance (FMM), unfactored strength values (FSM), and $(FS)_r = 1$ (NPPM).

10. Determine the section modulus, $S_x = M_{max}/f_a$, where $M_{max}$ is the maximum bending moment and $f_a$ is the allowable bending stress of the wall material.

### 10.10.6 Anchored Sheet Pile Walls

There are two methods used to analyze anchored sheet pile walls. One is the free earth method, the other is the fixed earth method. We will be discussing the free earth method, because it is frequently used in design practice.

In the free earth method, it is assumed that (1) the depth of embedment of the wall is insufficient to provide fixity at the bottom end of the wall and (2) rotation takes place about the point of attachment of the anchor, $O$ (Fig. 10.24a). The expected bending moment diagram, based on the above assumptions, is depicted in Fig. 10.24b.

The procedure for analyzing an anchored sheet pile wall is as follows:

1. Assume a depth of embedment, $d$.

2. Calculate the active and passive pressures using the FMM or FSM or NPPM.

3. Calculate the net pore water pressure ($u$) distribution and the seepage force per unit volume ($j_s$). The effective unit weight is increased by $j_s$ in the active zone and is decreased by $j_s$ in the passive zone.

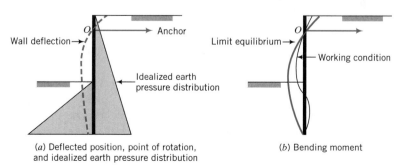

(a) Deflected position, point of rotation, and idealized earth pressure distribution

(b) Bending moment

**FIGURE 10.24** Free earth conditions for anchored retaining walls (Padfield and Mair, 1984).

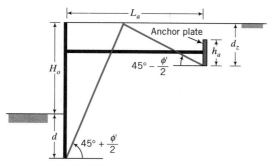

**FIGURE 10.25** Location of anchor plates.

**4.** Determine $d$ by taking moments about the point of attachment of the anchor, $O$. Usually, you will get a cubic equation, which you can solve by iteration or by using a goal seek option in a spreadsheet program or by using a polynomial function on a calculator.

**5.** Recalculate $d$ using unfactored passive resistance (FMM), unfactored strength (FSM), and $(FS)_r = 1$ (NPPM). Use this recalculated depth to determine the anchor force and the maximum bending moment.

**6.** Determine the anchor force per unit length of wall, $T_a$, by summing forces in the horizontal direction. The anchor force, $T_a$, is multiplied by a factor of safety $(FS)_a$, usually 2.

**7.** Determine the location of the anchor plates or deadmen. Let $d_z$ be the depth of the bottom of the anchor plate from the ground surface (Fig. 10.25). The force mobilized by the anchor plate must balance the design anchor force, that is,

$$\tfrac{1}{2}\gamma'\, d_z^2(K_p - K_a) = T_a \times (FS)_a$$

Solving for $d_z$, we get

$$d_z = \sqrt{\frac{2T_a \times (FS)_a}{\gamma'(K_p - K_a)}} \tag{10.58}$$

A passive wedge develops in front of the anchor plate and an active wedge develops behind the retaining wall. The anchor plate must be located outside the active slip plane. The minimum anchor length $(L_a)$ of the anchor rod, with reference to Fig. 10.25, is

$$L_a = (H_o + d)\tan(45° - \phi'/2) + d_z \tan(45° + \phi'/2) \tag{10.59}$$

**8.** Calculate the spacing of the anchors. Let $s$ be the longitudinal spacing of the anchors and $h_a$ be the height of the anchor plate. If $h_a \geq d_z/2$, the passive resistance of the anchor plate is assumed to be developed over the full depth $d_z$. From static equilibrium of forces in the horizontal direction, we obtain

$$s = \frac{\gamma' d_z^2 L_a}{2T_a(FS)_a}\,(K_p - K_a) \tag{10.60}$$

9. Calculate the maximum bending moment ($M_{max}$) using the embedment depth at limit equilibrium (unfactored passive resistance, unfactored strength values, or $(FS)_r = 1$).

10. Determine the section modulus, $S_x = M_{max}/f_a$, where $f_a$ is the allowable bending stress of the wall material.

## EXAMPLE 10.5

Determine the depth of embedment required for stability of the cantilever sheet pile wall shown in Fig. E10.5a. Compare the results of the three methods—FMM, FSM, and NPPM—using $(FS)_p = 2.0$, $F_\phi = 1.25$, and $(FS)_r = 1.5$. Calculate the maximum bending moment for each of these methods. Groundwater is below the base of the wall.

***Strategy*** You should use the Caquot and Kerisel (1948) passive pressures (see Appendix C) and either the Caquot and Kerisel (1948) or Coulomb active pressures. The key is to determine the lateral forces and then find $d_o$ (an arbitrarily selected embedment depth at which rotation is presumed to occur) using moment equilibrium. Since you have to find the depth for unfactored values to calculate the maximum bending moment, you should determine $K_{ax}$ and $K_{px}$ for factored and unfactored values at the very beginning of your solution.

## Solution 10.5

**Step 1:**  Calculate $K_{ax}$ and $K_{px}$.

$$\text{For FSM:} \quad \phi'_{design} = \frac{\phi'_{cs}}{F_\phi} = \frac{30}{1.25} = 24°$$

$$\text{For FMM and NPPM:} \quad \phi'_{design} = \phi'_{cs}$$

Use the Caquot–Kerisel $K_{ax}$ and $K_{px}$ (Appendix C).

FSM:  $K_{ax} = 0.36$    $(\phi'_{design} = 24°, \delta = \frac{2}{3}\phi'_{design}, \beta/\phi'_{design} = 0)$
$\quad\quad K_{px} = 3.3$    $(\phi'_{design} = 24°, \delta = \frac{1}{2}\phi'_{design}, \beta/\phi'_{design} = 0)$

FMM and NPPM:  $K_{ax} = 0.28$    $(\phi'_{cs} = 30°, \delta = \frac{2}{3}\phi'_{cs}, \beta/\phi'_{cs} = 0)$
$\quad\quad\quad\quad K_{px} = 4.6$    $(\phi'_{cs} = 30°, \delta = \frac{1}{2}\phi'_{design}, \beta/\phi'_{cs} = 0)$

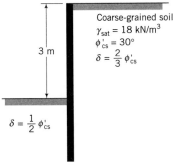

**FIGURE E10.5a**

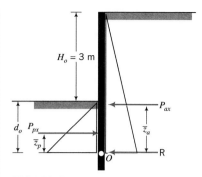

**FIGURE E10.5b**

**Step 2:**    Determine the lateral earth pressure distributions.
The lateral pressure distributions for the FMM and FSM have the same shape but different magnitudes because of the different lateral earth pressure coefficients (Fig. E10.5b). The lateral pressure distribution for the NPPM is shown in Fig. E10.5c. Since groundwater is not within the depth of the retaining wall, $\gamma' = \gamma_{sat} = \gamma$.
With reference to Fig. E10.5b:

**Active case**

$$P_{ax} = \tfrac{1}{2}K_{ax}\gamma(H_o + d_o)^2 = \tfrac{1}{2} \times K_{ax} \times 18(3 + d_o)^2 = 9K_{ax}(3 + d_o)^2$$

$$\bar{z}_a = \frac{H_o + d_o}{3} = \frac{3 + d_o}{3}$$

$$(M_O)_a = P_{ax}\bar{z}_a = 9K_{ax}(3 + d_o)^2\left(\frac{3 + d_o}{3}\right) = 3K_{ax}(3 + d_o)^2$$

**Passive case**

$$P_{px} = \tfrac{1}{2}K_{px}\gamma d_o^2 = \tfrac{1}{2} \times K_{px} \times 18 \times d_o^2 = 9K_{px}d_o^2$$

$$\bar{z}_p = \frac{d_o}{3}$$

$$(M_O)_p = P_{px}\bar{z}_p = 9K_{px}\,d_o^2 \times \frac{d_o}{3} = 3K_{px}d_o^3$$

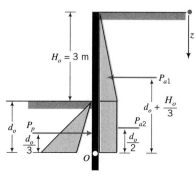

**FIGURE E10.5c**

**Step 3:** Find $d_o$.

All forces are calculated per meter length of wall.

**FMM**

The passive pressure is factored by $(FS)_p$.

$$K_{ax} = 0.28 \quad \text{and} \quad K_{px} = 4.6$$

$$(M_O)_a = 3 \times 0.28(3 + d_o)^3 = 0.84(3 + d_o)^3$$

$$(M_O)_p = \frac{3 \times 4.6d_o^3}{(FS)_p} = \frac{3 \times 4.6d_o^3}{2} = 6.9d_o^3$$

For equilibrium: $(M_O)_a = (M_O)_p$

$$\therefore \; 0.84(d_o^3 + 9d_o^2 + 27d_o + 27) = 6.9d_o^3$$

which simplifies to

$$7.21d_o^3 - 9d_o^2 - 27d_o - 27 = 0$$

By trial and error or by using the polynomial function on a calculator, $d_o = 2.95$ m.

**FSM**

$$K_{ax} = 0.36 \quad \text{and} \quad K_{px} = 3.3$$

$$(M_O)_a = 3 \times 0.36(3 + d_o)^3 = 1.08(3 + d_o)^3$$

$$(M_O)_p = 3 \times 3.3d_o^3 = 9.9d_o^3$$

For equilibrium: $(M_O)_a = (M_O)_p$

$$1.08(d_o^3 + 9d_o^2 + 27d_o + 27) = 9.9d_o^3$$

which simplifies to

$$8.17d_o^3 - 9d_o^2 - 27d_o - 27 = 0$$

By trial and error or by using the polynomial function on a calculator, $d_o = 2.75$ m.

**NPPM**

$$K_{ax} = 0.28 \quad \text{and} \quad K_{px} = 4.6$$

The pressure diagram for the NPPM is shown in Fig. E10.5c.

$$P_{a1} = \tfrac{1}{2} \times K_{ax}\gamma H_o^2 = \tfrac{1}{2} \times 0.28 \times 18 \times 3^2 = 22.7 \text{ kN}$$

$$P_{a2} = K_{ax}\gamma H_o d_o = 0.28 \times 18 \times 3 \times d_o = 15.1d_o$$

$$P_p = \tfrac{1}{2}(K_{px} - K_{ax})\gamma d_o^2 = \tfrac{1}{2}(4.5 - 0.28) \times 18d_o^2 = 38.9d_o^2$$

$$(M_O)_a = P_{a1}\left(d_o + \frac{H_o}{3}\right) + P_{a2}\frac{d_o}{2} = 7.6d_o^2 + 22.7d_o + 22.7$$

$$(M_O)_p = P_p\frac{d_o}{3} = 13d_o^3$$

$$(FS)_r = \frac{(M_O)_p}{(M_O)_a}$$

For $(FS)_r = 1.5$,

$$1.5(7.6d_o^2 + 22.7d_o + 22.7) = 13d_o^3$$

Rearranging, we get

$$7.67d_o^3 - 7.6d_o^2 - 22.7d_o - 22.7 = 0$$

By trial and error or by using the polynomial function on a calculator, $d_o = 2.41$ m.

**Step 4:**   Calculate the design depth.

$$\text{FMM:}\quad d = 1.2d_o = 1.2 \times 2.95 = 3.54 \text{ m}$$
$$\text{FSM:}\quad d = 1.2d_o = 1.2 \times 2.75 = 3.3 \text{ m}$$
$$\text{NPPM:}\quad d = 1.2d_o = 1.2 \times 2.41 = 2.89 \text{ m}$$

**Step 5:**   Determine $R$.

$R = P_{px} - P_{ax}$
FMM:  $d_o = 2.95$ m; $R = 9 \times 4.5 \times 2.95^2 - 9 \times 0.28 \times (3 + 2.95)^2 = 271.1$ kN/m
FSM:  $d_o = 2.75$ m; $R = 9 \times 3.3 \times 2.75^2 - 9 \times 0.36 \times (3 + 2.75)^2 = 109.4$ kN/m
NPPM:  $d_o = 2.41$ m; $R = 9 \times (4.6 - 0.28) \times 2.41^2$
$$- (9 \times 0.28 \times 3^2 + 0.28 \times 18 \times 3 \times 2.41) = 166.7 \text{ kN/m}$$

To calculate the net resistance below the assumed point of rotation, $O$, calculate the average passive pressure at the back of the wall and the average active pressure in front of the wall. Remember that below the point of rotation, passive pressure acts at the back of the wall and active pressure acts at the front of the wall. The mid-depth between $d_o$ and $1.2d_o$ is $1.1d_o$.

**FMM**

Average passive lateral pressure $= K_{px}\gamma(H_o + 1.1d_o) = 4.6 \times 18$
$$\times (3 + 1.1 \times 2.95) = 517.1 \text{ kPa}$$
Average active lateral pressure $= K_{ax}\gamma \times 1.1d_o = 0.28 \times 18 \times 1.1 \times 2.95$
$$= 16.4 \text{ kPa}$$
Net lateral pressure $= 517.1 - 16.4 = 500.7$ kPa
Net force $= 500.7 \times 0.2d_o = 295.4$ kN $> R$ ($= 271.1$ kN);
therefore, depth of penetration is satisfactory

**FSM**

Average passive lateral pressure $= K_{px}\gamma(H_o + 1.1d_o) = 3.3 \times 18$
$$\times (3 + 1.1 \times 2.75) = 357.9 \text{ kPa}$$
Average active lateral pressure $= K_{ax}\gamma \times 1.1d_o = 0.36 \times 18 \times 1.1 \times 2.75$
$$= 19.6 \text{ kPa}$$
Net lateral pressure $= 357.9 - 19.6 = 338.3$ kPa
Net force $= 338.3 \times 0.2d_o = 186.1$ kN $> R$ ($= 109.4$ kN);
therefore, depth of penetration is satisfactory

**NPPM**

Average passive lateral pressure $= K_{px}\gamma(H_o + 1.1d_o) = 4.6 \times 18$
$$\times (3 + 1.1 \times 2.41) = 467.9 \text{ kPa}$$
Average active lateral pressure $= K_{ax}\gamma \times 1.1d_o = 0.28 \times 18 \ \times 1.1 \times 2.41$
$$= 13.4 \text{ kPa}$$
Net lateral pressure $= 467.9 - 13.4 = 454.5$ kPa
Net force $= 454.5 \times 0.2d_o = 219.1$ kN $> R$ ($= 166.7$ kN);
therefore, depth of penetration is satisfactory

**Step 6:** Determine the maximum bending moment.

Maximum bending moment for $(FS)_p = F_\phi = 1$.

Let $z$ be the location of the point of maximum bending moment (point of zero shear) such that $z > H_o$.

$$M_z = \tfrac{1}{2} K_{ax}\gamma z^2 \times \frac{z}{3} - \tfrac{1}{2} K_{px}\gamma(z - H_o)^2 \times \frac{(z - H_o)}{3}$$

$$= \tfrac{1}{2} \times 0.28 \times 18 \times \frac{z^3}{3} - \tfrac{1}{2} \times 4.6 \times 18 \times \frac{(z - 3)^3}{3}$$

$$= 0.84z^3 - 13.8(z - 3)^3$$

To find $z$ at which the bending moment is maximum, we need to differentiate the above equation with respect to $z$ and set the result equal to zero.

$$\frac{dM_z}{dx} = 0 = 2.52z^2 - 41.4(z - 3)^2 = 38.9z^2 - 248.4z + 372.6 = 0$$

Solving for $z$, we get $z = 3.68$ m or $2.53$ m. The correct answer is 3.68 m since zero shear cannot occur above the excavation level in this problem (positive shear in the active zone only gets reduced below the excavation level).

$$M_z = 0.84 \times 3.68^3 - 13.8(3.68 - 3)^3 = 36 \text{ kN·m}$$

For this problem, it is easy to use calculus to determine the maximum bending moment. For most problems, you will have to find the shear force distribution with depth, identify or calculate the point of zero shear, and then calculate the maximum bending moment. ∎

## EXAMPLE 10.6

Determine the embedment depth and the anchor force of the tied-back wall shown in Fig. E10.6a using the FSM.

**Strategy**   You should use the Caquot and Kerisel (1948) passive pressures (see Appendix C) and either the Caquot and Kerisel (1948) or Coulomb active pressures. Groundwater level on both sides of the wall is the same, so seepage will not occur.

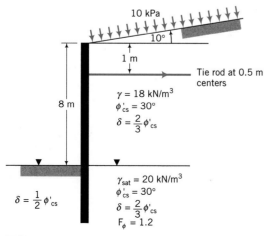

**FIGURE E10.6a**

## Solution 10.6

**Step 1:** Determine $K_{ax}$ and $K_{px}$.

$$\phi'_{design} = \frac{\phi'_{cs}}{F_\phi} = \frac{30}{1.2} = 25°; \quad \frac{\beta}{\phi'_{design}} = \frac{10}{25} = 0.4$$

From Appendix C,

$$K_{ax} = 0.42 \quad (\phi'_{design} = 25°, \; \beta/\phi'_{design} = 0.4, \; \delta = \tfrac{2}{3}\phi'_{desgn})$$
$$K_{px} = 3.4 \quad (\phi'_{design} = 25°, \; \beta/\phi'_{design} = 0, \; \delta = \tfrac{1}{2}\phi'_{design})$$

For unfactored strength values,

$$K_{ax} = 0.31 \quad (\phi'_{cs} = 30°, \; \beta/\phi'_{cs} = 10/30 = \tfrac{1}{3}, \; \delta = \tfrac{2}{3}\phi'_{cs})$$
$$K_{px} = 4.6 \quad (\phi'_{cs} = 30°, \; \beta/\phi'_{cs} = 0, \; \delta = \tfrac{1}{2}\phi'_{cs})$$

**Step 2:** Determine the lateral forces and moments.
Use a table to facilitate ease of computation and checking. (See Fig. E10.6b.) Below the groundwater level, $\gamma' = 20 - 9.8 = 10.2 \text{ kN/m}^3$.

| Part | Horizontal force (kN) | Moment arm from anchor (m) | Moment (kN · m) + ) |
|------|----------------------|----------------------------|---------------------|
| 1 | $K_{ax}q_s(H_o + d) =$ <br> $0.42 \times 10 \times (8 + d) = 4.2d + 33.6$ | $\left(\dfrac{H_o + d}{2}\right) - h =$ <br> $\left(\dfrac{8 + d}{2}\right) - 1 = 3 + \dfrac{d}{2}$ | $-(2.1d^2 + 29.4d + 100.8)$ |
| 2 | $\tfrac{1}{2}K_{ax}\gamma H_o^2 =$ <br> $\tfrac{1}{2} \times 0.42 \times 18 \times 8^2 = 241.9$ | $\tfrac{2}{3}H_o - h =$ <br> $\tfrac{2}{3} \times 8 - 1 = 4.33$ | $-1047.4$ |
| 3 | $K_{ax}\gamma H_o d =$ <br> $0.42 \times 18 \times 8 \times d = 60.5d$ | $H_o - h + \dfrac{d}{2} =$ <br> $7 + \dfrac{d}{2}$ | $-(30.3d^2 + 423.5d)$ |
| 4 | $\tfrac{1}{2}K_{ax}\gamma' d^2 =$ <br> $\tfrac{1}{2} \times 0.42 \times 10.2 \times d^2 = 2.1d^2$ <br> $\sum 2.1d^2 + 64.7d + 275.5$ | $H_o - h + \tfrac{2}{3}d =$ <br> $7 + \tfrac{2}{3}d$ | $-(1.4d^3 + 14.7d^2)$ |
| 5 | $\tfrac{1}{2} \times K_{px}\gamma' d^2 =$ <br> $\tfrac{1}{2} \times 3.4 \times 10.2 \times d^2 = 17.3d^2$ | $H_o - h + \tfrac{2}{3}d =$ <br> $7 + \tfrac{2}{3}d$ | $11.5d^3 + 121.1d^2$ |
| | | $\Sigma M =$ | $-(10.1d^3 + 74d^2 + 452.9d + 1148.2)$ |

**Step 3:** Determine $d$.
Equate the sum of moments to zero (simplify equation by dividing by the coefficient of $d^3$).

$$d^3 + 7.3d^2 - 44.8d - 113.7 = 0$$

By trial and error or by using the polynomial function on a calculator, $d = 5.38$ m.

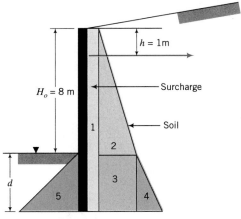

**FIGURE E10.6b**

**Step 4:** Determine $d$ for the unfactored strength values.

To calculate the new depth of penetration for unfactored strength values, use proportionality, for example,

Active moment = Active moment for factored strength values

$$\times \frac{\text{Unfactored } K_{ax}}{\text{Factored } K_{ax}}$$

$$(M_O)_A = -(1.4d^3 + 47.1d^2 + 452.9d + 1148.2) \times \frac{0.31}{0.42}$$

$$= -(d^3 + 34.8d^2 + 334.3d + 847.5)$$

Passive moment:  $(M_O)_p = (11.5d^3 + 121.1d^2) \times \dfrac{4.6}{3.4} = 15.5d^3 + 163.9d^2$

Sum of moments:  $(M_O)_p + (M_O)_a = 14.5d^3 + 129.1d^2 - 334.3d - 847.5$

Solving, we get $d = 3.38$ m.

**Step 5:** Determine the anchor force for $d = 3.38$ m.

$$\Sigma \text{ Active forces} = (2.1d^2 + 64.7d + 275.5) \times \frac{0.31}{0.42} = 381.9 \text{ kN}$$

$$\Sigma \text{ Passive forces} = 17.3d^2 \times \frac{4.6}{3.4} = 265.8 \text{ kN}$$

$$T_a = 381.9 - 265.8 = 116.1 \text{ kN}$$

$$(T_a)_{\text{design}} = (FS)T_a = 2 \times 116.1 = 232.2 \text{ kN}$$  ■

## EXAMPLE 10.7

Determine the embedment depth and the design anchor force required for stability of the sheet pile wall shown in Fig. E10.7a using the NPPM.

*Strategy* In the NPPM, you must use the unfactored strength values to calculate $K_{ax}$ and $K_{px}$ and then determine the net active and net passive lateral pressures. To calculate $d$, you have to do iterations. A simple approach to solve for $d$ is to set up the forces and moments in terms of the unknown $d$ and then assume values of $d$ until you find a $d$ value that gives the required factor of safety $((FS)_r \cong 1.5)$. A spreadsheet program or a programmable calculator is very helpful in solving this type of problem. In Excel, for example, you can use the Goal Seek function to find $d$. It is quite easy to make errors in calculations, so you

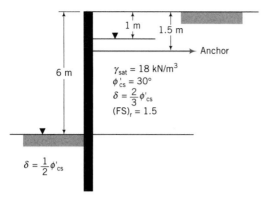

**FIGURE E10.7a**

should recheck your work and you must conduct a "hand" check when using outputs from computer programs. Since the groundwater levels are different in front and behind the wall, you need to consider seepage assuming a steady state seepage condition. To calculate the anchor force, you have to find $d$ for $(FS)_r = 1$ and then multiply the anchor force by 2 (factor of safety).

## Solution 10.7

**Step 1:** Determine $K_{ax}$ and $K_{px}$.
From Appendix C,

$$K_{ax} = 0.28 \quad (\phi'_{cs} = 30°, \beta/\phi'_{cs} = 0, \delta = \tfrac{2}{3}\phi'_{cs})$$
$$K_{px} = 4.6 \quad (\phi'_{cs} = 30°, \beta/\phi'_{cs} = 0, \delta = \tfrac{1}{2}\phi'_{cs})$$

**Step 2:** Determine the net lateral pressures. Make a table to do the calculations and draw a diagram of the lateral earth pressure distribution.
See Fig. E10.7b.

**Below groundwater level**

$$\text{Average seepage force/unit volume:} \quad j_s = \left(\frac{a}{a + 2d}\right)\gamma_w$$
$$= \left(\frac{5}{5 + 2d}\right)9.8 \text{ kN/m}^3$$

Active zone: $\gamma' = \gamma_{sat} - \gamma_w + j_s = 18 - 9.8 + j_s = 8.2 + j_s \text{ kN/m}^3$

Passive zone: $\gamma' = \gamma_{sat} - \gamma_w - j_s = 18 - 9.8 - j_s = 8.2 - j_s \text{ kN/m}^3$

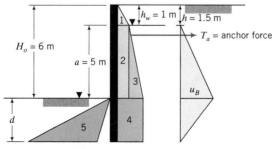

Lateral earth pressure from soil          Pore water pressure

**FIGURE E10.7b**

| Depth (m) | Active pressure (kPa) | Passive pressure (kPa) |
|---|---|---|
| 0 | 0 | 0 |
| 1 | $K_a\gamma h_w = 0.28 \times 18 \times 1 = 5$ | 0 |
| 6 | $K_a\gamma h_w + K_a(\gamma' + j_s)(H_o - h_w)$ | 0 |
| | $= 5.0 + 0.28 \times (8.2 + j_s) \times 5 = 16.5 + 1.4j_s$ | $K_p(8.2 - j_s)\, d - K_a(8.2 + j_s)\, d$ |
| $6 + d$ | $16.8 + 1.4j_s$ | $= 4.6(8.2 - j_s)\, d - 0.28(8.2 + j_s)\, d$ |
| | | $= (35.4 - 4.88j_s)\, d$ |
| Water | $u_B = \dfrac{2(ad)}{a + 2d}\gamma_w = \dfrac{98.1d}{5 + 2d}$ | |

**Step 3:** Calculate the lateral forces, the moment, and $(FS)_r$.

All forces and moments are per meter length of wall. The moment is the sum of moments about the anchor position and $R_x$ is the resultant active lateral force. In the first column under moment, a value of $d$ is guessed and $(FS)_r$ is calculated. In the second column under moment, the value of $d = 5.75$ m was obtained using a spreadsheet program (the actual value obtained from the spreadsheet program is $d = 5.73$ m for $(FS)_r = 1.5$). If you use a spreadsheet program, the first column under moment is not needed. In Excel, you use Tools → Goal Seek to find $d$ to satisfy the desired value of $(FS)_r$.

| | | | Moment (kN · m) | |
|---|---|---|---|---|
| Part | Forces (kN) | Moment arm (m) | $d = 7$ m | $d = 5.75$ m |
| 1 | $0.5 \times 5.0 \times 1 = 2.5$ | $h - h_w + \dfrac{h_w}{3} =$ $0.5 + 0.33 = 0.83$ | −2.1 | −2.2 |
| 2 | $5.0 \times 5 = 25$ | $\dfrac{a}{2} - (h - h_w) = 2.5 - 0.5 = 2$ | −50.0 | −50.0 |
| 3 | $0.5 \times (11.5 + 1.4j_s) \times 5$ $= 28.8 + 3.8j_s$ | $\dfrac{2a}{3} - (h - h_w) = \dfrac{10}{3} - 0.5$ $= 2.83$ | −109.2 | −113.4 |
| 4 | $(16.5 + 1.4j_s) \times d$ $= (16.5 + 1.4j_s)\, d$ Water: | $H_o - h - \dfrac{d}{2} = 4.5 + \dfrac{d}{2}$ | −1126.2 | −876.0 |
| | $\dfrac{ad(a + d)\gamma_w}{a + 2d} = \dfrac{49d(5 + d)}{5 + 2d}$ | $H_o + d - h - \bar{z}_w$ $= 4.5 + d - \dfrac{5 + 2d}{3} = \dfrac{8.5 + d}{3}$ | −1119.3 | −871.9 |
| | $R_x = 56.3 + 3.8j_s$ $+ (16.5 + 1.4j_s)\, d + \dfrac{49d(5 + d)}{5 + 2d}$ | $\Sigma M_d$ | −2402.6 | −1909.2 |
| 5 | $0.5 \times (35.4 - 4.88j_s)\, d^2$ $= (17.7 - 2.4j_s)\, d^2$ | $H_o - h + \dfrac{2d}{3} = 4.5 + \dfrac{2d}{3}\,\, \Sigma M_r =$ | 5179.8 | 2912.8 |
| | | $(FS)_r = \dfrac{\Sigma M_r}{\Sigma M_d}$ | 2.16 | 1.53 |

**Step 4:**    Calculate the anchor forces for $(FS)_r = 1$.
For $(FS)_r = 1$, $d = 4.62$ m. Substituting $d = 4.62$ m, we get

**Active zone**

$$j_s = \left(\frac{5}{5 + 2 \times 4.62}\right) 9.8 = 3.44 \text{ kN/m}^3$$

$$R_x = 56.3 + (3.8 \times 3.27) + (16.5 + 1.4 \times 3.44)4.62 + \frac{49 \times 4.62(5 + 4.62)}{5 + 2 \times 4.62}$$

$$= 320.1 \text{ kN}$$

**Passive**

Passive lateral force $= (17.7 - 2.4 \times 3.44)4.62^2 = 201.6$ kN

$T_a =$ Active lateral force $-$ Passive lateral force $= 320.1 - 201.6$

$$= 118.5 \text{ kN}$$

Design anchor force $= 2T_a = 237$ kN    ■

*What's next* . . .Sometimes, it is not possible to use an anchored sheet pile wall for an excavation. For example, an existing building near a proposed excavation can preclude the use of anchors. You may have to brace your sheet piles within the excavation using struts. The analysis of braced excavation (also called cofferdam) is presented in the next section.

## 10.11    BRACED EXCAVATION

Braced excavations consist of sheet piles driven into the soil to form the sides of an excavation (Fig. 10.26a) such as in the construction of bridge piers and abutments. As excavation proceeds within the area enclosed by the sheet piles, struts are added to keep the sheet piles in place.

The top struts are installed followed by others at lower depths. The wall displacements before the top struts are installed are usually very small but get larger as the excavation gets deeper. The largest wall displacement occurs at the base of the excavation (Fig. 10.26a). Wall displacements are inconsistent with all the earth pressure theories.

The critical design elements in a braced excavation are the loads on the struts, which are usually different because of different lateral loads at different depths, the time between excavations, and the installation procedure. Failure of a single strut can be catastrophic because it can lead to the collapse of the whole system. The analysis for the forces and deflection in braced excavation should ideally consider the construction sequence, and numerical methods such as the finite element method are preferred. Semi-empirical methods are often used for shallow braced excavations and in the preliminary design of deep braced excavations. The finite element method is beyond the scope of this book. We will only discuss a semi-empirical method.

Lateral stress distributions for use in the semi-empirical method are approximations from field measurements of strut loads in different types of soil. The lateral stress distributions used for coarse-grained and fine-grained soils are shown in Figs. 10.26b–d. These lateral stress distributions are not real but average

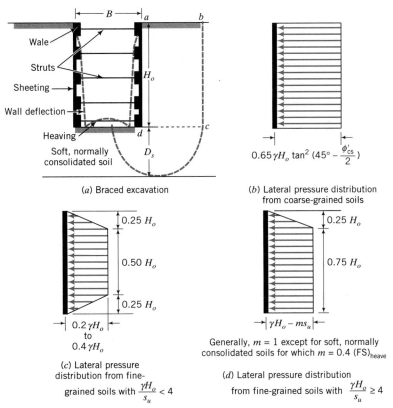

**FIGURE 10.26** Braced excavations.

approximate stress distributions. The lateral stress distribution for coarse-grained soils (Fig. 10.26b) was extrapolated from strut loads measured for dense sand adjacent to the excavation. The appropriate value of friction angle is $\phi_p'$ but because we cannot rely on dilation, the design friction angle should be $\phi_{cs}'$. For fine-grained soils, a total stress analysis is used and the lateral stress distribution depends on the stability number, $\gamma H_o/s_u$ (Peck, 1969). If the stability number is less than 4, the stress state of the soil adjacent to the excavation can be assumed to be elastic and the recommended lateral stress distribution is depicted in Fig. 10.26c. However, if the stability number is greater than or equal to 4, the stress state of the soil adjacent to the bottom of the excavation is expected to be plastic and the recommended lateral stress distribution is depicted in Fig. 10.26d.

If the soil below the base of the excavation is a soft, normally consolidated soil, it is possible that heaving can occur. The column of soil, *abcd* (Fig. 10.26a), above the base acts as a surcharge on the soil below the excavation level. This surcharge load may exceed the bearing capacity of the soft, normally consolidated soil, resulting in heaving. Bjerrum and Eide (1956) suggested that the excavation could be viewed as a footing of width $B$ and embedment depth $H_o$. They showed that the factor of safety against bottom heave is

$$(FS)_{heave} = N_c \frac{s_u}{\gamma H_o + q_s} \tag{10.61}$$

where $N_c$ is a bearing capacity coefficient given by Skempton (1951). The coefficient $N_c$ can be approximated, for practical purposes, by

$$N_c = 6\left(1 + 0.2\frac{H_o}{B}\right) \text{ for } \frac{H_o}{B} \leq 2.5 \quad \text{and} \quad N_c = 9 \text{ for } \frac{H_o}{B} > 2.5 \qquad (10.62)$$

where $H_o/B$ is the depth to width ratio.

If $(FS)_{heave} < 1.5$, the sheeting should be extended below the base of the excavation for stability. The value of the coefficient $m$ (Fig. 10.26d) is usually 1 except when the soil below the excavation is a deep, soft, normally consolidated soil, in which case $m \approx 0.4(FS)_{heave}$.

The strut loads at each level are found by assuming hinged connections of the struts to the sheet piles. A free-body diagram is drawn for each level and the forces imposed on the struts are determined using static equilibrium. Displacements of the walls are an important design consideration as adjacent structures may be affected. The method discussed above does not consider displacements. Analyses using numerical methods (e.g., finite element method) are better suited for the overall analysis of braced excavation.

The procedure for analysis of braced excavation is as follows:

1. Check the stability against bottom heave using Eq. (10.61). If $(FS)_{heave} < 1.5$, the walls should be extended below the base.

2. Determine the lateral stress on the walls for your soil type (Figs. 10.26b–d).

3. Treat the connections of the wall (sheet pile) to the struts as hinges.

4. Draw a free-body diagram at each level of the excavation.

5. Solve for the forces in the struts using the static equilibrium equations on each free-body diagram.

### EXAMPLE 10.8

Determine the forces on the struts for the braced excavation in soft, normally consolidated clay as shown in Fig. E10.8a.

**Strategy**  You need to determine the approximate lateral stress distribution by calculating $\gamma H_o/s_u$. To find the forces on the struts, draw free-body diagrams—one at each level—and use statics.

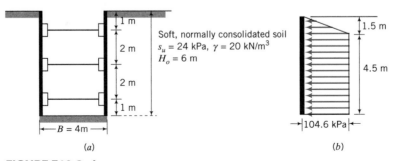

Soft, normally consolidated soil
$s_u = 24$ kPa, $\gamma = 20$ kN/m³
$H_o = 6$ m

(a)    (b)

**FIGURE E10.8a,b**

## Solution 10.8

**Step 1:** Check for stability against bottom heave.

$$\frac{H_o}{B} = \frac{6}{4} = 1.5, \quad N_c = 6\left(1 + 0.2\frac{H_o}{B}\right) = 6(1 + 2 \times 0.15) = 7.6,$$

$$q_s = 0$$

$$(FS)_{heave} = N_c \frac{s_u}{\gamma H_o + q_s} = 7.6 \frac{24}{20 \times 6} = 1.52 > 1.5;$$

therefore, excavation is safe against bottom heave

**Step 2:** Determine the lateral pressure diagram.

$$\frac{\gamma H_o}{s_u} = \frac{20 \times 6}{24} = 5; \quad \text{use Fig. 10.26d}$$

$$m = 0.4(FS)_{heave} = 0.4 \times 1.6 = 0.64$$

The maximum lateral pressure is

$$\gamma H_o - ms_u = 20 \times 6 - 0.64 \times 24 = 104.6 \text{ kPa}$$

**Step 3:** Draw the pressure diagram.
See Fig. E10.8b.

**Step 4:** Calculate the forces on the struts at each level.
All loads are per meter length of wall.

**Level 1 (Fig. E10.8c)**

$$\Sigma M_{B_1} = 0 = 2.0A$$

$$- \left[\frac{1}{2} \times 104.6 \times 1.5 \times \left(\frac{1.5}{3} + 1.5\right) + 1.5 \times 104.6 \times 0.75\right]$$

$$\therefore A = 137.3 \text{ kN/m}$$

$$\Sigma F_x = 0: \quad A + B_1 = \frac{1}{2} \times 104.6 \times 1.5 + 1.5 \times 104.6 = 235.4 \text{ kN/m}$$

$$\therefore B_1 = 235.4 - 137.3 = 98.1 \text{ kN/m}$$

**Level 2 (Fig. E10.8d)**

$$B_2 = C_1 = \frac{104.6 \times 2}{2} = 104.6 \text{ kN/m}$$

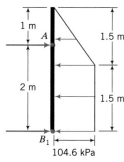

**FIGURE E10.8c**

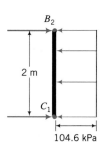

**FIGURE E10.8d**

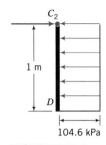

**FIGURE E10.8e**

**Level 3 (Fig. E10.8e)**

$$\Sigma F_x = 0: \quad C_2 = 104.6 \times 1 = 104.6 \text{ kN/m}$$

**Step 5:**    Calculate the forces on each strut.

$$A = 137.3 \text{ kN/m}$$
$$B = B_1 + B_2 = 98.1 + 104.6 = 202.7 \text{ kN/m}$$
$$C = C_1 + C_2 = 104.6 + 104.6 = 209.2 \text{ kN/m}$$ ∎

*What's next . . .*Soils reinforced by metal strips or geotextiles have become popular earth retaining structures because they are generally more economical than gravity retaining walls. These walls are called mechanical stabilized earth (MSE) walls. A brief introduction to MSE walls is presented next.

# 10.12   MECHANICAL STABILIZED EARTH WALLS

Mechanical stabilized earth (MSE) walls (Fig. 10.27a) are used for a variety of retaining structures. Metal strips (Fig. 10.27b), geotextiles (Fig. 10.27c), or geogrids (Fig. 10.27d) reinforce the soil mass. A geotextile is a planar, textile, polymeric product. A geogrid is a polymeric product formed by joining intersecting ribs. MSE walls are generally more economical than gravity walls. The basic mechanics of MSE walls is described in the next section.

## 10.12.1 Basic Concepts

You should recall from Chapter 5 that if a load is applied to a soil mass under axisymmetric undrained conditions, the lateral strain ($\varepsilon_3$) is one-half the axial strain ($\varepsilon_1$) as expressed by Eq. (5.42). If the undrained restriction is lifted, then

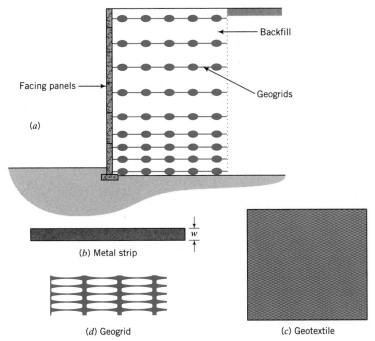

(a)

Backfill

Facing panels

Geogrids

(b) Metal strip

w

(d) Geogrid

(c) Geotextile

**FIGURE 10.27** (a) A geogrid reinforced wall, (b) metal strip, (c) geotextile, and (d) geogrid.

you can expect lateral strains greater than one-half the vertical strains. If we were to install strips of metal in the lateral directions of the soil mass, then the friction at the interfaces of the metal strips and the soil would restrain lateral displacements. The net effect is the imposition of a lateral resistance on the soil mass that causes Mohr's circle to move away from the failure line (Fig. 10.28). The lateral force imposed on the soil depends on the interface friction value between the reinforcing element and the soil mass and the vertical effective stress. For a constant interface friction value, the lateral frictional force would increase with depth. The reinforcing material will fail if the lateral stress exceeds its tensile strength.

The ***essential point*** in MSE walls is that the reinforcement serves as an internal lateral confinement that allows the soil to mobilize more shearing resistance.

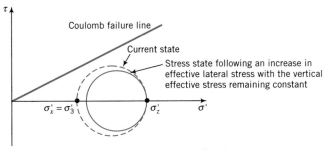

**FIGURE 10.28** Effects of increasing the lateral soil resistance by reinforcement.

## 10.12.2 Stability of Mechanical Stabilized Earth Walls

There are two sets of stability criteria to be satisfied for MSE walls. One is the internal stability; the other is the external stability. The external stability of a MSE wall is determined by analogy to a gravity retaining wall with a vertical face, $FG$, as illustrated in Fig. 10.29. The internal stability depends on the tensile strength of the reinforcing material and the slip at the interface of the reinforcing material and the soil. Tensile failure of the reinforcing material at any depth leads to progressive collapse of the wall, while slip at the interface of the reinforcing material and the soil mass leads to redistribution of stresses and progressive deformation of the wall.

Two methods of analysis are used to determine the internal stability. One method is based on an analogy with anchored flexible retaining walls and is generally used for reinforcing material with high extensibility, for example, polymerics such as geotextiles and geogrids. The Rankine active earth pressure theory is used with the active slip plane inclined at $\theta_a = 45° + \phi'_{cs}/2$ to the horizontal, as shown in Fig. 10.29. The frictional resistance develops over an effective length, $L_e$, outside the active slip or failure zone. At a depth $z$, the frictional resistance developed on both surfaces of the reinforcing material (Fig. 10.29) is

$$\boxed{P_r = 2wL_e(\sigma'_z + q_s)\tan\phi_i} \tag{10.63}$$

where $w$ is the width of the reinforcing material, $\sigma'_z$ is the vertical effective stress, $q_s$ is the surcharge, and $\phi_i$ is the friction angle at the soil–reinforcement interface. Consider a layer of reinforcement at a depth $z$. The tensile force is

$$\boxed{T = K_{aR}(\sigma'_z + q_s)S_zS_y} \tag{10.64}$$

where $S_z$ and $S_y$ are the spacing in the $Z$ and $Y$ directions and $T$ is the tensile force per unit length of wall. For geotextiles or geogrids, you would normally consider one unit length of wall and one unit width, so $S_y = 1$ and $w = 1$.

By setting $T = P_r$, we can find the effective length of the reinforcement required for limit equilibrium (factor of safety = 1). To find the design effective length, a factor of safety $(FS)_t$ is applied on the tensile force, $T$, and by solving for $L_e$ from Eqs. (10.63) and (10.64), we get

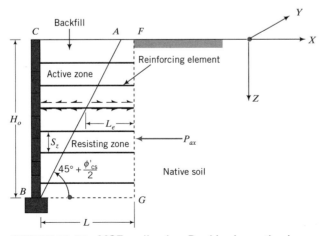

**FIGURE 10.29**  MSE wall using Rankine's method.

$$L_e = \frac{K_{aR}(\sigma_z' + q_s)S_zS_y(\text{FS})_t}{2w(\sigma_z' + q_s)\tan\phi_i} = \frac{K_{aR}S_zS_y(\text{FS})_t}{2w\tan\phi_i} \qquad (10.65)$$

where $(\text{FS})_t$ ranges from 1.3 to 1.5.

The total length of reinforcement is

$$L = L_e + L_R$$

where $L_R$ is the length of reinforcement within the active failure zone.

Because $L_R$ is zero at the base of the wall, the calculated length of reinforcement at the base is often the shortest. This calculated length, while adequate for internal stability, is often inadequate for translation or bearing capacity (external stability). You should check whether the calculated length of reinforcement at the base is adequate for translation or bearing capacity as follows:

1. Calculate the maximum lateral active earth force, $P_{ax}$.

$$P_{ax} = \tfrac{1}{2}K_{aC}\gamma H_o^2\cos\delta + K_{aC}q_sH_o\cos\delta = H_oK_{aC}\cos\delta(0.5\gamma H_o + q_s)$$

where $K_{aC}$ is the active lateral pressure coefficient using Coulomb's method with wall friction. The friction angle to use is $\phi' = (\phi_{cs}')_{\text{backfill}}$ and the wall friction is $\delta \approx \tfrac{2}{3}(\phi_{cs}')_{\text{native soil}}$ to $(\phi_{cs}')_{\text{native soil}}$. Coulomb's method is used because the interface between the reinforced backfill soil and the native soil (*FG*, Fig. 10.29) is frictional. If the interface friction is neglected, then you can use $K_{aR}$ instead of $K_{aC}$.

2. Calculate the length of reinforcement required for translation. For short-term loading in clays, the base resistance is $T = L_b s_w$, where $L_b$ is the length of reinforcement at the base and $s_w$ is the adhesion stress. By analogy with sheet pile retaining walls, we will use $s_w = 0.5s_u$ (maximum $s_w = 50$ kPa). For external stability against translation $T = P_{ax}(\text{FS})_T$, where $(\text{FS})_T$ is a factor of safety against translation; usual range 1.5–3.0. The required length at the base against translation for short-term loading in clays is then

$$L_b = \frac{P_{ax}(\text{FS})_T}{s_w} \qquad (10.66)$$

For long-term loading, $T = \sum_{i=1}^{n} W_i\tan\phi_b'$, where $W_i$ is the weight of soil layer $i$, $n$ is the number of layers, and $\phi_b'$ is the effective interfacial friction angle between the reinforcement and the soil at the base. Assuming a uniform soil unit weight throughout the height of the wall, then

$$T = \gamma H_o L_b \tan\phi_b'$$

The length of reinforcement at the base required to prevent translation under long-term loading is

$$L_b = \frac{P_{ax}(\text{FS})_T}{\gamma H_o\tan\phi_b'} = \frac{(K_{aC})_xH_o(0.5\gamma H_o + q_s)}{\gamma H_o\tan\phi_b'} = \frac{(K_{aC})_x(0.5H_o + q_s/\gamma)(\text{FS})_T}{\tan\phi_b'}$$

$$(10.67)$$

where $(K_{aC})_x$ is the horizontal component of Coulomb's active lateral earth pressure coefficient. You should use the larger value of $L_b$ obtained from Eqs. (10.66) and (10.67).

The procedure for analysis of a reinforced soil wall using polymeric materials is as follows:

1. Calculate the allowable tensile strength per unit width of the reinforcing polymeric material.

$$T_{all} = T_{ult} \frac{1}{(FS)_{ID} \times (FS)_{CR} \times (FS)_{CD} \times (FS)_{BD}} \qquad (10.68)$$

where $T_{all}$ is the allowable tensile strength, $T_{ult}$ is the ultimate tensile strength, FS is a factor of safety, and the subscripts have the following meaning:

ID—installation damage

CR—creep

CD—chemical degradation

BD—biological degradation

Typical values of the various factors of safety are shown in Table 10.2.

2. Calculate the vertical spacing at different wall heights.

$$S_z = \frac{T_{all}}{K_{aR}(\sigma'_z + q_s)(FS)_{sp}} \qquad (10.69)$$

where $(FS)_{sp}$ is a factor of safety between 1.3 and 1.5. It is customary to calculate the minimum vertical spacing using $\sigma'_z = \gamma' H_o$ and then check the vertical spacing required at $H_o/3$ and $2H_o/3$. For practical reasons, the vertical spacing either is kept at a constant value or varies by not more than three different values along the height.

3. Determine the length of reinforcement required at the base for external stability from Eqs. (10.66) and (10.67).

4. Determine the total length of reinforcement at different levels.

$$L = L_e + L_R$$

**TABLE 10.2  Typical Ranges of Factor of Safety**

| Factor of safety | Range |
| --- | --- |
| $(FS)_{ID}$ | 1.1 to 2.0 |
| $(FS)_{CR}$ | 2.0 to 4.0 |
| $(FS)_{CD}$ | 1.0 to 1.5 |
| $(FS)_{BD}$ | 1.0 to 1.3 |

where

$$L_R = (H_o - z) \tan\left(45° - \frac{\phi'_{cs}}{2}\right)$$ (10.70)

**5.** Determine the external stability (translation and bearing capacity). Remember that translation is already satisfied from item 3 above. Overturning is not crucial in MSE walls because these walls are flexible and cannot develop moment. However, you can verify that overturning is satisfied if $e < L_b/6$, where $e$ is the eccentricity of the vertical forces.

The other method is the coherent gravity method (Juran and Schlosser, 1978) and is applicable to low extensible reinforcing materials such as metal strips. Failure is assumed to occur progressively along a path defined by the maximum tensile strains at each level of the reinforcing material. The failure surface is a logarithm spiral that is approximated to a bilinear surface to simplify calculations (Fig. 10.30a). The lateral active pressure coefficient is assumed to vary linearly from $K_o$ at the top of the wall to $K_{aR}$ at a depth of 6 m and below (Fig. 10.30b).

The variation of lateral stress coefficient with depth (Fig. 10.30b) is

$$K = K_{aR}\frac{z}{6} + K_o\left(1 - \frac{z}{6}\right) \quad \text{for } z \leq 6 \text{ m}$$ (10.71)

$$K = K_{aR} \quad \text{for } z > 6 \text{ m}$$ (10.72)

The length of reinforcement within the failure zone (Fig. 10.30a) is

$$L_R = 0.2H_o + \left(0.1H_o - \frac{z}{6}\right) \quad \text{for } z \leq 0.6H_o$$ (10.73)

$$L_R = \frac{1}{2}H_o\left(1 - \frac{z}{H_o}\right) \quad \text{for } z > 0.6H_o$$ (10.74)

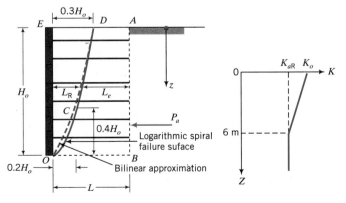

(a) Failure surface for the coherent gravity method

(b) Variation of active lateral stress coefficient with depth

**FIGURE 10.30** Coherent gravity method.

The procedure for analyzing a MSE wall using low extensible materials is as follows:

1. Select the spacing of the reinforcement in the $Z$ and $Y$ directions and the width of the reinforcement. Use a manufacturer's catalog to provide information on standard sizes.

2. Calculate the required maximum thickness of the reinforcement.

$$t_r = \frac{K_{aR}(\gamma' H_o + q_s)S_z S_y (FS)_{tr}}{w f_y} \tag{10.75}$$

where $K_{aR}$ is the Rankine active earth pressure coefficient, $f_y$ is the yield stress of the reinforcement, $w$ is the width of the reinforcement, and $(FS)_{tr}$ is a factor of safety, usually 3. The design thickness is the calculated thickness plus a corrosion thickness expected for the design life of the wall.

3. Determine the length of reinforcement required at the base from Eqs. (10.66) and (10.67).

4. Determine the total length of reinforcement at each level.

$$L = L_e + L_R$$

where $L_R$ is given by Eqs. (10.73) and (10.74) and $L_e$ is determined from Eq. (10.65) by using the appropriate value of the active lateral earth pressure coefficient from Eqs. (10.71) and (10.72).

5. Determine the external stability by assuming the wall is a gravity wall.

## EXAMPLE 10.9

Design a 4 m mechanical stabilized earth wall using a geotextile as the reinforcement. The backfill would be a compacted, coarse-grained soil with $\phi'_{cs} = 30°$ and $\gamma_{sat} = 18$ kN/m³. The surcharge is 15 kPa. The geotextile selected has an ultimate wide-width tensile strength of 58.5 kN/m and the soil–geotextile interface friction value is 20°. The native soil is a clay with parameters $\gamma_{sat} = 18.5$ kN/m³, $\phi'_{cs} = 28°$, $\sigma'_b = \frac{2}{3}\sigma'_{cs}$, and $s_u = 60$ kPa.

**Strategy** Follow the procedure for mechanical stabilized earth walls using a polymeric product.

## Solution 10.9

**Step 1:** Calculate the allowable tensile strength of the geotextile.
From Table 10.2, use $(FS)_{ID} = 1.5$, $(FS)_{CR} = 2$, $(FS)_{CD} = 1.3$, and $(FS)_{BD} = 1.3$.

$$T_{all} = \frac{58.5}{1.5 \times 2 \times 1.3 \times 1.3} = 11.5 \text{ N/mm} = 11.5 \text{ kN/m}$$

**Step 2:** Calculate the vertical spacing.

$$K_{aR} = \tan^2(45 - \phi'_{cs}/2) = \tan^2(45 - 30/2) = \tfrac{1}{3}$$

Lateral stress due to surcharge: $K_a q_s = \tfrac{1}{3} \times 15 = 5$ kPa

$$\sigma_x = K_{aR}\sigma'_z + K_{aR}q_s = \tfrac{1}{3}\gamma' z + 5 = \tfrac{1}{3} \times 18 \times z + 5 = 6z + 5$$

$$(\sigma_x)_{max} = 6 \times H_o + 5 = 6 \times 4 + 5 = 29 \text{ kPa}$$

From Eq. (10.69) with $(FS)_{sp} = 1.3$, we get

$$(S_z)_{min} = \frac{T_{all}}{K_{aR}(\sigma'_z + q_s)(FS)_{sp}} = \frac{11.5}{29 \times 1.3} = 0.305 \text{ m} = 305 \text{ mm}$$

Check spacing requirement at midheight ($z = 2$):

$$\sigma_x = 6 \times 2 + 5 = 17 \text{ kPa}$$

$$S_z = \frac{11.5}{17 \times 1.3} = 0.520 \text{ m} = 520 \text{ mm}$$

You should try to minimize the number of layers and keep the spacing to easily measurable values. Use $S_z = 250$ mm for the bottom half of the wall and 500 mm for the top half of the wall.

**Step 3:** Determine the length of reinforcement required at the base for translation. From Eq. (10.16),

$$K_{aC} = 0.3 \quad (\phi'_{cs} = 30°, \delta = 20°) \quad \text{and} \quad (K_{aC})_x = K_{aC} \cos \delta = 0.3 \times \cos 20 = 0.28$$

$$P_{ax} = \tfrac{1}{2}(K_{aC})_x \gamma H_o^2 + (K_{aC})_x q_s H_o = \tfrac{1}{2} \times 0.28 \times 18 \times 4^2 + 0.28 \times 15 \times 4$$
$$= 57.1 \text{ kN/m}$$

$$s_w = 0.5 s_u = 0.5 \times 60 = 30 \text{ kPa} < 50 \text{ kPa}; \quad \text{therefore use } s_w = 30 \text{ kPa}$$

Equation (10.66): $\quad L_b = \dfrac{P_{ax}(FS)_T}{s_w} = \dfrac{57.1 \times 1.5}{30} = 2.9 \text{ m}$

Equation (10.67): $\quad L_b = \dfrac{(K_{aC})_x(0.5 H_o + q_s/\gamma')(FS)_T}{\tan \phi'_b}$

$$= \frac{0.28(0.5 \times 4 + 15/18)1.5}{\tan(\tfrac{2}{3} \times 28)} = 3.5 \text{ m}$$

Use $L_b = 3.5$ m.

**Step 4:** Determine the total length of reinforcement at each level for internal stability.
Use a table to determine the total length as shown below.
Use $(FS)_t = 1.3$.

| z (m) | S_z (m) | $L_R$ (m) = $(H_o - z)$ $\tan(45° - \phi'_{cs}/2)$ | $L_e$ (m) = $\dfrac{K_{aR} S_z (FS)_t}{2 \tan \phi_i}$ | L (m) |
|---|---|---|---|---|
| 0.50 | 0.50 | 2.02 | 0.3 | 2.3 |
| 1.00 | 0.50 | 1.73 | 0.3 | 2.0 |
| 1.50 | 0.50 | 1.44 | 0.3 | 1.7 |
| 2.00 | 0.50 | 1.15 | 0.3 | 1.5 |
| 2.25 | 0.25 | 1.01 | 0.15 | 1.2 |
| 2.50 | 0.25 | 0.87 | 0.15 | 1.0 |
| 2.75 | 0.25 | 0.72 | 0.15 | 0.9 |
| 3.00 | 0.25 | 0.58 | 0.15 | 0.7 |
| 3.25 | 0.25 | 0.43 | 0.15 | 0.6 |
| 3.50 | 0.25 | 0.29 | 0.15 | 0.4 |
| 3.75 | 0.25 | 0.14 | 0.15 | 0.3 |
| 4.00 | 0.25 | 0.00 | 0.15 | 0.2 |

Since $L_b$ is greater than the total length required for internal stability at each level, use $L = L_b = 3.5$ m. The total length can be reduced toward the top of the wall but for construction purposes, it is best to use, in most cases, a single length of polymeric product.

**Step 5:** Check external stability.

Stability against translation is already satisfied in Step 3.

Check the bearing capacity.

$$(\sigma_z)_{max} = \gamma H_o = 18 \times 4 = 72 \text{ kPa}$$

**TSA**

Use Skempton's method [Eq. (7.14)].

$$q_{ult} = 5s_u = 5 \times 60 = 300 \text{ kPa}$$

$$(FS)_B = \frac{q_{ult}}{(\sigma_x)_{max}} = \frac{300}{72} = 4.2 > 3; \quad \text{okay}$$

**ESA**

Use Meyerhof's method [Eq. (7.16) with $D_f = 0$].

Equation (7.11):   $\phi' = 28°$:   $N_q = e^{\pi \tan 28} \tan^2(45 + 28/2) = 14.7$

Equation (7.19):   $N_\gamma = (14.7 - 1) \tan(1.4 \times 28) = 11.2$

$q_{ult} = \frac{1}{2}\gamma B N_\gamma = \frac{1}{2} \times 18.5 \times 7 \times 11.2 = 725.2 \text{ kPa}$

$(FS)_B = \dfrac{725.2}{72} = 10 > 3;$   therefore, bearing capacity is okay ■

## Example 10.10

Design a mechanical stabilized wall 6 m high using metal ties. The backfill is a coarse-grained soil with $\gamma_{sat} = 16.5$ kN/m³ and $\phi'_{cs} = 32°$. Galvanized steel ties are available with a yield strength $f_y = 2.5 \times 10^5$ kPa and a rate of corrosion of 0.025 mm/year. A factor of safety of 3 is desired for a design life of 50 years. The soil–tie interface friction value is 21°. The foundation soil is coarse-grained with parameters $\gamma_{sat} = 18$ kN/m³, $\phi'_{cs} = 28°$, and $\phi'_b = 20°$. The surcharge is 15 kPa.

**Strategy**   You have to guess the spacing of the ties. You can obtain standard widths and properties of ties from manufacturers' catalogs. Follow the procedure for the coherent gravity method.

## Solution 10.10

**Step 1:** Assume spacing and width of ties.

Assume $S_z = 0.5$ m, $S_y = 1$ m, and $w = 75$ mm

**Step 2:** Calculate required thickness of reinforcement.

$K_o = 1 - \sin \phi'_{cs} = 1 - \sin 32° = 0.47$

$K_{aR} = \tan^2(45 - \phi'_{cs}/2) = \tan^2(45 - 32/2) = 0.31$

$$t_r = \frac{K_{aR}(\gamma'H_o + q_s)S_zS_y(FS)_{tr}}{wf_y} = \frac{0.31(16.5 \times 6 + 15) \times 0.5 \times 1 \times 3}{0.075 \times 2.5 \times 10^5}$$

$$= 283 \times 10^{-5} \text{ m} = 2.8 \text{ mm}$$

$t_{corrosion}$ = Annual corrosion rate × design life = $0.025 \times 50 = 1.25$ mm

$t_{design}$ = Calculated thickness + Corrosion thickness = $2.8 + 1.25 = 4.05$ mm

You will select a tie thickness from standard sizes closest to 4.05 mm. Use $t = 5$ mm.

**Step 3:** Determine the length of reinforcement required at the base.

From Eq. (10.16): $K_{aC} = 0.28$     ($\phi'_{cs} = 32°$, $\delta = 28°$)

$(K_{aC})_x = K_{aC} \cos \delta = 0.28 \cos 20 = 0.26$

From Eq. (10.67): $L_b = L_b = \dfrac{0.26(0.5 \times 6 + 15/16.5)1.5}{\tan 20°} = 4.2$ m

For internal stability, the effective length at the wall base is

Equation (10.65): $L_e = \dfrac{0.31 \times 0.5 \times 1 \times 1.3}{2 \times 0.075 \times \tan 21°} = 3.5$ m $< 4.2$ m

**Step 4:** Determine the total length of reinforcement.
Use a table as shown below. $(FS)_t = 1.3$.

| $z$ (m) | $S_z$ (m) | K Eq. (10.71) | $L_r$ (m) Eq. (10.73) | $L_e$ (m) Eq. (10.65) | $L$ (m) | Recommended $L$ (m) |
|---|---|---|---|---|---|---|
| 0.50 | 0.50 | 0.49 | 5.49 | 1.72 | 7.2 | 7.5 |
| 1.00 | 0.50 | 0.47 | 5.33 | 1.63 | 7.0 | 7.5 |
| 1.50 | 0.50 | 0.46 | 5.17 | 1.55 | 6.7 | 7.5 |
| 2.00 | 0.50 | 0.44 | 5.02 | 1.47 | 6.5 | 6.5 |
| 2.50 | 0.50 | 0.43 | 4.86 | 1.38 | 6.2 | 6.5 |
| 3.00 | 0.50 | 0.42 | 4.70 | 1.30 | 6.0 | 6.5 |
| 3.50 | 0.50 | 0.40 | 4.55 | 1.22 | 5.8 | 6.5 |
| 4.00 | 0.50 | 0.39 | 4.39 | 1.00 | 5.4 | 6.5 |
| 4.50 | 0.50 | 0.38 | 4.23 | 0.75 | 5.0 | 5.0 |
| 5.00 | 0.50 | 0.36 | 4.08 | 0.50 | 4.6 | 5.0 |
| 5.50 | 0.50 | 0.35 | 3.92 | 0.25 | 4.2 | 5.0 |
| 6.00 | 0.50 | 0.33 | 3.76 | 0.00 | 3.8 | 5.0 |

**Step 5:** Check for external stability.
Translation is satisfied because $L$ used at the base is greater than $L_b$. Check bearing capacity.

$$(\sigma_z)_{max} = \gamma H_o = 16.5 \times 6 = 99 \text{ kPa}$$

**Using Terzaghi's method [Eq. (7.10) with $D_f = 0$]**

For $\phi'_{cs} = 28°$,   $N_\gamma = 16.7$

ESA: $q_{ult} = \frac{1}{2}B\gamma N_\gamma = \frac{1}{2} \times 8.4 \times 18 \times 16.7 = 1262.5$ kPa

$(FS)_B = \dfrac{q_{ult}}{(\sigma_z)_{max}} = \dfrac{1262.5}{99} = 12.8 > 3$;

therefore, bearing capacity is satisfactory ∎

## 10.13    SUMMARY

We have considered lateral earth pressures and their applications to several types of earth retaining structures in this chapter. Two earth pressure theories are in general use: one developed by Coulomb and the other by Rankine. Coulomb's equations for the lateral earth pressure coefficients are based on limit equilibrium and include the effects of soil–wall friction, wall slope, and backfill slope. Rankine's equations are based on stress states of the backfill and do not account for soil–wall friction. The failure planes in the Coulomb and Rankine methods are planar surfaces. Soil–wall friction causes the failure plane to curve, resulting in higher active lateral earth pressure coefficients and lower passive earth pressure coefficients than those found using Coulomb's equations.

Because of the flexibility of some earth retaining structures and construction methods used in practice, the "real" lateral earth pressures are different from either the Coulomb or Rankine theories. It is suggested in this book that the appropriate value of friction angle to use in the analysis of earth retaining structures is $\phi'_{cs}$. Three methods of analysis for flexible earth retaining walls were considered. The differences in the methods result mainly from how the lateral stresses are considered and how the factor of safety is applied.

*PRACTICAL EXAMPLES*

### EXAMPLE 10.11

Ore from a manufacturing plant is to be stored between two cantilever gravity retaining walls, as shown in Fig. E10.11a. A gantry crane will run on top of the walls to place the ore. The crane will apply a maximum vertical load of 24 kN and a horizontal load of ±4.5 kN on each wall. The base of the ore pile is only permitted to come within 0.5 m of the top of the walls. The ore surface should be the maximum admissible slope. The walls will be restrained from spreading outward by steel ties or rods at 1 m centers anchored to the base slab. A 2 m layer of the ore spoils of similar characteristics to the ore to be stored would be compacted to support the base of the wall. Determine (a) the stability of the wall

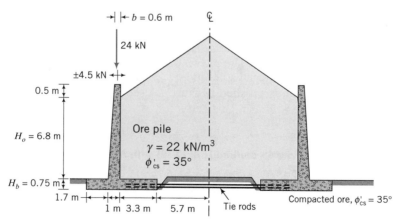

**FIGURE E10.11a**

for the geometry shown in Fig. E10.11a and (b) the force in the tie rods, assuming they resist all the horizontal loads. If the tie rods were not present, would the walls be safe? Do you expect any alignment problems with the gantry crane?

***Strategy*** The maximum slope angle would be the friction angle. Since the storage is symmetrical, each wall will carry identical loads. You will have to make an assumption regarding wall friction. You can assume $\delta = \frac{1}{2}\phi'_{cs}$ or $\delta = \frac{2}{3}\phi'_{cs}$ and use Coulomb's method to determine the lateral forces.

## Solution 10.11
**Step 1:** Determine the lateral forces.

The maximum admissible slope is $\beta = \phi'_{cs}$. Therefore, $\beta = 35°$.

$$H_o = 0.75 + 6.8 + 3.3 \tan 35° = 9.86 \text{ m}$$

From Eq. (10.16):  $K_a = 0.7$  for $\beta = 35°$, $\phi' = 35°$, $\delta = \frac{1}{2}\phi'$

$$P_a = \tfrac{1}{2}K_a\gamma H_o^2 = \tfrac{1}{2} \times 0.7 \times 22 \times 9.86^2 = 748.7 \text{ kN/m}$$
$$P_{ax} = P_a \cos \delta = 748.7 \times \cos 17.5° = 714.0 \text{ kN/m}$$
$$P_{az} = P_a \sin \delta = 748.7 \times \sin 17.5° = 225.1 \text{ kN/m}$$

**Step 2:** Calculate forces and moments. Draw a diagram (show only one-half because of symmetry, see Fig. E10.11b) and use a table to do the calculations.

| Part | Force (kN/m) | Moment arm from 0 (m) | Moment (kN · m) |
|---|---|---|---|
| *Wall* | *Vertical forces* | | |
| 1 | $(6.8 + 0.5) \times 0.6 \times 24 = 105.1$ | $1.7 + 0.4 + 0.6/2 = 2.4$ | +252.2 |
| 2 | $\frac{1}{2}(6.8 + 0.5) \times 0.4 \times 24 = 35$ | $1.7 + \frac{2}{3} \times 0.4 = 1.97$ | +69.0 |
| 3 | $0.75 \times 6 \times 24 = 108.0$ | $6/2 = 3.00$ | +324.0 |
| *Soil* | | | |
| 4 | $\frac{1}{2} \times 3.3 \times 2.31 \times 22 = 83.9$ | $1.7 + 1 + (\frac{2}{3} \times 3.3) = 4.9$ | +411.1 |
| 5 | $3.3 \times 6.8 \times 22 = 493.7$ | $1.7 + 1 + 3.3/2 = 4.35$ | +2147.6 |
| $P_{az}$ | 225.1 | 6 | +1350.6 |
| *Gantry* | | | |
| $P_z$ | 24 | 2.4 | +57.6 |
| | $R_z = \Sigma$ Vertical forces = 1074.8 | | |
| *Gantry* | *Lateral forces* | | |
| $P_x$ | 4.5 | $0.75 + 6.8 + 0.5 = 8.05$ | −36.2 |
| *Soil* | | | |
| $P_{ax}$ | 714 | $9.87/3 = 3.29$ | −2349.1 |
| | $R_x = \Sigma$ Lateral forces = 718.5 | $M = \Sigma$ Moments = 2226.8 | |

**Step 3:** Determine stability.

**Rotation**

$$\bar{x} = \frac{M}{R_z} = \frac{2226.8}{1074.8} = 2.07 \text{ m}$$

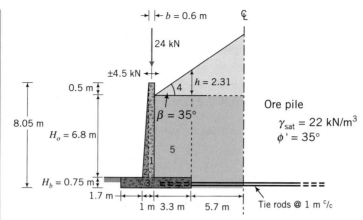

**FIGURE E10.11b**

$$e = \frac{B}{2} - \bar{x} = \frac{6}{2} - 2.07 = 0.93 \text{ m}$$

$$\frac{B}{6} = \frac{6}{6} = 1 \text{ m}; \quad e < \frac{B}{6}; \quad \text{therefore, wall is safe against rotation}$$

### Sliding

$$T = \text{sliding resistance at base} = R_z \tan \phi_b' = 1074.8 \tan 20° = 391.2 \text{ kN}$$

$$(FS)_T = \frac{T}{R_z} = \frac{391.2}{718.5} = 0.54 < 1.5$$

∴ Sliding would occur if tie rods were not present.

### Forces on tie rods

$$\text{For } (FS)_T = 1.5: \quad T_a + T = 1.5 R_x$$

where $T_a$ is the force on the tie rods/unit length of wall.

$$T_a = 1.5 \times 718.5 - 391.2 = 686.6 \text{ kN/m}$$

### Bearing capacity

Because the resultant force is eccentric, use Meyerhof's method.

$$B' = B - 2e = 6 - 2 \times 0.93 = 4.14 \text{ m}$$

Equation (7.11):  $N_q = e^{\pi \tan 35°} \tan^2(45° + 35°/2) = 33.3$

Equation (7.19):  $N_\gamma = (33.3 - 1) \tan (1.4 \times 35°) = 37.2$

$q_{ult} = 0.5 \gamma B' N_\gamma$

$$q_{ult} = 0.5 \times 18 \times 4.14 \times 37.2 = 1386.1 \text{ kPa}$$

$$\sigma_{max} = \frac{\Sigma R_z}{B \times 1} \left( 1 + \frac{6e}{B} \right) = \frac{1074.8}{6 \times 1} \left( 1 + \frac{6 \times 0.93}{6} \right) = 345.7 \text{ kPa}$$

$$(FS)_B = \frac{1386.1}{345.7} = 4 > 3; \quad \text{therefore, bearing capacity is satisfactory}$$

**Summary of Results**
1. The wall is unlikely to rotate.
2. Without the tie rods, the wall will translate.
3. The design tie rod force is ≈687 kN/m.
4. The soil bearing capacity is adequate.
5. Assuming that the base slab is rigid and the loading is symmetrical, there should be no alignment problem. ∎

### EXAMPLE 10.12

A retaining wall is required near a man-made lake in a housing scheme development. The site is a swamp and the topsoil consisting mainly of organic material will be removed up to the elevation of a deep deposit of clay with a silt and sand mixture. The wall is expected to retain a sand backfill of height 6 m. It is anticipated that a rapid drawdown condition could occur and the lake emptied for a long period. A flexible retaining wall is proposed as shown in Fig. E10.12a. Determine the embedment depth and the force for an anchor spacing of 3 m. A surcharge of 10 kPa should be considered.

***Strategy*** You can use either the FSM or NPPM or both. Because the clay layer has silt and sand, you should consider seepage forces for long-term conditions (assuming that the drawdown level would remain for some time).

### Solution 10.12
**Step 1:** Determine the lateral earth pressure coefficients.
Use Appendix C to find the lateral earth pressure coefficients.

$$\text{Sand:} \quad K_{ax} = 0.28 \quad (\phi'_{cs} = 30°, \delta = \tfrac{2}{3}\phi'_{cs})$$
$$\text{Clay:} \quad K_{ax} = 0.34 \quad (\phi'_{cs} = 27°, \delta = \tfrac{2}{3}\phi'_{cs});$$
$$K_{px} = 3.8 \quad (\phi'_{cs} = 27°, \delta = \tfrac{1}{2}\phi'_{cs})$$

**Step 2:** Determine the seepage forces and pore water pressures.

$$\text{Sand:} \quad \gamma' = 18.8 - 9.8 = 9 \text{ kN/m}^3$$
$$\text{Clay:} \quad \gamma' = 19.4 - 9.8 = 9.6 \text{ kN/m}^3$$

$$j_s = \frac{a}{a + 2d}\gamma_w = \left(\frac{3.5}{3.5 + 2d}\right)9.8 = \frac{34.3}{3.5 + 2d} \text{ kN/m}^3$$

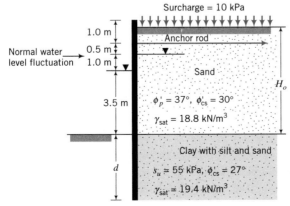

**FIGURE E10.12a**

**Active state**

$$\text{Sand:} \quad \gamma'_{ja} = \left(9 + \frac{34.3}{3.5 + 2d}\right) = \frac{65.8 + 18d}{3.5 + 2d}$$

$$\text{Clay:} \quad \gamma'_{ja} = \left(9.6 + \frac{34.3}{3.5 + 2d}\right) = \frac{67.9 + 19.2d}{3.5 + 2d}$$

**Passive state**

$$\text{Clay:} \quad \gamma'_{jp} = \left(9.6 - \frac{34.3}{3.5 + 2d}\right) = \frac{19.2d - 0.7}{3.5 + 2d}$$

$$u_B = \left(\frac{2ad}{a + 2d}\right)\gamma_w = \left(\frac{2 \times 3.5d}{3.5 + 2d}\right)9.8 = \frac{68.6d}{3.5 + 2d}$$

$$P_w = \frac{1}{2}(a + d)u_B = \frac{1}{2}(3.5 + d)\frac{68.6d}{3.5 + 2d} = \left(\frac{120d + 34.3d^2}{3.5 + 2d}\right)$$

$$\bar{z}_w = \frac{a + 2d}{3} = \frac{3.5 + 2d}{3}$$

$$H_o - h_w = 6 - 2.5 = 3.5 \text{ m}; \quad H_o - h = 6 - 1 = 5 \text{ m};$$

$$H_w - h = 2.5 - 1.0 = 1.5 \text{ m}$$

**Step 3:** Carry out the calculations.

Draw a diagram of the pressure distributions. Use a table (see below) or a spreadsheet program to carry out the calculations. See Fig. E10.12b for the lateral pressure distributions.

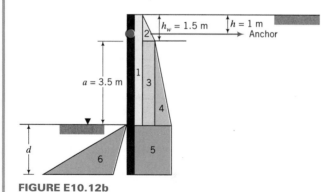

**FIGURE E10.12b**

**Step 4:** Calculate the anchor force.

$$T_a = 317.1 - 208.7 = 108.4 \text{ kN/m}$$

Assume a factor of safety of 2.

Design anchor force = $T_a(\text{FS}) \times$ Anchor spacing = $108.4 \times 2 \times 3 = 650.4$ kN

| Part | Pressure (kPa) | Force (kN/m) | Moment arm from 0 | Moment = Force × Moment arm | D = 5.57 m Moment (kN·m/m) | D = 4.39 m Moment (kN·m/m) | D = 4.39 m Force (kN/m) |
|---|---|---|---|---|---|---|---|
| 1 | $p_{a1} = K_{ax}q_s$ $= 0.3 \times 10 = 3$ | $p_{a1}H_o = 3 \times 6 = 18$ | $\frac{H_o}{2} - h = \frac{6}{2} - 1 = 2$ | $(-)33.6$ | $-33.6$ | $-33.6$ | $16.8$ |
| 2 | $p_{a2} = K_{ax}\gamma h_w$ $= 0.3 \times 18.8 \times 2.5 = 14.1$ | $\frac{1}{2}p_{a2}h_2 = \frac{1}{2} \times 14.1 \times 2.5 = 17.6$ | $\frac{2}{3}h_w - 1 = \frac{2}{3} \times 2.5 - 1 = 0.67$ | $(-)11.0$ | $-11.0$ | $-11.0$ | $16.5$ |
| 3 | $p_{a3} = p_{a2} = 14.1$ | $p_{a3}a = 14.1 \times 3.5 = 49.4$ | $\frac{a}{2} + (h_w - h) = \frac{3.5}{2} + 1.5 = 3.25$ | $(-)149.7$ | $-149.7$ | $-149.7$ | $46.1$ |
| 4 | $p_{a4} = K_{ax}\gamma_{ja}a$ $= 0.3\left(\dfrac{65.8 + 18d}{3.5 + 2d}\right)3.5$ $= \dfrac{69.1 + 18.9d}{3.5 + 2d}$ | $\frac{1}{2}p_{a4}a = \frac{1}{2}\left(\dfrac{69.1 + 18.9d}{3.5 + 2d}\right) \times 3.5$ $= \dfrac{120.9 + 33.1d}{3.5 + 2d}$ | $\frac{2}{3}a + (h_w - h) = \frac{2}{3} \times 3.5 + 1.5 = 3.83$ | $(-)\dfrac{463 + 126.8d}{3.5 + 2d}$ | $-74.6$ | $-77.5$ | $20.2$ |
| 5 | $p_{a5} = p_{a1} + p_{a2} + p_{a3} + p_{a4}$ $= 3 + 14.1 + 14.1 + \dfrac{69.1 + 18.9d}{3.5 + 2d}$ $= \dfrac{178.3 + 81.3d}{3.5 + 2d}$ | $p_{a5}d = \left(\dfrac{129d + 53.1d^2}{3.5 + 2d}\right)d$ | $\frac{d}{2} + (H_o - h) = \frac{d}{2} + 5 = \frac{d + 10}{2}$ | $(-)\left(\dfrac{26.5d^2 + 330d + 645}{3.5 + 2d}\right)d$ | $-1173.1$ | $-869.2$ | $120.8$ |
| Water | $u_b = \dfrac{68.6d}{3.5 + 2d}$ | $P_w = \dfrac{(120 + 34.3d)d}{3.5 + 2d}$ | $(H_o - h) + d - y_w = 5 + d - \frac{3.5 + 2d}{3}$ $= \frac{11.5 + d}{3}$ | $(-)\left(\dfrac{11.4d^2 + 171.4d + 460}{3.5 + 2d}\right)d$ | $-673.0$ | $-512.4$ | $96.7$ |
| 6 | $p_{a6} = K_{px}\gamma_{jp}d - K_{ax}\gamma_{ja}d$ $= \left[3.8\left(\dfrac{19.2d - 0.7}{3.5 + 2d}\right) - 0.34\left(\dfrac{67.9 + 19d}{3.5 + 2d}\right)\right]d$ $= \left(\dfrac{66.4d - 25.7}{3.5 + 2d}\right)d$ | $\frac{1}{2}p_{a6}d = \frac{1}{2}\left(\dfrac{66.4d - 25.7}{3.5 + 2d}\right)d$ $= \left(\dfrac{33.2d - 12.9}{3.5 + 2d}\right)d^2$ | $\frac{2}{3}d + (H_o - h) = \frac{2}{3}d + 5 = \frac{2d + 15}{3}$ | $(+)\left(\dfrac{22.1d^2 + 157.5d - 64.5)d^2}{3.5 + 2d}\right)$ | $+3172.9$ | $+1653.9$ | $208.7$ |
| | | | | | $\Sigma = -2114.9$ $+3172.9$ | $\Sigma = -1653.4$ $+1653.9$ | $\Sigma = 317.1$ |

# EXERCISES

## Theory

**10.1**   Show, using Mohr's circle, that the depth of a tension crack is $z_{cr} = 2s_u/\gamma_{sat}$ for a saturated clay.

**10.2**   Show that a tension crack will not appear in a saturated clay if a surface stress of $q_s \geq 2s_u$ is present.

## Problem Solving

**10.3**   Plot the variation of active and passive lateral pressures with depth for the soil profile shown in Fig. P10.3.

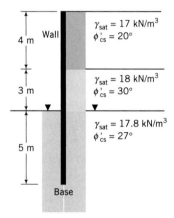

**FIGURE P10.3**

**10.4**   A retaining wall 8 m high supports a soil of saturated unit weight 18 kN/m³, $\phi'_{cs} = 30°$. The backfill is subjected to a surcharge of 15 kPa. Calculate the active force on the wall if (a) the wall is smooth and (b) the wall is rough ($\delta = 20°$). Groundwater is below the base of the wall.

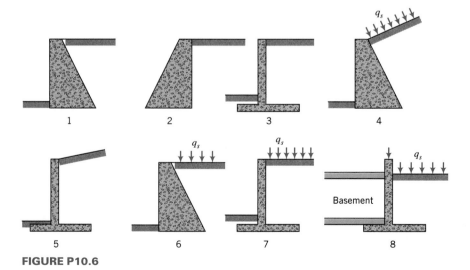

**FIGURE P10.6**

10.5 A retaining wall 5 m high was designed to stabilize a slope of 15°. The back of the wall is inclined 10° to the vertical and may be assumed to be rough with $\delta = 20°$. The soil parameters are $\phi'_{cs} = 30°$ and $\gamma_{sat} = 17.5$ kN/m³. After a flood, the groundwater level, which is usually below the base of the wall, rose to the surface. Calculate the lateral force on the wall. Neglect seepage effects.

10.6 Figure P10.6 shows rigid walls of height $H_o$ with different geometries. Sketch the distribution of lateral earth pressures on each wall; indicate the location and direction of the resultant lateral force. Show on your diagram what other forces act on the wall, for example, the weight of the wall ($W_w$) and the weight of the soil ($W_s$). You should consider two cases of soil–wall friction: (a) $\delta > 0$ and (b) $\delta = 0$.

10.7 Which of the two walls in Fig. P10.7 gives the larger horizontal force? (Show your calculations.)

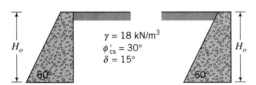

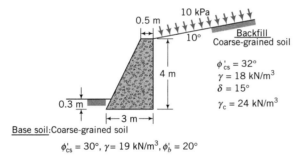

**FIGURE P10.7**

10.8 Determine the stability of the concrete gravity wall shown in Fig. P10.8.

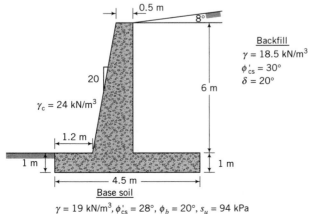

**FIGURE P10.8**

10.9 Determine the stability of the cantilever wall shown in Fig. P10.9 and sketch a drainage system to prevent buildup of pore water pressures behind the wall.

**FIGURE P10.9**

**10.10** The drainage system of a cantilever wall shown in Fig. P10.10 became blocked after a heavy rainstorm and the groundwater level, which was originally below the base, rose to 1.5 m below the surface. Determine the stability of the wall before and after the rainfall. Neglect seepage effects.

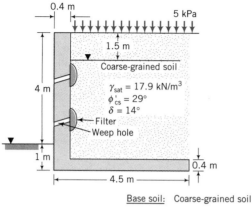

Base soil: Coarse-grained soil

$\gamma_{sat} = 19 \text{ kN/m}^3$, $\phi_{cs}' = 32°$, $\phi_b = 20°$

**FIGURE P10.10**

**10.11** Determine the embedment depth, $d$, and maximum bending moment for the cantilever sheet pile wall shown in Fig. P10.11. Use the factored strength method (FSM) with $F_\phi = 1.25$.

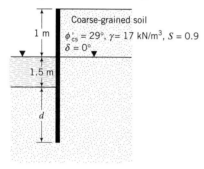

**FIGURE P10.11**

**10.12** Determine the embedment depth, $d$, and maximum bending moment for the cantilever sheet pile wall shown in Fig. P10.12 for long-term conditions. Use the FSM with $F_\phi = 1.25$, and the NPPM with $(FS)_r = 1.5$. Compare the results.

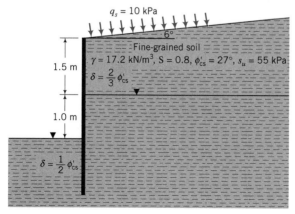

**FIGURE P10.12**

**10.13**   An anchored sheet pile wall is shown in Fig. P10.13. Determine the embedment depth, the maximum bending moment, and the force on the anchor per unit length of wall. Use either FSM ($F_\phi = 1.25$) or the NPPM with $(FS)_r = 1.5$. Assume the soil above the groundwater to be saturated.

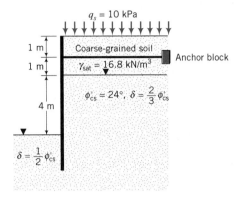

**FIGURE P10.13**

**10.14**   Determine the depth of embedment and the anchor force per unit length of wall for the retaining wall shown in Fig. P10.14 using the FMM. Assume the soil above the groundwater to be saturated.

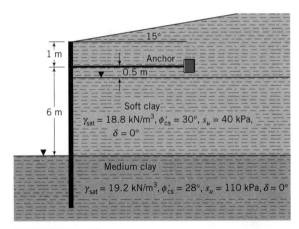

**FIGURE P10.14**

**10.15**   Calculate the strut loads per meter length for the braced excavation shown in Fig. P10.15.

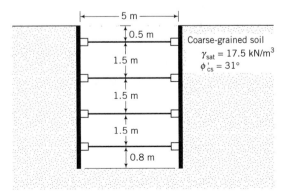

**FIGURE P10.15**

**10.16** A braced excavation is required in a soft clay as shown in Fig. P10.16. Determine the load on the struts per meter length and the factor of safety against bottom heave.

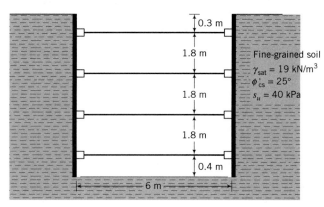

**FIGURE P10.16**

**10.17** A 6 m high geotextile wall is required to support a coarse-grained backfill with $\gamma_{sat} = 17.5$ kN/m³ and $\phi'_{cs} = 29°$. The foundation (base) soil is a clay with $\gamma_{sat} = 18$ kN/m³, $\phi'_{cs} = 22°$, $s_u = 72$ kPa, and $\phi'_b = 16°$. The ultimate strength of the geotextile is 45 kN/m and the soil–geotextile interface friction angle is 20°. The permanent surcharge is 15 kPa. Determine the spacing and length of geotextile required for stability.

**10.18** Redo Exercise 10.17 using galvanized steel ties 75 mm wide with a yield strength 2.5 × 10⁵ kPa, tie–soil interface friction of 20°, and rate of corrosion of 0.025 mm/yr. The design life is 50 years.

## Practical

**10.19** A section of an approach to a bridge is shown in Fig. P10.19. The sides of the approach are to be supported using MSE walls with 1 m × 1 m facing panels. Steel ties of width 300 mm are readily available. Determine the length and thickness of the ties required for stability. The design life is 50 years and the rate of corrosion is 0.025 mm/yr. The yield strength is 2.5 × 10⁵ kPa.

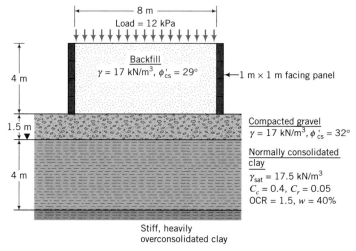

**FIGURE P10.19**

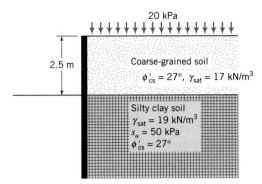

20 kPa

2.5 m

Coarse-grained soil
$\phi'_{cs} = 27°$, $\gamma_{sat} = 17$ kN/m³

Silty clay soil
$\gamma_{sat} = 19$ kN/m³
$s_u = 50$ kPa
$\phi'_{cs} = 27°$

**FIGURE P10.20**

10.20 A cantilever sheet pile wall is required to temporarily support an embankment for an access road as shown in Fig. P10.20. Determine the depth of penetration of the wall and the maximum bending moment. Select two methods from FMM, FSM, and NPPM, and compare the results.

## CHAPTER 11

# SLOPE STABILITY

## 11.0   INTRODUCTION

Slopes in soils and rocks are ubiquitous in nature and in man-made structures. Highways, dams, levees, canals, and stockpiles are constructed by sloping the lateral faces of the soil because slopes are generally less expensive than constructing walls. Natural forces (wind, water, snow, etc.) change the topography on Earth and other planets, often creating unstable slopes. Failure of natural slopes (landslides) and man-made slopes have resulted in much death and destruction. Some failures are sudden and catastrophic; others are insidious. Some failures are widespread; others are localized.

Geotechnical engineers have to pay particular attention to geology, surface drainage, groundwater, and the shear strength of soils in assessing slope stability. However, we are handicapped by the geological variability of soils and methods of obtaining reliable values of shear strength. The analyses of slope stability are based on simplifying assumptions and the design of a stable slope relies heavily on experience and careful site investigation.

In this chapter, we will discuss a few simple methods of analysis from which you should be able to:

* Estimate the stability of slopes with simple geometry and geological features
* Understand the forces and activities that provoke slope failures
* Understand the effects of geology, seepage, and pore water pressures on the stability of slopes

You would make use of the following:

* Shear strength of soils (Chapter 5)
* Effective stresses and seepage (Chapter 3)
* Flow through dams (Chapter 9)

*Sample Practical Situation*   A reservoir is required to store water for domestic use. Several sites were investigated and the top choice is a site consisting of clay soils (clay is preferred because of its low permeability—it is practically impervious). The soils would be excavated, forming sloping sides. You are required to determine the maximum safe slope for the reservoir. Slope failures occur frequently, one of which is shown in Fig. 11.1. Your job is to prevent such failure.

**FIGURE 11.1** Slope failure near a roadway. (Courtesy of Todd Mooney.)

## 11.1  DEFINITIONS OF KEY TERMS

*Slip* or *failure zone* is a thin zone of soil that reaches the critical state or residual state and results in movement of the upper soil mass.

*Slip plane* or *failure plane* or *slip surface* or *failure surface* is the surface of sliding.

*Sliding mass* is the mass of soil within the slip plane and the ground surface.

*Slope angle* $(\alpha_s)$ is the angle of inclination of a slope to the horizontal. The slope angle is sometimes referred to as a ratio, for example, 2:1 (horizontal:vertical).

*Pore water pressure ratio* $(r_u)$ is the ratio of pore water force on a slip surface to the total weight of the soil and any external loading.

## 11.2  QUESTIONS TO GUIDE YOUR READING

1. What types of slope failure are common in soils?
2. What factors provoke slope failures?
3. What is an infinite slope failure?
4. What methods of analysis are used to estimate the factor of safety of a slope?
5. What are the assumptions of the various methods of analysis?
6. How does seepage affect the stability of slopes?
7. What is the effect of rapid drawdown on slope stability?

## 11.3 SOME TYPES OF SLOPE FAILURE

Slope failures depend on the soil type, soil stratification, groundwater, seepage, and the slope geometry. We will introduce a few types of slope failure that are common in soils. Failure of a slope along a weak zone of soil is called a translational slide (Fig. 11.2a). The sliding mass can travel long distances before coming to rest. Translational slides are common in coarse-grained soils.

A common type of failure in homogeneous fine-grained soils is a rotational slide that has its point of rotation on an imaginary axis parallel to the slope. Three types of rotational failure often occur. One type, called a base slide, occurs by an arc engulfing the whole slope. A soft soil layer resting on a stiff layer of soil is prone to base failure (Fig. 11.2b). The second type of rotational failure is the toe slide, whereby the failure surface passes through the toe of the slope (Fig. 11.2c). The third type of rotational failure is the slope slide, whereby the failure surface passes through the slope (Fig. 11.2d).

A flow slide occurs when internal and external conditions force a soil to behave like a viscous fluid and flow down even shallow slopes, spreading out in several directions (Fig. 11.2e). The failure surface is ill defined in flow slides. Multiple failure surfaces usually occur and change continuously as flow proceeds. Flow slides can occur in dry and wet soils.

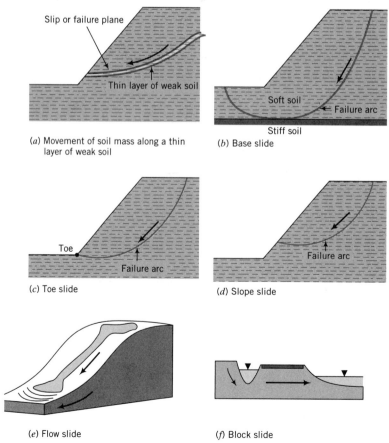

(a) Movement of soil mass along a thin layer of weak soil

(b) Base slide

(c) Toe slide

(d) Slope slide

(e) Flow slide

(f) Block slide

**FIGURE 11.2**  Some common types of slope failure.

Block or wedge slides occur when a soil mass is shattered along joints, seams, fissures, and weak zones by forces emanating from adjacent soils. The shattered mass moves as blocks and wedges down the slope (Fig. 11.2f).

*What's next . . .*What causes the slope failures that we briefly described above? The causes are many and varied. In the next section, we will describe some common causes of slope failure.

## 11.4 SOME CAUSES OF SLOPE FAILURE

Slope failures are caused, in general, by natural forces, human misjudgment and activities, and burrowing animals. We will describe below some of the main factors that provoke slope failures.

### 11.4.1 Erosion

Water and wind continuously erode natural and man-made slopes. Erosion changes the geometry of the slope (Fig. 11.3a), ultimately resulting in slope failure or, more aptly, a landslide. River and streams continuously scour their banks undermining their natural or man-made slopes (Fig. 11.3b).

### 11.4.2 Rainfall

Long periods of rainfall saturate, soften, and erode soils. Water enters into existing cracks and may weaken underlying soil layers, leading to failure, for example, mud slides (Fig. 11.3c).

### 11.4.3 Earthquakes

Earthquakes induce dynamic forces (Fig. 11.3d), especially dynamic shear forces that reduce the shear strength and stiffness of the soil. Pore water pressures in saturated coarse-grained soils could rise to a value equal to the total mean stress and cause these soils to behave like viscous fluids—a phenomenon known as dynamic liquefaction. Structures founded on these soils would collapse; structures buried within them would rise. The quickness (a few seconds) in which the dynamic forces are induced prevents even coarse-grained soils from draining the excess pore water pressures. Thus, failure in a seismic event often occurs under undrained conditions.

### 11.4.4 Geological Features

Many failures commonly result from unidentified geological features. A thin seam of silt (a few millimeters thick) under a thick deposit of clay can easily be overlooked in drilling operations or one may be careless in assessing borehole logs only to find later that the presence of the silt caused a catastrophic failure. Sloping, stratified soils are prone to translational slide along a weak layer(s) (Fig. 11.3e). You must pay particular attention to geological features in assessing slope stability.

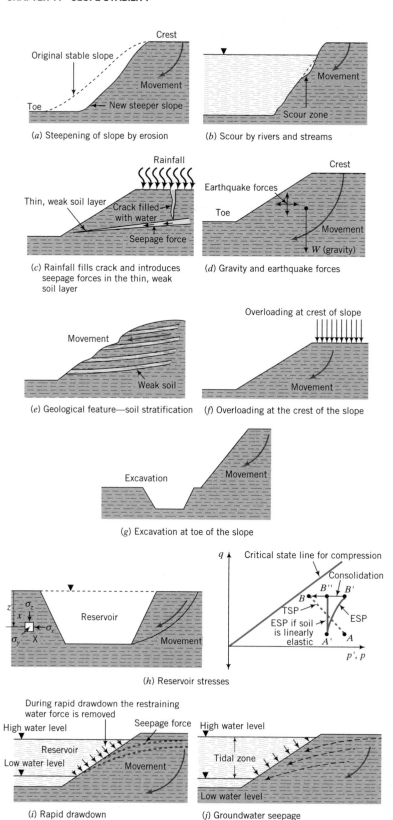

**FIGURE 11.3** Some causes of slope failure.

## 11.4.5 External Loading

Loads placed on the crest of a slope (the top of the slope) add to the gravitational load and may cause slope failure (Fig. 11.3f). A load placed at the toe, called a berm, will increase the stability of the slope. Berms are often used to remediate problem slopes.

## 11.4.6 Construction Activities

Construction activities near the toe of an existing slope can cause failure because lateral resistance is removed (Fig. 11.3g). We can conveniently divide slope failures due to construction activities into two cases. The first case is excavated slope and the second case is fill slope.

### 11.4.6.1 Excavated Slopes

When excavation occurs, the total stresses are reduced and negative pore water pressures are generated in the soil. With time the negative pore water pressures dissipate, causing a decrease in effective stresses and consequently lowering the shear strength of the soil. If slope failures were to occur, they would take place after construction is completed.

We can use our knowledge of stress paths (Chapter 3) to provide insight on the possible effects of excavation on slope stability. Let us consider a construction activity involving excavation of a normally consolidated fine-grained soil to construct a reservoir (Fig. 11.3h). Let us consider an element of soil, X, at a depth $z$ below the surface.

The initial vertical effective stress is $\sigma'_{zo} = \gamma'z$ and the lateral effective stresses are $\sigma'_{xo} = \sigma'_{yo} = K_o\sigma'_{zo}$. The initial pore water pressure is $u_o = \gamma_w z$. The stress invariants are

$$p'_o = \frac{\sigma'_{zo}}{3}(1 + 2K_o), \quad p_o = \frac{\sigma'_{zo}}{3}(1 + 2K_o) + u_o, \quad \text{and} \quad q_o = \sigma'_{zo}(1 - K_o)$$

In stress space $\{p'(p), q\}$, the initial total stresses are represented by point $A$ and the initial effective stresses are represented by point $A'$. The excavation will cause a reduction in $\sigma_x$ (i.e., $\Delta\sigma_x < 0$) but very little change in $\sigma_z$ (i.e., $\Delta\sigma_z \cong 0$) and $\sigma_y$ (i.e., $\Delta\sigma_y \cong 0$). The change in mean total stress is then $\Delta p = -\Delta\sigma_x/3$ and the change in deviatoric stress is $\Delta q = \Delta\sigma_x$. The total stress path (TSP) is depicted as $AB$ in Fig. 11.3h. Although $B$ is near the failure line, the soil is not about to fail because failure is dictated by effective not total stresses. In some cases, the slope fails because the undrained shear strength is exceeded. We will discuss slope failure under undrained condition later in this chapter.

If the soil were a linear, elastic material, the ESP would be $A'B''$ (recall that for elastic material $\Delta p' = 0$ under undrained condition). For an elastoplastic soil, the TSP will be nonlinear. Assuming our soil is elastoplastic, then $A'B'$ would represent our ESP. The ESP moves away from the failure line and the excess pore water pressure is negative. Therefore, failure is unlikely to occur during the excavation stage unless the undrained shear strength of the soil is exceeded.

After the excavation, the excess pore water pressure would dissipate with time. Since no further change in $q$ occurs, the ESP must move from $B'$ to $B$, that is, toward the failure line. The implication is that slope instability would occur under drained condition (after the excavation). The illustration of the excavation process using stress paths further demonstrates the power of stress paths to provide an understanding of construction events in geotechnical engineering.

***11.4.6.2 Fill Slopes***    Fill slopes are common in embankment construction. Fill (soil) is placed at the site and compacted to specifications, usually greater than 95% Proctor maximum dry unit weight. The soil is invariably unsaturated and negative pore water pressures develop. The soil on which the fill is placed, which we will call the foundation soil, may or may not be saturated. If the foundation soil is saturated, then positive pore water pressures will be generated from the weight of the fill and the compaction process. The effective stresses decease and consequently the shear strength decreases. With time the positive pore water pressures dissipate, the effective stresses increase and so does the shear strength of the soil. Thus, slope failures in fill slopes are likely to occur during or immediately after construction.

### 11.4.7 Rapid Drawdown

Reservoirs can be subjected to rapid drawdown. In this case the lateral force provided by the water is removed and the excess pore water pressure does not have enough time to dissipate (Fig. 11.3i). The net effect is that the slope can fail under undrained conditions. If the water level in the reservoir remains at low levels and failure did not occur under undrained conditions, seepage of groundwater would occur and the additional seepage forces can provoke failure (Fig. 11.3j).

> *The* **essential points** *are:*
> 1. *Geological features and environmental conditions (e.g., external loads and natural forces) are responsible for most slope failures.*
> 2. *The common modes of slope failure in soils are by translation, rotation, flow, and block movements.*

*What's next . . .*In the next section, we will discuss the methods of analyses used to evaluate the stability of slopes.

## 11.5   TWO-DIMENSIONAL SLOPE STABILITY ANALYSES

Slope stability can be analyzed using one or more of the following: the limit equilibrium method, limit analysis, finite difference method, and finite element method. Limit equilibrium is often the method of choice but the finite element method is more flexible and general. You should recall (Chapters 7 and 10) that in the limit equilibrium method, a failure mechanism must be postulated and then the equilibrium equations are used to solve for the collapse load. Several failure mechanisms must be investigated and the minimum load required for collapse is taken as the collapse load. The limit equilibrium method gives an upper bound solution (answer higher than the "true" collapse load) because a more efficient mechanism of collapse is possible than those postulated. The limit analysis makes use of the stress–strain characteristics and a failure criterion for the soil. The solution from a limit analysis is a lower bound (answer lower than the "true" collapse load). The finite element method requires the discretization of the soil domain, and makes use of the stress–strain characteristics of the soil and a failure criterion to identify soil regions that have reached the failure stress state. The finite element method does not require speculation on a possible failure surface. We will concentrate on the limit equilibrium method in this book because of its simplicity. We will develop the analyses using a generic friction angle, $\phi'$. Later, we will discuss the appropriate $\phi'$ to use. We will use an effective stress analysis (ESA) and a total stress analysis (TSA).

*What's next . . .* In the next section, we will study how to analyze a slope of infinite extent. We will make use of the limit equilibrium method of analysis and consider long-term and short-term conditions.

## 11.6  INFINITE SLOPES

Infinite slopes have dimensions that extend over great distances. The assumption of an infinite slope simplifies stability calculations. Let us consider a clean, homogeneous coarse-grained soil of infinite slope, $\alpha_s$. To use the limit equilibrium method, we must first speculate on a failure or slip mechanism. We will assume that slip would occur on a plane parallel to the slope. If we consider a slice of soil between the surface of the slope and the slip plane, we can draw a free-body diagram of the slice as shown in Fig. 11.4.

The forces acting on the slice per unit thickness are the weight $W_j = \gamma b_j z_j$, the shear forces $X_j$ and $X_{j+1}$ on the sides, the normal forces $E_j$ and $E_{j+1}$ on the sides, the normal force $N_j$ on the slip plane, and the mobilized shear resistance of the soil, $T_j$, on the slip plane. We will assume that forces that provoke failure are positive. If seepage is present, a seepage force $J_s = i\gamma_w b_j z_j$ develops, where $i$ is the hydraulic gradient. The sign of the seepage force depends on the seepage direction. In our case the seepage force is positive because seepage is in the direction of positive forces. For a uniform slope of infinite extent, $X_j = X_{j+1}$ and $E_j = E_{j+1}$.

To continue with the limit equilibrium method, we must now use the equilibrium equations to solve the problem. But before we do this, we will define the factor of safety of a slope. The factor of safety (FS) of a slope is defined as the

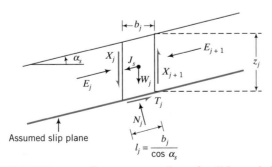

**FIGURE 11.4**  Forces on a slice of soil in an infinite slope.

ratio of the available shear strength of the soil, $\tau_f$, to the minimum shear strength required to maintain stability, $\tau_m$, that is,

$$FS = \frac{\tau_f}{\tau_m} \tag{11.1}$$

The shear strength of soils is governed by the Mohr–Coulomb failure criterion (Chapter 5): that is, $\tau_f = \sigma'_n \tan \phi'$ for an effective stress analysis and $\tau_f = s_u$ for a total stress analysis. The factor of safety is then

$$\text{ESA:} \quad FS = \frac{\sigma'_n \tan \phi'}{\tau_m} = \frac{N' \tan \phi'}{T_m} \tag{11.2}$$

$$\text{TSA:} \quad FS = \frac{s_u}{\tau_m} \tag{11.3}$$

where $N'$ is the normal effective force on the slip plane and $T_m$ is the mobilized shear force.

Let us now use statics to solve for the factor of safety. First, we will consider a **slope without seepage** and groundwater below the slip plane. Because the groundwater is below the slip plane, the effective stress is equal to the total stress. From statics,

$$N'_j = W'_j \cos \alpha_s \quad \text{and} \quad T_j = W'_j \sin \alpha_s$$

From the definition of factor of safety for the ESA [Eq. (11.2)], we get

$$\therefore FS = \frac{N'_j \tan \phi'}{T_j} = \frac{W'_j \cos \alpha_s \tan \phi'}{W'_j \sin \alpha_s} = \frac{\tan \phi'}{\tan \alpha_s} \tag{11.4}$$

At limit equilibrium, FS = 1. Therefore,

$$\alpha_s = \phi' \tag{11.5}$$

The implication of Eq. (11.5) is that the maximum slope angle of a coarse-grained soil cannot exceed $\phi'$.

We will now consider groundwater within the sliding mass and assume that **seepage is parallel to the slope**. The seepage force is

$$J_s = i \gamma_w b_j z_j$$

Since seepage is parallel to the slope, $i = \sin \alpha_s$. From statics,

$$N'_j = W'_j \cos \alpha_s = \gamma' b_j z_j \cos \alpha_s \tag{11.6}$$

and

$$T_j = W'_j \sin \alpha_s + J_s = \gamma' b_j z_j \sin \alpha_s + \gamma_w b_j z_j \sin \alpha_s = (\gamma' + \gamma_w) b_j z_j \sin \alpha_s$$
$$= (\gamma_{\text{sat}}) b_j z_j \sin \alpha_s$$

From the definition of factor of safety [Eq. (11.2)], we get

$$FS = \frac{N'_j \tan \phi'}{T_j} = \frac{\gamma' b_j z_j \cos \alpha_s \tan \phi'}{\gamma_{\text{sat}} b_j z_j \sin \alpha_s} = \frac{\gamma'}{\gamma_{\text{sat}}} \frac{\tan \phi'}{\tan \alpha_s} \tag{11.7}$$

At limit equilibrium, FS = 1. Therefore,

$$\tan \alpha_s = \frac{\gamma'}{\gamma_{\text{sat}}} \tan \phi' \tag{11.8}$$

For most soils, $\gamma'/\gamma_{\text{sat}} \approx \frac{1}{2}$. Thus, seepage parallel to the slope reduces the limiting slope of a clean, coarse-grained soil by about one-half.

The shear stress on the slip plane for a TSA, which is applicable to the short-term slope stability in fine-grained soils, is

$$\tau_j = \frac{T_j}{l_j} = \frac{W_j \sin \alpha_s}{l_j} = \frac{W_j \sin \alpha_s \cos \alpha_s}{b_j} = \frac{\gamma b_j z_j}{b_j} \sin \alpha_s \cos \alpha_s = \gamma z_j \sin \alpha_s \cos \alpha_s \tag{11.9}$$

The factor of safety [Eq. (11.3)] is

$$\text{FS} = \frac{s_u}{\Sigma \tau_j} = \frac{s_u}{\gamma z \sin \alpha_s \cos \alpha_s} = \frac{2s_u}{\gamma z \sin(2\alpha_s)} \tag{11.10}$$

At limit equilibrium, FS = 1. Therefore,

$$\boxed{\alpha_s = \tfrac{1}{2} \sin^{-1}(2s_u/\gamma z)} \tag{11.11}$$

and

$$z = \frac{2s_u}{\gamma \sin(2\alpha_s)} \tag{11.12}$$

The critical value of $z$ occurs when $\alpha_s = 45°$, that is,

$$\boxed{z = z_{\text{cr}} = 2s_u/\gamma} \tag{11.13}$$

which is the depth of tension cracks. Therefore, the maximum slope of fine-grained soils under short-term loading, assuming an infinite slope failure mechanism, is 45°. For slopes greater than 45° and depths greater than $2s_u/\gamma$, the infinite slope failure mechanism is not tenable. The infinite slope failure mechanism is more relevant to coarse-grained soils than fine-grained soils because most slope failures observed in fine-grained soils are finite and rotational.

> ***The* essential points *are:***
> 1. ***The maximum stable slope in a coarse-grained soil, in the absence of seepage, is equal to the friction angle.***
> 2. ***The maximum stable slope in coarse-grained soils, in the presence of seepage parallel to the slope, is approximately one-half the friction angle.***
> 3. ***The critical slope angle in fine-grained soils is 45° for an infinite slope failure mechanism. The critical depth is the depth of tension cracks, that is, 2s_u/γ.***

## EXAMPLE 11.1

Dry sand is to be dumped from a truck on the side of a roadway. The properties of the sand are $\phi' = 30°$, $\gamma = 17$ kN/m³, and $\gamma_{sat} = 17.5$ kN/m³. Determine the maximum slope angle of the sand in (a) the dry state, (b) the saturated state, without seepage and (c) the saturated state if groundwater is present and seepage were to occur parallel to the slope. What is the safe slope in the dry state for a factor of safety of 1.25?

**Strategy** The solution to this problem is a straightforward application of Eqs. (11.5) and (11.8).

## Solution 11.1

**Step 1:** Sketch a diagram.
See Fig. E11.1.

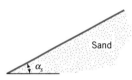

**FIGURE E11.1**

**Step 2:** Determine the maximum slope under the dry condition.

$$\text{Equation (11.5):} \quad \alpha_s = \phi' = 30°$$

You should note that any small disturbance would cause this slope to fail since it is at limit equilibrium.
For FS = 1.25, the safe slope is

$$\alpha_s = \frac{\tan \phi'}{FS} = \frac{\tan 30°}{1.25} = 24.8°$$

**Step 3:** Determine the maximum slope for the saturated condition.
We see that $\phi'$ is not significantly affected by whether the soil is dry or wet. Therefore, $\alpha_s = 30°$.

**Step 4:** Determine the maximum slope for seepage parallel to the slope.

$$\text{Equation (11.8):} \quad \tan \alpha_s = \frac{\gamma'}{\gamma_{sat}} \tan \phi'$$

$$\gamma' = \gamma_{sat} - \gamma_w = 17.5 - 9.8 = 7.7 \text{ kN/m}^3$$

$$\tan \alpha_s = \frac{\gamma'}{\gamma_{sat}} \tan \phi' = \frac{7.7}{17.5} \tan 30° = 0.25$$

$$\alpha_s = 14°$$ ∎

## EXAMPLE 11.2

A trench was cut in a clay slope (Fig. E11.2) to carry TV and telephone cables. When the trench reached a depth of 2 m, the top portion of the clay suddenly failed, engulfing the trench and injuring several workers. On investigating the failure, you observed a slip plane approximately parallel to the original slope. Determine the undrained shear strength of the clay.

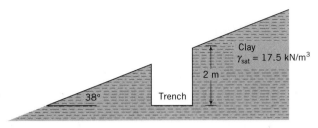

**FIGURE E11.2**

***Strategy*** The failure observed can be analyzed as an infinite slope failure in a fine-grained soil under undrained conditions. So you should consider a TSA.

**Solution 11.2**
**Step 1:** Determine $s_u$.

$$\text{From Eq. (11.10):} \quad s_u = \frac{\gamma z (\sin 2\alpha_s)}{2} \text{ FS}$$

Since the slope failed, FS $= 1$.

$$\therefore s_u = \frac{17.5 \times 2 \times \sin(2 \times 38°)}{2} = 17 \text{ kPa} \quad \blacksquare$$

**What's next . . .**The infinite slope failure mechanism is reasonable for coarse-grained soils. Homogeneous fine-grained soils have been observed to fail through a rotational mechanism. Next, we will consider rotational failures.

## 11.7 ROTATIONAL SLOPE FAILURES

We will continue to use the limit equilibrium method but instead of a planar slip surface of infinite extent we will assume circular (Fig. 11.5a) and noncircular (Fig. 11.5b) slip surfaces of finite extent. We will assume the presence of a phreatic surface within the sliding mass.

A free-body diagram of the postulated circular failure mechanism would show the weight of the soil within the sliding mass acting at the center of mass. If seepage is present, then the seepage forces, $J_s$, which may vary along the flow path, are present. The forces resisting outward movement of slope are the shearing forces mobilized by the soil along the slip surface.

We must now use statics to determine whether the disturbing forces and moments created by $W$ and $J_s$ exceed the resisting forces and moments due to the shearing forces mobilized by the soil. However, we have several problems in determining the forces and moments. Here is a list.

- It is cumbersome, if not difficult, to determine the location of the center of mass especially when we have layered soils and groundwater.
- The problem is statically indeterminate.
- We do not know how the mobilized shear strength, $\tau_m$, of the soil varies along the slip surface.

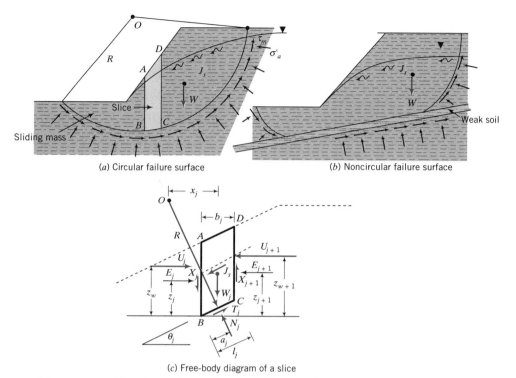

(a) Circular failure surface   (b) Noncircular failure surface

(c) Free-body diagram of a slice

**FIGURE 11.5**   Circular and noncircular failure surfaces and the forces on a slice of soil.

- We do not know how the normal effective stress, $\sigma'_n$, varies along the slip surface.
- We do not know how the seepage forces vary within the soil mass and along the failure surface.
- Even the weight of sliding mass is difficult to calculate because of soil layering (different unit weights of the soils) and complex geometry of some slopes.

One approach to solve our problem is to divide the sliding mass into an arbitrary number of slices and then sum the forces and moments of each slice. Of course, the larger the number of slices, the better the accuracy of our answer. Dividing the area inside the sliding mass into slices presents new problems. We now have to account for the internal forces or interfacial forces between the slices.

Let us consider an arbitrary slice, $ABCD$ (Fig. 11.5a), and draw a free-body diagram of the forces acting on the slices as illustrated in Fig. 11.5c. The forces have the following meaning:

- $W_j$ is the total weight of a slice including any external load.
- $E_j$ is the interslice lateral effective force.
- $(J_s)_j$ is the seepage force on the slice.
- $N'_j$ is the normal effective force along the slip surface.

- $T_j$ is the mobilized shear force along the slip surface.
- $X_j$ is the interslice shear force.
- $U_j$ is the force from the pore water pressure.
- $z_j$ is the location of the interslice lateral effective force.
- $z_w$ is the location of the pore water force.
- $a_j$ is the location of the normal effective force along the slip surface.
- $b_j$ is the width of the slice.
- $l_j$ is the length of slip surface along the slice.
- $\theta_j$ is the inclination of the slip surface within the slice to the horizontal plane.

The side $BC$ is assumed to be a straight line.

We now have to obtain the values of 13 parameters. We can find $W_j$, $U_j$, $(J_s)_j$, $b_j$, $l_j$, $z_w$, and $\theta_j$ from the geometry of the slice, the unit weights of the soils, and the location of the phreatic surface. We have six unknowns for each slice and only three equilibrium equations; our problem is then statically indeterminate. To solve our problem, we have to make assumptions regarding three of the unknown parameters. Several solution methods have evolved depending on the assumptions made about the unknown parameters and which equilibrium condition (force, moment, or both) is satisfied. Table 11.1 provides a summary of methods that have been proposed.

We will describe the methods developed by Bishop (1955) and by Janbu (1973) because they are popular methods and require only a calculator or a spreadsheet program. Computer programs are commercially available for all the methods listed in Table 11.1. Before you use these programs, you should under-

## TABLE 11.1 Slope Analyses Methods Based on Limit Equilibrium

| Method | Assumption | Failure surface | Equilibrium equation satisfied | Solution by |
|---|---|---|---|---|
| Swedish method (Fellenius, 1927) | Resultant of interslice force is zero; $J_s = 0$ | Circular | Moment | Calculator |
| Bishop's simplified method (Bishop, 1955) | $E_j$ and $E_{j+1}$ are collinear $X_j - X_{j+1} = 0$, $J_s = 0$ | Circular | Moment | Calculator |
| Bishop's method (Bishop, 1955) | $E_j$ and $E_{j+1}$ are collinear; $J_s = 0$ | Circular | Moment | Calculator/computer |
| Morgenstern and Price (1965) | Relationship between $E$ and $X$ of the form $X = \lambda f(x)E$, $f(x)$ is a function $\approx 1$, $\lambda$ is a scale factor, $J_s = 0$ | Any shape | All | Computer |
| Spencer (1967) | Interslice forces are parallel; $J_s = 0$ | Any shape | All | Computer |
| Bell's method (Bell, 1968) | Assumed normal stress distribution along failure surface; $J_s = 0$ | Any shape | All | Computer |
| Janbu (1973) | $X_j - X_{j+1}$ replaced by a correction factor, $f_o$; $J_s = 0$ | Noncircular | Horizontal forces | Calculator |
| Sarma (1976)[a] | Assumed distribution of vertical interslice forces; $J_s = 0$ | Any shape | All | Computer |

[a]Sarma's method includes seismic forces.

stand the principles employed and the assumptions made in their development. The methods of Bishop (1955) and Janbu (1973) were developed by assuming that soil is a cohesive–frictional material. We will modify the derivation of the governing equations of these methods by considering soil as a dilatant–frictional material. We will also develop separate governing equations for an effective stress analysis and a total stress analysis. However, we will retain the names of original developers. For long-term or drained condition, we have to conduct an effective stress analysis. For short-term or undrained condition in fine-grained soils, we have to conduct a total stress analysis.

## 11.8 METHOD OF SLICES

### 11.8.1 Bishop's Method

Bishop (1955) assumed a circular slip surface as shown in Fig. 11.5a. Let us apply the equilibrium equations to the forces on the slice shown in Fig. 11.5c, assuming that $E_j$ and $E_{j+1}$, $U_j$ and $U_{j+1}$ are collinear, $N_j$ acts at the center of the arc length, that is, at $l_j/2$, and $(J_s)_j = 0$.

Summing forces vertically, we get

$$N_j \cos \theta_j + T_j \sin \theta_j - W_j - X_j + X_{j+1} = 0 \tag{11.14}$$

The force due to the pore water pressure ($U_j$) is $U_j = u_j l_j$. From the principle of effective stress,

$$N_j' = N_j - u_j l_j \tag{11.15}$$

Combining Eqs. (11.14) and (11.15), we get

$$N_j' \cos \theta_j = W_j + X_j - X_{j+1} - T_j \sin \theta_j - u_j l_j \cos \theta_j \tag{11.16}$$

For convenience, let us define the force due to the pore water as a function of $W_j$ as

$$r_u = \frac{u_j b_j}{W_j} \tag{11.17}$$

where $r_u$ is called the pore water pressure ratio. Substituting Eq. (11.17) into (11.16) yields

$$N_j' \cos \theta_j = W_j(1 - r_u) - T_j \sin \theta_j + (X_j - X_{j+1}) \tag{11.18}$$

Bishop (1955) considered only moment equilibrium such that, from Fig. 11.5c,

$$\Sigma W_j x_j = \Sigma T_j R \tag{11.19}$$

where $x_j$ is the horizontal distance from the center of the slice to the center of the arc of radius $R$, and $T_j$ is the mobilized shear force. Solving for $T_j$ from Eq. (11.19) and noting that $x_j = R \sin \theta_j$, we get

$$\Sigma T_j = \Sigma \frac{W_j x_j}{R} = \Sigma W_j \sin \theta_j \tag{11.20}$$

Recall from Eq. (11.1) that the factor of safety is defined as

$$FS = \frac{\tau_f}{\tau_m} = \frac{(T_f)_j}{T_j} \tag{11.21}$$

where $T_f$ is the soil shear force at failure. In developing the governing equation for FS, we will, firstly, consider effective stress and later total stress. For an ESA,

$$FS = \frac{N_j' \tan(\phi')_j}{T_j} \tag{11.22}$$

By rearranging Eq. (11.22), we get

$$T_j = \frac{N_j' \tan(\phi')_j}{FS} \tag{11.23}$$

Substituting Eq. (11.23) into (11.18) yields

$$N_j' \cos \theta_j = W_j(1 - r_u) - \frac{N_j' \tan(\phi')_j \sin \theta_j}{FS} + (X_j - X_{j+1}) \tag{11.24}$$

Solving for $N_j'$, we get

$$N_j' = \frac{W_j(1 - r_u) + (X_j - X_{j+1})}{\cos \theta_j + \dfrac{\tan(\phi')_j \sin \theta_j}{FS}} \tag{11.25}$$

Putting

$$m_j = \frac{1}{\cos \theta_j + \dfrac{\tan(\phi')_j \sin \theta_j}{FS}} \tag{11.26}$$

we can write $N_j'$ as

$$N_j' = [W_j(1 - r_u) + (X_j - X_{j+1})]m_j \tag{11.27}$$

Substituting Eq. (11.22) into (11.20) gives

$$\Sigma \frac{N_j' \tan(\phi')_j}{FS} = \Sigma W_j \sin \theta_j \tag{11.28}$$

Combining Eqs. (11.27) and (11.28) yields

$$FS = \frac{\Sigma[W_j(1 - r_u) + (X_j - X_{j+1})] \tan(\phi')_j m_j}{\Sigma W_j \sin \theta_j} \tag{11.29}$$

Equation (11.29) is Bishop's equation for an ESA. Bishop (1955) showed that neglecting $(X_j - X_{j+1})$ resulted in about 1% of error. Therefore, neglecting $(X_j - X_{j+1})$, we get

$$FS = \frac{\Sigma[W_j(1 - r_u)] \tan(\phi')_j m_j}{\Sigma W_j \sin \theta_j} \tag{11.30}$$

Equation (11.30) is Bishop's simplified equation for an ESA. If groundwater is below the slip surface, $r_u = 0$ and

$$FS = \frac{\Sigma W_j \tan(\phi')_j m_j}{\Sigma W_j \sin \theta_j} \qquad (11.31)$$

Let us now consider a TSA. The mobilized shear force on the slip surface is

$$T_j = \frac{(s_u)_j l_j}{FS} \qquad (11.32)$$

where $(s_u)_j$ is the undrained shear strength of the soil along the slip surface within the slice. Combining Eqs. (11.20) and (11.32) yields

$$FS = \frac{\Sigma (s_u)_j l_j}{\Sigma W_j \sin \theta_j} \qquad (11.33)$$

Since $b_j = l_j \cos \theta_j$, Eq. (11.33) becomes

$$FS = \frac{\Sigma (s_u)_j \dfrac{b_j}{\cos \theta_j}}{\Sigma W_j \sin \theta_j} \qquad (11.34)$$

## 11.8.2 Janbu's Method

Janbu (1973) assumed a noncircular slip surface (Fig. 11.6a). The forces acting on a slice are as shown in Fig. 11.6b. Janbu considered equilibrium of horizontal forces and assumed that $E_j - E_{j+1} = 0$.

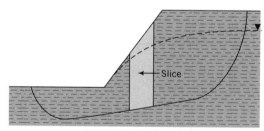

(a) Noncircular slip surface

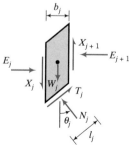

(b) Forces on a slice

**FIGURE 11.6** Failure surface proposed by Janbu and forces on a slice of soil.

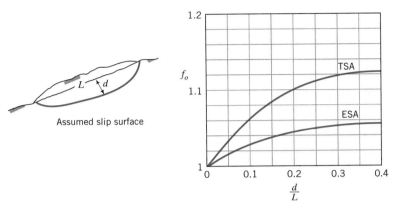

**FIGURE 11.7**  Correction factor for Janbu's method.

The factor of safety, defined with respect to equilibrium of horizontal forces, is

$$\text{FS} = \frac{\Sigma \text{ Resisting forces}}{\Sigma \text{ Disturbing forces}} = \frac{\Sigma (T_f)_j \cos \theta_j}{\Sigma [W_j + (X_j - X_{j+1})] \tan \theta_j} \tag{11.35}$$

Noting that $(T_f)_j = T_j(\text{FS}) = N_j \tan(\phi')_j \text{ FS}$, we can combine Eqs. (11.35) and (11.27) to yield, for an ESA,

$$\text{FS} = \frac{\Sigma [W_j(1 - r_u) + (X_j - X_{j+1})] m_j \tan \theta_j \cos \theta_j}{\Sigma [W_j + (X_j - X_{j+1})] \tan \theta_j} \tag{11.36}$$

Janbu (1973) then replaced the interslice shear forces ($X_j$ and $X_{j+1}$) by a correction factor $f_o$ as shown in Fig. 11.7. The simplified form of Janbu's equation for an ESA is

$$\text{FS} = \frac{f_o \Sigma W_j (1 - r_u) m_j \tan \phi'_j \cos \theta_j}{\Sigma W_j \tan \theta_j} \tag{11.37}$$

If the groundwater is below the slip surface, $r_u = 0$, and

$$\text{FS} = f_o \frac{\Sigma W_j m_j \tan \phi'_j \cos \theta_j}{\Sigma W_j \tan \theta_j} \tag{11.38}$$

For a TSA,

$$\text{FS} = \frac{\Sigma (s_u)_j b_j}{\Sigma [W_j + (X_j - X_{j+1})] \tan \theta_j} \tag{11.39}$$

Replacing $(X_i - X_{i+1})$ by a correction factor $f_o$ (Fig. 11.7), we get

$$\text{FS} = f_o \frac{\Sigma (s_u)_j b_j}{\Sigma W_j \tan \theta_j} \tag{11.40}$$

---

*The* **essential points** *are:*

*1. Bishop (1955) assumed a circular slip plane and considered only moment equilibrium. He neglected seepage forces and assumed that the lateral normal forces are collinear. In Bishop's simplified method, the resultant interface shear is assumed to be zero.*

> **2. *Janbu (1973) assumed a noncircular failure surface and considered equilibrium of horizontal forces. He made similar assumptions to Bishop (1955), except that a correction factor is applied to replace the interface shear.***
>
> **3. *For slopes in fine-grained soils, you should conduct both an ESA and a TSA for long-term loading and short-term loading respectively. For slopes in coarse-grained soils, only an ESA is necessary for short-term and short-term loading provided the loading is static. In most problems, you would find that an effective stress analysis would yield the minimum factor of safety.***

## 11.9  APPLICATION OF THE METHOD OF SLICES

The shear strength parameters are of paramount importance in slope stability calculations. The soils at the slip surface are at or near the critical or the residual state. You should use $\phi' = \phi'_{cs}$ in all slope stability calculations except for fissured overconsolidated clays. Progressive failure usually occurs in fissured overconsolidated clays. The appropriate value of $\phi'$ to use is $\phi'_r$—the residual friction angle. The measured undrained shear strength is often unreliable. You should use conservative values of $s_u$ for a TSA.

Tension cracks in fine-grained soils tend to develop on the crest and the face of slopes in fine-grained soils. There are three important effects of tension cracks. First, they modify the slip surface. The slip surface does not intersect the ground surface but stops at the base of the tension crack (Fig. 11.8). Recall from Chapter 10 that the depth of a tension crack is $z_{cr} = 2s_u/\gamma$.

Second, the tension crack may be filled with water. In this case, the critical depth is $z'_{cr} = 2s_u/\gamma'$ and a hydrostatic pressure is applied along the depth of the crack. The net effect is a reduction in the factor of safety because the disturbing moment is increased. The additional disturbing moment from the hydrostatic pressure is $\frac{1}{2}\gamma_w z_{cr}^2(z_s + \frac{2}{3}z_{cr})$, where $z_s$ is the vertical distance from the top of the

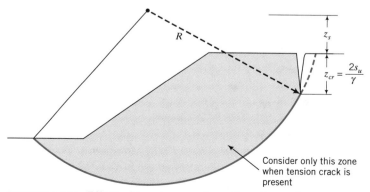

**FIGURE 11.8**  Effect of tension crack on the slip surface.

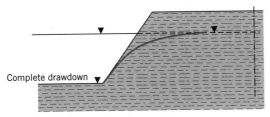

**FIGURE 11.9**  Drawdown in a reservoir.

tension crack to the center of rotation. The factor of safety using Bishop's simplified method becomes

$$
\text{ESA:} \quad FS = \frac{\Sigma W_j (1 - r_u)(\tan \theta'_j) m_j}{\Sigma W_j \sin \theta_j + \dfrac{\frac{1}{2}\gamma_w z_{cr}^2 (z_s + \frac{2}{3}z_{cr})}{R}} \tag{11.41}
$$

$$
\text{TSA:} \quad FS = \frac{\Sigma (s_u)_j \dfrac{b_j}{\cos \theta_j}}{\Sigma W_j \sin \theta_j + \dfrac{\frac{1}{2}\gamma_w z_{cr}^2 (z_s + \frac{2}{3}z_{cr})}{R}} \tag{11.42}
$$

Third, the tension crack provides a channel for water to reach underlying soil layers. The water can introduce seepage forces and weaken these layers. The locations of the tension crack and the critical slip plane are not sensitive to the location of the phreatic surface.

Dams and cuts supporting reservoirs can be subjected to rapid drawdown. Consider the earth dam shown in Fig. 11.9. When the reservoir is full, the groundwater level within the dam will equilibrate with the reservoir water level. If water is withdrawn rapidly, the water level in the reservoir will drop but very little change in the groundwater level in the dam will occur. In fine-grained soils, a few weeks of drawdown can be rapid because of the low permeability of these soils. Because the restraining lateral force of the water in the reservoir is no longer present and the pore water pressure in the dam is high, the FS will be reduced. The worst case scenario is rapid, complete drawdown. If a partial drawdown occurs and is maintained, then the phreatic surface will keep changing and seepage forces (resulting from pore water pressure gradients) are present in addition to the pore water pressures.

# 11.10  PROCEDURE FOR THE METHOD OF SLICES

The procedure to determine the factor of safety of slopes using the method of slices, with reference to Fig. 11.10, is as follows:

1. Draw the slope to scale and note the positions and magnitudes of any external loads.
2. Draw a trial slip surface and identify its point of rotation.
3. Draw the phreatic surface, if necessary (Chapter 9).

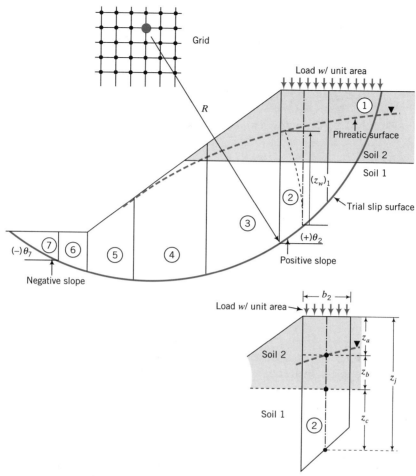

**FIGURE 11.10**   Method of slices.

4. If the soil is fine-grained, calculate the depth of the tension crack and sketch in a possible location of the tension crack. You can start at the crest and locate a point on your slip surface where the depth matches the depth of the tension crack.

5. Divide the soil mass above the slip surface into a convenient number of slices. More than five slices are needed for most problems.

6. For each slice:
    (a) Measure the width, $b_j$.
    (b) Determine $W_j$—the total weight of a slice including any external load. For example, for the two-layer soil profile shown in Fig. 11.10, the weight of slice ② ($j = 2$) is $W_2 = b_2(q_s + z_a(\gamma)_{s2} + z_b(\gamma_{sat})_{s2} + z_c(\gamma_{sat})_{s1})$, where $s_1$ and $s_2$ denote soil layers 1 and 2, $q_s$ is the surface load per unit area, and $z_a, z_b, z_c$ are the mean heights as shown in Fig. 11.10.
    (c) Measure the angle $\theta_j$ for each slice or you can calculate it if you measure the length, $l_j$ [$\theta_j = \cos^{-1}(b_j/l_j)$]. The angle $\theta_j$ can be negative. Angles left of the center of rotation are negative. For example, the value of $\theta$ for slice ⑦ is negative but for slice ②, $\theta$ is positive.

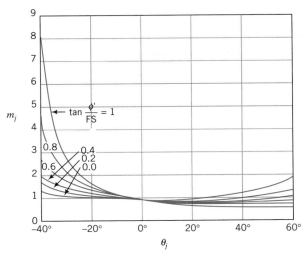

**FIGURE 11.11**  The $m_j$ to be used in Bishop's method.

Alternatively, negative values of the slope of the slip surface in a slice give negative values of $\theta_j$.

(d) Sketch an equipotential line starting from the intersection of the vertical center line and the slip surface to intersect the phreatic surface at $\sim 90°$. The vertical projection of the equipotential line is the pore water pressure head, $(z_w)_j$.

(e) Calculate $r_u = \gamma_w (z_w)_j / \gamma_j z_j$.

7. You now have values for all the required parameters to calculate the factor of safety. Prepare a table or use a spreadsheet program to carry out the calculations. To facilitate calculations, using a nonprogrammable calculator, a chart for $m_j$ is shown in Fig. 11.11. You have to guess a value of FS and then iterate until the guessed value of FS and the calculated value of FS are the same or within a small tolerance ($\approx 0.01$). If a tension crack is present, set the term $W_j(1 - r_u)(\tan \phi'_j)m_j$ to zero but keep the term $W \sin \theta_j$ for the slices above the tension crack when you are considering an ESA; for a TSA, set $s_u = 0$ for the slices above the tension crack but keep the term $W_j \sin \theta_j$.

8. Repeat the procedure from item 2 to item 7 until the smallest factor of safety is found. There are several techniques that are used to reduce the number of trial slip surfaces. One simple technique is to draw a grid and selectively use the nodal points as centers of rotation. Commercially available programs use different methods to optimize the search for the slip plane with the least factor of safety.

*The essential points are:*

1. *The appropriate value of $\phi'$ to use is $\phi'_{cs}$, except for fissured overconsolidated fine-grained soils, where you should use $\phi' = \phi'_r$.*

2. *Tension cracks in fine-grained soils reduce the factor of safety of a slope. Tension cracks may also provide channels for water to introduce seepage forces and weaken underlying soil layers.*

3. *For slopes adjacent to bodies of water, you should consider the effects of operating and environmental conditions on their stability.*

**4**

## EXAMPLE 11.3

Use Bishop's simplified method for the factors of safety of the slope shown in Fig. E11.3a. Assume the soil above the phreatic surface to be saturated. Consider three cases: Case 1—no tension crack; Case 2—tension crack; and Case 3—the tension crack in Case 2 is filled with water.

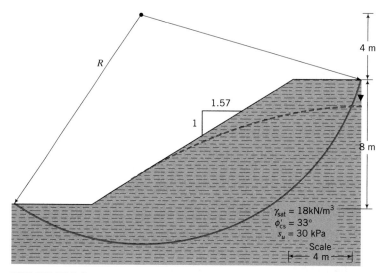

**FIGURE E11.3a**

***Strategy*** Set up a table to carry out the calculations following the procedure in Section 11.10. If you have access to a spreadsheet program, use it. For the solution of this example, the author used Excel. Before dividing the sliding mass into slices, determine the depth of the tension crack and locate it on the crest of the slope as a side of a slice.

## Solution 11.3

**Step 1:** Redraw the figure to scale.
See Fig. E11.3b.

**Step 2:** Find the depth of the tension crack.

$$z_{cr} = \frac{2s_u}{\gamma} = \frac{2 \times 30}{18} = 3.33 \text{ m}$$

**Step 3:** Divide the sliding mass into slices.
Find a height from the crest to the failure surface that equals $z_{cr}$ and sketch in the tension crack. Use this location of the tension crack as a side of a slice. In Fig. E11.3b, the sliding mass is divided into 9 slices.

**Step 4:** Set up a spreadsheet.
See Table E11.3.

**Step 5:** Extract the required values.
Follow the procedure in Section 11.10. The weight $W = \gamma_{sat}bz$. When the tension crack is considered, the shear resistance of slice ⑨ is

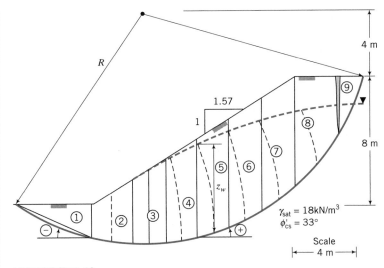

**FIGURE E11.3b**

neglected. When water fills the tension crack, the moment of the hydrostatic pressure in the tension crack is

$$\text{TCM} = \tfrac{1}{2}\gamma_w z_{cr}^2(z_s + \tfrac{2}{3}z_{cr})$$

For an ESA, assume a value of FS and then change this value until it becomes equal to the calculated value.

## TABLE E11.3   Bishop's Simplified Method

Homogeneous soil
$s_u$    30 kPa
$\phi'$    33°
$\gamma_w$    9.8 kN/m³
$\gamma_{sat}$    18 kN/m³
$z_{cr}$    3.33 m
FS    1.05 assumed

**No tension crack**

| Slice | $b$ (m) | $z$ (m) | $W = \gamma bz$ (kN) | $z_w$ (m) | $r_u$ | $\theta$ (deg) | $m_j$ | $W\sin\theta$ | ESA $W(1 - r_u)(\tan\phi')m_j$ | TSA $s_u b/\cos\theta$ |
|---|---|---|---|---|---|---|---|---|---|---|
| 1 | 4.9 | 1 | 88.2 | 1 | 0.54 | −23 | 1.47 | −34.5 | 38.3 | 159.7 |
| 2 | 2.5 | 3.6 | 162.0 | 3.6 | 0.54 | −10 | 1.14 | −28.1 | 54.6 | 76.2 |
| 3 | 2 | 4.6 | 165.6 | 4.6 | 0.54 | 0 | 1.00 | 0.0 | 49.0 | 60.0 |
| 4 | 2 | 5.6 | 201.6 | 5 | 0.49 | 9 | 0.92 | 31.5 | 62.1 | 60.7 |
| 5 | 2 | 6.5 | 234.0 | 5.5 | 0.46 | 17 | 0.88 | 68.4 | 72.2 | 62.7 |
| 6 | 2 | 6.9 | 248.4 | 5.3 | 0.42 | 29 | 0.85 | 120.4 | 80.1 | 68.6 |
| 7 | 2 | 6.8 | 244.8 | 4.5 | 0.36 | 39.5 | 0.86 | 155.7 | 87.6 | 77.8 |
| 8 | 2.5 | 5.3 | 238.5 | 2.9 | 0.30 | 49.5 | 0.90 | 181.4 | 97.5 | 115.5 |
| 9 | 1.6 | 1.6 | 46.1 | 0.1 | 0.03 | 65 | 1.02 | 41.8 | 29.6 | 113.6 |
| | | | | | | | Sum | 536.6 | 570.9 | 794.8 |
| | | | | | | | FS | | **1.06** | **1.48** |

**TABLE E11.3 (continued)**

**Tension crack**

FS = 1 assumed

| Slice | b (m) | z (m) | W = γbz (kN) | $z_w$ (m) | $r_u$ | θ (deg) | $m_j$ | W sin θ | ESA $W(1 - r_u)(\tan \phi')m_j$ | TSA $s_u b/\cos \theta$ |
|---|---|---|---|---|---|---|---|---|---|---|
| 1 | 4.9 | 1 | 88.2 | 1 | 0.54 | −23 | 1.50 | −34.5 | 39.1 | 159.7 |
| 2 | 2.5 | 3.6 | 162.0 | 3.6 | 0.54 | −10 | 1.15 | −28.1 | 55.0 | 76.2 |
| 3 | 2 | 4.6 | 165.6 | 4.6 | 0.54 | 0 | 1.00 | 0.0 | 49.0 | 60.0 |
| 4 | 2 | 5.6 | 201.6 | 5 | 0.49 | 9 | 0.92 | 31.5 | 61.8 | 60.7 |
| 5 | 2 | 6.5 | 234.0 | 5.5 | 0.46 | 17 | 0.87 | 68.4 | 71.5 | 62.7 |
| 6 | 2 | 6.9 | 248.4 | 5.3 | 0.42 | 29 | 0.84 | 120.4 | 78.9 | 68.6 |
| 7 | 2 | 6.8 | 244.8 | 4.5 | 0.36 | 39.5 | 0.84 | 155.7 | 85.8 | 77.8 |
| 8 | 2.5 | 5.3 | 238.5 | 2.9 | 0.30 | 49.5 | 0.87 | 181.4 | 95.1 | 115.5 |
| 9 | 1.6 | 1.6 | 46.1 | 0.1 | 0.03 | 65 | 0.99 | 41.8 | 0.0 | 0.0 |
| | | | | | | | Sum | 536.6 | 536.2 | 681.2 |
| | | | | | | | FS | | 1.00 | 1.27 |

**Tension crack filled with water**

R = 14.3 m
TCM/R = 23.7 kN
FS = 0.95 assumed

| Slice | b (m) | z (m) | W = γbz (kN) | $z_w$ (m) | $r_u$ | θ (deg) | $m_j$ | W sin θ | ESA $W(1 - r_u)(\tan \phi')m_j$ | TSA $s_u b/\cos \theta$ |
|---|---|---|---|---|---|---|---|---|---|---|
| 1 | 4.9 | 1 | 88.2 | 1 | 0.54 | −23 | 1.53 | −34.5 | 39.9 | 159.7 |
| 2 | 2.5 | 3.6 | 162.0 | 3.6 | 0.54 | −10 | 1.15 | −28.1 | 55.3 | 76.2 |
| 3 | 2 | 4.6 | 165.6 | 4.6 | 0.54 | 0 | 1.00 | 0.0 | 49.0 | 60.0 |
| 4 | 2 | 5.6 | 201.6 | 5 | 0.49 | 9 | 0.91 | 31.5 | 61.5 | 60.7 |
| 5 | 2 | 6.5 | 234.0 | 5.5 | 0.46 | 17 | 0.86 | 68.4 | 70.9 | 62.7 |
| 6 | 2 | 6.9 | 248.4 | 5.3 | 0.42 | 29 | 0.83 | 120.4 | 77.8 | 68.6 |
| 7 | 2 | 6.8 | 244.8 | 4.5 | 0.36 | 39.5 | 0.83 | 155.7 | 84.3 | 77.8 |
| 8 | 2.5 | 5.3 | 238.5 | 2.9 | 0.30 | 49.5 | 0.86 | 181.4 | 93.0 | 115.5 |
| 9 | 1.6 | 1.6 | 46.1 | 0.1 | 0.03 | 65 | 0.96 | 41.8 | 0.0 | 0.0 |
| | | | | | | | Sum | 536.6 | 531.7 | 681.2 |
| | | | | | | | FS | | 0.95 | 1.22 |

**Step 6:** Compare the factors of safety.

| | FS | |
|---|---|---|
| Condition | ESA | TSA |
| Without tension crack | 1.06 | 1.48 |
| With tension crack | 1.00 | 1.27 |
| Tension crack filled with water | 0.95 | 1.22 |

The smallest factor of safety occurs using an ESA with the tension crack filled with water. The slope, of course, fails because FS < 1. ■

## EXAMPLE 11.4

Determine the factor of safety of the slope shown in Fig. E11.4a. Assume no tension crack.

***Strategy*** Follow the same strategy as described in Example 11.3.

### Solution 11.4

**Step 1:** Redraw the figure to scale.
See Fig. E11.4b.

**Step 2:** Divide the sliding mass into slices as shown in Fig. E11.4b.

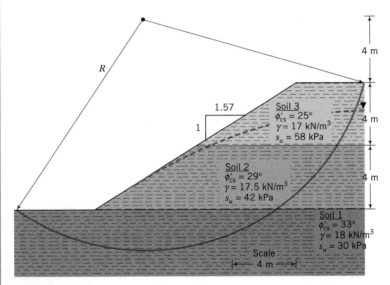

**FIGURE E11.4a**

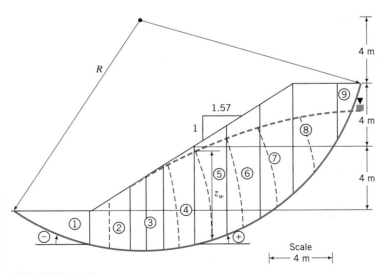

**FIGURE E11.4b**

**TABLE E11.4 Three Soil Layers**

|     | Soil 1 | Soil 2 | Soil 3 |        |
| --- | ------ | ------ | ------ | ------ |
| $s_u$ | 30 | 42 | 58 | kPa |
| $\phi'$ | 33° | 29° | 25° | |
| $\gamma_w$ | 9.8 | | | kN/m³ |
| $\gamma_{sat}$ | 18 | 17.5 | 17 | kN/m³ |
| FS | 1.01 | assumed | | |

|       |        |         |         |         |            |         |       |       |       | | ESA | TSA |
| ----- | ------ | ------- | ------- | ------- | ---------- | ------- | ----- | ----- | ------- | --- | --- | --- |
| Slice | $b$ (m) | $z_1$ (m) | $z_2$ (m) | $z_3$ (m) | $W = \gamma bz$ (kN) | $z_w$ (m) | $r_u$ | $\theta$ (deg) | $m_j$ | $W \sin \theta$ | $W(1 - r_u)(\tan \phi')m_j$ | $s_u b/\cos \theta$ |
| 1 | 4.9 | 1   | 0   | 0   | 88.2  | 1   | 0.54 | −23  | 1.49 | −34.5 | 39.0 | 159.7 |
| 2 | 2.5 | 2.3 | 1.3 | 0   | 160.4 | 3.6 | 0.55 | −10  | 1.15 | −27.8 | 53.7 | 76.2 |
| 3 | 2   | 2.4 | 2.2 | 0   | 163.4 | 4.6 | 0.55 | 0    | 1.00 | 0.0   | 47.6 | 60.0 |
| 4 | 2   | 2   | 3.6 | 0   | 198.0 | 5   | 0.49 | 9    | 0.92 | 31.0  | 59.7 | 60.7 |
| 5 | 2   | 0.9 | 4.1 | 1.5 | 226.9 | 5.5 | 0.48 | 17   | 0.87 | 66.3  | 67.6 | 62.7 |
| 6 | 2   | 0.8 | 4.1 | 2   | 240.3 | 5.3 | 0.43 | 29   | 0.84 | 116.5 | 74.7 | 68.6 |
| 7 | 2   | 0   | 3.7 | 3.1 | 234.9 | 4.5 | 0.38 | 39.5 | 0.89 | 149.4 | 72.6 | 108.9 |
| 8 | 2.5 | 0   | 1.5 | 3.8 | 227.1 | 2.9 | 0.31 | 49.5 | 0.94 | 172.7 | 81.1 | 161.7 |
| 9 | 1.6 | 0   | 0   | 1.6 | 43.5  | 0.1 | 0.04 | 65   | 1.19 | 39.4  | 23.3 | 219.6 |
|   |     |     |     |     |       |     |      | Sum  |      | 513.1 | 519.1 | 978.1 |
|   |     |     |     |     |       |     |      | FS   |      |       | **1.01** | **1.91** |

**Step 3:** Set up a spreadsheet.
See Table E11.4.

**Step 4:** Extract the required values and perform the calculations. Let $z_1$, $z_2$, and $z_3$ be the heights of soil 1, 2, and 3, respectively, in each slice. Slice ①, for example, only contains soil 1 while slice ⑥ contains each soil. Use the appropriate value of $\phi'_{cs}$ and $s_u$ for each of the slices. For example, $\phi'_{cs} = 33°$ and $s_u = 30$ kPa for soil 1 are applicable to slices ① through ⑥, while $\phi'_{cs} = 24°$ and $s_u = 58$ kPa are applicable to slice ⑨. ∎

**④** **EXAMPLE 11.5**

A coarse-grained fill was placed on saturated clay. A noncircular slip surface was assumed, as shown in Fig. E11.5a. Determine the factor of safety of the slope using an ESA. The groundwater level is below the assumed slip surface.

**Strategy** Since a noncircular slip surface is assumed, you should use Janbu's method. Groundwater is below the slip surface; that is, $r_u = 0$.

**Solution 11.5**

**Step 1:** Redraw the figure to scale.
See Fig. E11.5b.

**Step 2:** Divide the sliding mass into a number of slices.
In this case, four slices are sufficient.

**Step 3:** Extract the required parameters.

**Step 4:** Carry out the calculations.
Use a spreadsheet as shown in Table E11.5.

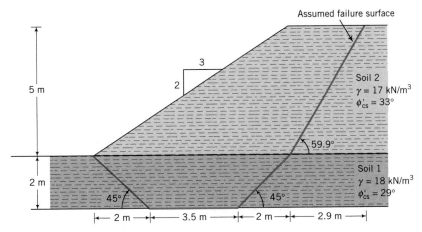

FIGURE E11.5a

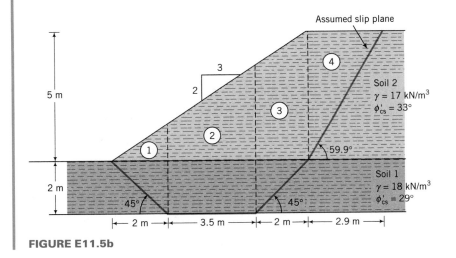

FIGURE E11.5b

## TABLE E11.5 Janbu's Method

|  | Soil 1 | Soil 2 |  |
|---|---|---|---|
| $\phi'$ | 29° | 33.5° |  |
| $\gamma_w$ | 9.8 |  | kN/m³ |
| $\gamma_{sat}$ | 18 | 17 | kN/m³ |
| $d$ | 4.5 |  | m |
| $l$ | 11.5 |  |  |
| $d/l$ | 0.39 | $f_o$ | 1.06 |
| FS | 1.04 | assumed |  |

|  |  |  |  |  |  |  |  | ESA |
|---|---|---|---|---|---|---|---|---|
| Slice | $b$ (m) | $z_1$ (m) | $z_2$ (m) | $W = \gamma b z$ (kN) | $\theta$ (deg) | $m_j$ | $W \tan \theta$ | $W \tan \phi' (\cos \theta) m_j$ |
| 1 | 2 | 1 | 0.7 | 59.8 | −45 | 3.03 | −59.8 | 71.0 |
| 2 | 3.5 | 2 | 2.5 | 274.8 | 0 | 1.00 | 0.0 | 152.3 |
| 3 | 2 | 1 | 4.3 | 182.2 | 45 | 0.92 | 182.2 | 65.9 |
| 4 | 2.9 | 0 | 2.5 | 123.3 | 59.9 | 0.95 | 212.6 | 38.9 |
|  |  |  |  |  |  | Sum | 335.0 | 328.0 |
|  |  |  |  |  |  | FS |  | 1.04 |

*What's next . . .*Charts can be prepared to allow you to quickly estimate the stability of slopes with simple geometry in homogeneous soils. In the next section, we present some of the popular charts.

# 11.11 STABILITY OF SLOPES WITH SIMPLE GEOMETRY

### 11.11.1 Taylor's Method

Let us reconsider the stability of a slope using a TSA as expressed by Eq. (11.34). We can rewrite Eq. (11.34) as

$$FS = N_o \frac{\Sigma(s_u)_j}{\Sigma(\gamma z)_j} \tag{11.43}$$

where $N_o$ is called stability number and depends mainly on the geometry of the slope. Taylor (1948) used Eq. (11.43) to prepare a chart to determine the stability of slopes in a homogeneous deposit of soil underlain by a much stiffer soil or rock. He assumed no tension crack, failure occurring by rotation, no surcharge or external loading, and no open water outside the slope.

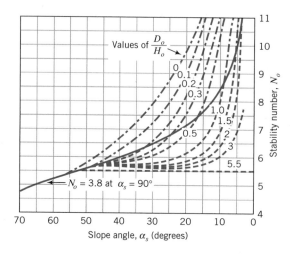

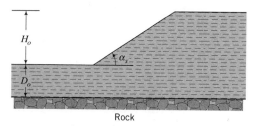

**FIGURE 11.12** Taylor's curves for determining the stability of simple slopes.

The procedure to use Taylor's chart to determine the safe slope in a homogeneous deposit of soil using a TSA, with reference to Fig. 11.12, is as follows:

1. Calculate $n_d = D_o/H_o$, where $D_o$ is the depth from the toe to the top of the stiff layer and $H_o$ is the height of the slope.
2. Calculate $N_o = FS(\gamma H_o/s_u)$.
3. Read the value of $\alpha_s$ at the intersection of $n_d$ and $N_o$.

If you wish to check the factor of safety of an existing slope or a desired slope, the procedure is as follows:

1. Calculate $n_d = D_o/H_o$.
2. Read the value of $N_o$ at the intersection of $\alpha_s$ and $n_d$.
3. Calculate $FS = N_o s_u/\gamma H_o$.

### 11.11.2 Bishop–Morgenstern Method

Bishop and Morgenstern (1960) prepared a number of charts for homogeneous soil slopes with simple geometry using Bishop's simplified method. Equation (11.30) was written as

$$FS = m - nr_u \tag{11.44}$$

where $m$ and $n$ are stability coefficients (Fig. 11.13) that depend on the friction angle and the geometry of the slope.

The procedure to use the Bishop–Morgenstern method is as follows:

1. Assume a circular slip surface.
2. Draw the phreatic surface (Chapter 9).
3. Calculate $r_u = \gamma_w(z_w)_j/\gamma_j z_j$ (see Fig. 11.10). Use a weighted average value of $r_u$ within the sliding mass. A practical range of values of $r_u$ is $\frac{1}{3}$ to $\frac{1}{2}$.
4. With $\phi' = \phi'_{cs}$ and the assumed slope angle, determine the values of $m$ and $n$ from Fig. 11.13.
5. Calculate FS using Eq. (11.44).

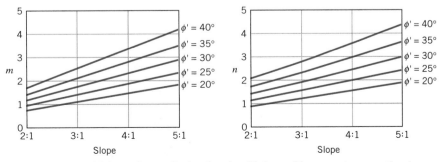

**FIGURE 11.13**  Values for $m$ and $n$ for the Bishop–Morgenstern method.

## EXAMPLE 11.6

Determine the factor of safety of the slope shown in Fig. E11.6.

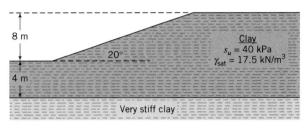

**FIGURE E11.6**

***Strategy***   Follow the procedures in Section 11.11.

### Solution 11.6
**Step 1:**   Calculate $n_d$.
$H_o = 8$ m, $D_o = 4$ m.

$$n_d = \frac{D_o}{H_o} = \frac{4}{8} = 0.5$$

**Step 2:**   Determine $N_o$.
From Fig. 11.12, for $\alpha_s = 20°$ and $n_d = 0.5$, we get
$N_o = 6.8$.

**Step 3:**   Calculate FS.

$$\text{FS} = \frac{N_o s_u}{\gamma H_o} = \frac{6.8 \times 40}{17.5 \times 8} = 1.94$$

■

## EXAMPLE 11.7

Determine the factor of safety of the slope shown in Fig. E11.7.

***Strategy***   Use the Bishop and Morgenstern (1960) charts and equations. Follow the procedures in Section 11.11. Since you are given $r_u$, you only need to do Steps 4 and 5.

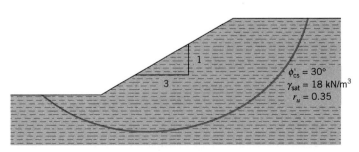

**FIGURE E11.7**

**Solution 11.7**

**Step 1:**   Determine $m$ and $n$.
From Fig. 11.13, $m = 1.73$ and $n = 1.92$ for a slope of $3:1$ and $\phi'_{cs} = 30°$.

**Step 2:**   Calculate FS.

$$\text{FS} = m - nr_u = 1.73 - 1.92 \times 0.35 = 1.06$$   ■

## 11.12   SUMMARY

In this chapter, we examined the stability of simple slopes. Slope failures are often catastrophic and may incur extensive destruction and deaths. Slopes usually fail from natural causes (erosion, seepage, and earthquakes) and by construction activities (excavation, change of land surface, etc.). The analyses we considered were based on limit equilibrium, which requires simplifying assumptions. Careful judgment and experience are needed to evaluate slope stability. The geology of a site is of particular importance in determining slope stability. You should consider both an effective stress analysis and a total stress analysis for slopes in fine-grained soils and an effective stress analysis for slopes in coarse-grained soils. The main sources of errors in slope stability analysis are the shear strength parameters, especially $s_u$, and the determination of the pore water pressures.

## EXERCISES

### Theory

**11.1**   A slope fails as shown in Fig. P11.1. Derive an equation for the undrained shear strength of the soil.

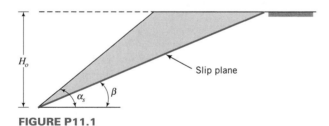

**FIGURE P11.1**

**11.2**   Derive an equation for the factor of safety of the slope in Fig. P11.2 using the mechanism shown.

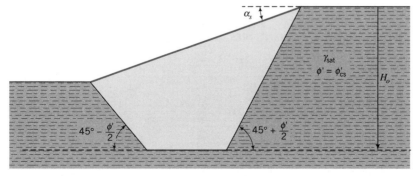

**FIGURE P11.2**

**11.3** Figure P11.3 shows the profile of a beach on a lake. It is proposed to drawdown the lake by 2 m. Determine the slope angle of the beach below the high water level after the drawdown. You may assume an infinite slope failure mechanism.

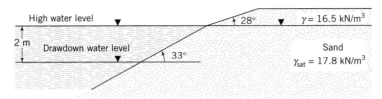

**FIGURE P11.3**

## Problem Solving

**11.4** A cut for a highway is shown in Fig. P11.4. Determine the factor of safety of the slope using an ESA and a TSA. Assume a center of rotation, $O$, such that the slip surface passes through the toe of the slope.

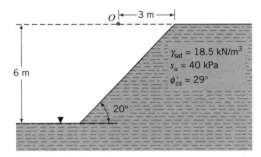

**FIGURE P11.4**

**11.5** Determine the factor of safety of the slope shown in Fig. P11.5 using an ESA and a TSA. The point of rotation is indicated by $O$ and the line representing the top of the stiff soil is a tangent to the slip plane.

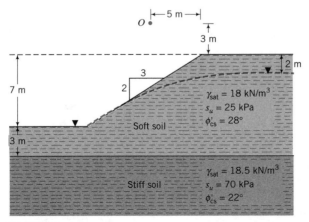

**FIGURE P11.5**

**11.6** A compacted earth fill is constructed on a soft, saturated clay (Fig. P11.6). The fill was compacted to an average dry unit of 19 kN/m³ and water content of 15%. The shearing strength of the fill was determined by CU tests on samples compacted to representative field conditions. The shear strength parameters are $s_u = 35$ kPa, $\phi'_p = 34°$, and $\phi'_{cs} = 28°$. The variation of undrained shear strength of the soft clay with depth as determined by simple shear tests is shown in Fig. P11.6 and the friction angle at the critical state is $\phi'_{cs} = 30°$. The average water content of the soft clay is 55%. Compute the factor of safety using Bishop's simplified method. Assume that a tension crack will develop in the fill.

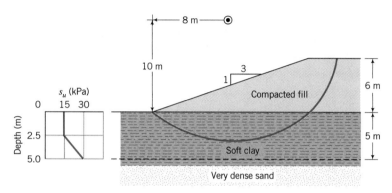

**FIGURE P11.6**

**11.7** A cross section of a canal is shown in Fig. P11.7. Determine the factor of safety for (a) the existing condition and (b) a rapid drawdown of the water level in the canal. Use Bishop's method. The center of rotation of the sliding mass is at coordinates $x = 113$ m and $y = 133$ m. The rock surface is tangent to the slip plane. The properties of the soil are as follows:

| Soil | Description | $\gamma_{sat}$ (kN/m³) | $s_u$ (kPa) | $\phi'_{cs}$ |
|------|-------------|------------------------|-------------|--------------|
| A | Clay | 17.8 | 34 | 30 |
| B | Clay | 18.0 | 21 | 28 |

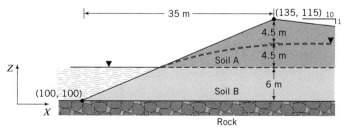

**FIGURE P11.7**

**11.8** Use Janbu's method to determine the factor of safety of the slope shown in Fig. P11.8.

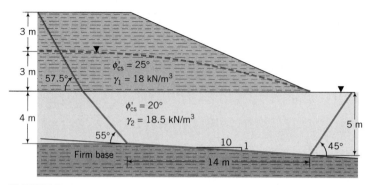

**FIGURE P11.8**

**11.9** Use Taylor's method to determine the factor of safety of the slope shown in Fig. P11.9.

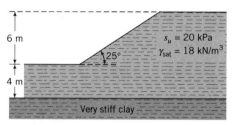

**FIGURE P11.9**

**11.10** Use Taylor's method to determine the slope in Fig. P11.10 for FS = 1.25.

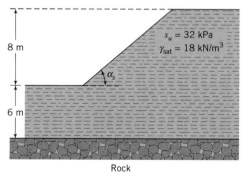

**FIGURE P11.10**

**11.11** Determine the factor of safety of a 2:1 slope with $r_u = 0.25$, $\phi'_{cs} = 27°$ using the Bishop–Morgenstern method.

## Practical

11.12    The soil at a site is shown in Fig. P11.12. A cut was made as shown and one possible slip plane is shown. Determine the factor of safety.

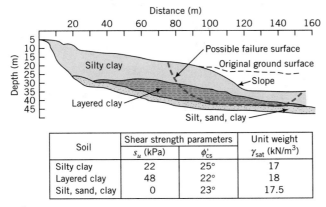

| Soil | Shear strength parameters | | Unit weight |
|---|---|---|---|
| | $s_u$ (kPa) | $\phi'_{cs}$ | $\gamma_{sat}$ (kN/m³) |
| Silty clay | 22 | 25° | 17 |
| Layered clay | 48 | 22° | 18 |
| Silt, sand, clay | 0 | 23° | 17.5 |

**FIGURE P11.12**

# A COLLECTION OF FREQUENTLY USED SOIL PARAMETERS AND CORRELATIONS

**TABLE A.1  Typical Values of Unit Weight for Soils**

| Soil type | $\gamma_{sat}$ (kN/m$^3$) | $\gamma_d$ (kN/m$^3$) |
| --- | --- | --- |
| Gravel | 20–22 | 15–17 |
| Sand | 18–20 | 13–16 |
| Silt | 18–20 | 14–18 |
| Clay | 16–22 | 14–21 |

**TABLE A.2  Description Based on Relative Density**

| $D_r$ (%) | Description |
| --- | --- |
| 0–15 | Very loose |
| 15–35 | Loose |
| 35–65 | Medium dense |
| 65–85 | Dense |
| 85–100 | Very dense |

**TABLE A.3  Soil Types, Description, and Average Grain Size According to USCS**

| Soil type | Description | Average grain size |
| --- | --- | --- |
| Gravel | Rounded and/or angular bulky hard rock | Coarse:  75 mm to 19 mm<br>Fine:   19 mm to 4 mm |
| Sand | Rounded and/or angular bulky hard rock | Coarse:  4 mm to 1.7 mm<br>Medium: 1.7 mm to 0.380 mm<br>Fine: 0.380 mm to 0.075 mm |
| Silt | Particles smaller than 0.075 mm exhibit little or no strength when dried | 0.075 mm to 0.002 mm |
| Clay | Particles smaller than 0.002 mm exhibit significant strength when dried; water reduces strength | <0.002 mm |

**TABLE A.4   Description of Soil Strength Based on Liquidity Index**

| Values of $I_L$ | Description of soil strength |
|---|---|
| $I_L < 0$ | Semisolid state—high strength, brittle (sudden) fracture expected |
| $0 < I_L < 1$ | Plastic state—intermediate strength, soil deforms like a plastic material |
| $I_L > 1$ | Liquid state—low strength, soil deforms like a viscous fluid |

**TABLE A.5   Atterberg Limits for Typical Soils**

| Soil type | $w_{LL}$ (%) | $w_{PL}$ (%) | $I_p$ (%) |
|---|---|---|---|
| Sand | | Nonplastic | |
| Silt | 30–40 | 20–25 | 10–15 |
| Clay | 40–150 | 25–50 | 15–100 |

**TABLE A.6   Coefficient of Permeability for Common Soil Types**

| Soil type | $k_z$ (cm/s) |
|---|---|
| Clean gravel | $>1.0$ |
| Clean sands, clean sand and gravel mixtures | $1.0$ to $10^{-3}$ |
| Fine sands, silts, mixtures comprising sands, silts, and clays | $10^{-3}$ to $10^{-7}$ |
| Homogeneous clays | $<10^{-7}$ |

**TABLE A.7   Typical Values of Poisson's Ratio**

| Soil type | Description | $v'$ |
|---|---|---|
| Clay | Soft | 0.35–0.4 |
| | Medium | 0.3–0.35 |
| | Stiff | 0.2–0.3 |
| Sand | Loose | 0.15–0.25 |
| | Medium | 0.25–0.3 |
| | Dense | 0.25–0.35 |

### TABLE A.8 Typical Values of *E* and *G*

| Soil type | Description | $E^a$ (MPa) | $G$ (MPa) |
| --- | --- | --- | --- |
| Clay | Soft | 1–15 | 0.4–5 |
| | Medium | 15–30 | 5–11 |
| | Stiff | 30–100 | 11–38 |
| Sand | Loose | 10–20 | 4–8 |
| | Medium | 20–40 | 8–16 |
| | Dense | 40–80 | 16–32 |

[a]These are secant values at peak deviatoric stress for dense and stiff soils, and when the maximum deviatoric stress is attained for loose, medium, and soft soils.

### TABLE A.9 Ranges of Friction Angles (degrees) for Soils

| Soil type | $\phi'_{cs}$ | $\phi'_p$ | $\phi'_r$ |
| --- | --- | --- | --- |
| Gravel | 30–35 | 30–50 | |
| Mixture of gravel and sand with fine-grained soils | 28–33 | 30–40 | |
| Sand | 27–37[a] | 32–50 | |
| Silt or silty sand | 24–32 | 27–35 | |
| Clays | 15–30 | 20–30 | 5–15 |

[a]Higher values (32° to 37°) in the range are for sands with significant amount of feldspar (Bolton, 1986). Lower values (27° to 32°) in the range are for quartz sands.

### TABLE A.10 $A_f$ Values

| Type of Clay | $A_f$ |
| --- | --- |
| Normally consolidated | $\frac{1}{2}$ to 1 |
| Compacted sandy clay | $\frac{1}{4}$ to $\frac{3}{4}$ |
| Lightly overconsolidated clays | 0 to $\frac{1}{2}$ |
| Compacted clay–gravel | $-\frac{1}{4}$ to $\frac{1}{4}$ |
| Heavily overconsolidated clays | $-\frac{1}{2}$ to 0 |

SOURCE: After Skempton (1954).

### TABLE A.11 Correlation of *N*, $N_{60}$, $\gamma$, $D_r$, and $\phi'$ for Coarse-Grained Soils

| *N* | $N_{60}$ | Description | $\gamma$ (kN/m³) | $D_r$ (%) | $\phi'$ (degrees) |
| --- | --- | --- | --- | --- | --- |
| 0–5 | 0–3 | Very loose | 11–13 | 0–15 | 26–28 |
| 5–10 | 3–9 | Loose | 14–16 | 16–35 | 29–34 |
| 10–30 | 9–25 | Medium | 17–19 | 36–65 | 35–40[a] |
| 30–50 | 25–45 | Dense | 20–21 | 66–85 | 38–45[a] |
| >50 | >45 | Very dense | >21 | >86 | >45[a] |

[a]These values correspond to $\phi'_p$.

**TABLE A.12   Correlation of $N_{60}$ and $s_u$ for Saturated Fine-Grained Soils**

| $N_{60}$ | Description | $s_u$ (kPa) |
|---|---|---|
| 0–2 | Very soft | $<10$ |
| 3–5 | Soft | 10–25 |
| 6–9 | Medium | 25–50 |
| 10–15 | Stiff | 50–100 |
| 15–30 | Very stiff | 100–200 |
| $>30$ | Extremely stiff | $>200$ |

**TABLE A.13   Empirical Soil Strength Relationships**

| Soil type | Equation | Reference |
|---|---|---|
| Normally consolidated clays | $\left(\dfrac{s_u}{\sigma_z'}\right)_{nc} = 0.11 + 0.0037 I_p$ | Skempton (1957) |
| Overconsolidated clays | $\dfrac{(s_u/\sigma_z')_{oc}}{(s_u/\sigma_z')_{nc}} = (OCR)^{0.8}$ | Ladd et al. (1977) |
| | $\dfrac{s_u}{\sigma_z'} = (0.23 \pm 0.04)OCR^{0.8}$ | Jamiolkowski et al. (1985) |
| All clays | $\dfrac{s_u}{\sigma_{zc}'} = 0.22$ | Mesri (1975) |
| Clean quartz sand | $\phi_p' = \phi_{cs}' + 3D_r(10 - \ln p_f') - 3$, where $p_f'$ is the mean effective stress at failure (in kPa) and $D_r$ is relative density. This equation should only be used if $12 > (\phi_p' - \phi_{cs}') > 0$. | Bolton (1986) |

# DISTRIBUTION OF SURFACE STRESSES WITHIN FINITE SOIL LAYERS

## B1  VERTICAL STRESSES IN A FINITE SOIL LAYER DUE TO A UNIFORM SURFACE LOAD ON A CIRCULAR AREA AND A RECTANGULAR AREA

### B1.1 Circular Area (Milovic, 1970)

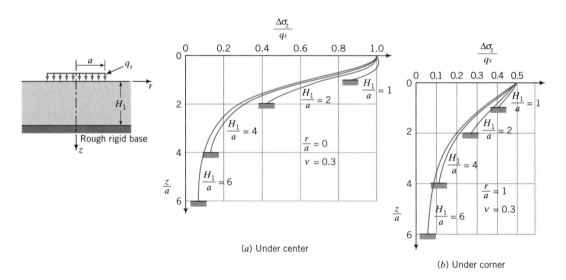

(a) Under center

(b) Under corner

## B1.2 Rectangular Area with Rough Rigid Base (Milovic and Tournier, 1971)

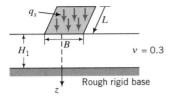

$v = 0.3$

Rough rigid base

| | | L/B | | | L/B | | |
|---|---|---|---|---|---|---|---|
| | | 1 | 2 | 5 | 1 | 2 | 5 |
| | | Center | | | Corner | | |
| $H_1/B$ | $z/B$ | $\Delta\sigma_z/q_s$ | | | $\Delta\sigma_z/q_s$ | | |
| 1 | 0.000 | 1.000 | 1.000 | 1.000 | 0.250 | 0.250 | 0.250 |
| | 0.100 | 0.974 | 0.992 | 0.996 | 0.250 | 0.250 | 0.250 |
| | 0.200 | 0.943 | 0.977 | 0.981 | 0.250 | 0.250 | 0.250 |
| | 0.400 | 0.842 | 0.924 | 0.922 | 0.250 | 0.250 | 0.250 |
| | 0.600 | 0.690 | 0.827 | 0.832 | 0.250 | 0.250 | 0.250 |
| | 0.800 | 0.570 | 0.734 | 0.751 | 0.238 | 0.244 | 0.244 |
| | 1.000 | 0.468 | 0.638 | 0.672 | 0.220 | 0.232 | 0.233 |
| 2 | 0.000 | 1.000 | 1.000 | 1.000 | 0.250 | 0.250 | 0.250 |
| | 0.100 | 0.970 | 0.985 | 0.990 | 0.250 | 0.250 | 0.250 |
| | 0.200 | 0.930 | 0.963 | 0.971 | 0.250 | 0.250 | 0.250 |
| | 0.400 | 0.802 | 0.878 | 0.890 | 0.244 | 0.249 | 0.247 |
| | 0.800 | 0.464 | 0.619 | 0.670 | 0.211 | 0.231 | 0.230 |
| | 1.200 | 0.286 | 0.441 | 0.528 | 0.172 | 0.207 | 0.208 |
| | 1.600 | 0.204 | 0.340 | 0.443 | 0.142 | 0.183 | 0.188 |
| | 2.000 | 0.155 | 0.269 | 0.377 | 0.117 | 0.160 | 0.168 |
| 3 | 0.000 | 1.000 | 1.000 | 1.000 | 0.250 | 0.250 | 0.250 |
| | 0.100 | 0.970 | 0.982 | 0.990 | 0.250 | 0.250 | 0.250 |
| | 0.200 | 0.930 | 0.962 | 0.969 | 0.249 | 0.250 | 0.249 |
| | 0.400 | 0.799 | 0.872 | 0.884 | 0.241 | 0.246 | 0.246 |
| | 0.800 | 0.453 | 0.599 | 0.650 | 0.203 | 0.222 | 0.225 |
| | 1.200 | 0.264 | 0.405 | 0.492 | 0.158 | 0.191 | 0.197 |
| | 1.600 | 0.172 | 0.289 | 0.395 | 0.122 | 0.163 | 0.174 |
| | 2.000 | 0.124 | 0.220 | 0.333 | 0.098 | 0.141 | 0.156 |
| | 2.500 | 0.093 | 0.171 | 0.281 | 0.078 | 0.120 | 0.138 |
| | 3.000 | 0.073 | 0.137 | 0.238 | 0.064 | 0.102 | 0.123 |
| 5 | 0.000 | 1.000 | 1.000 | 1.000 | 0.250 | 0.250 | 0.250 |
| | 0.100 | 0.970 | 0.981 | 0.990 | 0.250 | 0.250 | 0.250 |
| | 0.200 | 0.930 | 0.961 | 0.969 | 0.249 | 0.250 | 0.249 |
| | 0.400 | 0.798 | 0.870 | 0.881 | 0.241 | 0.245 | 0.245 |
| | 0.800 | 0.450 | 0.594 | 0.641 | 0.200 | 0.219 | 0.221 |
| | 1.200 | 0.258 | 0.394 | 0.475 | 0.153 | 0.184 | 0.191 |
| | 1.600 | 0.162 | 0.271 | 0.368 | 0.114 | 0.151 | 0.164 |
| | 2.000 | 0.111 | 0.195 | 0.296 | 0.087 | 0.125 | 0.143 |
| | 2.500 | 0.075 | 0.139 | 0.235 | 0.064 | 0.100 | 0.123 |
| | 3.000 | 0.056 | 0.105 | 0.193 | 0.050 | 0.082 | 0.108 |
| | 3.500 | 0.044 | 0.085 | 0.165 | 0.040 | 0.069 | 0.097 |
| | 4.000 | 0.037 | 0.071 | 0.144 | 0.034 | 0.060 | 0.089 |
| | 4.500 | 0.032 | 0.062 | 0.128 | 0.030 | 0.053 | 0.082 |
| | 5.000 | 0.027 | 0.053 | 0.113 | 0.026 | 0.047 | 0.075 |

## B1.3 Rectangular Area with Smooth Rigid Base (Sovinc, 1961)

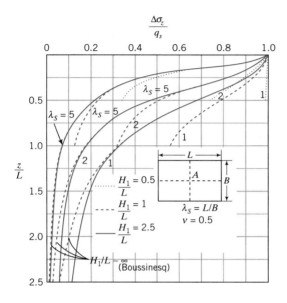

# B2  VERTICAL STRESSES IN A TWO-LAYER SOIL UNDER THE CENTER OF A UNIFORMLY LOADED CIRCULAR AREA (Fox, 1948)

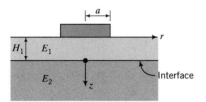

| a/H₁ | z | $E_1/E_2$ | | | | $E_1/E_2$ | | | |
|---|---|---|---|---|---|---|---|---|---|
| | | 1 | 10 | 100 | 1000 | 1 | 10 | 100 | 1000 |
| | | Rough interface | | | | Smooth | | | |
| | | $\Delta\sigma_z/q_s$ | | | | $\Delta\sigma_z/q_s$ | | | |
| | 0 | 0.284 | 0.101 | 0.0238 | 0.0051 | 0.31 | 0.105 | 0.0241 | 0.0051 |
| | $H_1$ | 0.087 | 0.047 | 0.0158 | 0.0042 | 0.141 | 0.063 | 0.0183 | 0.0045 |
| 1/2 | $2H_1$ | 0.0403 | 0.0278 | 0.0117 | 0.0035 | 0.064 | 0.0367 | 0.0136 | 0.0038 |
| | $3H_1$ | 0.023 | 0.0184 | 0.0091 | 0.0031 | 0.0346 | 0.0235 | 0.0105 | 0.0033 |
| | $4H_1$ | 0.0148 | 0.0129 | 0.0074 | 0.0028 | 0.0212 | 0.0161 | 0.0083 | 0.0029 |
| | 0 | 0.646 | 0.292 | 0.081 | 0.0185 | 0.722 | 0.305 | 0.082 | 0.019 |
| | $H_1$ | 0.284 | 0.168 | 0.06 | 0.0162 | 0.437 | 0.217 | 0.068 | 0.0172 |
| 1 | $2H_1$ | 0.145 | 0.105 | 0.046 | 0.0143 | 0.225 | 0.136 | 0.0525 | 0.0151 |
| | $3H_1$ | 0.087 | 0.07 | 0.036 | 0.0124 | 0.128 | 0.089 | 0.0409 | 0.0133 |
| | $4H_1$ | 0.057 | 0.05 | 0.029 | 0.011 | 0.081 | 0.062 | 0.0326 | 0.0117 |
| | 0 | 0.911 | 0.644 | 0.246 | 0.071 | 1.025 | 0.677 | 0.249 | 0.067 |
| | $H_1$ | 0.646 | 0.48 | 0.205 | 0.0606 | 0.869 | 0.576 | 0.225 | 0.063 |
| 2 | $2H_1$ | 0.424 | 0.34 | 0.165 | 0.0542 | 0.596 | 0.421 | 0.186 | 0.057 |
| | $3H_1$ | 0.284 | 0.244 | 0.133 | 0.048 | 0.396 | 0.302 | 0.15 | 0.051 |
| | $4H_1$ | 0.2 | 0.181 | 0.108 | 0.0428 | 0.271 | 0.22 | 0.122 | 0.0454 |

# LATERAL EARTH PRESSURE COEFFICIENTS (KERISEL AND ABSI, 1970)

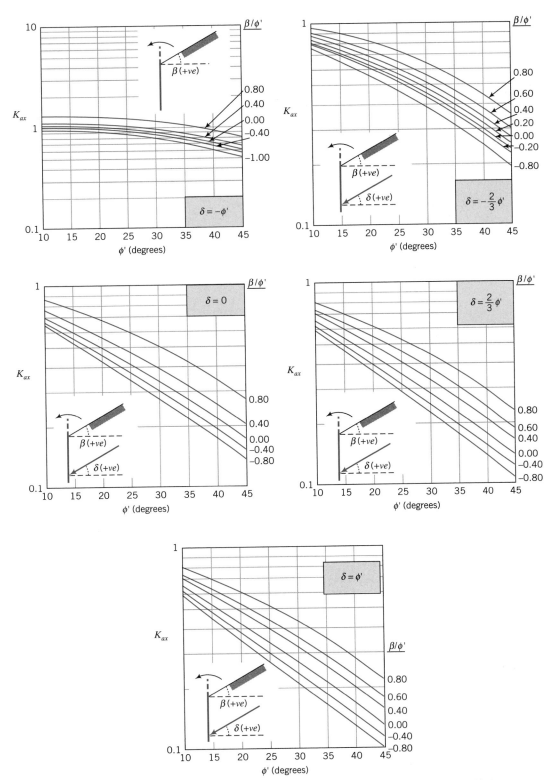

**FIGURE C.1**   Horizontal component of the active lateral pressure coefficient. (Plotted from data published by Kerisel and Absi, 1990.)

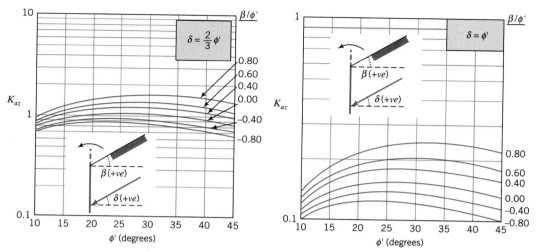

**FIGURE C.2** Vertical component of the active lateral pressure coefficient. (Plotted from data published by Kerisel and Absi, 1990.)

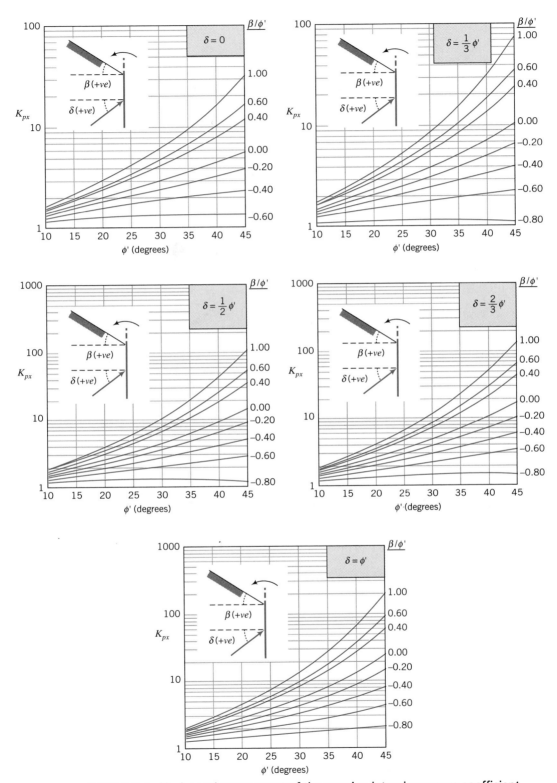

**FIGURE C.3** Horizontal component of the passive lateral pressure coefficient. (Plotted from data published by Kerisel and Absi, 1990.)

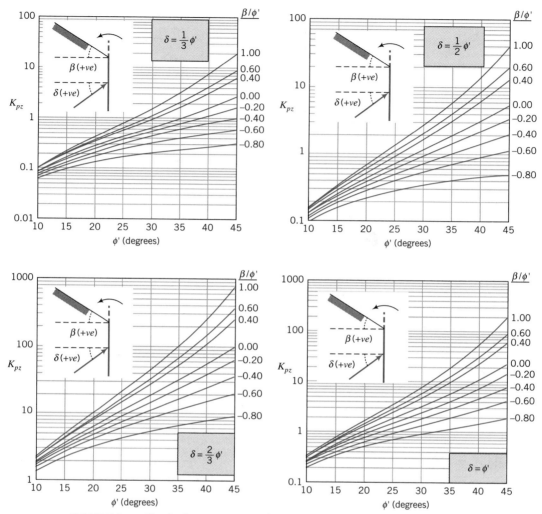

**FIGURE C.4** Vertical component of the passive lateral pressure coefficient. (Plotted from data published by Kerisel and Absi, 1990.)

# ANSWERS to selected problems

## Chapter 2

**2.6** (a) 28%    (b) 0.76    (c) 19.2 kN/m$^3$    (d) 9.4 kN/m$^3$

**2.8** (a) 38%    (b) 20.1 kN/m$^3$

**2.10** $D_r = 39\%$, $S = 1$

**2.12** SP

**2.14** (a) 40%    (b) 0.88    (c) Since $0 < I_L < 1$, then soil deforms like a plastic material.

**2.16** A: $h_p = 0.71$ m, $h_z = 0.25$ m, $H = 0.96$ m
B: $h_p = 1.02$ m, $h_z = 0.9$ m, $H = 1.92$ m
C: $h_p = 0$, $h_z = 0$, $H = 0$

**2.18** $k = 3 \times 10^{-2}$ cm/sec, $v = 0.045$ cm/sec, $v_s = 0.082$ cm/sec, $k = 0.08$ cm/sec

**2.20** $2.0 \times 10^{-6}$ cm/sec

**2.22** (a) 16%, 17.7 kN/m$^3$    (c) 86%

## Chapter 3

**3.6** $\epsilon_1 = 22.4 \times 10^{-4}$    $\epsilon_3 = -11.6 \times 10^{-4}$

**3.8** (a) 880 kPa, 314 kPa    (b) 23°    (c) 283 kPa    (d) 626 kPa, 280 kPa

**3.10**

| Depth (m) | $\sigma_z$ (kPa) | $u$ (kPa) | $\sigma'_z$ (kPa) |
|---|---|---|---|
| 0 | 0 | 0 | 0 |
| 4.5 | 85.5 | 0 | 85.5 |
| 9.5 | 182 | 49 | 133.0 |

**3.12** 19.6 kPa

**3.14**

| Position | $\sigma'_z$ (kPa) | $u$ (kPa) | $\sigma'_x$ (kPa) | $\sigma_x$ (kPa) |
|---|---|---|---|---|
| A | 17.3 | 39.2 | 8.7 | 47.9 |
| B | 28.1 | 53.9 | 14.1 | 68.0 |
| C | 42.4 | 73.6 | 21.2 | 94.8 |

**3.16** 134 kPa

**3.18** TSP: $\dfrac{\Delta q}{\Delta p} = -\dfrac{3}{2}$    ESP: $\dfrac{\Delta q}{\Delta p^1} \to \infty$    $\Delta u = 133.3$ kPa

## Chapter 4

**4.8** 30 mm, 57 mm

**4.10** 135 mm

**4.12** (b) 0.207, 0.017 (c) $4.1 \times 10^{-4}$ m²/kN

**4.14** (a) 0.61 (b) 0.665, 0.638, 0.611, 0.596, 0.575, 0.61
(c) 31.02 mm (d) $1.2 \times 10^{-4}$ m²/kN

**4.16** $t = 0.09$ yr., $\rho_{pc} = 6$mm, $t = 4.52$ yr, $\rho_{pc} = 42$ mm
$t = 10$ yr., $\rho_{pc} = 54.5$ mm

## Chapter 5

**5.4** Soil will not fail. FS = 1.55

**5.6** $\phi'_{cs} = 22°$ $\phi'_p = 45.6°$

**5.8** $\varepsilon_1 = 0.02$, $\varepsilon_p = 0.004$, $\phi = 28.3°$, $\theta = 59.1°$

**5.10** (c) Test 1: $\phi' = 25.3°$ Test 2: $\phi' = 20.4°$ (d) $s_u = 182.5$ kPa
(e) 144.7 kPa, 171 kPa

**5.12** (b) 22.7°, 22.9°, 22.5° (c) 132.9 kPa using $\phi' = 22.7°$ (d) 56.4°

**5.14** (a) 22.6°, 23.1°, 22.9° (b) 103.8 kPa
(c) Maximum shear stress: 43.2 kPa, 103.8 kPa, 172.5 kPa
Failure shear stress: 39.9 kPa, 95.4kPa, 158.9 kPa

**5.16** 31.1°

**5.18** (a) $\sigma'_1 = 311.8$ kPa, $\sigma'_3 = 88.2$ kPa, $\sigma'_1$ is inclined at 58.3° to the
horizontal plane.
(b) 34° (c) 111.8 kPa

## Chapter 6

**6.12** 151.3 kPa

# REFERENCES

Adamson, A. W. (1982). *Physical Chemistry of Surfaces.* Wiley, New York.

API (1984). *API Recommended Practice for Planning, Designing and Constructing Fixed Offshore Platforms.* American Petroleum Institute, Washington, D.C.

Atterberg, A. (1911). "Uber die Physikalishe Bodennuntersuchung und uber die Plastizitat der Tone." *Int. Mitt. Boden*, **1**, pp. 10–43.

Azzoz, A. S., Krizek, R. J., and Corotis, R. B. (1976). "Regression analysis of soil compressibility." *Soils Found.*, **16**(2), 19–29.

Bell, J. M. (1966). "Dimensionless parameters for homogeneous earth slopes." *J. Soil Mech. Found. Eng. Div. ASCE*, **95**(SM6), 1253–1270.

Berezartzev, V. G., Khristoforov, V. S., and Golubkov, V. N. (eds.) (1961). "Load bearing capacity and deformation of piled foundations." *Proceedings of the 5th International Conference on Soil Mechanics and Foundation Engineering*, Paris, Vol. 2, pp. 11–15.

Bishop, A. W. (1955). "The use of the slip circle in the stability analysis of slopes." *Geotechnique*, **5**(1), 7–17.

Bishop, A. W., Alan, I., Blight, G. E., and Donald, I. B. (eds.) (1960). "Factors controlling the strength of partially saturated cohesive soils." *Proceedings of the Research Conference on Shear Strength of Cohesive Soils*, American Society of Civil Engineers, VA, pp. 500–532.

Bishop, A. W., and Morgenstern, N. R. (1960). "Stability coefficients for earth slopes." *Geotechnique*, **10**(4), 129–150.

Bjerrum, L. (1954). "Geotechnical properties of Norwegian marine clays." *Geotechnique*, **4**(2), 49–69.

Bjerrum, L., and Eggestad, A. (eds.) (1963). "Interpretation of loading tests on sand." *Proceedings of the European Conference on Soil Mechanics and Foundation Engineering*, Wiesbaden, Vol. 1, pp. 199–203.

Bolton, M. D. (1986). "The strength and dilatancy of sands." *Geotechnique*, **36**(1), 65–78.

Boussinesq, J. (1885). *Application des Potentiels a l'Etude de le l'Equilibre et du Mouvement des Solides Elastiques.* Gauthier-Villars, Paris.

Bozozuk, M. (1976). "Tower Silo Foundation." CBd-177, Canadian Building Digest, Ottawa.

Budhu, M. (1979). *Simple Shear Deformation of Sands.* Ph.D. Thesis, Cambridge University, Cambridge, England.

Burland, J. B. (1973). "Shaft friction piles in clay—a simple fundamental approach." *Ground Eng.*, **6**(3), 30–42.

Burland, J. B., and Burbrige, M. C. (eds.) (1985). "Settlement of foundations on sand and gravel." *Proc. Inst. Civ. Eng. Part 1*, **78**, 1325–1381.

Caquot, A., and Kerisel, J. (1948). *Tables for the Calculation of Passive Pressure, Active Pressure and Bearing Capacity of Foundations.* Gauthier-Villars, Paris.

Casagrande, A. (1932). "Research on the Atterberg limits of soils." *Public Roads*, **13**(8), 121–136.

Casagrande, A. (ed.) (1936). "The determination of the preconsolidation load and its practical significance." *1st International Conference on Soil Mechanics and Foundation Engineering, Cambridge, Massachusetts*, Vol. 3, pp. 60–64.

Casagrande, A. (1937). "Seepage through dams." *J. N. Engl. Water Works Assoc.*, **L1**(2), 131–172.

Casagrande, A., and Fadum, R. E. (1940). "Notes on soil testing for engineering purposes." *Soil Mechanics Series*, Graduate School of Engineering, Harvard University, Cambridge, MA, Vol. 8(268), p. 37.

Chang, M. F. (ed.) (1988). "Some experience with the dilatometer test in Singapore." *Penetration Testing, Orlando*, **1**, 489–496.

Chu, S. (1991). "Rankine analysis of active and passive pressures in dry sands." *Soils Found.*, **31**(4), 115–120.

Coulomb, C. A. "Essai sur une application des regles de maximia et minimis a quelques problemes de statique relatifs a l'architecture." *Memoires de la Mathematique et de Physique, presentes a l'Academic Royale des Sciences, par divers savants, et lus dans ces Assemblees.* L'Imprimerie Royale, Paris, pp. 3–8.

Darcy, H. (1856). *Les Fontaines Publiques de la Ville de Dijon.* Dalmont, Paris.

**573**

Datta, M., Gulhati, S. K., and Rao, G. V. (eds.) (1980). "An appraisal of the existing practice of determining the axial load capacity of deep penetration piles in calcareous sands." *Proceedings of the 12th Annual OTC, Houston, Texas*, Vol. 4, pp. 119–130.

DeBeer, E. E. (1970). "Experimental determination of the shape factors and the bearing capacity factors of sand." *Geotechnique*, **20**(4), 347–411.

Decourt, L. (ed.) (1982). "Prediction of the bearing capacity of piles based exclusively on N values of SPT." *Proceedings ESOPT2, Amsterdam*, Vol. 1, pp. 29–45.

Drnevich, V. P. (1967). *Effect of Strain History on the Dynamic Properties of Sand*. Ph.D. Thesis, University of Michigan.

Dumbleton, M. J., and West, G. (1970). "The suction and strength of remoulded soils as affected by composition." LR306, Road Research Laboratory, Crowthorne.

Dutt, R. N., and Ingra, W. B. (eds.) (1984). "Jackup rig sitting in calcareous soils." *Proceedings of the 16th Annual OTC, Houston, Texas*, pp. 541–548.

Eden, S. M. (1974). *Influence of Shape and Embedment on Dynamic Foundation Response*. Ph.D. Thesis, University of Massachusetts.

Fellenius, W. (1927). *Erdstatische berechnungen*. W. Ernst und Sohn, Berlin.

Findlay, J. D. (1984). "Discussion: piling and ground treatment." *Piling and Ground Treatment*. Institution of Civil Engineers, Thomas Telford, London, pp. 189–190.

Fleming, W. G. K., Weltman, A. J., Randolp, M. F., and Elson, W. K. (1985). *Piling Engineering*. Halsted Press, New York.

Fletcher, M. S., and Mizon, D. H. (1984). "Piles in chalk for Orwell bridge." *Piling and Ground Treatment*. Institution of Civil Engineers, Thomas Telford, London, pp. 203–209.

Gazetas, G., and Hatzikonstantinou, E. (1988). "Elastic formulae for lateral displacement and rotation of arbitrarily-shaped embedded foundations." *Geotechnique*, **38**(3), 439–444.

Gazetas, G., and Stokoe, K. H. (1991). "Free vibration of embedded foundations: theory versus experiment." *J. Geotech. Eng.*, **117**(9), 1362–1381.

Gazetas, G., Tassoulas, J. L., Dobry, R., and O'Rourke, M. J. (1985). "Elastic settlement of arbitrarily shaped foundations embedded in half-space." *Geotechnique*, **35**(2), 339–346.

Giroud, J. P. (1968). "Settlement of a linearly loaded rectangular area." *J. Soil Mech. Found. Div. ASCE*, **94**(SM4), 813–831.

Graham, J., and Houlsby, G. T. (1983). "Elastic anisotropy of a natural clay." *Geotechnique*, **33**(2), 165–180.

Hansen, B. (ed.) (1961). "The bearing capacity of sand, tested by loading circular plates." *5th International Conference on Soil Mechanics and Foundation Engineering, Paris, France*, Vol. 1, pp. 659–664.

Hansen, J. B. (1970). *A Revised and Extended Formula for Bearing Capacity*, No. 28. Danish Geotechnical Institute, Copenhagen.

Hazen, A. (1930). *Water Supply. American Civil Engineers Handbook*. Wiley, New York, pp. 1444–1518.

Hobbs, N. B. (1977). "Behaviour and design of piles in chalk—an introduction to the discussion of the papers on chalk." *Piles in Weak Rock*. Institution of Civil Engineers, London, pp. 149–175.

Hull, T. S. (ed.) (1987). *The Static Behaviour of Laterally Loaded Piles*. Ph.D. Thesis, University of Sydney, Sydney, Australia.

Ignold, T. S. (1979). "The effects of compaction on retaining walls." *Geotechnique*, **29**(3), 265–283.

Jaky, J. (1944). "The coefficient of earth pressure at rest." *J. Soc. Hungarian Architects Eng.*, **7**, 355–358.

Jaky, J. (ed.) (1948). "Pressure in silos." *2nd International Conference on Soil Mechanics and Foundation Engineering, Rotterdam*, Vol. 1, pp. 103–107.

Jamiolkowski, M., Ladd, C. C., Germaine, J. T., and Lancellotta, R. (eds.) (1985). "New developments in field and laboratory testing of soils." *XIth International Conference on Soil Mechanics and Foundation Engineering, San Francisco*, Vol. 1, pp. 57–154.

Janbu, N. (ed.) (1957). "Earth pressure and bearing capacity by generalized procedure of slices." *Proceedings of the 4th International Conference on Soil Mechanics and Foundation Engineering, London*, Vol. 2, pp. 207–212.

Janbu, N. (ed.) (1973). "Slope stability computations." *Embankment Dam Engineering, Casagrande Memorial Volume*. Wiley, New York, pp. 47–86.

Janbu, N. (ed.) (1976). "Static bearing capacity of friction piles." *Proceedings of the 6th European Conference on Soil Mechanics and Foundation Engineering*, Vol. 1.2, pp. 479–488.

Jarquio, R. (1981). "Total lateral surcharge pressure due to a strip load." *J. Geotech. Eng. Div. ASCE*, **107**(10), 1424–1428.

Juran, I., and Schlosser, F. (eds.) (1978). "Theoretical analysis of failure in reinforced earth structures." *Symposium on Earth Reinforcement, ASCE Convention, Pittsburgh*. American Society of Civil Engineers, New York, pp. 528–555.

Kenney, T. C., Lau, D., and Ofoegbu, G. I. (1984). "Permeability of compacted granular materials." *Can. Geotech. J.*, **21**(4), 726–729.

Kerisel, J., and Absi, E. (1990). *Active and Passive Earth Pressure Tables*. Balkema, Rotterdam.

Kraft, L. M., and Lyons, C. G. (1974). "State of the art—ultimate axial capacity of grouted piles." *Proceedings of the 6th Annual OTC, Houston, Texas*, pp. 485–504.

Ladd, C. C., Foot, R., Ishihara, K., Schlosser, F., and Poulos, H. G. (1977). "Stress-deformation and strength characteristics." *Proceedings of the 9th International Conference on Soil Mechanics and Foundation Engineering, Tokyo*, pp. 421–494.

Liao, S. S. C., and Whitman, R. V. (1985). "Overburden correction factors for SPT in sand." *J. Geotech. Eng. Div. ASCE*, **112**(3), 373–377.

Love, A. E. H. (1927). *The Mathematical Theory of Elasticity*. Cambridge University Press, Cambridge.

MacDonald, D. H., and Skempton, A. W. (eds.) (1955). "A survey of comparisons between calculated and observed settlements of structures on clay." *Conference on Correlation of Calculated and Observed Stresses and Displacements*. Institution of Civil Engineers, London, pp. 318–337.

Martin, R. E., Seli, J. J., Powell, G. W., and Bertoulin, M. (1987). "Concrete pile design in Tidewater, Virginia." *J. Geotech. Eng. Div. ASCE*, **113**(6), 568–585.

McClelland, B. (1974). "Design of deep penetration piles for ocean structures." *J. Geotech. Eng. Div. ASCE*, **100**(GT7), 705–747.

Mesri, G. (1975). "Discussion: new design procedure for stability of soft clays." *J. Geotech. Eng. Div. ASCE*, **101**(GT4), 409–412.

Meyerhof, G. G. (1951). "The ultimate bearing capacity of foundations." *Geotechnique*, **2**(4), 301–331.

Meyerhof, G. G. (1953). "The bearing capacity of foundations under eccentric and inclined loads." *3rd International Conference on Soil Mechanics and Foundation Engineering, Zurich, Switzerland*, pp. 440–445.

Meyerhof, G. G. (1956). "Penetration tests and bearing capacity of cohesionless soils." *J. Soil Mech. Found. Eng. Div. ASCE*, **82**(SM1), 1–19.

Meyerhof, G. G. (ed.) (1957). "The ultimate bearing capacity of foundations of slopes." *4th International Conference on Soil Mechanics and Foundation Engineering, London*, Vol. 1, pp. 384–386.

Meyerhof, G. G. (1963). "Some recent research on the bearing capacity of foundations." *Can. Geotech. J.*, **1**(1), 16–26.

Meyerhof, G. G. (1965). "Shallow foundations." *J. Soil Mech. Found. Div. ASCE*, **91**(SM2), 21–31.

Meyerhof, G. G. (1976). "Bearing capacity and settlement of pile foundations." *J. Geotech. Eng. Div. ASCE*, **102**(GT3), 195–228.

Meyerhof, G. G., and Koumoto, T. (1987). "Inclination factors for bearing capacity of shallow footings." *J. Geotech. Eng. Div. ASCE*, **113**(9), 1013–1018.

Milovic, D. M., and Tournier, J. P. (1971). "Stresses and displacements due to a rectangular load on a layer of finite thickness." *Soils Found.*, **II**(1), 1–27.

Mindlin, R. D. (1936). "Force at a point in the interior of a semi-infinite solid." *J. Appl. Phys.*, **7**(5), 195–202.

Mitchell, J. K. (1993). *Fundamentals of Soil Behavior*. Wiley, New York.

Morgenstern, N. R., and Price, V. E. (1965). "The analysis of the stability of general slip surfaces." *Geotechnique*, **15**(1), 79–93.

Nagaraj, T. S., and Srinivasa Murthy, B. R. (1985). "Prediction of the preconsolidation pressure and recompression index of soils." *Geotech. Testing J. ASTM*, **8**(4), 199–202.

Nagaraj, T. S., and Srinivasa Murthy, B. R. (1986). "A critical reappraisal of compression index." *Geotechnique*, **36**(1), 27–32.

Newmark, N. M. (1942). *Influence Charts for Computation of Stresses in Elastic Foundations*, No. 338. University of Illinois Engineering Experiment Station, Urbana.

Packshaw, S. (1969). "Earth pressures and earth resistance." *A Century of Soil Mechanics*. The Institution of Civil Engineers, London, pp. 409–435.

Padfield, C. J., and Mair, R. J. (1984). *Design of Retaining Walls Embedded in Stiff Clay*. CIRIA, London.

Peck, R. B. (ed.) (1969). "Deep excavation and tunneling in soft ground." *7th International Conference on Soil Mechanics and Foundation Engineering, State-of-the-Art, Mexico City*, Vol. 3, pp. 147–150.

Peck, R. B., and Byrant, F. G. (1953). "The bearing capacity failure of the Transcona Grain Elevator." *Geotechnique*, **111**, 210–208.

Peck, R. B., Hanson, W. E., and Thornburn, T. H. (1974). *Foundation Engineering*. Wiley, New York.

Poncelet, J. V. (1840). "Memoire sur la stabilite des revetments et de leurs fondations. Note additionelle sur les relations analytiques qui lient entre elles la poussee et la butee de la terre." *Memorial de l'officier du genie*, 13.

Poulos, H. G. (ed.) (1988). "The mechanics of calcareous sediments." *5th Australia–New Zealand Geomechanics Conference, Australia Geomechanics*, pp. 8–41.

Poulos, H. G. (1989). "Pile behavior—theory and application." *Geotechnique*, **39**(3), 365–415.

Poulos, H. G., and Davis, E. H. (1974). *Elastic Solutions for Soil and Rock Mechanics*. Wiley, New York.

Prandtt, L. (1920). *Über die Härte plastischer Körper*, Nachrichten von der Königlichen Gesellschaft der Wissenschaften zu Göttingen (Mathematisch-physikalische Klasse aus dem Jahre 1920), 1920, Berlin, pp. 74–85.

Rankine, W. J. M. (1857). "On the stability of loose earth." *Philos. Trans. R. Soc. London*, **1**, 9–27.

Reese, L. C., and O'Neill, M. W. (1988). *Drilled Shafts: Construction Procedures and Design Methods*. Federal Highway Administration, Washington, DC.

Reynold, O. (1885). "On the dilatancy of media composed of rigid particles in contact; with experimental illustrations." *Philos. Mag.*, **20** (5th Series), 469–481.

Richart, F. E. (1959). "Review of the theories for sand drains." *Trans. ASCE*, **124**, 709–736.

Robertson, P. K., and Campanella, R. E. (1983). "Interpretation of cone penetration tests. Part I: Sand." *Can. Geotech. J.*, **22**(4), 718–733.

Roscoe, K. H., and Burland, J. B. (1968). "On the generalized stress–strain behavior of 'wet' clay." *Engineering Plasticity*, J. Heyman and F. Leckie (eds.). Cambridge University Press, Cambridge, pp. 535–609.

Rosenfard, J. L., and Chen, W. F. (1972). "Limit analysis solutions of earth pressure problems." Fritz Engineering Laboratory Report 35514, Lehigh University, Bethlehem.

Rowe, P. W. (1957). "Anchored sheet pile walls." *Proc. Inst. Civ. Eng., Part 1*, **1**, 27–70.

Samarasinghe, A. M., Huang, Y. H., and Drnevich, V. P. (1982). "Permeability and consolidation of normally consolidated soils." *J. Geotech. Eng. Div. ASCE*, **108**(GT6), 835–850.

Sarma, S. K. (1979). "Stability analysis of embankments and slopes." *J. Geotech. Eng. Div. ASCE*, **105**(GT12), 1511–1524.

Schmertmann, J. (1978). *Use the SPT to Measure Dynamic Soil Properties?—Yes, But...!* Dynamic Geotechnical Testing, ASTM Spec. Tech. Publ. 654.

Schmertmann, J. H. (1953). "The undisturbed consolidation behavior of clay." *Trans. ASCE*, **120**, pp. 1201.

Schmertmann, J. H. (1970). "Static cone to compute static settlement over sand." *J. Soil Mech. Found. Div. ASCE*, **96**(SM3), 1011–1043.

Schofield, A., and Wroth, C. P. (1968). *Critical State Soil Mechanics*. McGraw-Hill, London.

Scott, R. E. (1963). *Principles of Soil Mechanics*. Addison-Wesley, Reading, MA.

Seed, H. B., and Idriss, I. M. (1970). "Soil moduli and damping factors for dynamic response analyses." Report EERC 70-10, Earthquake Engineering Research Center, University of California, Berkeley.

Semple, R. M., and Rigden, W. J. (1984). "Shaft capacity of driven piles in clay." *Analysis and Design of Pile Foundations*. American Society of Civil Engineers, New York, pp. 59–79.

Shioi, Y., and Fukui, J. (eds.) (1982). "Application of *N*-value to design of foundations in Japan." *Proceedings ESOPT2, Amsterdam*, pp. 159–164.

Sibley, E. A. (ed.) (1967). "Backfill adjacent to structures." *Proceedings of the Montana Conference on Soil Mechanics and Foundation Engineering, Bozeman*, Montana State University.

Skempton, A. W. (1944). "Notes on the compressibility of clays." *Q. J. Geol. Soc. London*, **100**(C: parts 1 & 2), 119–135.

Skempton, A. W. (1951). *The Bearing Capacity of Clay.* Building Research Congress, London.

Skempton, A. W. (ed.) (1953). "The colloidal activity of clays." *Proceedings, Third International Conference on Soil Mechanics and Foundation Engineering, Zurich,* Vol. 1, pp. 57–61.

Skempton, A. W. (1954). "The pore water coefficients A and B." *Geotechnique,* **4,** 143–147.

Skempton, A. W. (1959). "Cast-in-situ bored piles in London clay." *Geotechnique,* **9**(4), 153–173.

Skempton, A. W. (1970). "The consolidation of clays by gravitational compaction." *Q. J. Geol. Soc. London,* **125**(3), 373–411.

Skempton, A. W., and Bjerrum, L. (1957). "A contribution to the settlement analysis of foundation on clay." *Geotechnique,* **7**(4), 168–178.

Skempton, A. W., and MacDonald, D. H. (1956). "The allowable settlement of buildings." *Proc. Inst. Civ. Eng. Part III,* **5,** 727–768.

Sovinc, I. (ed.) (1961). "Stresses and displacements in a limited layer of uniform thickness, resting on a rigid base, and subjected to a uniformly distributed flexible load of rectangular shape." *5th International Conference on Soil Mechanics and Foundation Engineering, Paris,*Vol. 1, pp. 823–827.

Spencer, E. (1967). "A method of analysis of embankments assuming parallel interslices technique." *Geotechnique,* **17**(1), 11–26.

Stas, C. V., and Kulhawy, F. H. (1984). "Critical evaluation of design methods for foundations under axial uplift and compression loading." EL-3771, Cornell University (report for EPRI), Ithaca.

Taylor, D. W. (1942). "Research on consolidation of clays." Serial No. 82, Massachusetts Institute of Technology, Cambridge.

Taylor, D. W. (1948). *Fundamentals of Soil Mechanics.* Wiley, New York.

Terzaghi, K. (1925). *Erdbaumechanik.* Franz Deuticke, Vienna.

Terzagi, K. (1943). *Theoretical Soil Mechanics.* Wiley, New York.

Terzaghi, K., and Peck, R. B. (1948). *Soil Mechanics in Engineering Practice.* Wiley, New York.

Thorburn, S., and MacVicar, S. L. (1971). "Pile load tests to failure in the Clyde alluvium." *Behaviour of Piles.* Institution of Civil Engineers, London, pp. 1–7, 53–54.

Vesic, A. S. (1973). "Analysis of ultimate loads of shallow foundations." *J. Soil Mech. Found. Div. ASCE,* **99**(SM1), 45–73.

Vesic, A. S. (1975). "Bearing capacity of shallow foundations." *Foundation Engineering Handbook,* H. F. Winkerton and H. Y. Fang, eds. Van Nostrand-Reinhold, New York, p. 121.

White, L. S. (1953). "Transcona elevator failure: eye witness account." *Geotechnique,* **111**(5), 209–215.

Wong, P. K., and Mitchell, R. J. (1975). "Yielding and plastic flow of sensitive cemented clay." *Geotechnique,* **25**(4), 763–782.

Wood, D. M. (1990). *Soil Behavior and Critical State Soil Mechanics.* Cambridge University Press, Cambridge.

Wood, D. M., and Wroth, C. P. (1978). "The correlation of index properties with some basic engineering properties of soils." *Can. Geotech. J.,* **15**(2), 137–145.

Wright, S. J., and Reese, L. C. (1979). "Design of large diameter bored piles." *Ground Eng.,* **Nov.,** 17–50.

Wroth, C. P. (1984). "The interpretation of in situ soil tests." *Geotechnique,* **34**(4), 449–489.

Wroth, C. P., and Hughes, O. J. M. (eds.) (1973). "An instrument for the in-situ measurement of the properties of soft clays." *8th International Conference on Soil Mechanics and Foundation Engineering, Moscow,* Vol. 1, pp. 487–494.

Yamashita, K., Tomono, M., and Kakurai, M. (1987). "A method for estimating immediate settlement of piles and pile groups." *Soils Found.,* **27**(1), 61–76.

Youssef, M. S., El Ramli, A. H., and El Demery, M. (eds.) (1965). "Relationship between shear strength, consolidation, liquid limit, and plastic limit for remolded clays." *6th International Conference on Soil Mechanics and Foundation Engineering.* Toronto University Press, Montreal, Vol. 1, pp. 126–129.

Zadroga, B. (1994). "Bearing capacity of shallow foundations on noncohesive soils." *J. Geotech. Eng.,* **120**(11), 1991–2008.

# INDEX

**578**

# SOIL MECHANICS AND FOUNDATIONS CD-ROM

The accompanying CD-ROM complements this textbook and contains multimedia interactive animation of the essential concepts of soil mechanics and foundations, virtual laboratories, a glossary, notation, quizzes, notepads, interactive problem solving, spreadsheet links and computer program utilities. Not all sections of the textbook are covered in the CD-ROM. The author has selected only those sections of the textbook that can be enhanced by the application of multimedia. You should refer to the CD-ROM section of "NOTES FOR INSTRUCTORS" on page ix and the section on SUGGESTIONS FOR USING TEXTBOOK AND CD-ROM in "NOTES FOR STUDENTS AND INSTRUCTORS" on page xiii for further details.

## Installing the Soil Mechanics and Foundations CD

To install the CD
1. Start Windows if it is not already running on your computer.
2. Insert the CD into your CD-ROM drive.
3. Locate the file, **setup.exe**, on your CD.
4. Double click on **setup.exe**
*Alternatively*
From Windows, Click **Start ➤Run** and enter **D:/setup.exe** where D is the drive letter of your CD-ROM drive. If your CD-ROM drive has a different letter, for example E, then you should enter **E:/setup.exe**.
5. Follow the instructions presented by the installation program.
6. An Application shortcut named *SoilMechanics* will be created in the start menu.

## Starting Soil Mechanics and Foundations CD

1. Click the start program SoilMechanics.
2. For the first time that you start the Soil Mechanics and Foundations CD, you will be prompted for the following:
(i)Your name.
(ii) A password. *The password is a word randomly selected from the textbook. You must have your textbook available to obtain the password.*
3. Soil Mechanics and Foundations CD require that you have certain auxiliary programs such as dynamic link libraries (dll) available on your computer. If you do not have these files, Soil Mechanics and Foundations setup will install them in your computer. Follow the instructions presented by the installation program. This is a one-time installation.
4. A main menu page should appear. Click the button on the chapter that you would like to go to.

## Troubleshooting

| Problems | Possible causes and fixes. |
| --- | --- |
| Sentences cut off, word overlap and words outside of defined area. | Soil Mechanics and Foundations CD requires that you have the **Verdana** font installed in your Windows environment. Check that this font is present in your Windows/Fonts directory. If this font is not present, install it from your Windows CD. |
| Missing MSVBVM50.DLL, Run-time error 380, Run-time error 6, Run-time error 13 | A required dynamic link library (dll) is missing. Open the **Vbdll** directory on the CD and double click **Setup.exe**. This setup program will try to reinstall all the required DLL. Follow the instructions presented by the installation program. |
| Video and animations encroach on other workspaces. | Some media pieces (video, animations, etc.) require that you close them before moving to another section, page or chapter. Cultivate a habit to close all media pieces before proceeding to other parts of the CD. |
| Double clicking the application icon, SoilMechanics, does not navigate to the main menu. | Double click the file **SMF.exe** on the CD-ROM drive to manually start the program. |
| Presentation window is not centered on screen or part of it off-screen. | Try changing the setting on your display to a higher resolution or lower resolution. |
| A white bar appears on the main menu when a chapter is highlighted. | Try changing the setting on your display to a higher resolution or lower resolution. Change also the Color palette to True Color (24 bit). |
| When I quit or exit the CD, a white page appears on the Desktop. | The program was not exited correctly. Press Ctrl-Alt-Del keys simultaneously. Make sure that SoilMechanics is highlighted in the window that appears and then click on End task. |

## Minimum System requirements

Windows-compatible PC running Windows 95, 98 or NT 4.0
Pentium 90, 16MB RAM, 10MB free hard disk space, VGA Graphics, 4X CD-ROM drive, sound card and speakers.
*Technical Support*
Wiley Customer Technical Support
Voice: 212-850-6753
Email: techhelp@wiley.com
Web: http://www.wiley.com/techsupport/